Fluorescence Methods for Investigation of Living Cells and Microorganisms

Edited by Natalia Grigoryeva

Published in London, United Kingdom

IntechOpen

Supporting open minds since 2005

Fluorescence Methods for Investigation of Living Cells and Microorganisms
http://dx.doi.org/10.5772/intechopen.83296
Edited by Natalia Grigoryeva

Contributors
Minchul Kang, Natalia Yu. Grigoryeva, Robert Thomas Youker, Danielle Voet, Anton Shakhov, Irina Alieva, Evgeny Kurashov, Larisa Kapustina, Galina Mitrukova, Julia Krylova, Alexey Moshkov, Mirya Kuranova, Aleksandra Nosdracheva, Nadezhda Pleskach, Daria Pasekova, Aleksey Manikhas, Maria Istomina, Vyacheslav Moshnikov, Anton Belorus, Dmitry V. Korolev, Dmitry Sonin, Artem Brovko, Galina Shulmeyster, Natalia Evreinova, Robson Faria, Guilherme Teixeira, Svetlana Averina, Ekaterina Senatskaya, Alexander Pinevich, Dmitrii Mazing, Lev Matyushkin, Anna Drobintseva, Igor Kvetnoy, Yuliya Krylova, Boris Nikolaev, Olga Aleksandrova, Sergey Musikhin, Yaroslav Marchenko, Victoriya Polyakova, Maxim Shevtsov, Ludmila Yakovleva, Nikolai Vekshin, Galina Glushchenko, Viktoriya Gerasyuk, Alexey Kleshchenkov, Sergey Berdnikov, Yuriy Yurasov, Pavel Plyaka, Margarita Shevchenko, Georgy Valov, Gayane Popovyan, Igor Tron, Klim Grigorenko, Alexander Dubonosov, Vladimir Bren, Ivo Crnolatac, Antonio Jesus Castro, Alfonso Clemente, Juan de Dios Alche, Stefano Santabarbara, William Remelli, Anastasia Petrova, Anna Paola Casazza, Helena Cristina Vasconcelos, Joao A. Lopes, Afonso Silva Pinto, Maria João Pereira, Benjamin Gantenbein, Marie Larraillet, Andreas Shaun Croft, Kalyan Sundar Ghosh, Anchal Sharma

First published in London, United Kingdom, 2020 by IntechOpen
IntechOpen is the global imprint of INTECHOPEN LIMITED, registered in England and Wales, registration number: 11086078, 5 Princes Gate Court, London, SW7 2QJ, United Kingdom
Printed in Croatia

British Library Cataloguing-in-Publication Data
A catalogue record for this book is available from the British Library

Additional hard and PDF copies can be obtained from orders@intechopen.com

Fluorescence Methods for Investigation of Living Cells and Microorganisms
Edited by Natalia Grigoryeva
p. cm.
Print ISBN 978-1-83968-039-7
Online ISBN 978-1-83968-040-3
eBook (PDF) ISBN 978-1-83968-041-0

An electronic version of this book is freely available, thanks to the support of libraries working with Knowledge Unlatched. KU is a collaborative initiative designed to make high quality books Open Access for the public good. More information about the initiative and links to the Open Access version can be found at www.knowledgeunlatched.org

For EU product safety concerns:
IN TECH d.o.o., Prolaz Marije Krucifikse Kozulić 3, 51000 Rijeka, Croatia, info@intechopen.com or visit our website at intechopen.com.

Meet the editor

Dr. Grigoryeva Natalia is currently a senior research fellow at Saint-Petersburg Scientific Research Centre for Ecological Safety of Russian Academy of Sciences, Russia. She is also an official scientific consultant of the Department of Nanoelectronics and the Department of Software Engineering and Computer Applications at Saint-Petersburg Electrotechnical University. In the last six years, she was involved in studying cyanobacteria and developing fluorescence methods for their investigation. She developed an in vivo technique for investigating weak external actions on the physiological state of cyanobacteria. Dr. Grigoryeva recently elaborated on the base of a neural network algorithm an original method of ataxonomic classification of cyanobacteria using single-cell self-fluorescence spectra. Her research interests include both environmental and biotechnological aspects of cyanobacterial applications.

Contents

Preface

Nowadays, methods of fluorescence analysis play a leading role in the investigation of biological objects. Actually, they are the only non-destructive methods for investigating living cells and microorganisms in vivo. Using intrinsic fluorescence as well as artificial fluorescence gives us deep insight into mechanisms underlying physiological and biochemical processes. Over the last several years, a large variety of novel research techniques based on intrinsic and artificial fluorescence have been elaborated. This book covers a wide range of modern experimental techniques such as fluorescence microscopy, fluorescence spectroscopy, confocal laser scanning microscopy (FRAP, FRET, microscopic spectroscopy, etc.), flow cytofluorometry, delayed fluorescence excitation spectroscopy, pulse-amplitude-modulation fluorometry, and methods of fluorescent dye staining.

The goal of this book is to give a clear picture of the current state-of-the-art of fluorescence methods in biological studies across two main areas of interest: algology and medicine. Worldwide ecological problems are closely related with phytoplankton variation, which is a main indicator of environmental health and fresh water quality. Algae physiology and dynamics can be easily evaluated via fluorescence methods due to self-fluorescence of their photosynthetic pigments. From this point of view, fluorescence methods provide natural facilities for online environmental monitoring. The most valuable part of fluorescence techniques deals with artificial fluorescence (staining with dyes, antibodies, nanoagents, etc.) and the main field of its application lies in medicine. The investigation of cancer cells is one of the most important medical applications of fluorescence methods. As such, the second part of the book is mostly devoted to this theme. Overall, this book illustrates the latest achievements of fluorescence analysis, which help to achieve better understanding of living cell processes. Its main objective is to summarize the advantages and specific features of different fluorescence method applications in rather different fields.

This book is useful for students and researchers in any fields dealing with the investigation of living cells. It contains twenty chapters divided into three sections: "Fluorescence Methods in Algology," "Fluorescence Methods in Medicine," and "Special Theories, Protocols and Techniques." The chapters concerned with the application of fluorescence methods in algology include investigations based on self-fluorescence of living microalgae cells. The chapters concerned with the use of fluorescence methods in medicine include a description of the techniques based on artificial fluorescence, produced by dyes and nanoagents, and coupled to different proteins in mammalian cells. The final section includes some special fluorescence techniques and protocols that may be interesting for biologists.

The editor would like to express thanks to the chapter authors for their invaluable contributions. Especially I want to convey my gratitude to Knowledge Unlatched GmbH for financial support of this project.

Dr. Natalia Grigoryeva
Laboratory of Biological Methods of Environmental Safety,
Scientific-Research Center for Ecological Safety,
Russian Academy of Sciences,
Saint-Petersburg, Russia

Section 1

Fluorescence Methods in Algology

Chapter 1

Studying Cyanobacteria by Means of Fluorescence Methods: A Review

Natalia Yu. Grigoryeva

Abstract

Self-fluorescence is a powerful tool for investigation of living photosynthetic microorganisms. Since the physiological state of such microorganisms is closely related to the operation and activity of photosynthetic system; thus, any variations in spectroscopic properties of their self-fluorescence indicate the changes in their physiological state. Cyanobacteria (or blue green alga) are one of the most widespread photosynthetic organisms in nature, and the ecological aspect in their investigation is quite valuable. On the other hand, thousands of strains belonging to different cyanobacterial species are cultivated in biolaboratories all over the world for different biotechnological applications such as biofuel cells, food production, pharmaceuticals, and fertilizers. Thus, the novel noninvasive methods of their investigation are quite important for on-line monitoring of cyanobacterial cultures. In this chapter, several fluorescence techniques are presented for investigation of living cyanobacterial cells and cultures.

Keywords: cyanobacteria, photosynthesis, fluorescence spectroscopy, confocal microscopic spectroscopy, fluorescence induction, flow cytometry, remote sensing

1. Introduction

In the last years, the investigation of cyanobacteria attracts a considerable attention due to their environmental and biotechnological importance. New applications of cyanobacterial communities appear in biotechnology (i.e., production of biofuel, agricultural fertilizers, pharmaceutics, and food additives ultraviolet protection compounds) [1–5] and biosensing [6–9]. Large-scale industrial production of the cyanobacterial products requires optimization and more detailed control of incubation conditions in order to increase productivity. On the other hand, the ecological problem of harmful bloom monitoring and control makes an important contribution in this rising interest to cyanobacteria, as to one of the main indicator organisms of the environment pollution and to methods of their investigation [10–18]. Obviously, future research in this field will be focused on the improvement of different treatments that inhibit harmful growth; thus, the development of novel noninvasive methods for on-line monitoring of cyanobacterial communities *in vivo* is quite needed. The presence of a photosynthetic apparatus similar to higher plants makes microalgae and cyanobacteria suitable for use in biosensors based on monitoring changes in photosynthesis and other bioenergetic processes in the cell [19–21].

A vast amount of work was done to achieve a present insight into the physiological processes, which rules cyanobacteria life, and their genetic background [10–18]. Nowdays, it was clearly shown that there is a universal genetic response to most abiotic stressors (heat, osmotic, saline, acid stress, strong light, and ultraviolet radiation) [22]. Moreover, this universal response is closely related to the photosynthetic activity of cyanobacterial cells [23]. This fact gives a unique opportunity for operational in-line monitoring (including remote sensing) of the physiological state of photosynthetic microorganisms in variable environmental conditions. Since most photosynthetic pigments possess self-fluorescent properties, thus, the emission spectral composition and fluorescence kinetics can provide important information about photosynthetic activity of living cells and reveal any variations in their physiological state under the influence of external factors.

All fluorescence methods can be divided in to two groups by their degree of organism impact. First group deals with *in vivo* investigations of living cells. Such methods should be noninvasive and very sensitive to study the life as it is, in dynamics of growth and adaptation. The second group deals with dead material. By these methods, chemical, morphological and genetic structure of the cell can be studied in different soluble and nonsoluble chemical components. In this chapter, we will be interested in the first ones. The most naturally suitable and appropriate methods of investigation of photosynthetic organisms are to use the self-fluorescence of photosynthetic apparatus. In this sense, following *in vivo* and *in situ* methods can be suggested: fluorescence spectroscopy (steady-state fluorescence) [24, 25], chlorophyll fluorescence induction analysis (fast and slow kinetics) [26, 27], pulse amplitude-modulated fluorometry (chlorophyll fluorescence quenching) [28, 29], delayed fluorescence [30, 31], flow cytometry [32, 33], confocal microscopic spectroscopy [34, 35], and remote sensing (passive and active, e.g., fluorescence LIDAR technique) [36, 37].

Two main aspects in studying cyanobacteria usually are interested both ecologists and biotechnologists: physiological state of cyanobacteria on population and cell level, and biological diversity of the considered community. Fluorescence methods can successfully solve both problems.

2. Peculiarities of cyanobacterial photosynthetic apparatus

Photosynthetic system of cyanobacteria, in contrast to green algae and higher plants, contains the external membrane light-harvesting complexes. The antenna complex for photosystem II (PSII) and photosystem I (PSI) is extrinsic and is formed as a large multiprotein organelle (phycobilisome), which is located on the stromal side of the thylakoid membrane. The detailed description of the morphology, structure, chemical, and optical properties of light-harvesting complex of cyanobacteria, detached phycobilisomes, and phycobilins can be found in numerous publications [10–15, 38–40]. Here, we only pointed out several main features of its self-fluorescence spectrum that are essential for further discussion.

Phycobilisomes (PBSs) are assembled from 12 to 18 different types of polypeptides, which may be grouped into three classes: (1) phycobiliproteins, (2) linker polypeptides, and (3) PBS-associated proteins. And only first ones bear chromophores and form fluorescence response together with chlorophyll *a* (Chl *a*) of PSII and PSI. PBSs are constructed from two main structural elements: a core substructure and peripheral rods that are arranged in a hemidiscoidal fashion around that core (**Figure 1**). During the energy transfer process through PBS to PSII and PSI, the occasional quenching of the absorbed light by fluorescence

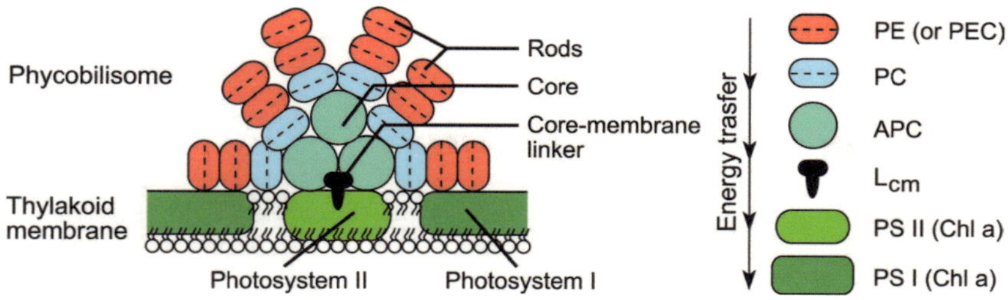

Figure 1.
Schematic drawing of photosynthetic apparatus and energy transfer in cyanobacteria.

can occur and this is the essential property for fluorescence spectroscopy. Both steady-state and time-resolved fluorescence measurements are used for probing the structure and functioning of photosynthetic system of cyanobacteria via this specific fluorescent response.

The fluorescence of intact living cyanobacterial cells is originated from the efficiency of the energy transfer between all components of the energy transfer chain including the final step, the delivery to PSII or PSI (**Figure 1**). Each transfer step results in the spectrum shape as a peak or shoulder [41, 42]. This is due to the fact that in the course of the energy transfer from the initially photoexcited phycobiliprotein to the reaction center of photosystems PSI and PSII; fluorescence is emitted from almost every type of pigment and can be used as a probe to examine the mechanism of energy transfer within the light-harvesting system and between two photosystems [30, 38–40]. To monitor this energy transfer process, a sample should be irradiated with the light that selectively absorbs by one set of pigments and then the fluorescence, originated from another set of pigments, should be recorded. This type of fluorescence excitation experiment can be used to measure the efficiency of energy transfer from one set of pigments to another, which reflects the species-specific features, if cyanobacteria are in the optimal physiological state. On the other hand, under any external actions this fluorescence "finger prints" will change according to mechanisms of the cyanobacterial response on this action, and the changes in steady-state fluorescence and its kinetics finally can reveal these mechanisms.

As it was mentioned previously, a vast amount of fluorescence techniques were elaborated to study physiological state and biological diversity of photosynthetic organisms, but not all of them can be directly applied to cyanobacteria. Several crucial features of their fluorescence response lead to the necessity of changes in conventional techniques; otherwise, it will lead to insufficient or wrong results. To demonstrate these features of cyanobacteria, we present here several simple but representative examples, which show specific fluorescence response of cyanobacteria compared to green microalgae *Chlorella ellipsoidea Gerneck CALU 3*. It should be mentioned here that the fluorescence response of green microalgae is quite similar to higher plants one, due to the similarity of the structure of their photosynthetic apparatus.

In **Figure 2**, three sets of fluorescence spectra, recorded via standard lambda-scanning procedure of confocal laser scanning microscope (CLSM) Leica TCS-SP5, and microphotographs of corresponding species, obtained simultaneously in fluorescence and transmission detection channels, are presented. The excitation wavelengths, corresponding to the emission lines of three CLSM lasers, are indicated over the curves (405, 458, 476, 488, 496, 514, 543, and 633 нм). It is well seen that depending on the excitation wavelength, the room temperature fluorescence emission spectrum of the intact cyanobacterial cells exhibits various extents of

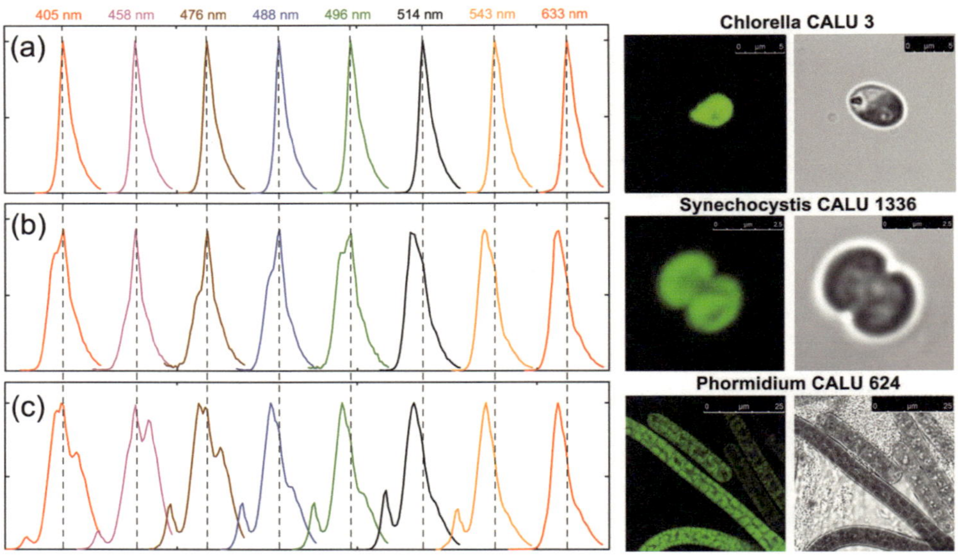

Figure 2.
Three characteristic sets of CLSM single-cell fluorescence spectra and microphotographs of corresponding strains, obtained simultaneously in fluorescence and transmission detection channels. (a) Chlorella ellipsoidea Gerneck CALU 3*; (b)* Synechocystis aquatilis CALU 1336*; (c)* Phormidium favosum str. Marvan CALU 624*. All spectra are normalized to the maximum intensity and shifted along x-axis for convenience of observation. Corresponding excitation wavelengths are given over the curves. Dashed lines indicate wavelength of Chl* a *fluorescence maximum at 682 nm. The white bar corresponds to 5, 2.5, and 25 μm, respectively.*

contribution of phycobilisome emission to the spectrum, opposite to green algae. If one exclusively excites chlorophyll *a* fluorescence, using a 458-nm excitation wavelength, the emission spectrum of cyanobacterial cells shows no appreciable emission of phycocyanin or allophycocyanin. It is due to the fact that, in cyanobacteria, the 458-nm excitation wavelength is preferentially absorbed by PSI and PSII, bypassing photosynthetic pigments of light harvesting complex (LHC). The excitation by intermediate (blue and green) wavelengths (405, 476, 488, and 496 nm) reveals fluorescent maxima of all photosynthetic pigments, belonging to the corresponding species, since the light in this range is absorbed by all pigment-protein complexes almost in equal portions and fluorescence emits by all steps of energy transfer chain. The direct excitation of cyanobacterial cells in the phycocyanin absorption region at 514 and 543 nm results in emission spectrum with two main peaks at 580 and 656 nm, which are due to phycoerythrin, phycocyanin, and allophycocyanin emission. The spectra for 633-nm excitation wavelength directly give a prominent emission band at 656 nm that originates from phycocyanin, omitting the band at 580 nm, which cannot be excited by 633 nm. Other small emission bands, corresponding to fine pigment structure of antenna complex, are not resolved at the room temperature investigation.

Opposite to cyanobacterial species, green algae demonstrate poor diversity in the emission spectra (**Figure 2(a)**). Comparative analysis of the series of fluorescence spectra for different cyanobacterial species and strains (**Figure 2(b)** and **(c)**) reveals visible variations in their shape, and thus, species/strains differentiation (so-called ataxonomic analysis) could be carried out [41, 42]. Moreover, this diversity in fluorescence responses on different excitation wavelength in cyanobacterial cells shows that all fluorescence techniques based on the analysis of chlorophyll *a* fluorescence, developed for higher plants (pulse amplitude modulation technique, flow cytometry, remote sensing, etc.), should be strongly modified for studying cyanobacteria.

Another important difference in the *in vivo* fluorescence response of cyanobacteria and green algae lies in stress-state investigations. **Figure 3** illustrates the temporal changes of the *in vivo* fluorescence spectrum in one living cell of cyanobacterial strain *Synechocystis aquatilis CALU 1336* and one living cell of *Chlorella ellipsoidea Gerneck CALU 3* under light and heat stress. The series of spectra were recorded by means of CLSM lambda-scanning technique at the excitation wavelength 488 nm. Comparison of **Figure 3(a)** and **(b)** demonstrates the difference in stress fluorescence response for these two cases. Thus, the conclusions about stress state of cyanobacteria, made on the basis of the techniques elaborated for green algae and higher plants, are sometimes meaningless and sometimes absolutely incorrect. The presence of additional photosynthetic pigments in LHC of cyanobacteria and some specific features in their photoprotection mechanisms makes them principally different from other photosynthetic taxa.

It should be noted here that this temporal set of fluorescence emission spectra illustrates all stages of cyanobacterial cell degradation, that is, all possible physiological states of the considered speci. Thus, during the growth evolution of the culture and aging of each cell, all these stages will be presented simultaneously in the volume samples of the whole population. This means that the analysis of the population growth and the community viability by means of fluorescence methods should also be corrected for cyanobacteria compared to green algae ones.

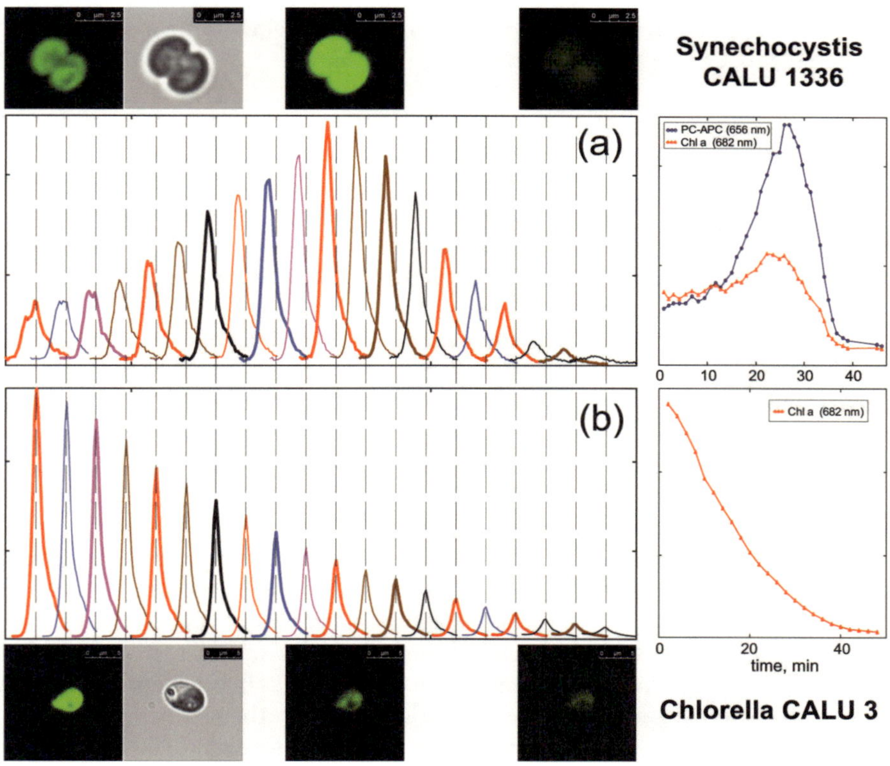

Figure 3.
Time degradation of living cell of (a) cyanobacterial strain Synechocystis aquatilis CALU 1336 *and (b)* Chlorella ellipsoidea Gerneck CALU 3 *under light and heat stress. Spectra were recorded at the excitation wavelength 488 nm and with the time step 2 min. Spectra are shifted along x-axis relative to each other for convenience of observation. Over the spectra, there are several fluorescent images taken at corresponding physiological states and recorded via CLSM. Right panels represent time dependence of the fluorescence intensity at 656 nm (PC-APC fluorescence) and at 682 nm (Chl* a *fluorescence) for cyanobacteria and at 682 nm for green microalgae. Dashed lines indicate the wavelength of Chl* a *fluorescence maximum at 682 nm.*

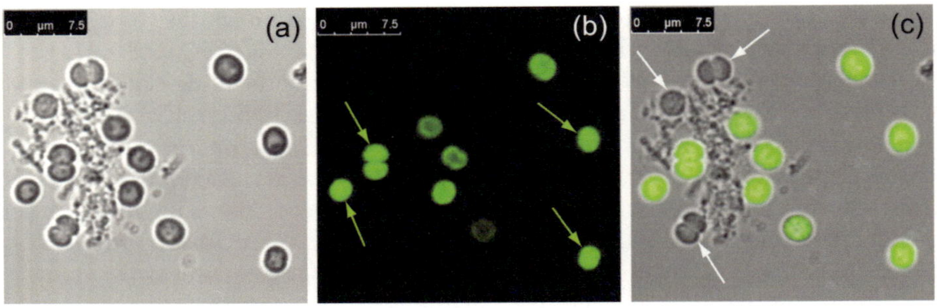

Figure 4.
CLSM micrographs of cyanobacterium Synechocystis aquatilis CALU 1336, *visualized with CLSM. (a) Transmission image, (b) fluorescence false-green channel represents the response from phycobilins and Chl* a *at 650–680 nm, (c) represents an overlap of fluorescence and transmission channels. Scale bar = 7.5 µm.*

Figure 4 illustrates several cells of unicelluar cyanobacterial culture *Synechocystis aquatilis CALU 1336*. Diverse physiological states are observed and indicated with arrows. Small fluorescent signal corresponds to healthy cells. Cells in frustrated physiological state (dying cells) have a high fluorescence and are indicated by green arrows. Dead cells have no fluorescence and can be observed only in transmission channel. They are indicated with white arrows.

Here, we have demonstrated only two examples of the differences between cyanobacterial and green-algae fluorescent response. Below some more cases will be considered concerning specific fluorescence techniques.

3. Fluorescence methods to study cyanobacteria

As it was mentioned previously, two main properties of cyanobacterial communities are of great interest during the investigation and control: physiological state (or viability) and biological diversity. And the fluorescence measurements contribute greatly to the study of these issues.

Historically, since the 1960s, researches had noted that changes in physiological state of cyanobacterial cell, occurring when it is grew old or is treated with different external actions, manifest themselves as corresponding changes in the intrinsic self-fluorescence spectrum [43]. Several studies have shown that changes in the pigment fluorescence are associated with the changes in the enzymatic activity of the cell and an increase in the permeability of the cell membrane, which can be used as an indicator of aging for cyanobacterial species [41, 44]. Today, there is no doubt that the *in vivo* analysis of fluorescence characteristics of photosynthetic apparatus is a powerful tool for studying the impact of a wide variety of environmental factors on cyanobacteria [24, 45]. The changes in the steady-state and variable fluorescencereflect the in-time physiological state of cyanobacterial communities, exposed to different external factors [23]. The environmental changes cause the changes in bioenergetic processes occurring in cyanobacterial cells; they significantly affect the kinetics parameters and spectral features of the intrinsic fluorescence spectra. Since early 1980s, chlorophyll fluorescence kinetics has been widely used for investigation of viability of eukaryotic plants [28, 46]; however, the specific features of cyanobacteria only recently began taking into account. Therefore, in some earlier works, it was reported that the results of chlorophyll fluorescence measurements in cyanobacteria did not reflect the condition of photosynthesis at all [29, 47–49].

The fluorescence spectra have been widely used to classify phytoplankton populations since approximately early 1970s [50]. However, because of the generally low number of available excitation wavelengths in the conventional devices, the rate of species discrimination was relatively low. Researches can separate only algal groups that differ greatly in the structure of the photosynthetic apparatus (e.g., cryptophytes, chlorophytes, and cyanobacteria) but cannot separate groups that are more similarly pigmented (e.g., among cyanobacterial species) [51, 52]. Discrimination between similarly pigmented taxa or even between species within a taxon requires high-resolution and multi-wavelength spectral analysis to reveal small peculiarities in configuration and functioning of photosynthetic system. The rigorous discrimination is possible if the inter-species differences are greater than those within a species. For cyanobacteria, the spectral features of fluorescence, emitted by single cells *in vivo*, depends only on the structure and operational effectiveness of photosynthetic apparatus; thus, they reflect the individual characteristic of cyanobacterial strain and can be used for confident differentiation up to strain level [42]. The possible contribution of environmental adaptation effects to the resulting fluorescence spectra can be minimized by an accurate definition of the corresponding spectral regions of the excitation and emission wavelengths. Recent progress in CLSM methods allows to get spectroscopic data from single cells, instead of the whole population (like in conventional fluorescence spectroscopy); thus, the external interference has a minimal effect on the resultant spectrum.

Several fluorescent methods for *in vitro* and *in situ* investigations will be considered in this part, and the specific features of their use for studying cyanobacteria will be outlined.

3.1 Fluorescence spectroscopy

Steady-state fluorescence spectroscopy can solve many routine problems of limnology and oceanology in a noninvasive manner. For example, the gravimetrically measured dry weight is an invasive method that is unsuitable for monitoring growth over time in the same sample; however, fluorescence methods can successfully replace it by applying calibration curves for *in vivo* chlorophyll *a* fluorescence of different cyanobacteria species [53]. This is the only one simple example of growth rate measurement. In more complicated cases, like monitoring of physiological state or biological diversity of cyanobacterial communities, fluorescence spectroscopy can become the only possible noninvasive method of investigation.

Opposite to the absorption spectra, the *in vivo* fluorescence spectra are much more informative. Moreover, it was shown in [54, 55] that the fluorescence assay would be a preferred alternative to absorbance in quantifying cyanobacterial growth because the fluorescence measures only viable cells and absorbance measures total cell biomass (including dead cells). Fluorescence spectroscopy is a highly sensitive, nearly instantaneous, noninvasive way to study various components and processes *in situ* and *in vivo*. Although the fluorescence spectra contain the information only about photosynthetic apparatus of different algal groups, they include the information about the chemical structure of light harvesting complex (LHC) and accessory pigment-proteins, as well as about the character of links between pigment-protein complexes and the efficiency of energy transfer in the light-harvesting process. When compared with absorption, fluorescence is affected by the excitation wavelength and energy. Thus, the use of different excitation wavelengths can provide more detailed information for the study of single-cell composition (**Figure 5**).

As it was mentioned previously, the specific features of fluorescence measurements of cyanobacterial communities, based on their initial fluorescence, lay mostly in the peculiarities of their photosynthetic apparatus, in particular, of the light-harvesting antennae. The distinct set of photopigments makes cyanobacteria

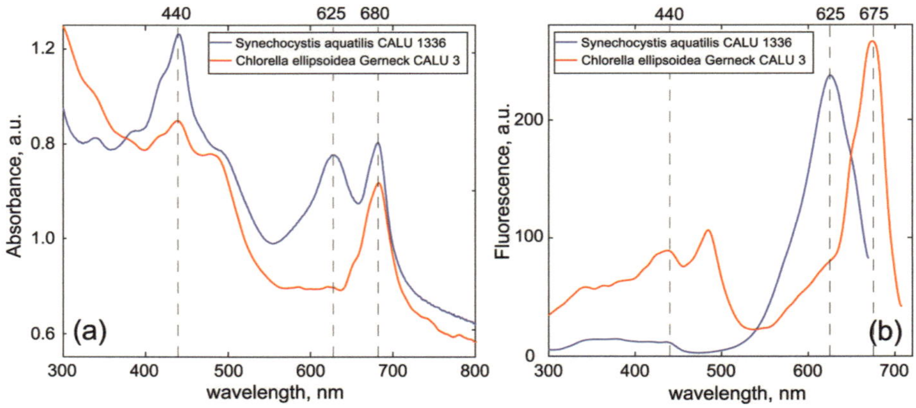

Figure 5.
Absorption (a) and fluorescence excitation (action) spectra (b) for Synechocystis aquatilis CALU 1336 *(blue line) and* Chlorella ellipsoidea Gerneck CALU 3 *(red line).*

an ideal laboratory object for various biological investigations. Fast adaptation and understandable fluorescence response on the environmental changes allow using cyanobacteria as an indicator organism in ecological studies and as a laboratory object in biotechnological applications. However, the structure and functioning of their photosynthetic apparatus differ drastically from eukaryotic plants (**Figure 5**). Thus, the attention should be paid while applying fluorescence methods, elaborated for higher plants, to cyanobacteria and *vice versa,* during the application of the results of cyanobacterial studies to green algae. Let us mention several specific features that should be kept in mind while working with cyanobacteria.

The problem of registration and control of the physiological state of natural communities and the viability of cultures during incubation is a primary task in both ecological and biotechnological fields of application of cyanobacteria. In cyanobacteria, the main pigments responsible for photosynthetic activity are phycobilins; therefore, the methods associated with the analysis of only chlorophyll fraction do not give satisfactory results in the study of cyanobacterial cultures [56–58]. Moreover, the analysis of the fraction of water-soluble phycobilins may give the ambiguous information, due to debris of disrupted cells in the culture medium. The former gives underestimated results, and the latter, overestimated results. Thus, the analysis of the viability of living cyanobacterial cells should be carried out considering a set of fluorescence spectra with different excitation wavelength, to elucidate the impact of each photosynthetic pigment [41, 44, 49]. The examples of one-excitation-wavelength studies are also present in the literature [59, 60]. However, the conclusions made on the basis of such investigations are questionable.

The investigation of phytoplankton communities and the identification of individual algae species, in the presence of different photosynthetic pigments, are completely impossible via one or two fluorescence spectra. Here, only a set of fluorescence spectra with different excitation wavelength can give a sufficient accuracy (see e.g., [42, 61]).

In the last years, so-called excitation-emission fluorescence matrices are often used in investigations of the effect of pollutants on natural algae communities [24], as well as in biological diversity studies [25, 62, 63]. Some researches even develop submersible devices to carry out such fluorescence study *in situ* [64].

3.2 Chlorophyll fluorescence kinetics

A vast amount of works are devoted to the investigation of chlorophyll fluorescence induction (fast and slow kinetics) and chlorophyll fluorescence quenching analysis via saturation pulse method in green algae and higher plants, but only few of them concerns studies of cyanobacteria [28, 29, 47, 65–67]. Here, we will be interested in the peculiarities of using this technique for studying cyanobacteria. Therefore, we will not discuss here theoretical basics but only pointed out some important features of fluorescence kinetics in cyanobacterial photosynthetic apparatus.

In numerous reviews, the theoretical investigations of chlorophyll *a* (Chl *a*) fluorescence induction in different photosynthetic organisms are presented [26, 27, 65–69]. Basically, three classes of processes modulate Chl *a* fluorescence: photochemical quenching [q(P)], nonphotochemical quenching [q(N)], and fluorescence lowering or fluorescence increase due to state 1 to 2 and state 2 to 1 transitions, respectively. In addition at high actinic light excitation, Chl *a* fluorescence can be subject to various photoinhibitory quenching processes q(I), which are much slower than the previous three.

Differences in the interpretation of cyanobacterial fluorescence induction start from the first step. First of all, the dark adaptation shifts plants and algae toward light state 1 (enlarged PS II antenna, high fluorescence state), whereas in cyanobacteria, dark adaptation shifts toward light state 2 (enlarged PS I antenna, low fluorescence state). This leads to discrepancy in interpretation of the JIP-test results like very small rise of OJIP fluorescence, compared to eukaryotic plants (see **Figure 6**) and very large rise of PSM fluorescence instead of decay for green algae and higher plants [26] Here and hereafter, the notation of main points in polyphasic chlorophyll a fluorescence curve (OJIP and PSMT) is taken from the main works of Strasser and Stamatakis [65, 66].

The first recorded signal in fluorescence induction (FI) experiment, F_0, is considered usually as the maximally oxidized Q_A and as the maximal rate of primary charge separation in the RCIIs in dark-adapted green algae and higher plants. In cyanobacteria, the O level of fluorescence induction curve is higher (see **Figure 6**) because of C-phycocyanin and allophycocyanin fluorescence contributions at 710 nm and also because Q_A is partly reduced in the dark due to equilibration with the PQ pool, which is reduced in darkness by respiratory substrates. This happens because in cyanobacteria, the PQ pool is shared both by photosynthetic and respiratory electron transport (PSET and RSET), contrary to

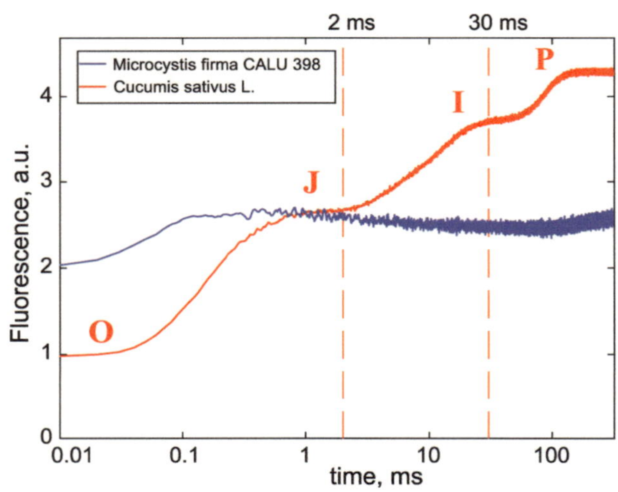

Figure 6.
Fast OJIP chlorophyll fluorscence kinetics for Microcystis firma CALU 398 *(blue line) and* Cucumis sativus L. *(red line). Characteristics were recorded with PAM-2500, Walz GmbH, Effeltrich, Germany.*

other photosynthetic organisms, where PSET and RSET functions are segregated in different organelles, plastoquinones are oxidized enzymically in darkness, and their fluorescence quenching effect is enhanced. Also, in cyanobacteria, Chl *a* is more abundant in PSI than in PSII compared to eukaryotic plants. Besides this, a dark-reduced PQ pool shifts cyanobacteria to state 2, in which the excitation which PBS donates to PSI is maximized, increasing the contribution of PSI to F_0. In other words, in cyanobacteria, F_0 contains a larger fraction of the redox-silent PSI fluorescence than it does in higher plants and algae. Finally, the OP rise is the integrated response of excited Chls *a* to redox signals and electrostatic fields that PSET generates within the thylakoid membrane.

Compared to the OP rise, the PSMT decay (in plans and rise in cyanobacteria) is far more complex (see e.g., **Figure 2** in [47]). In cyanobacteria, lumen acidity-dependent q(N) works in a characteristically different way than in eukaryotic plants. While in the latter, q(N) dissipates primarily the excess Chl *a* excitation in PSII; in cyanobacteria, it dissipates primarily the excess Chl *a* excitation in PSI, which is obtained from PBS when the cyanobacterium is in state 2 (e.g., after dark adaptation). In other words, in cyanobacteria, q(N) quenching occurs as a sequel to the fluorescence lowering during the state 1 to 2 transition [70]. Moreover, in contrast to algae and higher plants, cyanobacteria have the maximum fluorescence signal in point M between O and T, instead of point P for other photosynthetic organisms. This is due to the fact that in cyanobacteria, the PQ pool is reduced by respiratory substrates during dark adaptation, as it is shared by both the PSET and the RSET chains [71, 72]. Thereby, an extensive, and often monotonous, P to T fluorescence decay in higher plants reflects a state 1 to 2 transition. And the opposite situation occurs in cyanobacteria, where the SMT transient is dominant in the total fluorescence induction pattern, with SM rise corresponding to a state 2 to 1 transition. Thus, it is obvious that the F_m parameter for cyanobacteria should be obtained at another point (M), than for higher plants (P). The SM rise in cyanobacteria displays interesting phenomenology, which cannot always be explained adequately. In contrast to green algae, noncyclic and cyclic PSET inhibitors do not block the SM rise in cyanobacteria, although the fluorescence induction kinetic details are changed. When the MT fluorescence decay takes place, a cyanobacterium cell is at state 1 and most likely stays there during the MT decay because it is illuminated continuously. The MT decay in cyanobacteria is observed only under strong actinic illumination, raising the possibility of q(I) (as that part of a light-induced depression of Chl *a* fluorescence that is not reversed by dark adaptation) contributions to it.

Concerning fluorescence quenching analysis based on pulse amplitude modulation technique [73], it should be noted that this method can give detailed information of the main processes in photosynthetic system of cyanobacteria, but inherent simplicity of this method there is a risk of uncritical use which in certain cases may lead to questionable results and conclusions. In several reviews [28, 29, 67, 74, 75], a lot of peculiarities are pointed out, which distinguish cyanobacteria from eukaryotic plants. They are difference in metabolic interactions (i.e., reduced PQ pool in dark-acclimated cyanobacterial cells due to respiratory electron transport), the problems arise from pigment composition (i.e., the impact of phycobilins fluorescence into the overall chlorophyll fluorescence and the dependence of the result on the wavelength of excitation light), the high rate of development of nonphotochemical quenching, the problem with the estimation of electron transport rate (ETR) (due to the difference in distribution of light energy between PSI and PSII in cyanobacteria), discrepancy in the behavior of q(N) (nonphotochemical quenching) and q(P) (the redox state of PQ pool) (because the regulation of state transition in cyanobacteria is due to not only redox control, but in some way to component in cytochrome b_6/f complex), the role of the blue light intensity in nonphotochemical quenching (nonphotochemical OCP quenching), etc. Some of them were already described previously.

The main problem of chlorophyll fluorescence analysis in cyanobacteria is related with the determination of the actual F_0 and F_m, since cyanobacteria have highly fluorescent phycobilisome and high PSI/PSII ratio; thus, the contribution of "basal" fluorescence should be far higher than in land plants. The method of calculation of the "true" value of F_v/F_m in the absence of the "basal" fluorescence was suggested in [76], and the corrections that are due to phycobilin fluorescence only are presented in [77, 78]. Another problem with right determination of F_m deals with the reduction of PQ pool in the dark-acclimated cyanobacterial cells, since PQ pool and electron transport are shared by photosynthetic and respiration systems. This problem can be solved by using weak blue light illumination, which oxidizes PQ pool, instead of dark. The determination of F_0 also is not so simple. In the dark, cyanobacterial cells are in state 2; thus in order to determine the true F_0 level, they should be brought to state 1. Again appropriate blue light acts as PSI-excitation light and oxidizes PQ pool as well as Q_A in many cyanobacteria [28]. Thus, we should be especially cautious to interpret the differences in chlorophyll fluorescence parameters determined in the dark or under low light in the analysis of cyanobacteria. To demonstrate the difference in chlorophyll fluorescence kinetics between cyanobacteria and higher plants in **Figure 7**, several characteristics for cyanobacteria (*Microcystis firma CALU 398*) and cucumber leaves (*Cucumis sativus L.*) are presented. The characteristics were recorded at blue actinic light for cyanobacteria and at red actinic light for cucumber leaves. Obviously, the fluorescence emission trace for cyanobacterial and cucumber quenching analysis are quite different, as well as ETR, photochemical [Y(II)) and nonphotochemical (Y(NPQ)] quenching dependencies.

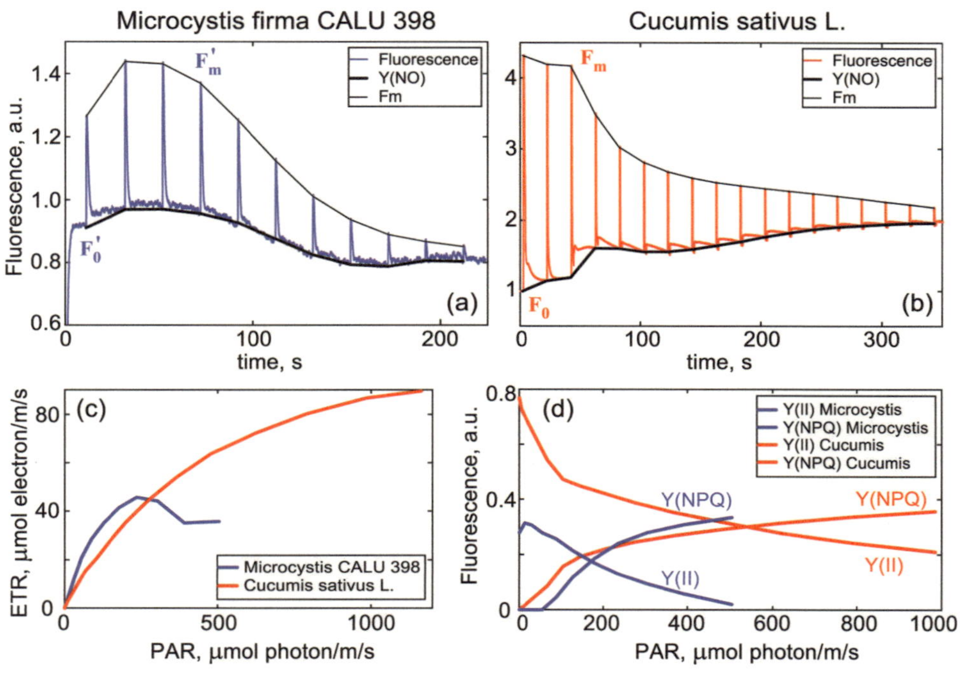

Figure 7.
Fluorescence emission trace for quenching analysis (a) Microcystis firma CALU 398 *(blue lines) and (b)* Cucumis sativus L. *(red lines), (c) ETR, (d) Effective quantum yield Y(II) and nonphotochemical quenching Y(NPQ) dependencies. Characteristics were recorded with PAM-2500, Walz GmbH, Effeltrich, Germany.*

The instrumental support of these two fluorescence methods, the fluorometers, that are currently in use for kinetic fluorescence studies, is based on different approaches to measure the variable Chl *a* fluorescence differing in the manner by which the photochemistry is saturated: shutterless and LED-based instruments for direct fluorometry, as plant efficiency analyzer (HandyPEA, M-PEA, PocketPEA; Hansatech Instruments, Kings Lynn Norfolk, UK) [79], pulse amplitude modulation, PAM fluorometry (Phyto-PAM Phytoplankton Analyzer; PAM-2000, Walz GmbH, Germany) [80], the pump and probe fluorometry [81], the fast repetition rate fluorometry [82], the pump during probe fluorometry [83], and several others that are functionally similar, such as the fluorescence induction and relaxation technique [84], the background irradiance gradient single turnover fluorometry [85], and advanced laser fluorometry [86]. In the last years, a lot of new modifications of these instruments appeared, including multicolor PAM (MULTI-COLOR-PAM and PHYTO-PAM Phytoplankton Analyzer, Waltz GmbH, Effeltrich, Germany) [87–89] and imaging PAM (IMAGING-PAM M-Series, Waltz GmbH, Effeltrich, Germany) [73]. Recently developed multi-color-PAM (Waltz GmbH, Effeltrich, Germany) [87] provides six different colors of measure light and six different colors of actinic light, all of which qualify for highly accurate measurements of fast induction kinetics and assessment of wavelength-dependent F_v/F_m and functional absorption cross-section of PSII.

3.3 Delayed fluorescence analysis

Delayed fluorescence (DF) occurs in photosynthetically active cells and is emitted between 450 and 750 nm with decay times from milliseconds to minutes [30, 31, 52, 90]. DF excitation spectroscopy is a real-time, nondestructive method for rapid estimation of the photosynthetically active pigments—chlorophylls, accessory pigments and phycobiliproteins. It can be used to determine chlorophyll concentration and phytoplankton composition [91–95], as well as to study the changes in electron transport [96], different stress states [97–100], the response to the diurnal light cycle [101], etc. The DF measurement distinguishes between cyanobacteria, cryptophyta, green algae, and chromophyte algae because they have different chlorophylls, accessory pigments, and phycobiliproteins. DF has some specific features in cyanobacteria, due to their specific structural arrangement of the photosynthetic membranes. In particular, in contrast to other algae, some parameters of delayed fluorescence from cyanobacteria show specific changes within the temperature range at which membrane lipids undergo thermophilic phase transitions [102].

The origin of DF is as follows. The accessory pigments, absorbing photons, provide charge separation at the reaction centers of PSII and PSII after the excitonic energy transfer in light-harvesting complex. Then electrons are transported through the electron transport chain to the Calvin cycle. Stopping the illumination initiates processes of the reverse light phase, that is, electrons in the electron transport chain flow back to the oxidized reaction center $P680^+$, leading to an excited state $P680^*$. This excited state $P680^*$ decays to the ground state emitting a delayed fluorescence in 580–720 nm wavelength range. DF can be observed during several minutes until a charge equilibrium between donor side (water splitting complex) and acceptor side (plastoquinone, PSI) is reached by recombination. The resulting decay curve includes the fast decay, produced by recombination of electrons and holes, which are located near $P680^+$, and the long-lasting decay, caused by electron-hole pairs, which are located at greater distances from $P680^+$ (e.g., electrons near PSI together with holes at the Mn^{n+} complex) [96]. Thus, the cellular components, which absorb light, but do not lead to charge separation (like dead cells, or photoprotective pigments—carotenoids), are not detected by DF spectroscopy.

The peculiarities of DF spectroscopy for cyanobacteria are associated with the difference from plants in cellular metabolism, and structure and functioning of their photosynthetic apparatus, which are pointed out in the previous section. The differences in kinetics and spectral characteristics of DF can be easily traced from higher plants taking into account the relation between photosynthetic and respiratory systems in sharing of electron transport chain and the presence of external light-harvesting complex (phycobilisome). The former influences the DF kinetics, and the latter, DF spectral features. For example, the temporal DF dependence (DF decay curve) shows the transient peak with variable, species-specific heights and timing, in contrast to green algae, which have no any peak (see, e.g., **Figure 2** in [100]). The height and timing of this peak is affected differently by different toxicants [100]. Thus, DF can be used as a rapid, reliable, and sensitive phytoplankton toxicity test. In the laboratory experiments, the comparison of four measuring techniques (delayed fluorescence, prompt fluorescence, photometric absorbance, and microscopic cell count) to assess growth inhibition freshwater algae and cyanobacteria shows very high sensitivity of delayed fluorescence excitation spectroscopy [90]. DF as a parameter is used to investigate the phytoplankton community in lakes [95], its seasonal aspect, and vertical distribution [95, 103], as well as continuous biomonitoring of surface waters [94]. Delayed fluorescence excitation spectra, obtained by measuring the intensity of delayed fluorescence at different excitation wavelengths and reflecting the photosynthetic pigment composition, is used to determine chlorophyll concentration and phytoplankton composition [30, 94, 95].

Finally, it should be noted that the delayed fluorescence is the only method that can diverse live and dead cells during on-line monitoring. Its major advantage is that it is emitted only from cells that are photosynthetically active, that is, alive; thus, the additional signals from dead cell debris do not interfere with the measurements. Long-term DF emission also prevents interference problems with fluorescent backgrounds in natural samples [94]. Furthermore, DF can measure nano- and pico-plankton, which may be lost during filtration or may be unaccounted in direct microscopic analysis.

Since 1995, DF excitation spectroscopy has been used in addition to microscopic biovolume estimation to monitor the phytoplankton composition in Germany [104]. Within an EC project "Phytoplankton On-Line," started in 2000, an on-line version of the DF excitation spectrometer has been developed to monitor continuously the composition of phytoplankton species in Lake Erken, Sweden, Lake Balaton, Hungary and Lake Kinneret, Israel.

3.4 Flow cytometry

Flow cytometry (FCM) is a rapid technique for the quantitative measurement of the properties of individual cells [105, 106]. It is a method for conducting a multi-parameter analysis of cells suspended in liquid and passing through a laser beam. In FCM, the excitation and emission properties of individual cells at one or more wavelength are measured together with the light scatter parameters, such as perpendicular light scatter, forward light scatter, and time of flight. Usually in order to detect any fluorescence, suspended cells require staining with organic dyes. In cyanobacteria, self-fluorescence of photopigments (chlorophylls and phycobilins) not only allows cells to be counted in a rapid manner but also enables a wide range of potential applications in ecological and biochemical studies, such as definition of biomass, distribution, and dynamics of phytoplankton [33, 107–109], discrimination and identification of algae species [110–112], and the investigation of the responses to contaminants [113–116]. It should be noted that the presence of these photopigments limits the possibility to stain cells with synthetic dyes as their excitation and emission spectra can greatly overlap. Thus, the attention should be paid while choosing appropriate fluorescent dyes with respect to the particular microorganism and the available lasers with which the FCM instrument is equipped [117].

For example, recently, an FCM protocol for quantitative and rapid analysis of *Microcystis* cells and colonies in lake sediment was developed [118]. Because of its many advantages, including rapidity, sensitivity, multiparametric analysis, and cell sorting capacity, FCM has been increasingly applied to environmental studies with aquatic microorganisms.

Various FCM instruments are also coupled with cell sorters [106, 119]. In this case, individual cells can be diverted from the fluid stream based on their properties such as cell morphology, taxonomic position, surface and intracellular protein expression, gene expression, and cellular physiology and collected into homogenous fractions for further analyses [120]. In spite of different cell viability after such a process (especially for fragile species), FCM featuring a sorting device has been successfully applied to isolate different phytoplankton taxa from mixed cultures at a success rate of 80% [121].

Initially, most of the commercially available flow cytometers cannot be directly applied to the study of aquatic biota, due to the large size range and low densities of the field samples. However, in 1989, the so-called Optical Plankton Analyzer has been designed by a group of American researchers [122]. Later, several companies in the United Kingdom and United States began to produce new modifications of flow cytometers for limnological and oceanographical applications [123]. Currently, the major producers in this field include Beckman Coulter, Becton-Dickinson, Thermo Fisher Scientific, and DakoCytomation provide a number of devices equipped with several lasers with different emitted wavelengths that allow studying cells stream flowing at a speed of over 100,000 cells per second. One of the advantages of FCM is the possibility of rapid and automated measurement of very small volume of the tested samples (1 cm^3 1 × 10^6 cells).

The particularity of studying cyanobacteria via FCM lays in the right choice of excitation and emission wavelengths for recording fluorescent signal, as it was described in previous sections. Measurement of chlorophyll and phycobilin fluorescence via FCM can provide the information on changes in the efficiency of photosynthetic apparatus, and indirectly, it can impart information on the cell's condition and growth state. It can be used to test the impact of different factors on cyanobacteria development, that is pH, toxic metals, herbicides, and other pollutants, under controlled laboratory conditions [114, 115, 124]. In addition to chlorophyll fluorescence, the fluorescence emission of phycocyanin pigment-protein complexes that participate in transferring signals also can be used in environmental stress studies of cyanobacteria, similar to those based on chlorophyll *a* measurement. Thus, cyanobacteria are well suited to flow cytometric analysis.

Considering that FCM equipped with a dual laser is able to measure chlorophyll *a* and phycocyanin, this method could also be of value in ecological studies of cyanobacteria. Once coupled with cell sorting, FCM can be successfully applied to isolate cyanobacteria from mixed phytoplankton samples. The recent emergence of advanced microfluidics, the availability of sophisticated digital image processing techniques, the advent of various optical imaging modalities, and image-guided cell sorting extend the capabilities of FCM from the measurements of one-dimensional signal intensities to multidimensional information-rich images, which can be used to analyze the spatial architecture of single cells in flow in addition to the analysis of the volume samples [125]. However, the presence of autofluorescence in cyanobacterial cells limits the use of staining cells with synthetic fluorochromes due to signal overlap. Moreover, the colonial forms and filamentous cyanobacteria (especially in freshwater species) are still rather difficult to study using FCM due to their different colony size and trichome lengths. In this case, the light scattering does not allow individual cells to be distinguished.

3.5 Confocal microscopic spectroscopy

Self-fluorescence of the photosynthetic apparatus of cyanobacteria originates from excited states of light-harvesting complex and PSII/PSI that were lost before photochemistry took place. It usually represents a small fraction of the excited state decay in photosynthetic complex of healthy cells. Nevertheless, this small fraction can be easily detected by confocal laser scanning microscopy (CLSM) [126, 127]. With CLSM, a very small excitation and detection areas can be investigated, so that single cells under nondamage conditions can be studied *in vivo*. In addition to above-mentioned techniques that now can be carried out on the single-cell level, CLSM provides some additional facilities. The investigation of self-fluorescence of single living cells reveals the relation between the physiological state and the operational activity of photosynthetic system. The investigation of self-fluorescence gives the information about single-cell processes as well as about the collaboration in cell communities. The changes in spectral characteristics of living photosynthetic cells indicate changes in their physiological state and can be applied for the studies of the results of stress states and external actions. Light-, heat-, ultrasound-, and toxin-induced changes can be distinguished by means of confocal microscopic spectroscopy since all these external actions are stress factors affecting photosynthetic process [44].

Modern CLSMs are unique tools for visualizing cellular structures and analyzing dynamic processes inside single cells. They exceed classical light microscopes in their axial resolution, which enables to acquire optical sections (slices) of a specimen. Thus, the object can be imaged in three dimensions and subsequently visualized as a 3D computer model. CLSM is designed for the quantification and analysis of image-coded information. Among other things, they allow easy determination of fluorescence intensities, distances, areas, and their changes over time. They are capable of quickly detecting and quantitatively unmixing the spectral signatures of fluorescent objects. Many useful parameters as the degree of colocalization of labeled structures or the ion concentration in a specimen can be analyzed [128]. The most interesting feature of CLSM is the capability for single-cell microscopic spectroscopy, which allows obtaining spectroscopic information inside small regions and single cells [35]. Another group of facilities deals with the quantitative analysis of dynamic processes in living cells using techniques such as fluorescence recovery after photobleaching (FRAP) [129, 130], fluorescence resonance energy transfer (FRET), fluorescence lifetime imaging microscopy (FLIM) [131], photoactivation, and photoconversion. Recently, a lot of additional techniques appear in modern CLSMs, which open new perspectives for single-cell investigation, such as white laser, which provides the ability to obtain not only fluorescence emission spectra, but also single-cell excitation and absorption spectra [132]; hyperspectral CLSM that allows more precise fluorescence spectra through the cell thickness and gives more detailed fluorescent pigments location [34, 133]; and STED and multiphotonic techniques, which extend the CLSM abilities to single-molecular studies [134, 135].

However, while working with living cells, the caution must be taken. The studies of the photosynthetic cells require some additional specific skills and techniques to perform measurements and to carry out data processing [35, 41, 42, 44]. The efficiency of photosynthesis and photosynthetic rate are highly dependent on irradiance. This can be seen in the light dependency of various photosynthetic parameters. Moreover, not only light quantity, but also light "quality" (wavelength) is an important factor [38]. Thus, special spectroscopic methods are required to study the physiology of phototrophic microorganisms [136]. These organisms employ light-dependent photosynthesis as the main energy source for their metabolism, and the detected self-fluorescence finally reflects the diversity in morphological and physiological states of their photosynthetic cells.

The main problem of *in vivo* investigations is to protect the object under consideration from light and heat damage. Since in CLSM the laser light density in the focus point is high, the dwell time and the intervals between the illuminations may influence photodamage and saturation of photosynthesis. It is well known that most chromophores and natural pigments bleach under the high laser excitation energies; thus, a bleach test should be performed [35]. It was shown experimentally that especially phycoerythrin and phycocyanin, as accessory pigments, were very sensitive to photobleaching, while the fluorescence of chlorophyll *a* and allophycocyanin remained stable in the intact cells [137]. Thus, during fluorescence detection of the main accessory pigments of cyanobacteria, the changes in their fluorescence intensity should be controlled during the recording period. The power of individual laser lines should be chosen according to the photodamage they cause. In some experiments, the repeated spectra were obtained under selected excitation power, and in this case, the excitation power would affect the cells (see **Figure 3**). In each case, the optimal zoom and excitation energy parameters should be selected, so the fluorescence spectra did not vary within the experimental error. When excitation energy is increased, both the intensity and shape of fluorescence spectra can change because of photodamage or structure breakdown in photosynthetic systems.

In the experiments, where several laser lines were involved in the investigation (see **Figure 2**), it is useful to record the first spectrum again at the end of each series to control the initial state of the cell. Moreover, the cell fluorescence intensities depend on the sensitivity setting of the photomultiplier, distance from the sample, and the excitation energy (which varies for different laser lines of CLSM); thus, the only qualitative analysis of spectral characteristics is acceptable in multispectra regime.

CLSM gives the opportunity to truly understand the cellular function; thus, we need to extend our imaging capabilities in ways that enable us to follow sequential events in real time, monitor the kinetics of dynamic processes, and record sensitive or transient events. Fluorescence spectra of cyanobacterial photosynthetic pigments are easily recorded by spectral CLSM. The fluorescence shares of individual phycobiliproteins can be reliably determined by spectral unmixing, showing that the spectral resolution of CLSM is well suited for this approach. In addition, CLSM technique allows the investigation not only cultivable but also the noncultivable cyanobacterial species.

3.6 Remote sensing

Methods of fluorescence spectroscopy suit very well for large-scale monitoring and remote sensing. They can be used in ecological studies, water quality control and on-line monitoring of cyanobacterial occurrence, distribution, and spatial and temporal dynamics in the environment [138–140]. Obviously, the presence of cyanobacteria and their photopigments influence reflectance, which is the ratio of radiation leaving the water (reflected) to the total amount of radiation incident on the water, is detected by remote sensing devices. Reflectance can be viewed by remote sensors called spectroradiometers, which can be either multispectral (sampling the electromagnetic spectrum over a relatively wide range of wavelengths in each discrete band) or hyperspectral (collecting and processing information from many narrow contiguous wavelength intervals across the electromagnetic spectrum). Remote sensing of chlorophyll *a* using its absorbance in the blue and red electromagnetic wavelengths has been successfully applied in the monitoring of phytoplankton blooms, and many single-band or band-ratio-type algorithms have been developed for chlorophyll *a* concentration retrieval from both multispectral and hyperspectral data obtained from ground-based, airborne, or satellite sensors [36]. However, the selective determination of cyanobacteria in the presence of other photoautotrophs requires the use of targeted algorithms (baseline, band-ratio, and semi-analytical nested band-ratio) for retrieval of phycocyanin concentration. Two spectral characteristics should be taken into account: a pronounced absorption peaks near 570 and 630 nm and two fluorescent peaks near 580 and 660 nm. Unfortunately, only in some instances, retrieval of phycocyanin concentrations at high cyanobacterial densities is possible from multispectral data, for example, provided by the MERIS satellite imaging spectrometer [141–143]. More sensitive hyperspectral remote sensing of phycocyanin is applied in cyanobacterial bloom monitoring using shipboard [144, 145] or airborne sensors [146, 147] and is eventually possible with the satellite Hyperion Hyperspectral Imager, which is capable of acquiring hyperspectral data [36, 148]. The methods are being continuously improved in their sensitivity, selectivity, resolution, robustness, and applicability, including new instrument designs or derivation of new algorithms for pigment concentration derivation [146, 149–153].

The most interesting instrument, providing active remote sensing of phytoplankton and dissolved organic matter (DOM) fluorescence, is a laser remote sensing spectrometer (LIDAR) used in shipborne laboratories [154–159]. A LIDAR fluorosensor for marine and freshwater applications is remotely excite and detect laser-induced fluorescence emissions by chromophore groups belonging to different species present in the water body. Four out of 12 available spectral channels, corresponding to water Raman backscattering (402 nm), DOM (450 nm), and phytoplankton pigments (phycocyanin, 650 nm; chlorophyll, 690 nm), were selected in wavelength by placing suitable interference filters. The water Raman peak is used to measure the water turbidity. This allows comparing different seawaters. Some laser transmitters of the LIDAR are also capable of operating in the dual pulse mode, to remotely measure the Chl *a* fluorescence yield according to the pump-and-probe excitation scheme [154].

Thus, the question of the applicability of the remote sensing to monitor the spatial distribution of cyanobacteria can be now answered affirmatively. Recent progress in spatial and spectral resolution of remote sensing allows distinction of cyanobacteria and various groups of algae on the basis of their photosynthetic pigments at a detailed spatial scale. In particular, quantification of cyanobacterial biomass and the recognition of cyanobacterial surface blooms by means of optical detection become true, and these advances allow field studies at the population level [160].

4. Conclusion

In the last years, fluorescence methods became a strong alternative to traditional methods of environmental research of cyanobacteria. They allow selective and sensitive quantification of cyanobacteria to be done quickly and easily in a nondestructive manner *in situ* with real-time data acquisition using fixed, portable, or remote-sensing devices, giving a unique opportunity for automated and continual monitoring of cyanobacterial communities by handheld instruments, shipboard sensors, or airborne and satellite platforms. Data obtained by fluorescence methods can valuably contribute to our understanding of the environmental distribution and dynamics of cyanobacteria, bloom formation, and cyanotoxin production and to water toxicity risk assessment. Large-scale remote sensing can give the information about the development of cyanobacterial blooms in oceans and freshwater bodies. On the other hand, the laboratory results of the single-cell spectroscopic investigations can give new information to improve remote sensing control. Airborne and satellite data should be also accompanied with shipboard (based phycocyanin fluorescence monitoring) and laboratory studies to provide deep insight to the cyanobacteria nature.

Conflict of interest

The authors declare no conflict of interest.

Author details

Natalia Yu. Grigoryeva
St. Petersburg Federal Research Center of Russian Academy of Science, Scientific-Research Centre for Ecological Safety, Saint-Petersburg, Russia

*Address all correspondence to: renes3@mail.ru

References

[1] Abed RM, Dobretsov S, Sudesh K. Applications of cyanobacteria in biotechnology. Journal of Applied Microbiology. 2009;**106**(1):1-12. DOI: 10.1111/j.1365-2672.2008.03918.x

[2] Vijayakumar S, Menakha M. Pharmaceutical applications of cyanobacteria—A review. Journal of Acute Medicine. 2015;**1**(5):15-23. DOI: 10.1016/j.jacme.2015.02.004

[3] Singh S, Kate BN, Banerjee UC. Bioactive compounds from cyanobacteria and microalgae: An overview. Critical Reviews in Biotechnology. 2005;**25**(3):73-95. DOI: 10.1080/07388550500248498

[4] Thajuddin N, Subramanian G. Cyanobacterial biodiversity and potential applications in biotechnology. Current Science. 2005;**89**(1):47-57

[5] Richmond A, editor. Handbook of Microalgal Culture: Biotechnology and Applied Phycology. Carlton: Blackwell Science Ltd.; 2004. p. 576. DOI: 10.1002/9780470995280

[6] Lei Y, Chen W, Mulchandani A. Microbial biosensors. Analytica Chimica Acta. 2006;**568**(1-2):200-210. DOI: 10.1016/j.aca.2005.11.065

[7] Reshetilov AN. Biosensor development in Russia. Biotechnology Journal: Healthcare Nutrition Technology. 2007;**2**(7):849-862. DOI: 10.1002/biot.200700021

[8] Ignatov SG, Ferguson JA, Walt DR. A fiber-optic lactate sensor based on bacterial cytoplasmic membranes. Biosensors & Bioelectronics. 2001;**16**(1-2):109-113. DOI: 10.1016/S0956-5663(00)00144-5

[9] Giardi MT, Rea G, Berra B, editors. Bio-farms for nutraceuticals: Functional food and safety control by biosensors. In: Advances in Experimental Medicine and Biology. Vol. 698. New York, USA: Springer Science & Business Media; 2011. p. 360. DOI: 10.1007/978-1-4419-7347-4

[10] Whitton BA, editor. Ecology of Cyanobacteria II: Their Diversity in Space and Time. Durham, UK: Springer Science & Business Media; 2012. p. 760. DOI: 10.1007/978-94-007-3855-3

[11] Gault PM, Marler HJ, editors. Handbook on Cyanobacteria: Biochemistry, Biotechnology and Applications. (Bacteriology Research Developments Series). New York: Nova Science Publishers; 2009. p. 538. ISBN: 978-1-60741-092-8

[12] Huisman J, Matthijs HC, Visser PM, editors. Harmful Cyanobacteria. Aquatic Ecology Series. Vol. 3. Dordrecht: Springer; 2005. p. 243. DOI: 10.1007/1-4020-3022-3

[13] Granéli E, Turner JT, editors. Ecology of Harmful Algae. Berlin: Springer Science & Business Media; 2006. p. 406. DOI: 10.1007/978-3-540-32210-8

[14] Seckbach J, editor. Algae and Cyanobacteria in Extreme Environments. Israel: Springer Science & Business Media; 2007. p. 786. DOI: 10.1007/978-1-4020-6112-7

[15] Bryant DA, editor. The Molecular Biology of Cyanobacteria (Advances in Photosynthesis 1). Dordrecht: Kluwer Academic Press; 1994. p. 881. DOI: 10.1007/978-94-011-0227-8

[16] Radix P, Léonard M, Papantoniou C, Roman G, Saouter E, Gallotti-Schmitt S, et al. Comparison of four chronic toxicity tests using algae, bacteria, and invertebrates assessed with sixteen chemicals. Ecotoxicology and Environmental Safety. 2000;**47**(2): 186-194. DOI: 10.1006/eesa.2000.1966

[17] Juneau P, Dewez D, Matsui S, Kim SG, Popovic R. Evaluation of different algal species sensitivity to mercury and metolachlor by PAM-fluorometry. Chemosphere. 2001;**45**(4-5):589-598. DOI: 10.1016/S0045-6535(01)00034-0

[18] Solé A, Gaju N, Méndez-Álvarez S, Esteve I. Confocal laser scanning microscopy as a tool to determine cyanobacteria biomass in microbial mats. Journal of Microscopy. 2001;**204**(3):258-262. DOI: 10.1046/j.1365-2818.2001.00951.x

[19] Khishamuddin NA, Shing WL, Kin CM, Niu VB. Fluorometric response of photosynthetic microorganism consortium as potential bioindicator for heavy metals detection in water. EnvironmentAsia. 2018;**11**(1):80-86. DOI: 10.14456/ea.2018.6

[20] Teo SC, Wong LS. Whole cell-based biosensors for environmental heavy metals detection. Annual Research & Review in Biology. 2014:2663-2674. DOI: 10.9734/ARRB/2014/9472

[21] Wong LS, Lee YH, Surif S. Performance of a cyanobacteria whole cell-based fluorescence biosensor for heavy metal and pesticide detection. Sensors. 2013;**13**(5):6394-6404. DOI: 10.3390/s130506394

[22] Los DA, Zorina A, Sinetova M, Kryazhov S, Mironov K, Zinchenko VV. Stress sensors and signal transducers in cyanobacteria. Sensors. 2010;**10**(3):2386-2415. DOI: 10.3390/s100302386

[23] Mironov KS, Sinetova MA, Shumskaya M, Los DA. Universal molecular triggers of stress responses in *Cyanobacterium synechocystis*. Lifestyles. 2019;**9**(3):67. DOI: 10.3390/life9030067

[24] Zung JB, Woodlee RL, Fuh MR, Warner IM. Preliminary investigation of the effect of select pollutants on marine phytoplankton using multidimensional fluorescence measurements. International Journal of Environmental Analytical Chemistry. 1990;**41**(3-4):149-158. DOI: 10.1080/03067319008027357

[25] Simis SG, Huot Y, Babin M, Seppälä J, Metsamaa L. Optimization of variable fluorescence measurements of phytoplankton communities with cyanobacteria. Photosynthesis Research. 2012;**112**(1):13-30. DOI: 10.1007/s11120-012-9729-6

[26] Stirbet A. On the relation between the Kautsky effect (chlorophyll a fluorescence induction) and photosystem II: Basics and applications of the OJIP fluorescence transient. Journal of Photochemistry and Photobiology, B: Biology. 2011;**104**(1-2):236-257. DOI: 10.1016/j.jphotobiol.2010.12.010

[27] Goltsev VN, Kalaji HM, Paunov M, Bąba W, Horaczek T, Mojski J, et al. Variable chlorophyll fluorescence and its use for assessing physiological condition of plant photosynthetic apparatus. Russian Journal of Plant Physiology. 2016;**63**(6):869-893. DOI: 10.1134/S1021443716050058

[28] Schreiber U, Endo T, Mi H, Asada K. Quenching analysis of chlorophyll fluorescence by the saturation pulse method: Particular aspects relating to the study of eukaryotic algae and cyanobacteria. Plant & Cell Physiology. 1995;**36**(5):873-882. DOI: 10.1093/oxfordjournals.pcp.a078833

[29] Ogawa T, Misumi M, Sonoike K. Estimation of photosynthesis in cyanobacteria by pulse-amplitude modulation chlorophyll fluorescence: Problems and solutions. Photosynthesis Research. 2017;**133**(1-3):63-73. DOI: 10.1007/s11120-017-0367-x

[30] Berden-Zrimec M, Drinovec L, Zrimec A. Delayed fluorescence. In:

Suggett D, Prášil O, Borowitzka M, editors. Chlorophyll a Fluorescence in Aquatic Sciences: Methods and Applications. Dordrecht: Springer; 2010. pp. 293-309. DOI: 10.1007/978-90-481-9268-7_14

[31] Goltsev V, Zaharieva I, Chernev P, Strasser RJ. Delayed fluorescence in photosynthesis. Photosynthesis Research. 2009;**101**(2-3):217-232. DOI: 10.1007/s11120-009-9451-1

[32] Ormerod MG, Imrie PR. Flow cytometry. Methods in Molecular Biology. 1990;**5**:543-558. DOI: 10.1385/0-89603-150-0:543

[33] Jonker RR, Meulemans JT, Dubelaar GB, Wilkins MF, Ringelberg J. Flow cytometry: A powerful tool in analysis of biomass distributions in phytoplankton. Water Science and Technology. 1995;**32**(4):177-182. DOI: 10.1016/0273-1223(95)00696-6

[34] Vermaas WF, Timlin JA, Jones HD, Sinclair MB, Nieman LT, Hamad SW, et al. *In vivo* hyperspectral confocal fluorescence imaging to determine pigment localization and distribution in cyanobacterial cells. Proceedings of the National Academy of Sciences. 2008;**105**(10):4050-4055. DOI: 10.1073/pnas.0708090105

[35] Grigoryeva N, Chistyakova L. Confocal laser scanning microscopy for spectroscopic studies of living photosynthetic cells. In: Zeng L-W, Cao S-L, editors. Color Detection. UK: IntechOpen; 2020. pp. 39-64. DOI: 10.5772/intechopen.84825

[36] Kutser T. Passive optical remote sensing of cyanobacteria and other intense phytoplankton blooms in coastal and inland waters. International Journal of Remote Sensing. 2009;**30**(17):4401-4425. DOI: 10.1080/01431160802562305

[37] Churnside JH, Shaw JA. Lidar remote sensing of the aquatic environment. Applied Optics. 2020;**59**(10):C92-C99. DOI: 10.1364/AO.59.000C92

[38] Papageorgiou GC. Fluorescence emission from the photosynthetic apparatus. In: Eaton-Rye J, Tripathy B, Sharkey T, editors. Photosynthesis. Advances in Photosynthesis and Respiration. Vol. 34. Dordrecht: Springer; 2012. pp. 415-443. DOI: 10.1007/978-94-007-1579-0_18

[39] Green BR, Parson WW, editors. Light-Harvesting Antennas in Photosynthesis. Advances in Photosynthesis and Respiration, Vol. 13. Dordrecht: Springer Netherlands Kluwer Academic Publishers; 2003. p. 514. DOI: 10.1007/978-94-017-2087-8

[40] Blankenship RE. Molecular Mechanisms of Photosynthesis. 2nd ed. Chichester, UK: John Wiley & Sons; 2014. p. 314. DOI: 10.1002/9780470758472

[41] Grigoryeva N, Chistyakova L. Fluorescence microscopic spectroscopy for investigation and monitoring of biological diversity and physiological state of cyanobacterial cultures. In: Tiwari A, editor. Cyanobacteria. Rijeka: IntechOpen; 2018. pp. 11-44. DOI: 10.5772/intechopen.78044

[42] Grigoryeva N. Self-fluorescence of photosynthetic system—A powerful tool for investigation of microalgal biological diversity. In: Vítová M, editor. Microalgae—From Physiology to Application. Rijeka: IntechOpen; 2019. pp. 3-26. DOI: 10.5772/intechopen.88785

[43] Sirenko LA, Sakevich AI, Osipovich LF. Methods for Physiological and Biochemical Researching Algae in Hydrobiological Practice. Kiev: Naukova Dumka; 1975. [in Russian]

[44] Grigoryeva NY, Chistyakova LV, Liss AA. Spectroscopic techniques for

estimation of physiological state of blue-green algae after weak external action. Oceanology. 2018;**58**(6):896-904. DOI: 10.1134/S0001437018060061

[45] Mineeva NM, xKorneva LG, Solovyova VV. Influence of environemntal factors on phytoplankton photosynthetic activity in the Volga River reservoirs. Inland Water biology. 2016;**9**(3):258-267. DOI: 10.1134/S1995082916030160

[46] Krause GH, Weis E. Chlorophyll fluorescence as a tool in plant physiology. Photosynthesis Research. 1984;**5**(2):139-157. DOI: 10.1007/BF00028527

[47] Papageorgiou GC, Tsimilli-Michael M, Stamatakis K. The fast and slow kinetics of chlorophyll a fluorescence induction in plants, algae and cyanobacteria: A viewpoint. Photosynthesis Research. 2007;**94**(2-3):275-290. DOI: 10.1007/s11120-007-9193-x

[48] Schuurmans RM, van Alphen P, Schuurmans JM, Matthijs HC, Hellingwerf KJ. Comparison of the photosynthetic yield of cyanobacteria and green algae: Different methods give different answers. PLoS One. 2015;**10**(9):e0139061. DOI: 10.1371/journal.pone.0139061

[49] Remelli W, Santabarbara S. Excitation and emission wavelength dependence of fluorescence spectra in whole cells of the cyanobacterium Synechocystis sp. PPC6803: Influence on the estimation of photosystem II maximal quantum efficiency. Biochimica et Biophysica Acta (BBA)-Bioenergetics. 2018;**1859**(11):1207-1222. DOI: 10.1016/j.bbabio.2018.09.366

[50] Yentsch CS, Yentsch CM. Fluorescence spectral signatures: The characterization of phytoplankton populations by the use of excitation and emission spectra. Journal of Marine Research. 1979;**37**(3):471-483

[51] Richardson TL, Lawrenz E, Pinckney JL, Guajardo RC, Walker EA, Paerl HW, et al. Spectral fluorometric characterization of phytoplankton community composition using the algae online analyser. Water Research. 2010;**44**(8):2461-2472. DOI: 10.1016/j.watres.2010.01.012

[52] Suggett DJ, Borowitzka MA, Prášil O. Editors. Chlorophyll a Fluorescence in Aquatic Sciences: Methods and Applications (Developments in Applied Phycology 4). Dordrecht, The Netherlands: Springer; 2010. p. 326. DOI: 10.1007/978-90-481-9268-7

[53] Karsten U, Klimant I, Holst G. A new *in vivo* fluorimetric technique to measure growth of adhering phototrophic microorganisms. Applied and Environmental Microbiology. 1996;**62**(1):237-243

[54] Johnson TJ, Hildreth MB, Gu L, Baldwin EL, Zhou R, Gibbons WR. Evaluating viable cell indicators for filamentous cyanobacteria and their application. Journal of Microbiology, Biotechnology and Food Sciences. 2019;**2019**:886-893. DOI: 10.15414/jmbfs.2016/17.6.3.886-893

[55] Wong LS, Lee YH, Surif S. The fluorometric response of cyanobateria to short exposure of heavy metal. Advances in Environmental Biology. 2012;**6**(1):103-108

[56] Millan-Almaraz JR, Guevara-Gonzalez RG, de Jesus R-TR, Osornio-Rios RA, Torres-Pacheco I. Advantages and disadvantages on photosynthesis measurement techniques: A review. African Journal of Biotechnology. 2009;**8**(25):7340-7349. DOI: 10.5897/AJB2009.000-9569

[57] Eilola K, Meier HM, Almroth E. On the dynamics of oxygen, phosphorus and cyanobacteria in the Baltic Sea; a model study. Journal of Marine Systems.

2009;**75**(1-2):163-184. DOI: 10.1016/j.jmarsys.2008.08.009

[58] DeEll JR, Toivonen PM, editors. Practical Applications of Chlorophyll Fluorescence in Plant Biology. New York, USA: Springer Science & Business Media; 2012. p. 271. DOI: 10.1007/978-1-4615-0415-3

[59] Babu NG, Sarma PA, Attitalla IH, Murthy SD. Effect of selected heavy metal ions on the photosynthetic electron transport and energy transfer in the thylakoid membrane of the cyanobacterium, *Spirulina platensis*. Acad J Plant Sci. 2010;**3**(1):46-49

[60] Zayadan BK, Akmuhanova NR, Sadvakasova AK, Kirbaeva DK, Bolatkhan K, Bauyenova MO. Influence of heavy metals on fluorescence activity of perspective strains of microalgae and cyanobacteria. International Journal of Biology and Chemistry. 2016;**9**(1):42-45. DOI: 10.26577/2218-7979-2016-9-1-42-45

[61] MacIntyre HL, Lawrenz E, Richardson TL. Taxonomic discrimination of phytoplankton by spectral fluorescence. In: Suggett D, Prášil O, Borowitzka M, editors. Chlorophyll a Fluorescence in Aquatic Sciences: Methods and Applications. Dordrecht: Springer; 2010. pp. 129-169. DOI: 10.1007/978-90-481-9268-7_7

[62] Zhang F, Su R, He J, Cai M, Luo W, Wang X. Identifying phytoplankton in seawater based on discrete excitation-emission fluorescence spectra. Journal of Phycology. 2010;**46**(2):403-411. DOI: 10.1111/j.1529-8817.2009.00805.x

[63] Zhang QQ, Lei SH, Wang XL, Wang L, Zhu CJ. Discrimination of phytoplankton classes using characteristic spectra of 3D fluorescence spectra. Spectrochimica Acta Part A: Molecular and Biomolecular Spectroscopy. 2006;**63**(2):361-369. DOI: 10.1016/j.saa.2005.05.041

[64] Beutler M, Wiltshire KH, Meyer B, Moldaenke C, Lüring C, Meyerhöfer M, et al. A fluorometric method for the differentiation of algal populations *in vivo* and *in situ*. Photosynthesis Research. 2002;**72**(1):39-53. DOI: 10.1023/A:1016026607048

[65] Strasser RJ, Srivastava A, Govindjee. Polyphasic chlorophyll a fluorescence transient in plants and cyanobacteria. Photochemistry and Photobiology. 1995;**61**(1):32-42. DOI: 10.1111/j.1751-1097.1995.tb09240.x

[66] Stamatakis K, Tsimilli-Michael M, Papageorgiou GC. Fluorescence induction in the phycobilisome-containing cyanobacterium *Synechococcus* sp PCC 7942: Analysis of the slow fluorescence transient. Biochimica et Biophysica Acta (BBA)-Bioenergetics. 2007;**1767**(6):766-772. DOI: 10.1016/j.bbabio.2007.02.009

[67] Campbell D, Hurry V, Clarke AK, Gustafsson P, Öquist G. Chlorophyll fluorescence analysis of cyanobacterial photosynthesis and acclimation. Microbiology and Molecular Biology Reviews. 1998;**62**(3):667-683. DOI: 10.1128/MMBR.62.3.667-683.1998

[68] Genty B, Briantais J-M, Baker NR. The relationship between the quantum yield of photosynthetic electron transport and quenching of chlorophyll fluorescence. Biochimica et Biophysica Acta (BBA) - General Subjects. 1989;**990**(1):87-92. DOI: 10.1016/s0304-4165(89)80016-9

[69] Van Kooten O, Snel JF. The use of chlorophyll fluorescence nomenclature in plant stress physiology. Photosynthesis Research. 1990;**25**(3):147-150. DOI: 10.1007/BF00033156

[70] Campbell D, Oquist G. Predicting light acclimation in cyanobacteria from nonphotochemical quenching

of photosystem II fluorescence, which reflects state transitions in these organisms. Plant Physiology. 1996;**111**(4):1293-1298. DOI: 10.1104/pp.111.4.1293

[71] Hirano M, Satoh K, Katoh S. Plastoquinone as a common link between photosynthesis and respiration in a blue-green alga. Photosynthesis Research. 1980;**1**(3):149-162. DOI: 10.1007/BF00020594

[72] Scherer S. Do photosynthetic and respiratory electron transport chains share redox proteins? Trends in Biochemical Sciences. 1990;**15**(12):458-462. DOI: 10.1016/0968-0004(90)90296-N

[73] Demmig-Adams B, Garab G, Adams W III, editors. Non-photochemical Quenching and Energy Dissipation in Plants, Algae and Cyanobacteria. vol. 40. Dordrecht: Springer Netherlands; 2014. p. 650. DOI: 10.1007/978-94-017-9032-1

[74] Bailey S, Grossman A. Photoprotection in cyanobacteria: Regulation of light harvesting. Photochemistry and Photobiology. 2008;**84**(6):1410-1420. DOI: 10.1111/j.1751-1097.2008.00453.x

[75] Kirilovsky D, Kerfeld CA. The orange carotenoid protein in photoprotection of photosystem II in cyanobacteria. Biochimica et Biophysica Acta. 1817;**2012**:158-166. DOI: 10.1016/j.bbabio.2011.04.013

[76] Ogawa T, Sonoike K. Dissection of respiration and photosynthesis in the cyanobacterium *Synechocystis* sp. PCC6803 by the analysis of chlorophyll fluorescence. Journal of Photochemistry and Photobiology, B: Biology. 2015;**144**:61-67. DOI: 10.1016/j.jphotobiol.2015.02.005

[77] Misumi M, Katoh H, Tomo T, Sonoike K. Relationship between photochemical quenching and non-photochemical quenching in six species of cyanobacteria reveals species difference in redox state and species commonality in energy dissipation. Plant & Cell Physiology. 2016;**57**(7):1510-1517. DOI: 10.1093/pcp/pcv185

[78] Acuña AM, Snellenburg JJ, Gwizdala M, Kirilovsky D, Van Grondelle R, Van Stokkum IH. Resolving the contribution of the uncoupled phycobilisomes to cyanobacterial pulse-amplitude modulated (PAM) fluorometry signals. Photosynthesis Research. 2016;**127**(1):91-102. DOI: 10.1007/s11120-015-0141-x

[79] Strasser RJ. Govindjee. The Fo and the O–J–I–P fluorescence rise in higher plants and algae. In: Argyroudi-Akoyunoglou JH, editor. Regulation of Chloroplast Biogenesis. Nato ASI Series (Series a: Life Sciences). Vol. 226. Boston: Springer; 1992. pp. 423-426. DOI: 10.1007/978-1-4615-3366-5_60

[80] Schreiber U. Pulse-amplitude-modulation (PAM) fluorometry and saturation pulse method: An overview. In: Papageorgiou GC, Govindjee, editors. Chlorophyll a Fluorescence. Advances in Photosynthesis and Respiration. Vol. 19. Dordrecht: Springer; 2004. pp. 279-319. DOI: 10.1007/978-1-4020-3218-9_11

[81] Falkowski PG, Wyman K, Ley AC, Mauzerall DC. Relationship of steady-state photosynthesis to fluorescence in eucaryotic algae. Biochimica et Biophysica Acta (BBA)-Bioenergetics. 1986;**849**(2):183-192. DOI: 10.1016/0005-2728(86)90024-1

[82] Kolber ZS, Prášil O, Falkowski PG. Measurements of variable chlorophyll fluorescence using fast repetition rate techniques: Defining methodology and experimental protocols. Biochimica et Biophysica Acta (BBA)-Bioenergetics.

1998;(1-3):1367, 88-1106. DOI: 10.1016/S0005-2728(98)00135-2

[83] Olson RJ, Chekalyuk AM, Sosik HM. Phytoplankton photosynethetic characteristics from fluorescence induction assays of individual cells. Limnology and Oceanography. 1996;**41**(6):1253-1263. DOI: 10.4319/lo.1996.41.6.1253

[84] Gorbunov MY, Falkowski PG. Fluorescence induction and relaxation (FIRe) technique and instrumentation for monitoring photosynthetic processes and primary production in aquatic ecosystems. In: Proceedings of 13th International Congress of Photosynthesis. 29-30 August 2004. Vol. 2. Montreal, QC: Allen Press; 2004. pp. 1029-1031

[85] Johnson ZI. Description and application of the background irradiance gradient-single turnover fluorometer (BIG-STf). Marine Ecology Progress Series. 2004;**283**:73-80. DOI: 10.3354/meps283073

[86] Chekalyuk A, Hafez M. Advanced laser fluorometry of natural aquatic environments. Limnology and Oceanography: Methods. 2008;**6**(11):591-609. DOI: 10.4319/lom.2008.6.591

[87] Schreiber U, Klughammer C, Kolbowski J. Assessment of wavelength-dependent parameters of photosynthetic electron transport with a new type of multi-color PAM chlorophyll fluorometer. Photosynthesis Research. 2012;**113**(1-3):127-144. DOI: 10.1007/s11120-012-9758-1

[88] Lysenko V, Lazar D, Varduny T. A method of a bicolor fast-Fourier pulse-amplitude modulation chlorophyll fluorometry. Photosynthetica. 2018;**56**(4):1447-1452. DOI: 10.1007/s11099-018-0848-y

[89] Chazalon F, Rabouille S, Hartmann P, Sciandra A, Bernard O. A dynamical model to study the response of microalgae to pulse amplitude modulated fluorometry. IFAC Proceedings Volumes. 2014;**47**(3):7152-7157. DOI: 10.3182/20140824-6-ZA-1003.02162

[90] Breuer F, Dören L, Ebke KP. Comparison of four measuring techniques to assess growth inhibition in standardized tests with seven freshwater algae and cyanobacteria. Toxicological and Environmental Chemistry. 2016;**98**(8):848-859. DOI: 10.1080/02772248.2016.1139116

[91] Bodemer U. Variability of phycobiliproteins in cyanobacteria detected by delayed fluorescence excitation spectroscopy and its relevance for determination of phytoplankton composition of natural water samples. Journal of Plankton Research. 2004;**26**(10):1147-1162. DOI: 10.1093/plankt/fbh105

[92] Gerhardt V, Bodemer U, Eckert W. Determination of phytoplankton colour classes: Online excitation spectroscopy of the delayed fluorescence (DF) of algae. Internationale Vereinigung für theoretische und angewandte Limnologie: Verhandlungen. 2005;**29**(1):306-309. DOI: 10.1080/03680770.2005.11902019

[93] Drinovec L, Flander-Putrle V, Knez M, Beran A, Berden-Zrimec M. Discrimination of marine algal taxonomic groups using delayed fluorescence spectroscopy. Environmental and Experimental Botany. 2011;**73**:42-48. DOI: 10.1016/j.envexpbot.2010.10.010

[94] Istvánovics V, Honti M, Osztoics A, Shafik HM, Padisak J, Yacobi Y, et al. Continuous monitoring of phytoplankton dynamics in Lake Balaton (Hungary) using on-line delayed fluorescence excitation spectroscopy. Freshwater Biology.

2005;**50**(12):1950-1970. DOI: 10.1111/j.1365-2427.2005.01442.x

[95] Yacobi YZ, Gerhardt V, Gonen-Zurgil Y, Sukenik A. Delayed fluorescence excitation spectroscopy: A rapid method for qualitative and quantitative assessment of natural population of phytoplankton. Water Research. 1998;**32**(9):2577-2582. DOI: 10.1016/s0043-1354(98)00032-3

[96] Gerhardt V, Bodemer U. Delayed fluorescence: An *in vivo* method for electron transport studies and on line applications in limnology. In: Proceedings of the 12th International Congress on Photosynthesis. 18-23 August 2001. Vol. 3, No. 1. Queensland, Australia: Science Access; 2001. S14-009. DOI: 10.1071/SA0403352

[97] Zhang L, Xing D. Rapid determination of the damage to photosynthesis caused by salt and osmotic stresses using delayed fluorescence of chloroplasts. Photochemical & Photobiological Sciences. 2008;**7**(3):352-360. DOI: 10.1039/B714209A

[98] Omagari R, Katsumata M, Arizono K, Ishibashi Y. Toxicity evaluation of 4-Nonylphenol isomers with ecotoxicological bioassay using delayed luminescence in the green alga *Pseudokirchneriella subcapitata*. Journal of Environment and Safety. 2017;**8**(2):57-63. DOI: 10.11162/daikankyo.E16RP0801

[99] Yamagishi T, Katsumata M, Yamaguchi H, Shimura Y, Kawachi M, Koshikawa H, et al. Rapid ecotoxicological bioassay using delayed fluorescence in the marine cyanobacterium *Cyanobium* sp. (NIES-981). Ecotoxicology. 2016;**25**(10):1751-1758. DOI: 10.1007/s10646-016-1718-7

[100] Leunert F, Grossart HP, Gerhardt V, Eckert W. Toxicant induced changes on delayed fluorescence decay kinetics of cyanobacteria and green algae: A rapid and sensitive biotest. PLoS One. 2013;**8**(4):e63127. DOI: 10.1371/journal.pone.0063127

[101] Kurzbaum E, Eckert W, Yacobi YZ. Delayed fluorescence as a direct indicator of diurnal variation in quantum and radiant energy utilization efficiencies of phytoplankton. Photosynthetica. 2007;**45**(4):562-567. DOI: 10.1007/s11099-007-0096-z

[102] Kaurov Y, Aksyonova GE, Lovyagina ER, Ivanov II, Rubin AB. Thermally-induced delayed fluorescence of photosystem I and II chlorophyll in thermophilic cyanobacterium. General Physiology and Biophysics. 1992;**11**:229-239

[103] Greisberger S, Teubner K. Does pigment composition reflect phytoplankton community structure in differing temperature and light conditions in a deep alpine lake? An approach using HPLC and delayed fluorescence techniques. Journal of Phycology. 2007;**43**(6):1108-1119. DOI: 10.1111/j.1529-8817.2007.00404.x

[104] Bodemer U, Gerhardt V, Yacobi YZ, Zohary Z, Friedrich G, Pohlmann M. Phytoplankton abundance and composition of freshwaters systems determined by DF excitation spectroscopy and conventional methods. Archiv für Hydrobiologie Special Issues Advances in Limnology. 2000;**55**:87-100

[105] Cunningham A. Analysis of microalgae and cyanobacteria by flow cytometry. In: Lloyd D, editor. Flow Cytometry in Microbiology. London: Springer; 1993. pp. 131-142. DOI: 10.1007/978-1-4471-2017-9_10

[106] Poniedziałek B, Falfushynska HI, Rzymski P. Flow cytometry as a valuable tool to study cyanobacteria: A mini-review. Limnological Review.

2017;**17**(2):89-95. DOI: 10.1515/limre-2017-0009

[107] Olson RJ, Chisholm SW, Zettler ER, Altabet MA, Dusenberry JA. Spatial and temporal distributions of prochlorophyte picoplankton in the North Atlantic Ocean. Deep Sea Research Part A. Oceanographic Research Papers. 1990;**37**(6):1033-1051. DOI: 10.1016/0198-0149(90)90109-9

[108] Moreira-Turcq PF, Cauwet G, Martin JM. Contribution of flow cytometry to estimate picoplankton biomass in estuarine systems. Hydrobiologia. 2001;**462**(1-3):157-168. DOI: 10.1023/a:1013138317897

[109] Mukhanov VS, Rylkova OA, Churilova TY, Sakhon EG, Pimenov NV. Structure and seasonal trophodynamics of picophytoplankton in Sevastopol bay and adjacent waters (the Black Sea). Microbiology. 2016;**85**(5):553-561. DOI: 10.1134/S002626171605012X

[110] Mukhanov VS, Naidanova OG, Lopukhina OA, Kemp RB. Cell-, biovolume-and biosurface-specific energy fluxes through marine picoplankton as a function of the assemblage size structure. Thermochimica Acta. 2007;**458**(1-2): 23-33. DOI: 10.1016/j.tca.2007.02.024

[111] Boddy L, Morris CW, Wilkins MF, Al-Haddad L, Tarran GA, Jonker RR, et al. Identification of 72 phytoplankton species by radial basis function neural network analysis of flow cytometric data. Marine Ecology Progress Series. 2000;**195**:47-59. DOI: 10.3354/meps195047

[112] Hofstraat JW, De Vreeze ME, Van Zeijl WJ, Peperzak L, Peeters JC, Balfoort HW. Flow cytometric discrimination of phytoplankton classes by fluorescence emission and excitation properties. Journal of Fluorescence. 1991;**1**(4):249-265. DOI: 10.1007/BF00865249

[113] Dudkowiak A, Olejarz B, Łukasiewicz J, Banaszek J, Sikora J, Wiktorowicz K. Heavy metals effect on cyanobacteria *Synechocystis aquatilis* study using absorption, fluorescence, flow cytometry, and photothermal measurements. International Journal of Thermophysics. 2011;**32**(4):762-773. DOI: 10.1007/s10765-010-0852-3

[114] Stauber JL, Franklin NM, Adams MS. Applications of flow cytometry to ecotoxicity testing using microalgae. Trends in Biotechnology. 2002;**20**(4):141-143. DOI: 10.1016/S0167-7799(01)01924-2

[115] Franklin NM, Stauber JL, Lim RP. Development of flow cytometry-based algal bioassays for assessing toxicity of copper in natural waters. Environmental Toxicology and Chemistry: An International Journal. 2001;**20**(1): 160-170. DOI: 10.1002/etc.5620200118

[116] Li WK. Primary production of prochlorophytes, cyanobacteria, and eucaryotic ultraphytoplankton: Measurements from flow cytometric sorting. Limnology and Oceanography. 1994;**39**(1):169-175. DOI: 10.4319/lo.1994.39.1.0169

[117] Karo O, Wahl A, Nicol SB, Brachert J, Lambrecht B, Spengler HP, et al. Bacteria detection by flow cytometry. Clinical Chemistry and Laboratory Medicine (CCLM). 2008;**46**(7):947-953. DOI: 10.1515/CCLM.2008.156

[118] Zhou Q, Chen W, Zhang H, Peng L, Liu L, Han Z, et al. A flow cytometer based protocol for quantitative analysis of bloom-forming cyanobacteria (Microcystis) in lake sediments. Journal of Environmental Sciences. 2012;**24**(9):1709-1716. DOI: 10.1016/S1001-0742(11)60993-5

[119] Davey HM, Kell DB. Flow cytometry and cell sorting of heterogeneous microbial populations: The importance of single-cell

analyses. Microbiological Reviews. 1996;**60**(4):641-696

[120] Ibrahim SF, van den Engh G. Flow cytometry and cell sorting. Advances in Biochemical Engineering/ Biotechnology. 2007;**106**:19-39. DOI: 10.1007/10_2007_073

[121] Cellamare M, Rolland A, Jacquet S. Flow cytometry sorting of freshwater phytoplankton. Journal of Applied Phycology. 2010;**22**(1):87-100. DOI: 10.1007/s10811-009-9439-4

[122] Dubelaar GB, Groenewegen AC, Stokdijk W, Van Den Engh GJ, Visser JW. Optical plankton analyser: A flow cytometer for plankton analysis, II: Specifications. Cytometry: The Journal of the International Society for Analytical Cytology. 1989;**10**(5):529-539. DOI: 10.1002/cyto.990100508

[123] Boddy L, Wilkins MF, Morris CW, Tarran GA, Burkill PH, Jonker RR. Techniques for neural network identification of phytoplankton for the EurOPA flow cytometer. In: Proceedings of OCEANS'94; 1994 Sep 13-16. Vol. 1. Brest, France: IEEE. p. I-565. DOI: 10.1109/oceans.1994.363967

[124] Rzymski P, Poniedzialek B, Niedzielski P, Tabaczewski P, Wiktorowicz K. Cadmium and lead toxicity and bioaccumulation in *Microcystis aeruginosa*. Frontiers of Environmental Science & Engineering. 2014;**8**(3):427-432. DOI: 10.1007/ s11783-013-0566-4

[125] Goda K, Filby A, Nitta N. In flow cytometry, image is everything. Cytometry Part A. 2019;**95**(5):475-477. DOI: 10.1002/cyto.a.23778

[126] Pawley JB, editor. Handbook of Biological Confocal Microscopy. Boston, MA: Springer; 1995. p. 346. DOI: 10.1007/978-1-4757-5348-6

[127] Wilson T. Confocal Microscopy. London, etc: Academic Press; 1990. p. 426. DOI: 10.1007/978-1-4615-7133-9

[128] Yang S, Su Z, Li H, Feng J, Xie J, Xia A, et al. Demonstration of phycobilisome mobility by the time-and space-correlated fluorescence imaging of a cyanobacterial cell. Biochimica et Biophysica Acta (BBA)-Bioenergetics. 2007;**1767**(1):15-21. DOI: 10.1016/j. bbabio.2006.11.012

[129] Mullineaux CW. FRAP analysis of photosynthetic membranes. Journal of Experimental Botany. 2004;**55**(400):1207-1211. DOI: 10.1093/ jxb/erh106

[130] Liu LN, Aartsma TJ, Thomas JC, Zhou BC, Zhang YZ. FRAP analysis on red alga reveals the fluorescence recovery is ascribed to intrinsic photoprocesses of phycobilisomes than large-scale diffusion. PLoS One. 2009;**4**(4):e5295. DOI: 10.1371/journal. pone.0005295

[131] Vitali M, Reis M, Friedrich T, Eckert HJ. A wide-field multi-parameter FLIM and FRAP setup to investigate the fluorescence emission of individual living cyanobacteria. In: Proceedings of SPIE Conference on Laser Applications in Life Sciences (Vol. 7376); 9-11 June 2010; Oulu, Finland: SPIE; 2010. p. 737610-737610-6. DOI: 10.1117/12.871520

[132] Borlinghaus R. Colours count: how the challenge of fluorescence was solved in confocal microscopy. Modern research and educational topics in microscopy. In: Méndez-Vilas A, Díaz J, editors. Colours Count: How the Challenge of Fluorescence Was Solved in Confocal Microscopy. Vol. 2007. FORMATEX; 2007. pp. 890-899. DOI: 10.1.1.564.5122

[133] Sinclair MB, Haaland DM, Timlin JA, Jones HD. Hyperspectral confocal microscope. Applied Optics.

2006;**45**(24):6283-6291. DOI: 10.1364/AO.45.006283

[134] Li D, Shao L, Chen BC,zZhang X, Zhang M, Moses B, et al. Extended-resolution structured illumination imaging of endocytic and cytoskeletal dynamics. Science. 2015;**349**(6251):aab3500. DOI: 10.1126/science.aab350

[135] van de Linde S, Heilemann M, Sauer M. Live-cell super-resolution imaging with synthetic fluorophores. Annual Review of Physical Chemistry. 2012;**63**:519-540. DOI: 10.1146/annurev-physchem-032811-112012

[136] Yokono M, Takabayashi A, Akimoto S, Tanaka A. A megacomplex composed of both photosystem reaction centres in higher plants. Nature Communications. 2015;**6**:6675. DOI: 10.1038/ncomms7675

[137] Wolf E, Schussler A. Phycobiliprotein fluorescence of *Nostoc punctiforme* changes during the life cycle and chromatic adaptation: Characterization by spectral confocal laser scanning microscopy and spectral unmixing. Plant, Cell & Environment. 2005;**28**(4):480-491. DOI: 10.1111/j.1365-3040.2005.01290.x

[138] Férard JF, Blaise C. Encyclopedia of Aquatic Ecotoxicology. Netherlands: Springer; 2013. p. 805

[139] Seppälä J, Ylöstalo P, Kaitala S, Hällfors S, Raateoja M, Maunula P. Ship-of-opportunity based phycocyanin fluorescence monitoring of the filamentous cyanobacteria bloom dynamics in the Baltic Sea. Estuarine, Coastal and Shelf Science. 2007;**73**(3-4):489-500. DOI: 10.1016/j.ecss.2007.02.015

[140] Tyler AN, Hunter PD, Carvalho L, Codd GA, Elliott JA, Ferguson CA, et al. Strategies for monitoring and managing mass populations of toxic cyanobacteria in recreational waters: A multi-interdisciplinary approach. Environmental Health. 2009;**8**(S1):S11. DOI: 10.1186/1476-069X-8-S1-S11

[141] Metsamaa L, Kutser T, Strömbeck N. Recognising cyanobacterial blooms based on their optical signature: A modelling study. Boreal Environment Research. 2006;**11**(6):493-506

[142] Matthews MW, Bernard S, Winter K. Remote sensing of cyanobacteria-dominant algal blooms and water quality parameters in Zeekoevlei, a small hypertrophic lake, using MERIS. Remote Sensing of Environment. 2010;**114**(9):2070-2087. DOI: 10.1016/j.rse.2010.04.013

[143] Simis SG, Ruiz-Verdú A, Domínguez-Gómez JA, Peña-Martinez R, Peters SW, Gons HJ. Influence of phytoplankton pigment composition on remote sensing of cyanobacterial biomass. Remote Sensing of Environment. 2007;**106**(4):414-427. DOI: 10.1016/j.rse.2006.09.008

[144] Randolph K, Wilson J, Tedesco L, Li L, Pascual DL, Soyeux E. Hyperspectral remote sensing of cyanobacteria in turbid productive water using optically active pigments, chlorophyll a and phycocyanin. Remote Sensing of Environment. 2008;**112**(11):4009-4019. DOI: 10.1016/j.rse.2008.06.002

[145] Zhang Y, Feng L, Li J, Luo L, Yin Y, Liu M, et al. Seasonal–spatial variation and remote sensing of phytoplankton absorption in Lake Taihu, a large eutrophic and shallow lake in China. Journal of Plankton Research. 2010;**32**(7):1023-1037. DOI: 10.1093/plankt/fbq039

[146] Hunter PD, Tyler AN, Carvalho L, Codd GA, Maberly SC. Hyperspectral

remote sensing of cyanobacterial pigments as indicators for cell populations and toxins in eutrophic lakes. Remote Sensing of Environment. 2010;**114**(11):2705-2718. DOI: 10.1016/j.rse.2010.06.006

[147] Li L, Sengpiel RE, Pascual DL, Tedesco LP, Wilson JS, Soyeux E. Using hyperspectral remote sensing to estimate chlorophyll-a and phycocyanin in a mesotrophic reservoir. International Journal of Remote Sensing. 2010;**31**(15):4147-4162. DOI: 10.1080/01431161003789549

[148] Kutser T, Metsamaa L, Strömbeck N, Vahtmäe E. Monitoring cyanobacterial blooms by satellite remote sensing. Estuarine, Coastal and Shelf Science. 2006;**67**(1-2):303-312. DOI: 10.1016/j.ecss.2005.11.024

[149] Beutler M, Wiltshire KH, Arp M, Kruse J, Reineke C, Moldaenke C, et al. A reduced model of the fluorescence from the cyanobacterial photosynthetic apparatus designed for the in situ detection of cyanobacteria. Biochimica et Biophysica Acta (BBA)-Bioenergetics. 2003;**1604**(1):33-46. DOI: 10.1016/S0005-2728(03)00022-7

[150] Lee Z, Carder KL. Absorption spectrum of phytoplankton pigments derived from hyperspectral remote-sensing reflectance. Remote Sensing of Environment. 2004;**89**(3):361-368. DOI: 10.1016/j.rse.2003.10.013

[151] Vernet M, Iturriaga R. Contribution of Phycoerythrin-Containing Phytoplankton to Remotely Sensed Signals in the Ocean. Final Report, NASA project #10-70414, 1997. USA, CA. p. 14

[152] Subramaniam A, Carpenter EJ, Falkowski PG. Bio-optical properties of the marine diazotrophic cyanobacteria Trichodesmium spp. II. A reflectance model for remote sensing. Limnology and Oceanography. 1999;**44**(3):618-627. DOI: 10.4319/lo.1999.44.3.0618

[153] Fadeev VV, Maslov DV, Matorin DN, Reuter R, Zavyalova TI. Some peculiarities of fluorescence diagnostics of phytoplankton in coastal waters of the Black Sea. In: Proceedings of EARSeL-SIG-Workshop LIDAR, June 16-17. Dresden:Earsel: FRG; 2000. pp. 205-213

[154] Barbini R, Colao F, Fantoni R, Palucci A, Ribezzo S. Differential lidar fluorosensor system used for phytoplankton bloom and seawater quality monitoring in Antarctica. International Journal of Remote Sensing. 2001;**22**(2-3):369-384. DOI: 10.1080/014311601449989

[155] Babichenko S, Kaitala S, Leeben A, Poryvkina L, Seppälä J. Phytoplankton pigments and dissolved organic matter distribution in the Gulf of Riga. Journal of Marine Systems. 1999;**23**(1-3):69-82. DOI: 10.1016/s0924-7963(99)00051-2

[156] Lednev VN, Grishin MY, Pershin SM, Bunkin AF, Kapustin IA, Molkov AA, et al. Laser remote probing of freshwater reservoir with high phytoplankton concentration. Laser. 2016;**13**(1):119-134. DOI: 10.21046/2070-7401-2016-13-1-119-134

[157] Bazzani M, Breschi B, Cecchi G, Pantani L, Tirelli D, Valmori G, et al. Phytoplankton monitoring by laser induced fluorescence. EARSeL Adv. Remote Sens. 1992;**1**(2-II):106-110

[158] Moore TS, Churnside JH, Sullivan JM, Twardowski MS, Nayak AR, McFarland MN, et al. Vertical distributions of blooming cyanobacteria populations in a freshwater lake from LIDAR observations. Remote Sensing of Environment. 2019;**225**:347-367. DOI: 10.1016/j.rse.2019.02.025

[159] Chekalyuk AM, Gorbunov MY. Laser remote sensing of phytoplankton

photosynthetic activity and chlorophyll by using a shipboard lidar system. In: Korppi-Tommola JE, editor. Laser Study of Macroscopic Biosystems. Vol. 1922. Finland: SPIE; 1992. pp. 391-400. DOI: 10.1117/12.146164

[160] Gons HJ, Hakvoort H, Peters SW, Simis SG. Optical detection of cyanobacterial blooms. In: Huisman J, Matthijs HC, Visser PM, editors. Harmful Cyanobacteria. Aquatic Ecology Series. Dordrecht: Springer; 2005. pp. 177-199. DOI: 10.1007/1-4020-3022-3_8

Chapter 2

Far-Red Light Absorbing Photosynthetic Pigments in Cyanobacteria: Steady-State Fluorescence Detection, Time-Resolved Fluorescence Spectroscopy, and Confocal Laser Scanning Microscopy

Svetlana Averina, Ekaterina Senatskaya and Alexander Pinevich

Abstract

The use of far-red light (FRL) is observed in some cyanobacteria, as well as in some marine and freshwater algae. While algae mobilize FRL absorbing antenna, which contains common chlorophyll *a* (Chl *a*), cyanobacteria produce red-shifted Chl *d* and/or Chl *f*. These pigments are synthesized either irrespective of ambient light or synthesized during FRL photoacclimation (FaRLiP), or adaptive remodeling of photosynthetic apparatus induced by relative enrichment with FRL quanta. The presence of red-shifted chlorophylls as well as their functions and topography are registered with various methods based on fluorescence measurement, such as: (1) steady-state fluorescence detection in live cells, cell fractions, and photosynthetic apparatus constituents; (2) time-resolved fluorescence spectroscopy, which traces energy transfer between individual pigments; (3) confocal laser scanning microscopy (CLSM), which helps to localize photosynthetic pigments in situ. This chapter describes photosynthetic apparatus in cyanobacteria and their photoacclimation phenomena. Over past decades, FRL photoacclimation has been studied in a small number of cyanobacteria. Novel Chl *f*-producing strains *Chlorogloeopsis* sp. CALU 759 and *Synechocystis* sp. CALU 1173 would represent promising model objects. Importantly, although they belong to alternative morphotypes and distant phylogenetic lineages, fluorescence pattern of their FRL-grown cells similarly falls within general FaRLiP response.

Keywords: cyanobacteria, chlorophylls *d* and *f*, steady-state fluorescence detection, time-resolved fluorescence spectroscopy, confocal laser scanning microscopy

1. Introduction

Cyanobacteria are the only up-to-date known prokaryotes capable of oxygenic photosynthesis. They assimilate light energy via electron transfer from water to oxidized ferredoxin, and excrete triplet dioxygen as a waste product. The majority of cyanobacteria possess chlorophyll *a* (Chl *a*), bicyclic carotenoids, and phycobiliproteins (PBP). In common, these pigments absorb light quanta within a whole visible area (400–700 nm). At the same time, some cyanobacteria have been shown to use far-red light (FRL) with wavelengths of more than 700 nm [1]. For this purpose, red-shifted Chl *d* and/or Chl *f* are produced [2, 3]. In search of novel representatives of FRL-adapting cyanobacteria, various analytical methods are employed, especially those based on fluorescence detection.

This chapter includes four sections, which describe: photosynthetic apparatus in cyanobacteria; photoacclimation phenomena in cyanobacteria; fluorescence methods employed in the study of FRL-adapting cyanobacteria; FRL photoacclimating strains in CALU collection (St. Petersburg University, St. Petersburg, Russia).

2. Photosynthetic apparatus in cyanobacteria

Light energy assimilation machinery in cyanobacteria is usually localized in/on intracytoplasmic membrane structures termed thylakoids. Among principal constituents are: reaction centers (RC) of two photosystems (PSs), a main light-harvesting complex (LHC), and electron transfer chain (ETC). The essential component of PSII is unique light-dependent enzyme—H_2O dehydrogenase, or water oxidizing complex (WOC).

In the majority of cyanobacteria, Chl *a* represents the only chlorophyll while in rare cases other chlorophylls are additionally produced [4, 5].

Both PSs are supplied with their respective light-harvesting antennae, which contain Chl *a*. LHC is usually represented by phycobilisome (PBS), supramolecular aggregate of phycobiliproteins (PBP) stabilized by colorless linker polypeptides, and anchored to thylakoid surface with high-molecular mass colored polypeptide [6, 7]. PBP apoprotein moiety is represented by α or β subunits that covalently bind non-porphyrin tetrapyrrole chromophores [8, 9]. Depending on chromophore type and number, four main PBP types are distinguished: phycocyanin (PC), phycoerythrin (PE), allophycocyanin (APC), and phycoerythrocyanin (PEC) [8]. Standard PBS consists of the central core and peripheral rods. The former contains two to three "cylinders" assembled of several $(\alpha\beta)_3$ trimers each, while the latter includes several $[2(\alpha\beta)_3]$ hexamers each. Due to anisotropic structure of rods and core, a waste-less channeling of excitation energy to RC chlorophyll is achieved [10, 11].

3. Photoacclimation phenomena in cyanobacteria

Cyanobacteria can acclimate themselves to light quantity and quality, that is, they adaptively respond to the shifts in ambient light color and intensity. For this purpose, they can modulate: (1) PS content, (2) the interaction between PS and PBS, and (3) LHC structure, PBP content in particular.

In the first case, antenna size varies inversely with a flood of light quanta. This strategy is performed via the changes in thylakoid surface and PS packing density [12].

In the second case (given that ambient light is enriched with either short- or long-wavelength quanta), excitation energy equally distributes between long-wavelength PSI and short-wavelength PSII. This phenomenon is termed State 1 ↔ State 2 transition. State 1 is achieved in response to relative over-excitation of PSII [13]. Here, PBS behaves as a mobile antenna, and it laterally moves from PSII to PSI. In the opposite situation (State 2), PBS detaches from PSI and comes back to PSII [14]. State transition scenario is as follows: up- or downshifts in ETC reduction level → redox-sensitive (de)phosphorylation of proteins within the PBS baseplate → coulomb attraction/repulsion between PBS and PS [15].

In the third case, PBS absorbance peak adjusts to ambient light color (preferentially green or red). This adaptation is performed via the shifts in PE (green light absorbing PBP) and PC (red light absorbing PBP) content. The knowledge on this phenomenon termed complementary chromatic adaptation (CA) has been comprehensively reviewed [16, 17].

In agreement with a response to green or red light, cyanobacteria have been ascribed to three CA groups [18]: group CA1 (steady PE and PC content), group CA2 (PE content varies), and group CA3 (PE and PC contents vary).

Groups CA2 and CA3 are specified according to the presence of regulatory photoreceptors CcaS and RcaE, which belong to the "cyanobacteriochromes" phytochrome family [19–21]. These receptors contain one and the same bilin-binding domain GAF (cyclic guanosine monophosphate phosphodiesterase/adenylyl cyclase/FhlA), which regulates green or red light-triggered photocycle [22, 23]. In the case of CA2, under green light, CcaS (signal transducer) phosphorylates transcription factor CcaR (response regulator) that induces the production of PE [22]. In the case of CA3, under red light, RcaE phosphorylates transcription factors RcaF and RcaC, which regulate numerous participants of PC and PE biosynthesis pathway [23, 24].

Group CA4 is represented by marine *Synechococcus* strains grown under blue or green light [25–27]. Under blue light, PE phycoerythrobilin chromophores replace phycourobilin chromophores, which adapt PBS to the use of smaller wavelength light. Although corresponding photoreceptor is unknown, light signal was shown to be transmitted to transcription factors FciA and FciB with a participation of chromophore liases MpeZ and MpeW [26, 28].

Group CA5 is typical of *Acaryochloris marina* [29]. Irrespective of light conditions, this cyanobacterium uses membrane-embedded LHC containing Chl *d*. Under red light (625 nm), atypical rod-like PC and APC containing PBS is additionally produced [30]. Such PBS disassembled in far-red light (720 nm), and corresponding PBP are destroyed [31, 32]. In other words, in these light conditions, *A. marina* uses two LHCs in common. Although the underlying regulatory mechanism is unknown, *A. marina* genome has been shown to contain a motif similar to that coding for FRL accepting phytochromes [33].

Group CA6 is represented by several strains of cyanobacteria, which adaptively produce Chl *f* and/or Chl *d* in FRL [16]. In more detail, this phenomenon is dwelled on below.

Group CA7 [16] has been described in the case of cyanobacteria which synthesize yellow-green light absorbing PEC, and regulate the amount of this PBP light dependently. Similar to CA2, the adaptive response is under control of two-component regulatory system CcaS/CcaR.

Another type of reaction to green or red light is typical for CA0 group [16]. In this case, PBP content is stable, while the amount of CpcG2 (CpcL) and CpcG1 linker polypeptides varies reciprocally [34]. In green light, the CcaS/CcaR system upgrades the production of CpcG2 (CpcL) linker involved in the biogenesis of APC lacking PBS [22]. The CpcG2 (CpcL) type PBS is suggested to be rod-like, and it possibly supplies energy to PSI under suboptimal conditions [16, 22].

Cyanobacterial response to FRL has been described in cyanobacteria only recently, and it is based on red-shifted Chl *d* and/or Chl *f* (697 and 706 nm maxima in methanol, respectively). First representative of Chl *d*-containing species was *A. marina* [1], while the initially described Chl *f*-containing species was *Halomicronema hongdechloris* [35].

Cyanobacteria can use two main strategies of FRL photoacclimation.

In the first strategy, Chl *d* represents the bulk chlorophyll, and it is constitutively produced in *A. marina* [36–39]. The pigment participates in PSI and PSII, as well as in LHC [40–43]. Such strategy is observed in cyanobacteria inhabiting marine or continental haline water bodies poor in visible light but rich in FRL.

In the second strategy, Chl *a* represents the bulk chlorophyll while Chl *f* (sometimes together with a small amount of Chl *d*) is produced adaptively in FRL [35, 44–46]. In this case, ~10% of Chl *a* in PSI and PSII is replaced with Chl *f*; special paralogue polypeptides substitute for PSI and PSII subunits; PBS is remodeled [45]. The latter strategy is termed FRL photoacclimation—FaRLiP [3], or CA6 (see above). In this strategy, FRL induces the expression of 21 genes of FaRLiP cluster including those coding for PSI and PSII subunit paralogues [2, 45, 46]. The products of these genes specifically bind Chl *a* together with red-shifted chlorophylls [2]. FaRLiP cluster also includes the genes coding for PBP subunit paralogues [45, 47]. Resulting PBS are optimal in new light climate because they are devoid of short-wavelength PBP (PE and PEC), and correspondingly lack rod periphery [46].

Leptolyngbya sp. JSC-1 grown in white light (WL) or red light (RL) contains the standard five-cylinder-core PBS, while in FRL a two-cylinder-core PBS is produced [45]. The latter has a 708-nm maximum (40 nm longer than in APCB, the top wavelength PBP previously reported). In its turn, FRL-grown *H. hongdechloris* produces a two-cylinder-core mini PBS, which contains APC with 653- and 712-nm maxima [48].

FRL-grown *Synechococcus* sp. PCC 7335 produces PBS of two types: three-cylinder-core PBS containing PC and APC, and two-cylinder-core PBS, which contains only APC with FaRLiP gene encoded subunits, and displays red-shifted 650- and 711-nm light absorbance and 730-nm fluorescence emission maxima [49].

FaRLiP response falls under the control of two-component phosphorelay system [47]. Sensory component (RfpA photoreceptor) represents a far-red light regulated cyanobacteriochrome with histidine kinase domain. Response regulators RfpB and RfpC have two CheY-like signal accepting domains, which flank the DNA-binding domain. RfpB acts as a transcription activator for FaRLiP genes [47, 50]. Within this phoshorelay, RfpA histidine kinase becomes (de)activated, and that influences the RfpB key response regulator. In its turn, RfpC is involved in transfer of phosphoryl group from RfpA to RfpB.

4. Fluorescence methods employed in the research of far-red light-adapting cyanobacteria

4.1 Steady-state fluorescence detection

Absorbed energy of light quanta brings photosynthetic pigments into excited state, which is relaxed by: (1) productive energy assimilation in the form of charge separation within RC, (2) counterproductive energy dissipation into heat, or with fluorescence quanta [51]. Since the photosynthetic apparatus is less than 100%

effective, the second mechanism is universally in action, although it depends on environmental and physiological regimes.

Steady-state fluorescence detection helps to identify photosynthetic pigments because they demonstrate individual fluorescence excitation and emission spectra. Additionally, this method can detect energy transfer between pigments. Fluorescence spectra can be obtained at either room or cryogenic temperature (most frequently, 77 K). Low temperatures are preferred because of lowered molecular mobility, and due to smaller intramolecular vibrations; as a result, peaks become higher and better resolved [51]. Importantly, low-temperature regime (4–77 K) helps to discriminate PSI chlorophyll (~ 720 nm) and PSII chlorophyll (~ 685 nm) emission peaks [51–53].

Current data on steady-state fluorescence of red-shifted chlorophylls are few. For instance, the spectra of WL-grown *H. hongdechloris* were compared with those of FRL culture [35]. In the case of WL cells, Chl *a*-specific 405-nm fluorescence excitation yielded fluorescence emission maxima at 640, 658, 682, and 730 nm of comparable height. In contrast, FRL cells exhibited a major 748-nm maximum (indicating the adaptive synthesis of red-shifted Chl *f*), and 682- and 720-nm minor maxima, respectively. Also, the emission at 440-nm excitation was compared in *Calothrix* sp. PCC 7507, *Chlorogloeopsis* sp. PCC 9212, *Chroococcidiopsis* sp. PCC 7203, *Fischerella* sp. PCC 7521, and *Synechococcus* sp. PCC 7335 grown under different light conditions [46]. In all these strains, WL cultures showed emission maxima of Chl *a* (683–684 nm, 693–695 nm, and 718–727 nm). At the same time, FRL spectra were strain specific, Chl *f* being most pronounced (*Calothrix* sp. PCC 7507—736 nm; *Chlorogloeopsis* sp. PCC 9212—739 nm; *Chroococcidiopsis* sp. PCC 7203—718, 736, and 750 nm; *Fischerella* sp. PCC 7521—749 nm; *Synechococcus* sp. PCC 7335—738 nm).

Apart from the experiments on cell suspensions, steady-state fluorescence of Chl *a* and Chl *f* has been detected with subcellular fractions, as well as with PSI, PSII, and PBS preparations. For example, *H. hongdechloris* emission was compared for WL and FRL thylakoid preparations [54]. In WL preparation, major 710-nm peak as well as 680- and 732-nm minor peaks were observed, while FRL preparation demonstrated a major 740-nm peak due to adaptively produced Chl *f*.

Cryogenic detection helped to monitor energy transfer in FRL-adapted *Synechococcus* sp. PCC 7335 [55]. Purified PSI or PSII preparations obtained from FRL-adapted cells showed one and the same well-expressed peak at 738–740 nm indicating similar effectiveness of energy coupling notwithstanding a distinction in PS structure.

Steady-state fluorescence emission was also detected in the experiments with PBP-specific 590-nm excitation of RL- or FRL-grown *Synechococcus* sp. PCC 7335. Besides 643-, 658-, and 679-nm peaks in the RL culture, FRL-grown cells acquired 717- and 737-nm maxima that could be explained by PBS rearrangement due to induction of FaRLiP gene cluster [50].

4.2 Time-resolved fluorescence spectroscopy

The method helps to analyze molecular processes within a picosecond/nanosecond timescale [56]. Because primary photosynthetic processes take several femtoseconds/nanoseconds, time-resolved fluorescence spectroscopy can trace corresponding rates and pathways of energy transfer [57]. Recently, this method has helped to elucidate the role(s) of red-shifted chlorophylls in cyanobacterial PS.

In the case of Chl *f*-producing *H. hongdechloris*, time-resolved fluorescence detection was performed at 77 K, with 425-nm excitation [41]. In WL cells, the excitation raised 685-nm (F685) and 730-nm (F730) emission peaks characteristic for PSII and PSI correspondingly. A shift from F685 to F695 occurred with time, and an extra F742 peak appeared and decayed. In FRL cells, F685 and F748 maxima were observed. The former quenched rapidly indicating energy transfer from PSII to PSI. F748 signified the presence of Chl *f*, and uphill energy transfer (from Chl *f* to Chl *a*) was proposed.

Time-resolved fluorescence detection within a picosecond/femtosecond timescale was also performed with *H. hongdechloris* thylakoid membrane preparations [54]. Similar to WL cells, WL membrane emission belonged to PSI and PSII Chl *a* (727 and 685 nm correspondingly). However, these peaks were attended with PBP-specific 678-nm peak. In the case of FRL membrane, main fluorescence emission maxima belonged to Chl *a* (685 nm) and Chl *f* (745 nm). In FRL membrane preparations, fluorescence rise and decay curves suggested Chl *f* to be energy donor to Chl *a*.

Later, the reality of uphill energy transfer from Chl *f H. hongdechloris* was confirmed in the experiments using TCSPC (time-correlated single photon counting) and DAS (decay-associated spectra) methods [58]. According to obtained data, energy was transferred within PSII on the route: PBS → Chl *f* → Chl *a*; only then charge separation began, and Chl *a* represented the P680 primary donor. Thus, Chl *f* in *H. hongdechloris* is PSII antenna pigment, and it does not participate in charge separation.

At the same time, the involvement of red-shifted chlorophylls in charge separation within PSI and PSII was proposed for *Chroococcidiopsis* sp. PCC 7203, which adaptively produces Chl *d* and Chl *f* [59, 60]. Data obtained using plenty of methods, TCSPC in particular, showed that RCI in the FRL-adapted cells contained 7–8 Chl *f* molecules, together with 90 Chl *a* molecules. Special pair (P_A and P_B) was represented by Chl *a* and Chl *a* epimer (*a'*). Primary acceptors (A_{0A} and A_{0B}) were also Chl *a* molecules. Chl *f* molecule absorbing 745-nm light was associated with primary donor (A_{-1A} and/or A_{-1B}). In its turn, RCII contained 4 Chl *f* molecules, 1 Chl *d* molecule, and 30 Chl *a* molecules. Primary donors (Chl_{D1}) were either Chl *d* or Chl *f* molecules absorbing 727-nm light.

Room temperature and 77 K fluorescence data in unicellular strain KC1 producing Chl *f* in FRL helped to obtain TRS (time-resolved) and DAS (decay-associated) emission spectra [61]. In the experiments with PSI exited with 405-nm laser, energy was rapidly transferred to Chl *f* (emission maxima 720–760 and 805 nm). In the case of 630-nm excitation (light absorbed by PBP associated with PSII), energy was transferred to Chl *f* (emission maxima 720–770 and 810 nm) and Chl *a* (emission maximum 694 nm). Distinctions in Chl *f* fluorescence show that PSI and PSII contained Chl *f* molecules with different properties, and possibly they represented antenna pigments.

Similarly, femtosecond pump-probe spectroscopy of *Fischerella thermalis* PCC 7521 [62] showed that Chl *f* represented PSI antenna pigment; it was also argued that energy was transferred uphill from Chl *f* to P700 Chl *a* special pair.

Time-resolved fluorescence spectroscopy also helped to analyze energy transfer in RL-or FRL-grown cells of *Synechococcus* sp. PCC 7335 [55], as well as in *A. marina* mini PBS [63].

4.3 Confocal laser scanning microscopy (CLSM)

The method is based on the auto fluorescence of photosynthetic pigments, chlorophylls in the first instance [64]. CLSM permits to distinguish individual pigments based on their emission peaks, to evaluate peak intensity, and to localize pigments in situ. Highly sensitive up-to-date CLSM microscopes help to obtain 3D images of cells and tissues, and to analyze dynamic physiological processes. Unfortunately, the potentiality of this method, at least with regard to FRL-adapting cyanobacteria, is underestimated, and thus corresponding data are few.

CLSM has been applied to the study of Chl *d* and PBP localization in *Candidatus* Acaryochloris bahamensis, an epibiont of colonial ascidian *Lissoclinum fragile* [65]. In 3D reconstruction, these pigments were asymmetrically distributed throughout cell interior and that possibly promoted optimal sunlight absorption within *L. fragile* tunic. *Acaryochloris*-like cells were also shown to inhabit the surface of *Cystodytes dellechiajei* ascidian body, and CLSM data indicated the presence of Chl *d* and PBP in this cyanobacterium [66].

Anisotropic distribution of pigments was also observed in *H. hongdechloris,* which adaptively produces accessory Chl *f* in FRL [67]. In WL culture, chlorophylls and PBP were shown to co-localize at cell periphery. However, in FRL culture, PBP fluorescence was detected only at cell poles. In their turn, FRL-grown *Synechococcus* sp. PCC 7335 demonstrated lower PBP fluorescence and higher chlorophyll fluorescence compared with WL cells. In FRL culture, PBP fluorescence was spatially isolated from chlorophyll fluorescence: the latter occupied cell center indicating that PBS detached from PS.

5. Far-red light photoacclimating strains in the CALU collection

5.1 Growth conditions

An ability of FRL-dependent synthesis of Chl *f* in trichome-forming strain *Chlorogloeopsis* sp. CALU 759 and unicellular strain *Synechocystis* sp. CALU 1173 has been deduced from whole-cell light absorbance spectra (**Figure 1**) as well as according to methanol extract spectrometry and HPLC results.

Liquid cultures were grown for a 2- to 3-week time in modified BG-11 medium [68], at 20–22°C, under permanent WL (500 lx) or FRL illumination (LED with ~750 nm emission maximum).

5.2 Steady-state fluorescence detection protocol

Cell pellet was washed with 100 mM HEPES buffer (pH 8.0), and resuspended in 50 mM HEPES (pH 8.0) with or without the addition of 25% glycerol. Room temperature fluorescence was detected in a Cary Eclipse scanning spectrophotometer/fluorimeter (Agilent Technologies, Santa Clara, CA, USA), and obtained data were treated with a Cary Eclipse Scan program. Emission spectra were detected for excitation at 440 nm (specific for chlorophyll) and 550 nm (specific for PBP). Excitation spectra (chlorophyll fluorescence at 745 nm) were also detected.

5.3 CLSM protocol

The method helped to reveal photosynthetic pigments topography. Life preparations of WL- or FRL-grown cells were observed in a Leica TCS SP5 MP STED confocal microscope (Leica Microsystems GmbH, Germany). Fluorescence was raised with a 458-nm-wavelength nitrogen laser. The images were obtained in XYZ scanning regime with a 0.5-μm step. The following emission channels were used: 560–600 nm (PE), 620–650 nm (PC), 670–700 nm (Chl *a*), and 705/710–790 nm (Chl *f*). The obtained images were analyzed with a LAS AF program.

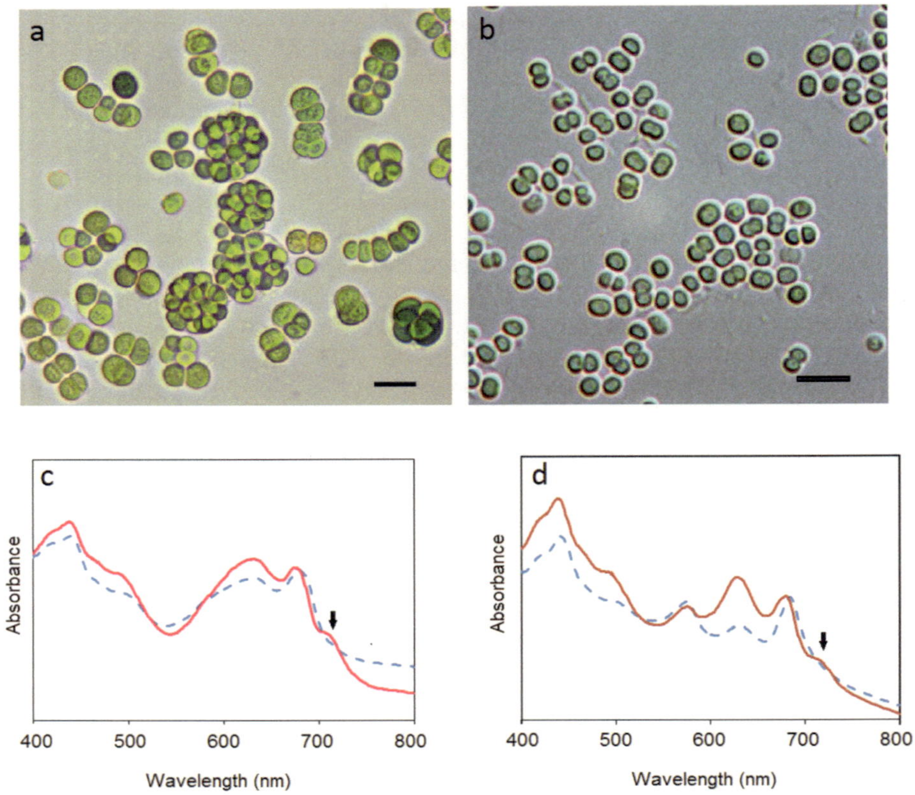

Figure 1.
Light micrographs and light absorbance spectra of Chlorogloeopsis *sp. CALU 759 (a, c) and* Synechocystis *sp. CALU 1173 (b, d) cells. Dashed line, white light-grown cells; solid line, far-red light-grown cells. Arrow, shoulder explained by absorbance of red-shifted chlorophyll. Scale bar, 10 μm (a) or 5 μm (b).*

5.4 Steady-state fluorescence detection of chlorophylls and phycobiliproteins in white light- or far-red light-grown cells

Fluorescence emission spectra (**Figures 2(a,b)** and **3(a,b)**) obtained at chlorophyll-specific 440-nm excitation demonstrated 670–740 nm maxima. In WL-grown *Chlorogloeopsis* sp. CALU 759 (**Figure 2(a)**), it was a single peak with a shoulder while in *Synechocystis* sp. CALU 1173 (**Figure 3(a)**) twin peaks were observed. In FRL-grown cells of both strains, a single 725–740 nm emission peak was observed (**Figures 2(b)** and **3(b)**). In agreement with the previous data obtained with FRL-grown cyanobacteria [35, 46], peaks with more than 720-nm wavelength (observed in light absorbance or fluorescence emission spectra) corresponded to Chl *f*.

Fluorescence emission spectra obtained with PBP-specific 590-nm excitation resulted in 645–660 nm emission peak with broad 690–715 nm shoulder. In WL-grown *Chlorogloeopsis* sp. CALU 759 (**Figure 2(c)**), the shoulder was steep while in *Synechocystis* sp. CALU 1173 (**Figure 3(c)**) it was gently sloping.

In FR-grown cells, larger size 645–660 nm peak was attended with smaller 715–725 nm peak (**Figures 2(d)** and **3(d)**). Modified emission spectra, as compared with WL-grown cells, could be explained by some changes in PBS arrangement and behavior during the FaRLiP response [45, 48, 49].

The comparison of emission spectra in **Figures 2** and **3** showed that, in the case of FRL-grown cells, direct excitation of chlorophylls (blue light excitation, 440-nm

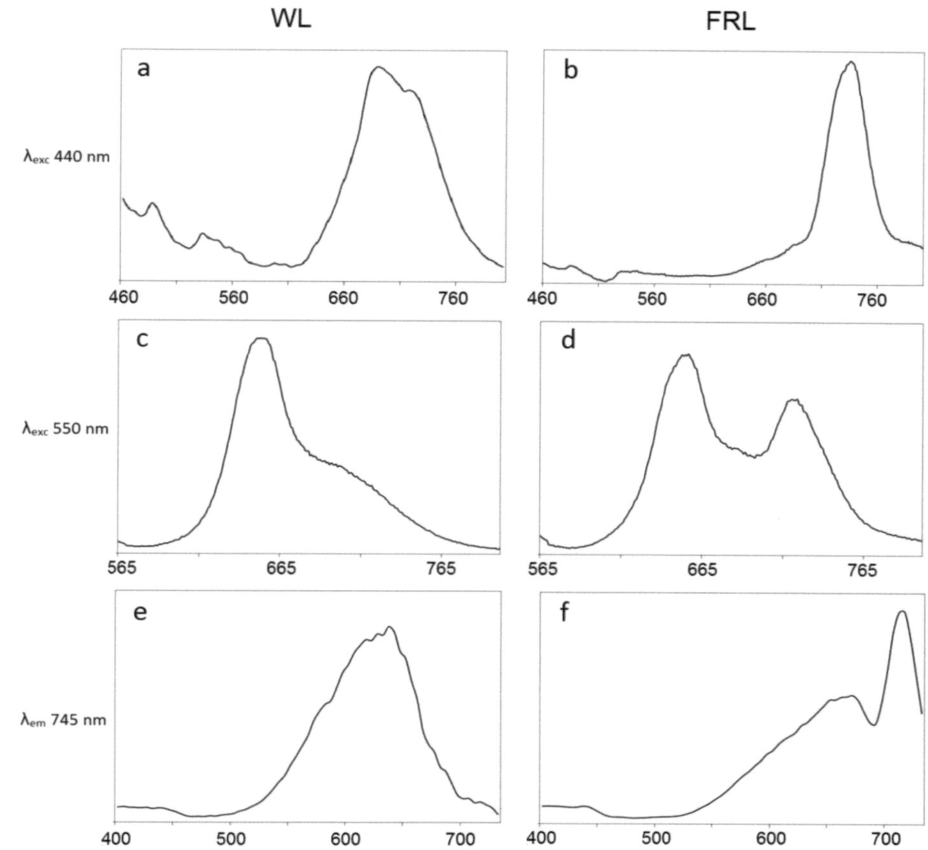

Figure 2.
Fluorescence spectra of white light (WL)-grown (a, c) or far-red light (FRL)-grown (b, d) Chlorogloeopsis *sp. CALU 759 cells. Emission spectra: 440 nm (a, b) and 550 nm (c, d) excitation. Excitation spectra: chlorophyll fluorescence at 745 nm (e, f).*

wavelength) was more efficient than via antenna PBP pigments (550-nm green light exciting mainly PE and PC). This effect can be explained by PBS uncoupling with PSII because bulk fluorescence at room temperature is known to issue from PSII. If not coupled, PBP excited at 550 nm should exhibit considerable fluorescence around 700 nm (that is the case here).

Fluorescence excitation spectra of WL-grown cells (**Figures 2(e)** and **3(e)**) demonstrated a single peak at 625–650 nm excitation wavelength. Negligible fluorescence during illumination with 400–500 nm light (which preferentially excited chlorophylls) can be explained by shading of the blue region with photosynthetically inactive carotenoids. At the same time, fluorescence during illumination with ~600 nm (which preferentially excited PBP) raised a maximum chlorophyll fluorescence that additionally indicated an efficient coupling of PBS with PSII. In contrast to WL-grown cells, in FRL-grown cells (**Figures 2(f)** and **3(f)**), peak chlorophyll fluorescence was observed at ~670 and ~ 720 nm excitation. High ratio of 720-nm peak/670-nm peak size additionally witnesses for a change of photosynthetic apparatus at FaRLiP response. It is noteworthy that unlike cryogenic method, room temperature fluorescence detection could not comment on the specificity of Chl *f* functioning in analyzed strains (that is the aim of our future research).

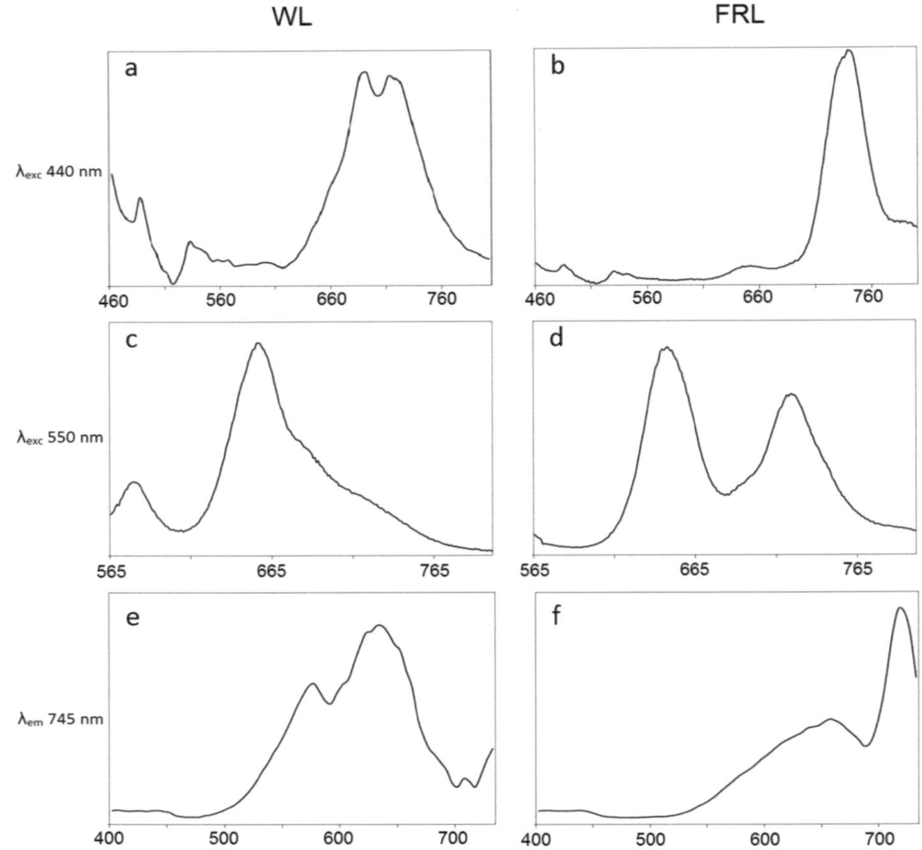

Figure 3.
Fluorescence spectra of white light (WL)-grown (a, c) or far-red light (FRL)-grown (b, d) Synechocystis *sp. CALU 1173 cells. Emission spectra: 440 nm (a, b) and 550 nm (c, d) excitation. Excitation spectra: Chlorophyll fluorescence at 745 nm (e, f).*

5.5 CLSM of chlorophylls and phycobiliproteins in white light- and far-red light-grown cells

In the case of WL-grown *Chlorogloeopsis* sp. CALU 759 (**Figure 4**), false-green 670–700 nm fluorescence typical for Chl *a* (**Figure 4(a)**) and false-blue 620–650 nm fluorescence typical for PC (**Figure 4(b)**) were observed not only at

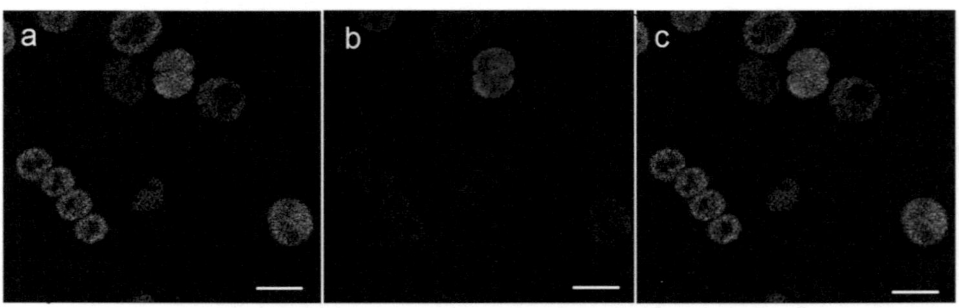

Figure 4.
Laser confocal scanning micrographs of white light-grown Chlorogloeopsis *sp. CALU 759 cells. Chlorophyll* a *fluorescence is shown with false-green (a); phycocyanin fluorescence is shown with false-blue (b); chlorophyll* a *and phycocyanin total fluorescence (c).*

cell periphery but also in cell center. Moreover, mutually overlapping fluorescence channels (**Figure 4(c)**) indicated that the pigments had a common localization.

Similar topology was observed in FRL-grown *Chlorogloeopsis* sp. CALU 759 (**Figure 5**). Importantly, Chl *a* false-green emission (**Figure 5(a)**), Chl *f*

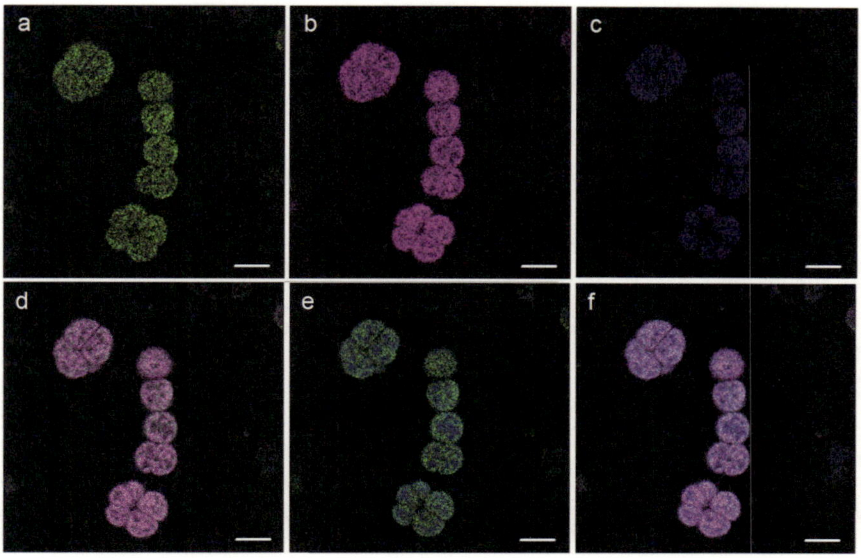

Figure 5.
Laser confocal scanning micrographs of far-red light-grown Chlorogloeopsis *sp. CALU 759 cells. Chlorophyll* a *fluorescence is shown with false-green (a); chlorophyll* f *fluorescence is shown with false-purple (b); phycocyanin fluorescence is shown with false-blue (c); chlorophyll* a *and chlorophyll* f *fluorescence (d); chlorophyll* a *and phycocyanin fluorescence (e); total pigment fluorescence (f). Scale bar, 5 μm.*

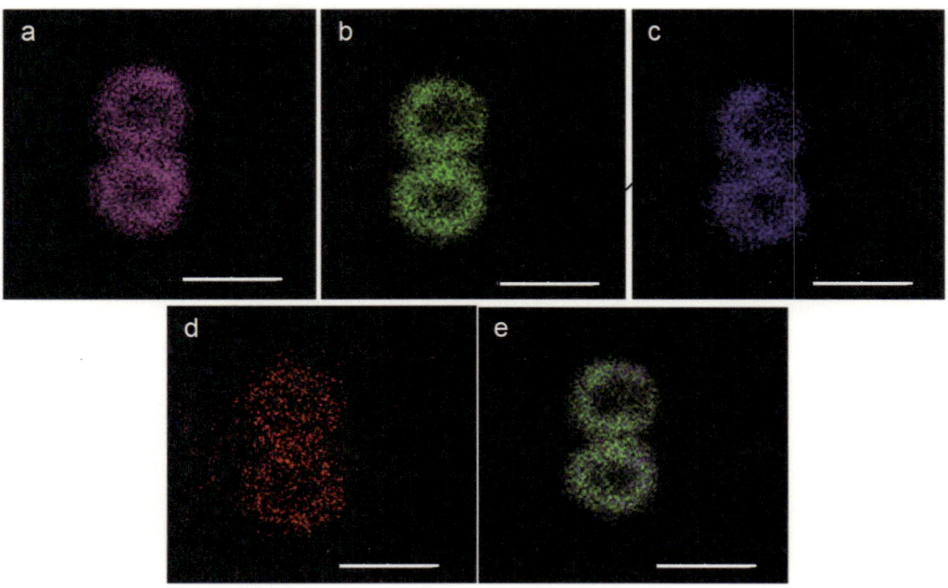

Figure 6.
Laser confocal scanning micrographs of far-red light-grown Synechocystis *sp. CALU 1173 cells. Chlorophyll* f *fluorescence is shown with false-purple (a); chlorophyll* a *fluorescence is shown with false-green (b); phycocyanin fluorescence is shown with false-blue (c); phycoerythrin fluorescence is shown with false-red (d); total pigment fluorescence (e). Scale bar, 2 μm.*

false-purple emission (**Figure 5(b)**), and Chl *a*/Chl *f* mixed emission (**Figure 5(d)**) coincided spatially, and that indicated a relative integrity of photosynthetic apparatus. At the same time, PC fluorescence showed a more homogeneous distribution (**Figure 5(c,e,f)**). Thus, PBS either partially detached from thylakoid membrane, or PBS coupling with PSII lowered (which is in agreement with steady-state fluorescence results). Similar images were obtained for FRL-grown *Synechococcus* sp. PCC 7335 [67].

In the case of FRL-grown *Synechocystis* sp. CALU 1173 (**Figure 6**), fluorescence topology was similar within the observed wavelength span. Namely, Chl *f* false-purple fluorescence (**Figure 6(a)**), Chl *a* false-green fluorescence (**Figure 6(b)**), PC false-blue fluorescence (**Figure 6(c)**), PE false-red fluorescence (**Figure 6(d)**), and pigment total fluorescence (**Figure 6(e)**) occupied cell periphery, and fluorescence channels mutually overlapped. The observed fluorescence topology matched with the electron microscopy data on peripheral thylakoid arrangement in *Synechocystis* spp. [69, 70].

6. Conclusion

Over past decades, FaRLiP photoacclimation has been studied in many respects, although in a small number of cyanobacterial strains [3]. In this connection, novel Chl *f*-producing strains *Chlorogloeopsis* sp. CALU 759 and *Synechocystis* sp. CALU 1173 would represent promising model objects. Importantly, although they belong to alternative morphotypes (trichome and single cell, respectively) and two distant phylogenetic lineages [71], fluorescence patterns of their FRL-grown cells similarly fall within the general FaRLiP response.

Acknowledgements

The authors are grateful to St. Petersburg University Research Center "Cultivation of microorganisms" (http://researchpark.spbu.ru/collection-ccem-rus/1628-ccem-kollekciya-calu-rus) for a help with strains maintenance. They also thank St. Petersburg University Research Centers "Molecular and Cell Technologies" and "Chromas" for research fellowship. The work was supported by the Russian Foundation for Fundamental Research, project no. 20-04-00020.

Author details

Svetlana Averina*, Ekaterina Senatskaya and Alexander Pinevich
Saint-Petersburg State University, St. Petersburg, Russia

*Address all correspondence to: s.averina@spbu.ru; spbu.svetik@rambler.ru

References

[1] Miyashita H, Ikemoto H, Kurano N, et al. Chlorophyll *d* as a major pigment. Nature. 1996;**383**:402. DOI: 10.1038.383402a0

[2] Gan F, Bryant DA. Adaptive and acclimative responses of cyanobacteria to far-red light. Environmental Microbiology. 2015;**17**:3450-3465. DOI: 0.1111/1462-2920.12992

[3] Averina S, Velichko N, Senatskaya E, Pinevich A. Far-red light photoadaptations in aquatic cyanobacteria. Hydrobiologia. 2018;**813**:1-17. DOI: 10.1007/s10750-018-3519-x

[4] Pinevich A, Velichko N, Ivanikova N. Cyanobacteria of the genus *Prochlorothrix*. Frontiers in Microbiology. 2012;**3**:173. DOI: 10.3389/fmicb.2012.00173

[5] Averina S, Velichko N, Senatskaya E, et al. Non *a*-chlorophylls in cyanbacteria. Photosynthetica. 2019;**57**:1109-1118. DOI: 10.32615/ps.2019.130

[6] Adir N. Elucidation of the molecular structures of components of the phycobilisome: Reconstructing a giant. Photosynthesis Research. 2005;**85**:15-32. DOI: 10.1007/s11120-004-2143-y

[7] Watanabe M, Ikeuchi M. Phycobilisome: Architecture of a light-harvesting supercomplex. Photosynthesis Research. 2013;**116**:265-276. DOI: 10.1007/s11120-013-9905-3

[8] MacColl R. Cyanobacterial phycobilisomes. Journal of Structural Biology. 1998;**124**:311-334. DOI: 10.1006/jsbi.1998.4062

[9] Scheer H, Zhao KH. Biliprotein maturation: The chromophore attachment. Molecular Microbiology. 2008;**68**:263-276. DOI: 10.1111/j.1365-2958.2008.06160.x

[10] Chang L, Liu X, Li Y, et al. Structural organization of an intact phycobilisome and its association with photosystem II. Cell Research. 2015;**25**:726-737. DOI: 10.1038/cr.2015.59

[11] Zhang J, Ma J, Liu D, et al. Structure of phycobilisome from the red alga *Griffithsia pacifica*. Nature. 2017;**551**:57-63. DOI: 10.1038/nature24278

[12] Muramatsu M, Hihara Y. Acclimation to high-light conditions in cyanobacteria: From gene expression to physiological responses. Journal of Plant Research. 2012;**125**:11-39. DOI: 10.1007/s10265-011-0454-6

[13] Mullineaux CW. Electron transport and light-harvesting switches in cyanobacteria. Frontiers in Plant Science. 2014;**5**:7. DOI: 10.3389/fpls.2014.00007

[14] McConnell MD, Koop R, Vasil'ev S, Bruce D. Regulation of the distribution of chlorophyll and phycobilin-absorbed energy in cyanobacteria. A structure-based model for the light state transition. Plant Physiology. 2002;**130**:1201-1212. DOI: 10.1104/pp.009845

[15] Mullineaux CW, Allen JF. State 1-state 2 transitions in the cyanobacterium *Synechococcus* 6301 are controlled by the redox state of electron carriers between photosystem I and II. Photosynthesis Research. 1990;**23**:297-311. DOI: 10.1007/BF00034860

[16] Hirose Y, Chihong S, Watanabe M, et al. Diverse chromatic acclimation processes regulating phycoerythrocyanin and rod-shaped phycobilisome in cyanobacteria.

Molecular Plant. 2019;**12**:715-725. DOI: 0.1016/jmolp.2019.02.010

[17] Sanfilippo JE, Garczarek L, Partensky F, Kehoe DM. Chromatic acclimation in cyanobacteria: A diverse and widespread process for optimizing photosynthesis. Annual Review of Microbiology. 2019;**73**:407-433. DOI: 10.1146/annurev-micro-020518-115738

[18] Tandeau de Marsac N. Occurrence and nature of chromatic adaptation in cyanobacteria. Journal of Bacteriology. 1977;**130**:82-91. PMCID: PMC235176

[19] Yoshihara S, Katayama M, Geng X, Ikeuchi M. Cyanobacterial phytochrome-like PixJ1 holoprotein shows novel reversible photoconversion between blue- and green-absorbing forms. Plant & Cell Physiology. 2004;**45**:1729-1737. DOI: 10.1093/pcp/pch214

[20] Ikeuchi M, Ishizuka T. Cyanobacteriochromes: A new superfamily of tetrapyrrole-binding photoreceptors in cyanobacteria. Photochemical & Photobiological Sciences. 2008;**7**:1159-1167. DOI: 10.1039/B802660M

[21] Rockwell NC, Lagarias JC. A brief history of phytochromes. ChemPhysChem. 2010;**11**:1172-1180. DOI: 10.1002/cphc.2009200900894

[22] Hirose Y, Shimada T, Narikawa R, et al. Cyanobacteriochrome CcaS is the green light receptor that induces the expression of phycobilisome linker protein. Proceedings of the National Academy of Sciences of the United States of America. 2008;**105**:9528-9533. DOI: 10.1073/pnas.0801826105

[23] Hirose Y, Rockwell NC, Nishiyama K, et al. Green/red cyanobacteriochromes regulate complementary chromatic acclimation via a protochromic photocycle. Proceedings of the National Academy of Sciences of the United States of America. 2013;**110**:4974-4979. DOI: 10.1073/pnas.1302909110

[24] Kehoe DM, Grossman AR. Similarity of a chromatic adaptation sensor to phytochrome and ethylene receptors. Science. 1996;**273**:1409-1412. DOI: 10.1126/science.273.5280.1409

[25] Everroad C, Six C, Partensky F, et al. Biochemical bases of type IV chromatic adaptation in marine *Synechococcus* spp. Journal of Bacteriology. 2006;**188**:3345-3356. DOI: 10.1128/JB.188.9.3345-3356.2006

[26] Shukla A, Biswas A, Blot N, et al. Phycoerythrin-specific bilin lyase-isomerase controls blue-green chromatic acclimation in marine *Synechococcus*. Proceedings of the National Academy of Sciences of the United States of America. 2012;**109**:20136-20141. DOI: 10.1073/pnas.1211777109

[27] Grebert T, Dore H, Partensky F, et al. Light color acclimation is a key process in the global ocean distribution of *Synechococcus* cyanobacteria. Proceedings of the National Academy of Sciences of the United States of America. 2018;**115**:E2010-E2019. DOI: 10.1073/pnas.1717069115

[28] Sanfilippo JE, Nguyen AA, Karty JA, et al. Self-regulating genomic island encoding tandem regulators confers chromatic acclimation to marine *Synechococcus*. Proceedings of the National Academy of Sciences of the United States of America. 2016;**113**:6077-6082. DOI: 10.1073/pnas.1600625113

[29] Chen M, Floetenmeyer M, Bibby TS. Supramolecular organization of phycobiliproteins in the chlorophyll *d*-containing cyanobacterium *Acaryochloris marina*. FEBS Letters. 2009;**583**:2535-2539. DOI: 10.1016/j.febslett.2009.07.012

[30] Hu Q, Marquardt J, Iwasaki I, et al. Molecular structure, localization and function of biliproteins in the chlorophyll *a*/*d* containing oxygenic photosynthetic prokaryote *Acaryochloris marina*. Biochimica et Biophysica Acta - Bioenergetics. 1999;**1412**:250-261. DOI: 10.1016/S0005-2728(99)0067-5

[31] Gloag RS, Ritchie RJ, Chen M, et al. Chromatic photoacclimation, photosynthetic electron transport and oxygen evolution in the chlorophyll *d*-containing oxyphotobacterium *Acaryochloris marina*. Biochimica et Biophysica Acta - Bioenergetics. 2007;**1767**:127-135. DOI: 10.1007/s11120-009-9466

[32] Duxbury Z, Schliep M, Ritchie RJ, et al. Chromatic photoacclimation extends utilisable photosynthetically active radiation in the chlorophyll *d*-containing cyanobacterium, *Acaryochloris marina*. Photosynthesis Research. 2009;**101**:69-75. DOI: 10.1007/s11120-009-9466-7

[33] Loughlin PC, Duxbury Z, Mugerwa TTM, et al. Spectral properties of bacteriophytochrome AM15894 in the chlorophyll *d*-containing cyanobacterium *Acaryochloris marina*. Scientific Reports. 2016;**6**:27547. DOI: 10.1038/srep27547

[34] Kondo K, Geng XX, Katayama M, Ikeuchi M. Distinct roles of CpcG1 and CpcG2 in phycobilisome assembly in the cyanobacterium *Synechocystis* sp PCC 6803. Photosynthesis Research. 2005;**84**:269-273. DOI: 10.1007/s11120-004-7762-9

[35] Chen M, Li Y, Birch D, Willows RD. A cyanobacterium that contains chlorophyll *f*—A red-absorbing photopigment. FEBS Letters. 2012;**586**:3249-3254. DOI: 10.1016/j.febslett.2012.06.045

[36] Miyashita H, Adachi K, Kurano N, et al. Pigment composition of a novel oxygenic photosynthetic prokaryote containing chlorophyll *d* as the major chlorophyll. Plant & Cell Physiology. 1997;**38**:274-281. DOI: 0.1093/oxfordjournals.pcp.a029163

[37] Schiller H, Senger H, Miyashita H, et al. Light-harvesting in *Acaryochloris marina*: Spectroscopic characterization of a chlorophyll *d*-dominated photosynthetic antenna system. FEBS Letters. 1997;**410**:433-436. DOI: 10.1016/s0014-05793(97)00620-0

[38] Mimuro M, Akimoto S, Gotoh T, et al. Identification of the primary electron donor in PS II of the Chl *d* dominated cyanobacterium *Acaryochloris marina*. FEBS Letters. 2004;**556**:95-98. DOI: 10.1016/s0014-5793(03)-01383-8

[39] Lin Y, Crossett B, Chen M. Effects of anaerobic conditions on photosynthetic units of *Acaryochloris marina*. In: Proceedings of 15th International Conference on Photosynthesis. Berlin, Heidelberg: Springer Science + Business Media B.V.; 2013. pp. 121-124. DOI: 10.1007/978-3-642-32034-7_26

[40] Hu Q, Miyashita H, Iwasaki I, et al. A photosystem I reaction center driven by chlorophyll *d* in oxygenic photosynthesis. Proceedings of the National Academy of Sciences of the United States of America. 1998;**95**:13319-13323. DOI: 10.1073/pnas.95.22.13319

[41] Tomo T, Shinoda T, Chen M, et al. Energy transfer processes in chlorophyll *f*-containing cyanobacteria using time-resolved fluorescence spectroscopy on intact cells. Biochimica et Biophysica Acta - Bioenergetics. 2014;**1837**:1484-1489. DOI: 10.1016/jbbabio.2014.04.009

[42] Chen M, Hiller RG, Howe CJ, Larkum AWD. Unique origin and lateral transfer of prokaryotic chlorophyll-*b* and chlorophyll-*d* light-harvesting systems. Molecular Biology and

Evolution. 2005;**22**:21-28. DOI: 10.1093/molbev/msh250

[43] Chen M, Telfer A, Lin S, et al. The nature of the photosystem II reaction centre in the chlorophyll *d*-containing prokaryote, *Acaryochloris marina*. Photochemical & Photobiological Sciences. 2005;**4**:1060-1064. DOI: 10.1039/b507057k

[44] Akutsu S, Fujinuma D, Furukawa H, et al. Pigment analysis of a chlorophyll *f*-containing cyanobacterium strain KC1 isolated from Lake Biwa. Photomedicine and Photobiology. 2011;**33**:35-40

[45] Gan F, Zhang S, Rockwell NC, et al. Extensive remodeling of a cyanobacterial photosynthetic apparatus in far-red light. Science. 2014;**345**:1312-1317. DOI: 10.1126/science.1256963

[46] Gan F, Shen G, Bryant DA. Occurrence of far-red light photoacclimation (FaRLiP) in diverse cyanobacteria. Life. 2015;**5**:4-24. DOI: 10.3390/life501004

[47] Zhao C, Gan F, Shen G, Bryant DA. RfpA, RfpB and RfpC are the master control elements of far-red light photoacclimation (FaRLiP). Frontiers in Microbiology. 2015;**6**:1303. DOI: 10.3389/fmicb.2015.01303

[48] Li Y, Lin Y, Garvey CJ, et al. Characterization of red-shifted phycobilisomes isolated from the chlorophyll *f*-containing cyanobacterium *Halomicronema hongdechloris*. Biochimica et Biophysica Acta - Bioenergetics. 2016;**1857**:107-114. DOI: 10.1016/j.bbabio.2015.10.009

[49] Ho M-Y, Gan F, Shen G, Bryant DA. Far-red light photoacclimation (FaRLiP) in *Synechococcus* sp. PCC 7335. II. Characterization of phycobiliproteins produced during acclimation to far-red light. Photosynthesis Research. 2016;**131**:187-202. DOI: 10.1007/s11120-016-0303-6

[50] Ho M-Y, Gan F, Shen G, et al. Far-red light photoacclimation (FaRLiP) in *Synechococcus* sp. PCC 7335: I. Regulation of FaRLiP gene expression. Photosynthesis Research. 2016;**131**:173-186. DOI: 1o.1007/s11120-016-0309-z

[51] Lamb JJ, Røkke G, Hohmann-Marriott MF. Chlorophyll fluorescence emission spectroscopy of oxygenic organisms at 77 K. Photosynthetica. 2018;**56**:105-124. DOI: 10.1007/s11099-018-0791-y

[52] Cho F, Govindjee. Low-temperature (4-77°K) spectroscopy of *Anacystis*; temperature dependence of energy transfer efficiency. Biochimica et Biophysica Acta. 1970;**216**:151-161. DOI: 10.1016/0005-2728(70)90167-2

[53] Govindjee, Shevela D, Björn LO. Evolution of the Z-scheme of photosynthesis: A perspective. Photosynthesis Research. 2017;**133**:5-15. DOI: 10.1007/s11120-016-0333-z

[54] Akimoto S, Shinoda T, Chen M, Allakhverdiev SI, Tomo T. Energy transfer in the chlorophyll *f*-containing cyanobacterium, *Halomicronema hongdechloris*, analyzed by time-resolved fluorescence spectroscopies. Photosynthesis Research. 2015;**125**:115-122. DOI: 10.1007/s11120-015-0091-3

[55] Ho M-Y, Niedzwiedzki DM, MacGregor-Chatwin C, et al. Extensive remodeling of the photosynthetic apparatus alters energy transfer among photosynthetic complexes when cyanobacteria acclimate to far-red light. Biochimica et Biophysica Acta - Bioenergetics. 2020;**1861**:148064. DOI: 10.1016/j.bbabio.2019.148064

[56] Millar DP. Time-resolved fluorescence spectroscopy. Current Opinion in Structural Biology.

1996;**6**:637-642. DOI: 10.1016/s0959-440x(96)80030-3

[57] Chukhutsina VU, Holzwarth AR, Croce R. Time-resolved fluorescence measurements on leaves: Principles and recent developments. Photosynthesis Research. 2019;**140**:355-369. DOI: 10.1007/s11120-018-0607-8

[58] Schmitt FJ, Campbell ZY, Mai V, et al. Photosynthesis supported by a chlorophyll *f*-dependent, entropy-driven uphill energy transfer in *Halomicronema hongdechloris* cells adapted to far-red light. Photosynthesis Research. 2019;**139**:185-201. DOI: 10.1007/s11120-018-0556-2

[59] Kaucikas M, Nürnberg D, Dorlhiac G, et al. Femtosecond visible transient absorption spectroscopy of chlorophyll *f*-containing photosystem I. Biophysical Journal. 2017;**112**:234-249. DOI: 10.1016/j.bpj.201612.022

[60] Nürnberg DJ, Morton J, Santabarbara S, et al. Photochemistry beyond the red-limit in chlorophyll *f*-photosystems. Science. 2018;**360**: 1210-1213. DOI: 10.1026/science.aar8313

[61] Itoh S, Ohno T, Noji T, et al. Harvesting far-red light by chlorophyll *f* in photosystems I and II of unicellular cyanobacterium strain KC1. Plant & Cell Physiology. 2015;**56**:2024-2034. DOI: 10.1039/pcp/pcv122

[62] Cherepanov DA, Shelaev IV, Gostev FE, et al. Evidence that chlorophyll *f* functions solely as an antenna pigment in far-red-light photosystem I from *Fischerella thermalis* PCC 7521. Biochimica et Biophysica Acta - Bioenergetics. 2020;**1861**:148-184. DOI: 10.1016/j.bbabio.2020.148184

[63] Bar-Zvi S, Lahav A, Harris D, et al. Structural heterogeneity leads to functional homogeneity in *A. marina* phycocyanin. Biochimica et Biophysica Acta - Bioenergetics. 2018;**1859**:544-553. DOI: 10.1016/j.bbabio.2018.04.007

[64] Krause GH, Weis E. Chlorophyll fluorescence and photosynthesis: The basics. Annual Review of Plant Physiology and Plant Molecular Biology. 1991;**42**:313-349. DOI: 10.1146/annurev.pp.42.060191.001525

[65] López-Legentil S, Song B, Bosch M, et al. Cyanobacterial diversity and a new *Acaryochloris*-like symbiont from Bahamian sea-squirts. PLoS One. 2011;**6**:e23938. DOI: 10.1371/journal.pone.0023938

[66] Martinez-Garcia M, Koblizek M, Lopez-Legentil S, Anton J. Epibiosis of oxygenic phototrophs containing chlorophylls *a*, *b*, *c* and *d* on the colonial ascidian *Cystodytes dellechiajei*. Microbial Ecology. 2011;**61**:13-19. DOI: 10.1007/s00248-010-9694-6

[67] Majumder EL-W, Wolf BM, Liu H, et al. Subcellular pigment distribution is altered under far-red light acclimation in cyanobacteria that contain chlorophyll *f*. Photosynthesis Research. 2017;**134**:183-192. DOI: 10.1007/s11120-017-0428-1

[68] Pinevich AV, Matthijs HCP, Gavrilova OV, et al. New ultrastructural aspects of membranes and cell inclusions in *Prochlorothrix hollandica* (*Prochlorales*, *Cyanobacteria*). Microbios. 1996;**84**:217-225

[69] Korelusova J, Kaštovsky J, Komarek J. Heterogeneity of the cyanobacterial genus *Synechocystis* and description of a new genus, *Geminocystis*. Journal of Phycology. 2009;**45**:928-937. DOI: 10.1111/j.1529-8817.2009.00701.x

[70] Komarek J, Kaštovsky J, Mareš J, Johansen JR. Taxonomic classification of cyanoprokaryotes (cyanobacterial genera) using a polyphasic approach. Preslia. 2014;**86**:295-335

[71] Castenholz RW. General characterization of cyanobacteria. In: Boone DR, Castenholz RW, editors. Bergey's Manual of Systematic Bacteriology. New York: Springer-Verlag; 2001. pp. 474-487

Chapter 3

Influence of the Wavelength of Excitation and Fluorescence Emission Detection on the Estimation of Fluorescence-Based Physiological Parameters in Different Classes of Photosynthetic Organisms

Stefano Santabarbara, William Remelli,
Anastasia A. Petrova and Anna Paola Casazza

Abstract

Fluorescence-based methodologies are commonly employed to determine a wide spectrum of physiological parameters in intact photosynthetic organisms. These methods rely on the detection of Chlorophyll a fluorescent emission, which exhibits changes in its intensity due to the occurrence of quenching phenomena of either photochemical or non-photochemical nature, as well as in response of the absorption cross-section of the photosystems. At room temperature, it is generally considered that most of the emission stems from Photosystem II, and therefore, most of the physiological parameters rely on the assumption that contribution from Photosystem I is negligible. Moreover, it is often considered that the whole light-harvesting antenna is efficiently coupled to either of the photosystems and does not contribute, independently, to the detected emission. When these caveats are not realised, fluorescence-based indicators might be subjected to biases that tend to underestimate the extent of both photochemical and non-photochemical quenching. The contribution of Photosystem I and partially coupled/antenna components can be assessed through the analysis of the dependency of steady-state emission as a function of both the excitation and the emission wavelengths. On this basis, methods relying on using different combinations of excitation and emission wavelengths will be discussed in order to minimise the bias on the estimation of physiologically relevant parameters.

Keywords: chlorophyll fluorescence, maximal photochemical efficiency, photochemical quenching, non-photochemical quenching, Photosystem II, Photosystem I, light harvesting complexes, phycobilisomes

1. Introduction

Methodologies relying on the monitoring of Chlorophyll (Chl) *a* fluorescence emission have a very widespread application in photosynthesis research and have therefore been amply adopted by the scientific community (for a comprehensive survey, see the quite recent monography edited by [1]). Since Chls are fundamental pigments, naturally coordinated by the photosynthetic photosystems, these fluorescence-based methods do not require the use of exogenous dyes, or their coordination to specific biological targets. Thus, techniques based on Chl-fluorescence monitoring have become popular as they allow to address a spectrum of physiologically related questions, non-invasively, with relatively fast temporal responses and high sensitivity. Moreover, these methods are suitable for both *in vitro*, i.e. on isolated photosynthetic membranes or Chl-protein complexes, and *in vivo*, either on leaves or unicellular and multicellular algae. Furthermore, they can be used in combination with other analytical approaches, such as gas exchange and/or transmission/reflectance changes, without extensive cross-interference. Noteworthy, fluorescence detection strategies are often the method of choice for remote sensing as well, especially when, although not exclusively, sunlight can be used as the excitation source.

The most widely employed physiological measurements based on chlorophyll fluorescence detection are the determination of fluorescence quenching processes resulting either from photochemical reactions (photochemical quenching), or regulatory processes affecting the fluorescence emission yield, collectively referred to as non-photochemical quenching (NPQ).

1.1 Photochemical quenching

Photochemical quenching is the process through which the emission yield of protein-bound Chl is decreased in response to energy conversion reactions occurring in photochemically active pigments, the reaction centre (RC), of a photosystem. Historically, the occurrence of fluorescence quenching as a response to the redox state of the terminal electron acceptors, Q_A and Q_B (e.g. [2–4]) of PSII, and similarly in the reaction centre of purple bacteria which shares some functional similarities, paved the way for the now amply diffused utilisation of fluorescence measurements to determine physiological parameters. The fluorescence yield of PSII is low when the electron acceptors are oxidised (F_0, open centres) and reaches a maximal value when the redox centre Q_A is reduced (F_M, closed centres). The difference between the maximal fluorescence level and the minimal one is defined as variable fluorescence ($F_V = F_M - F_0$).

The maximal fluorescence level can be promoted either by brief, intense, light pulses, or flashes that cause the almost complete reduction of PSII acceptor pool (saturating pulses or flashes), so that Q_A is reduced as well, or using inhibitors, like 3-(3,4-dichlorophenyl)-1,1-dimethylurea (DCMU), that inhibit the oxidation of reduced Q_A^- by Q_B (e.g. [1, 3–5]). When inhibitors are used, the F_M condition is attained even when employing low actinic illumination, such as the measuring beam in a conventional fluorescence measurement, so that the use of additional saturating pulses could be avoided.

The conclusive demonstration that the differences in the emission intensity under F_0 and F_M conditions correspond to an effective change in fluorescence yield came from fluorescence lifetimes investigation, where it was shown that under open centre conditions the average fluorescence lifetimes, for the whole PSII supercomplex composed of its core complex and external antenna complement is in the order of 200–300 ps, whereas upon centre closures, it reaches values of

1.2–1.5 ns (e.g. [6–10]), that is, the same 3- to 5-fold increase observed by steady-state/pre-steady-state methods.

It can then be straightforwardly demonstrated, and it is therefore commonly accepted, that the maximal photochemical quantum efficiency of PSII, $\Phi_{PC,PSII}^{\max}$, could be estimated with sufficiently good approximation by the ratio of the variable and maximal fluorescence levels, i.e. by F_V/F_M (e.g. [1, 3–5]). The relation follows from considering that F_0 and F_M level could be described by the following expressions:

$$\begin{cases} F_0 = \alpha\phi_0 = \alpha\dfrac{k_f}{\sum_i k_i + K_{pc}\gamma_{Q,0}} \\ \\ F_M = \alpha\phi_M = \alpha\dfrac{k_f}{\sum_i k_i + K_{pc}\gamma_{Q,M}} \end{cases} \tag{1}$$

where α is a factor depending on the number of photons absorbed at the excitation wavelength and emitted at the detection wavelength, which is independent of the redox state of the acceptors in the RC, ϕ_0 and ϕ_M are the fluorescence emission yield at open and closed centres, respectively, which could be further expressed in terms of probabilities of the excited state de-excitation by fluorescence (k_f), with respect to the sum of all the so-called natural de-excitation processes, including fluorescence, $\sum_i k_i$ and quenching by photochemistry, here described in terms of a macroscopic, rather than molecular rate constant, K_{pc}, weighted by the molar fraction of oxidised electron acceptors at open ($\gamma_{Q,0}$) and closed centres ($\gamma_{Q,M}$). Considering that $\gamma_{Q,0} = 1$ and $\gamma_{Q,M} = 0$, that is, in the case of PSII, Q_A being completely oxidised and reduced, and substituting into Eq. (1), results in:

$$\frac{F_V}{F_M} = \frac{K_{pc}}{\sum_i k_i + K_{pc}} \approx \Phi_{PC,PSII}^{Max} \tag{2}$$

Although the above-presented derivation presents the yield in the form of the so-called "lake model", in which the quenching induced by photochemistry is shared over a large pool of photosystems, thereby approaching the classic Stern-Volmer description, for the F_0 and F_M level, identical conclusion would be reached by considering either separate units or partial connectivity, under the condition that *all* centres are either open or closed. On the other hand, these models do differ significantly in the description of any intermediate level between F_0 and F_M, that is, for as any value of γ_Q different from 0 or 1, which manifests either in the kinetics or rise to F_M (so-called induction curve) or in the steady-state levels, F_{ss}, determined by a given background illumination. It is also worth noticing that Eq. (2) is valid, provided that the *only* quenching process differing at F_0 and F_M is strictly due to a modulation of photochemical quenching. It is however being shown that the fluorescence level reached by single-turnover flashes can be slightly lower than those observed under the so-called multiple turnover pulses, or by continuous illumination, indicating that the estimation of $\Phi_{PC,PSII}^{\max}$ from F_V/F_M could be somewhat biased (e.g. [11, 12]). However, in general, these processes, whose molecular nature remains to be fully determined, lead to contained distortions in photochemical yield estimation. Therefore, since measurements of F_V/F_M are relatively rapid, they remain one of the most employed indicators to determine not only PSII quantum yield, but also its changes in response to stress, both in the laboratory and in the field. Measurements of variable fluorescence in the presence of actinic backgrounds also have become a widespread practise to gain information concerning photosynthetic electron transport. A number of indicators have been proposed in order to extract this information from fluorescence data and, at least for land plants, strong correlation with simultaneous gas-exchange measurements has been demonstrated [5, 13, 14]. This is because the variation in the amplitude of F_V in the presence of actinic illumination, in combination with changes in the F_M level, with respect to the dark-adapted state, could be related to the fraction of "open/closed" PSII centres as well as to their photochemical quantum yield under operation conditions. A detailed discussion of the parameters is beyond the scope of this chapter, but several approaches are reported in the recent monography [1] and extended review [5] on these topics. We limit to notice that the interpretation of F_V-related fluorescence indicators in the presence of an actinic illumination is less straightforward than for dark adapted conditions.

1.2 Non-photochemical quenching

On top of photochemical quenching processes, fluorescence monitoring finds its natural application in the study of regulative mechanisms of light harvesting efficiency, which manifest themselves principally as a quenching of the Chl fluorescence emission. The establishment of non-photochemical quenching (NPQ) results in the lowering of the F_M level in response to actinic illumination regimes close to or exceeding the saturation of the photosynthetic electron transport capacity (e.g., [15, 16]). The fluorescence level in the presence of a non-photochemical quenching, most commonly measured at closed centres by the application of multiple turn-over saturating pulse, is referred to as F'_M and can be described, similarly to F_M, within a lake model approximation as follows:

$$F'_M = \alpha \frac{k_f}{\sum_i k_i + [Q_{NP}] \cdot K_{Q,NP}} \tag{3}$$

where $K_{Q,NP}$ and Q_{NP} are the macroscopic quencher rate constant and the concentration of the non-photochemical quencher, respectively. These two quantities are however difficult to assess separately and are generally determined as the product $[Q_{NP}] \cdot K_{Q,NP}$ through the *NPQ* parameters (note in the following the italics will identify the parameters whereas upright capitals indicate the process), that is defined as $F_M/F'_M - 1$, with the definitions in Eqs. (1) and (3) as follows:

$$NPQ = \frac{F_M - F'_M}{F'_M} = [Q_{NP}] \cdot \frac{K_{Q,NP}}{\sum_i k_i} \tag{4}$$

NPQ has proven to be a very complex phenomenon, showing several phases of development, upon light exposure, and relaxation, upon return to darkness [15, 16], that despite the generality of its manifestation, depends strongly on the organism or at least the class of organisms in which it is investigated [17, 18]. This large variability is likely due to the generally accepted knowledge that NPQ processes occur principally, when not exclusively, in the external antenna complement of PSII [15–18]. Whereas the core complex, which serves as an internal light harvesting system as well as the photochemical reaction centre, is well conserved amongst different oxygenic photosynthetic organisms [19], the composition of the external antenna is instead very diversified, having evolved to optimise light harvesting in ecological niches possessing different spectral distributions either in terms of light spectrum or light intensity [20, 21]. Hence, it is not surprising that also adaptive responses to changes in light conditions vary significantly in different organisms.

1.3 Source of possible distortion due to emission from Photosystem I and decoupled light-harvesting antenna

One of the caveats behind Chl-based fluorescence indicators is the assumption that the monitored fluorescence is emitted exclusively or almost exclusively from PSII. This is, in general, a justified assumption since, particularly under closed centres conditions, and at a room temperature or temperatures covering the "physiological interval" of living organisms, the fluorescence yield of PSII could be several folds larger than the one of the other ubiquitous component of the electron transfer chain, Photosystem I (PSI). Measurements on PSI isolated from different species indicate that its average fluorescence lifetime is in the order of 20–40 ps [19, 22, 23], that is, 5–10 times faster than for the PSII-antenna supercomplex at open centres and 200–500 faster at closed centres, in the absence of non-photochemical quenching [6–11]. Similar results concerning PSI lifetimes are also retrieved for measurements when the photosystem is embedded in the thylakoid membranes [8, 9, 24]. Moreover, PSI does not show a significant change in emission yield upon oxidisation of the terminal donor, P_{700} [25]. Hence, differently from PSII, PSI does not exhibit any significant induction, so that, in practise, its contribution to the cell emission is expected to be larger under F_0 than F_M conditions. Direct comparison of lifetimes and steady state emission levels holds, however, provided that the photosystems are in close stoichiometric ratio, that they absorb almost the same number of photons at the excitation wavelength and also have the same emission bandwidth at the excitation wavelength. For instance, in most cases, PSI emission results red-shifted with respect to that of PSII (e.g. [19, 22, 23] and references therein), so that its contribution to the detected fluorescence emission from photosynthetic material is expected to be a function of the detection wavelength, which depends on the specific emission spectra of PSII and PSI in a given organism. Similarly, either the absorption or the relative stoichiometry, hence the fraction of excitation photons absorbed by each photosystem, are expected to be dependent on organism. Especially for the determination of F_0, contribution from PSI emission could become significant and not necessarily negligible [24, 26–28], and similarly for cases in which the F'_M level is quenched significantly below F_M.

Another possible source of distortion in the determination of these fluorescence levels is the contribution from populations of the external antenna not coupled in terms of energy transfer to the core of either PSII or PSI. Differently from PSI, the core-decoupled pigment-protein complexes of the antenna are very bright in terms of fluorescence, emitting with lifetimes comparable or longer than PSII at closed centres. Typically, the population of decoupled antenna tends to be vanishingly small in most organisms having transmembrane light-harvesting apparatus [27, 29], but it can be sizable in cyanobacteria, accounting for 1–5% of the external antenna [26, 30, 31] and therefore clearly contributing to the cellular emission [26–28, 32–34]. Thus, differences between organisms, which harbour diversified external antenna complements, are to be expected.

2. Spectral dependence of the F_V/F_M ratio

A strategy to verify the distortion originating from the emission due to either PSI or a fraction of decoupled/weakly coupled external antenna components to the determination of the F_V/F_M ratio is that of recording the steady-state fluorescence emission spectrum of photosynthetic material under open PSII RC conditions, by using a low-intensity excitation having minimal actinic effect thereby approaching the real dark-adapted F_0 level, hereafter these will be referred to as F'_0, and to compare it with the spectrum obtained for the same excitation regime in the presence of an inhibitor, such as DCMU, that promotes PSII centres closure, thereby reaching the F_M conditions. The expectation is that when the fluorescence spectra are effectively dominated by PSII emission, the F_V/F_M ratio (or analogously the F_M/F_0 ratio) will attain the same value, irrespectively on the detection wavelength. Similarly, recording the emission spectra under F'_0 and F_M conditions, but at different excitation wavelengths, the expected result in the absence of distortion is that the F_V/F_M ratio is not only independent of the observation wavelength but also from the one of excitation. Contributions from either PSII or decoupled antenna, or both, would result in an observation-wavelength dependency of the F_V/F_M value, which might also depend on the excitation [26–28, 32–36].

In the successive paragraphs, experiments performed in model green algae (*Chlorella sorokiniana* and *Chlamydomonas reinhardtii*) and model cyanobacteria (*Synechocystis sp.* PCC6803 and *Synechococcus sp.* PCC7942) are summarised, because these classes of organisms display significant differences [27].

Conceptually, these types of experiments are pretty simple, but technically they might be challenging as they require a rather sensitive apparatus in order to collect F'_0 spectra with sufficient signal-to-noise ratio, due to the low intensity of the excitation beam. On the other hand, recording the emission under F_M conditions is relatively straightforward and will therefore be suggested as an initial screening for possible sources of bias in the F_V/F_M, and thus $\Phi^{Max}_{PC,PSII}$, estimations.

2.1 Excitation and emission dependency in model green algae

The emission spectra recorded under F'_0 and F_M conditions in the green algae *C. sorokiniana* and *C. reinhardtii* for excitation at 435, 475, 520 and 570 nm are shown in **Figure 1**. These wavelengths were chosen as they represent preferential excitation of Chl *a* (435 nm) and Chl *b* (475 nm), which are the main pigments in green algae, where Chl *a* is present both in the core complexes and the external antenna, and Chl *b* is present only in the light-harvesting complexes associated with both PSII and PSI. Although the external antenna complement is composed by several Chl-protein complexes, it is convenient in this case in which the individual contributions cannot be discriminated, to refer to them collectively as LHCI and LHCII, for PSI and PSII, respectively. Excitations at 520 and 570 nm are relatively unselective in this class of organisms, but they represent preferential excitation for phycobilisomes (PBS), which represent the main, often the only, external antenna complement in cyanobacteria.

For ease of comparison, the spectra, at each excitation wavelength, are normalised to the maximum of the F_M spectrum, and the F'_0 one is scaled accordingly. The spectra reported in **Figure 1** demonstrate that in both model green algae, the emission bandwidth displays a very weak dependency on the excitation wavelength, and that, as better appreciable in the insets where each spectrum is normalised to its maximum, the overall bandwidth shows only limited differences between F'_0 and F_M conditions. Yet, the spectra recorded at F'_0 show a slight increase in the relative intensity in the 700–730 nm window, with respect to those recorded at F_M. This difference can be interpreted, dominantly, by a relatively

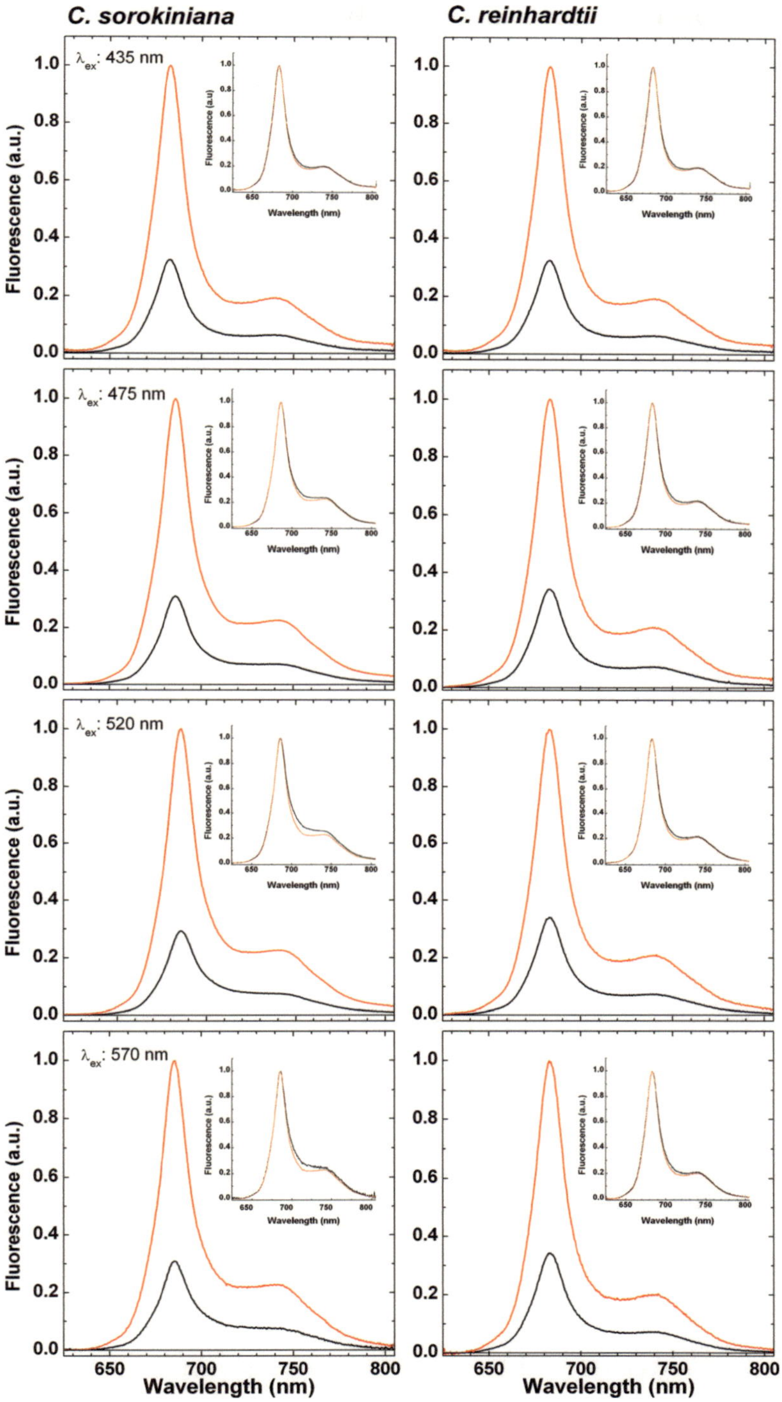

Figure 1.
Emission spectra recorded under F_0' (black lines) and F_M (red lines) conditions in the model green algae C. sorokiniana *(left) and* C. reinhardtii *(right), at room temperature, for excitation at 435, 475, 520 and 570 nm. The spectra in the main panels are normalised and scaled to the maximal emission at F_M. The insets show the spectra normalised each at its maximal emission. The samples were diluted to an optical density (scatter compensated) in between 0.05 and 0.1 OD at the maximal absorption in the red, at 680 nm. The excitation bandwidth was between 0.5 and 1 nm (FWHM); spectra were acquired with a resolution of 0.25 nm and are corrected for the sensitivity of the detector. Further details on the experimental set-up are described in Refs. [24, 26,27].*

larger contribution of PSI emission in this wavelength region, when PSII fluorescence is quenched photochemically [24, 27].

In **Figure 2** are shown the F_V/F_M spectra, obtained from data of **Figure 1**. For both model green algae, the F_V/F_M spectra are indeed quite flat across the whole emission band, showing however a trough, corresponding to minimal values, in the 700–740 window, in correspondence to the maximal *in vivo* emission of PSI. Maximal values are recorded close to maximal cellular emission, with values ranging in 0.66–0.72 interval, that is dominated by PSII instead, corresponding to an underestimation of F_V/F_M of 5–8% with respect to its maximal value. It is also worth remarking that the F_V/F_M values have a rather limited dependency on the excitation wavelength, which is probably for the results reported here, within the limits of the actinic effect of the excitation beam at F'_0, determining a condition which more (for larger F_V/F_M) or less (for lower F_V/F_M) closely approaches the actual F_0.

A strategy to extract the *in vivo* emission spectrum of PSII is that of calculating the F_V spectra, which are shown in **Figure 3**, each normalised to its maximal value. It is clear then that the bandshape of the F_V spectra does not depend on the excitation wavelength. This is a key indication, supported by parallel fluorescence lifetime evidences [24], that the observed spectral dependence of the F_V/F_M spectrum is not due to, and certainly not dominated by, an intrinsic spectral dependency of photochemical quenching in PSII.

Figure 3 also shows a comparison for the normalised F_M spectra, highlighting their very limited dependency on the excitation wavelength. As mentioned above, the recording of spectra at F_M is attained more straightforwardly than that at F'_0, as it does not suffer from possible biases from the actinic effect of the excitation wavelength. Although the spectra shown here were acquired using the same excitation intensity at F'_0 and F_M, in general, the latter can be collected with higher excitation fluxes henceforth in conventional instruments. Care shall however be taken to avoid the onset of non-photochemical quenching processes. The invariance of the emission bandshape of the F_M spectrum with respect to the excitation wavelength can be considered as a preliminary verification that this experimental condition does not affect the measured F_V/F_M ratio.

To better compare the effect of excitation and detection wavelengths in determining the F_V/F_M ratio in traditional set-ups where detection is commonly performed using interferential or band-pass filters, F'_0 and F_V/F_M were calculated convolving the emission spectra with either ideal interferential (FWHM 10 nm) or band-pass filters. The results obtained are listed in **Table 1**, all normalised to F_M. Maximal values of F_V/F_M (corresponding to minimal ones of F'_0) are obtained for detection centred at ~685 nm and minimal at 715 nm, for the conditions tested here. Similarly band-pass with cut-on filter having a threshold frequency above 700 nm yielded lower F_V/F_M values than for shorter-wavelengths, which integrate over a larger portion of the emission band.

2.2 Excitation and emission dependency in model cyanobacteria

The emission spectra recorded under F'_0 and F_M conditions in two model cyanobacteria, *Synechocystis sp.* PCC 6803 and *Synechococcus sp.* PCC7942 (from now on referred simply as Synechocystis and Sinechococcus), and at different excitation wavelengths are shown in **Figure 4**. Differently from what reported in **Figure 1** for model green algae, in both model cyanobacteria, the emission bandshapes display a strong dependency both on the excitation wavelength and on the redox conditions

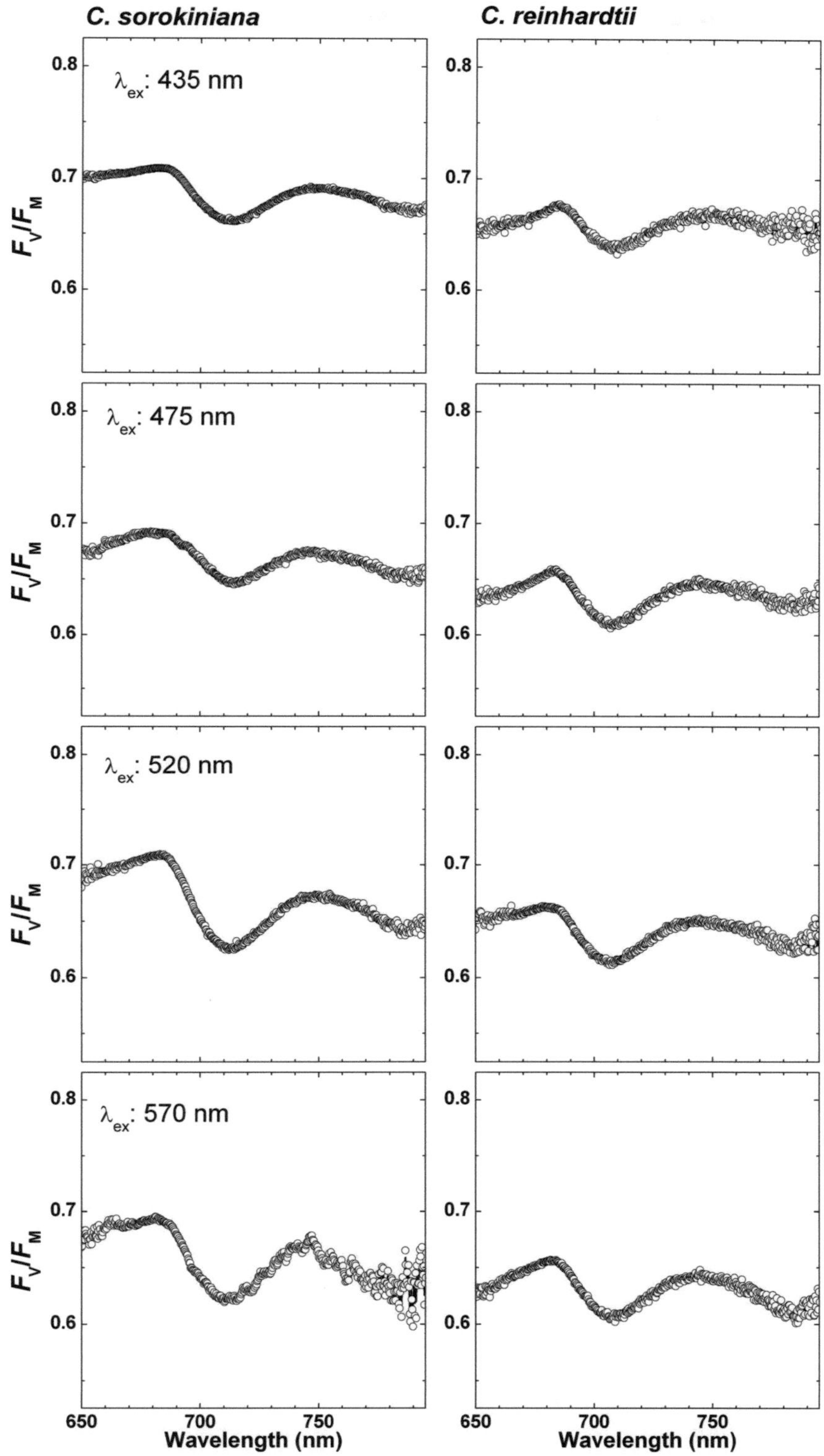

Figure 2.
Emission wavelength dependency of the F_V/F_M ratio in C. sorokiniana *(left) and* C. reinhardtii *(right), at room temperature, for excitation at 435, 475, 520 and 570 nm, calculated from the spectra of* ***Figure 1****.*

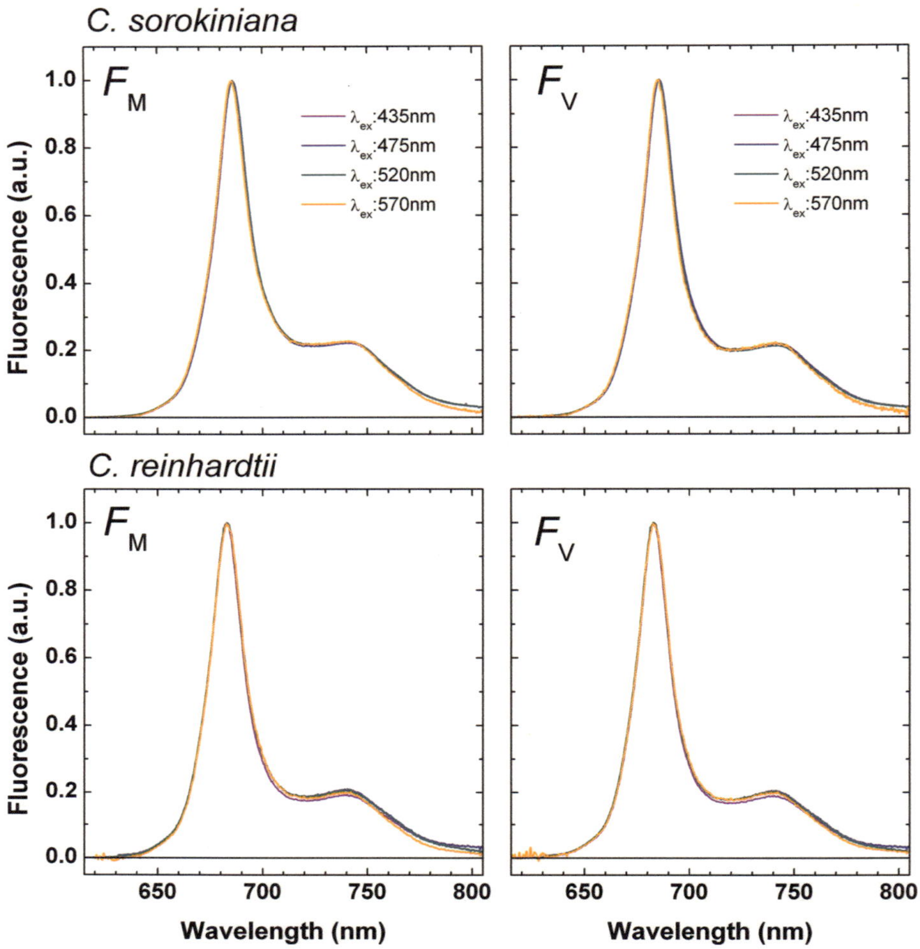

Figure 3.
Comparison of the normalised F_M and calculated $F_V = F_M - F'_0$ emission spectra, in C. sorokiniana *(top) and* C. reinhardtii *(bottom), at room temperature, for excitation at 435 nm (violet), 475 nm (blue), 520 nm (green) and 570 nm (orange). Each spectrum is normalised to its maximum value.*

of the PSII acceptors. Upon excitation at 435 and 475 nm, the main peak is centred at 684 nm and intense shoulders, more pronounced than what is observed in green algae, are observed on the long wavelength range above ~700 nm. Moreover, a relatively intense blue wing is clearly observed in the 610–660 nm range, which becomes larger and even dominant when the sample is excited at 520 and 570 nm, that is, for preferential excitation in the PBS antenna, where a clear satellite peak centred at 600 nm is discernible. It is immediately noticeable that the short wavelength emission displays negligible variation in intensity when PSII centres are closed with respect to when they are open.

This is further highlighted in **Figure 5**, where the F_V/F_M spectra derived from the data of **Figure 4** are reported. The results show that although the details depended on the species, in general, in both model cyanobacteria, the F_V/F_M value displays a significant dependence both on the wavelength of detection as well as on the wavelength of excitation. For excitation at 435 nm, that is preferential for Chl *a* (and carotenoids), maximal F_V/F_M values are obtained, in correspondence to the fluorescence emission peak (675–685 nm), in the 0.6 range. However, even when

	C. sorokiniana							
Excitation	435 nm		475 nm		520 nm		570 nm	
Interferential (Centre, nm)	F'_0	F_V/F_M	F'_0	F_V/F_M	F'_0	F_V/F_M	F'_0	F_V/F_M
650	0.302	0.698	0.325	0.675	0.310	0.690	0.319	0.681
670	0.292	0.708	0.312	0.688	0.298	0.702	0.311	0.689
680	0.292	0.708	0.309	0.691	0.293	0.707	0.308	0.692
690	0.304	0.696	0.315	0.685	0.305	0.695	0.319	0.681
700	0.327	0.673	0.330	0.670	0.342	0.658	0.353	0.647
720	0.347	0.653	0.349	0.651	0.366	0.634	0.370	0.630
Bandpass (cut-on, nm)								
655	0.314	0.686	0.325	0.675	0.323	0.677	0.334	0.666
675	0.317	0.683	0.326	0.674	0.326	0.674	0.336	0.664
695	0.331	0.669	0.336	0.664	0.347	0.653	0.354	0.646
720	0.327	0.673	0.335	0.665	0.341	0.659	0.348	0.652
	C. reinhardtii							
Inteferential (Centre, nm)	F'_0	F_V/F_M	F'_0	F_V/F_M	F'_0	F_V/F_M	F'_0	F_V/F_M
650	0.343	0.657	0.365	0.635	0.348	0.652	0.371	0.629
670	0.336	0.664	0.354	0.646	0.341	0.659	0.352	0.648
680	0.328	0.672	0.346	0.654	0.339	0.661	0.346	0.654
690	0.330	0.670	0.353	0.647	0.350	0.650	0.355	0.645
700	0.350	0.650	0.377	0.623	0.375	0.625	0.381	0.619
720	0.352	0.648	0.377	0.623	0.372	0.628	0.381	0.619
Bandpass (Cut-on, nm)								
655	0.337	0.663	0.359	0.641	0.353	0.647	0.361	0.639
675	0.337	0.663	0.360	0.640	0.355	0.645	0.362	0.638
695	0.345	0.655	0.369	0.631	0.365	0.635	0.374	0.626
720	0.340	0.660	0.361	0.639	0.357	0.643	0.367	0.633

Values of F'_0 and F_V/F_M determined from the F'_0 and F_M emission spectra of model green algae, after convolution with either ideal interferential filter (Gaussian, FWHM 10 nm) or long-pass filter (cut-on transitions 4 nm at the indicated 50% transmission). Values are normalised to F_M = 1 under each excitation/emission condition.

Table 1.
Estimated F'_0 and F_M detected through interferential and band-pass filters in green algae.

monitoring F_V/F_M at the emission maxima, its value depends on the excitation wavelength, decreasing to 0.45–0.5 for all other wavelengths tested. Moreover, in both organisms and for all excitations, the F_V/F_M ratio decreases sharply in the 700–730 nm window, to values of 0.2–0.3, corresponding to ~40% underestimation with respect to the maximal. Similarly to what was discussed for the case of green algae, this decrease can be ascribed to a significant contribution of PSI emission, especially at F'_0, in this wavelength region [26, 27]. The effect is however much larger than in green algae. This could be explained by differences in antenna properties for these two classes of organisms, as well as in the PSI:PSII stoichiometries. Whereas for excitation in the blue, PSI and PSII absorb almost equal fractions of photons, in cyanobacteria, the difference is larger as each

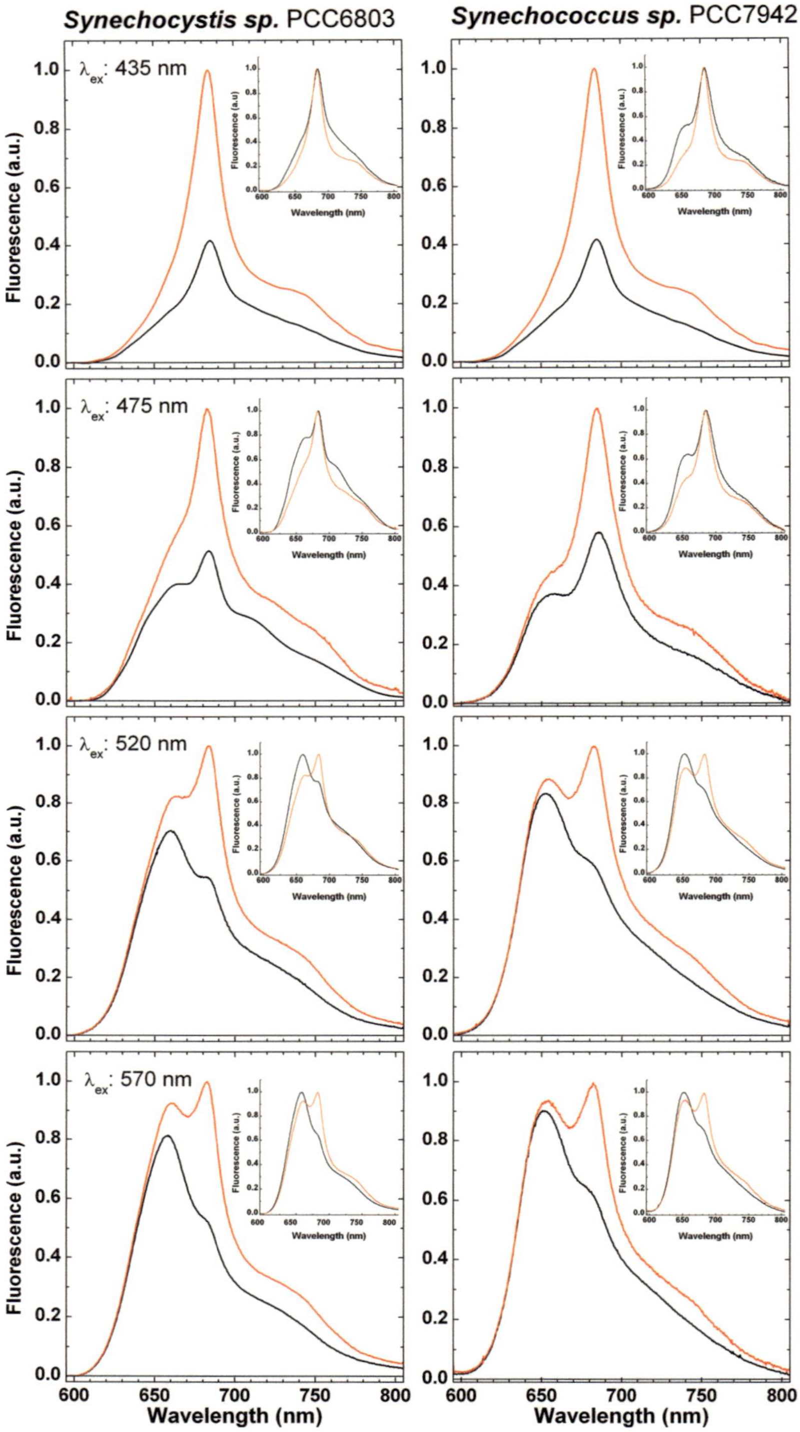

Figure 4.
Emission spectra recorded under F'_0 (black lines) and F_M (red lines) conditions in the model cyanobacteria Synechocystis sp. *PCC6803 (left) and* Synechococcus sp. *PCC7942 (right), at room temperature, for excitation at 435, 475, 520 and 570 nm. The spectra in the main panels are normalised and scaled to the maximal emission at F_M. The insets show the spectra normalised each at its maximal emission.*

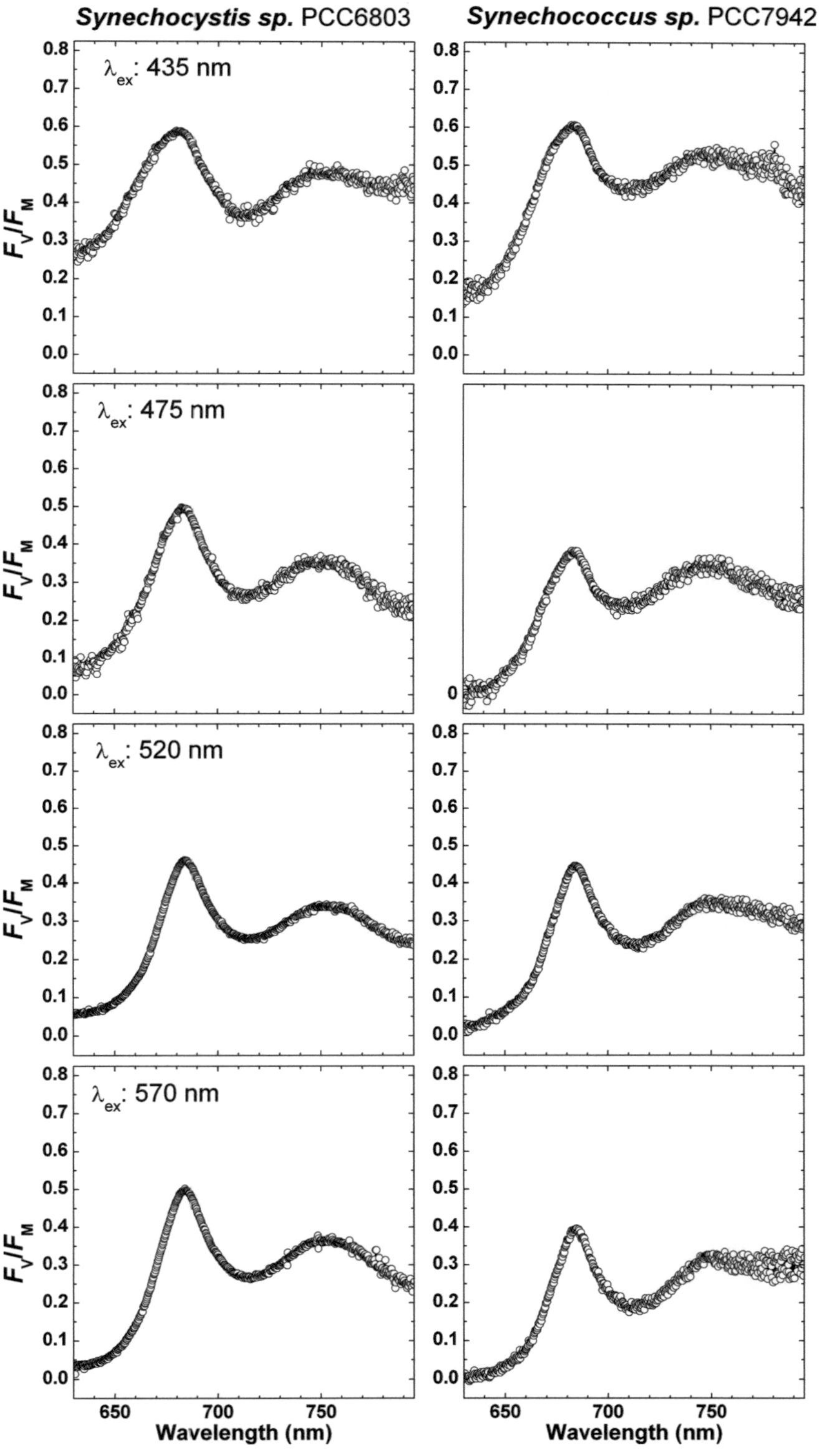

Figure 5.
Emission wavelength dependency of the F_V/F_M ratio in Synechocystis sp. *PCC6803 (left) and* Synechococcus sp. *PCC7942 (right), at room temperature, for excitation at 435, 475, 520 and 570 nm, calculated from the spectra of **Figure 4**. Please note the difference in the y-scale with respect to **Figure 2**.*

monomeric PSII core binds about 45 Chl *a* molecules, whereas PSI monomers bind about 90 Chl *a*. Furthermore, whereas PSII is a functional dimer (total ~ 90 Chls), PSI is, at least as the most abundant form, a trimer (270 Chls). Even for a 1:1 ratio of the two supercomplexes, PSI is expected to absorb about 3 times more photons than PSII. Moreover, it is often reported that the PSI:PSII stoichiometry in cyanobacteria could be much larger than 1, and thereby, PSI will result contributing significantly to the steady state emission despite its low fluorescence quantum yield.

Furthermore, the F_V/F_M spectra in model cyanobacteria display a very steep decrease to a minimal value on the blue emission wing, below ~665 nm, in correspondence to PBS emission. The decrease in F_V/F_M is much steeper when the external antenna is excited directly (520 and 570 nm, in **Figure 5**, and reaching negligible values well below 0.1). At least for the case of Synechocystis, this applies moreover to all wavelengths at which PBS absorption is dominant [27]. Hence, monitoring F_V/F_M on the short emission wing leads to profound underestimation of the maximal photochemical yield of PSII.

However, the bandshape of the F_V spectra does not depend significantly on the excitation wavelength, as also observed in model green algae. In the case of model cyanobacteria, a more pronounced wing extending down to 625–630 nm is visible (**Figure 6**) in the F_V spectra. This could be ascribed to variable fluorescence of the energetically coupled fraction of the PBS antenna, at least of its terminal emitters [27, 37, 38]. The relative invariance indicates that the changes in the F_V/F_M value are not dominated by an intrinsic variation of photochemical quenching over the PSII emission band (see discussion in ref. [27] for further detail), rather to the overlapped contributions of different chromophore-protein supercomplexes, nominally PSI and a fraction of the PBS antenna which is not energetically coupled to the core complexes of either PSI or PSII. The need for considering an energetically uncoupled antenna stems from the large increase in its emission upon direct excitation and the close-to-absent photochemical quenching, which is to be expected for a core-coupled antenna. It is also demonstrated by the long fluorescence lifetimes (1–4 ns, cite), even under open PSII RC conditions.

In **Figure 6** are also shown the normalised F_M spectra recorded at different excitation wavelengths. It is immediately noticeable that the bandshape of the F_M spectra, differently from the case of green algae and the F_V spectra in cyanobacteria, changes markedly with the excitation wavelength. As already mentioned, recording of emission spectra under F_M conditions, at room/physiologically relevant temperatures, is relatively straightforward. Observations of pronounced changes in the bandshape of the emission spectrum as a function of the excitation wavelength could then be used as a preliminary screening for indication of non-negligible contribution of pigment-protein complexes other than PSII to the cellular (or photosynthetic membrane) emission. If any independent knowledge is available concerning the spectral characteristics of the external antenna and/or of the PSI, which would help interpreting the F_M spectral excitation dependency, then appropriate pairs of excitation/detection wavelength can be chosen for classic fluorescence induction curve measurements.

To further highlight the importance of measurement conditions on the determination of the F_V/F_M ratio in cyanobacteria, in **Table 2** are reported the values of F'_0, F_M and F_V/F_M which would be obtained monitoring these levels though collection by either interference or band-pass filters. Although the general trend is similar to what was reported for green algae in **Table 1**, the variations are much larger in the case of cyanobacteria where values exceeding 0.5 are obtained only for detection close to emission maximum (interferential at 685 nm) and avoiding preferential PBS excitation.

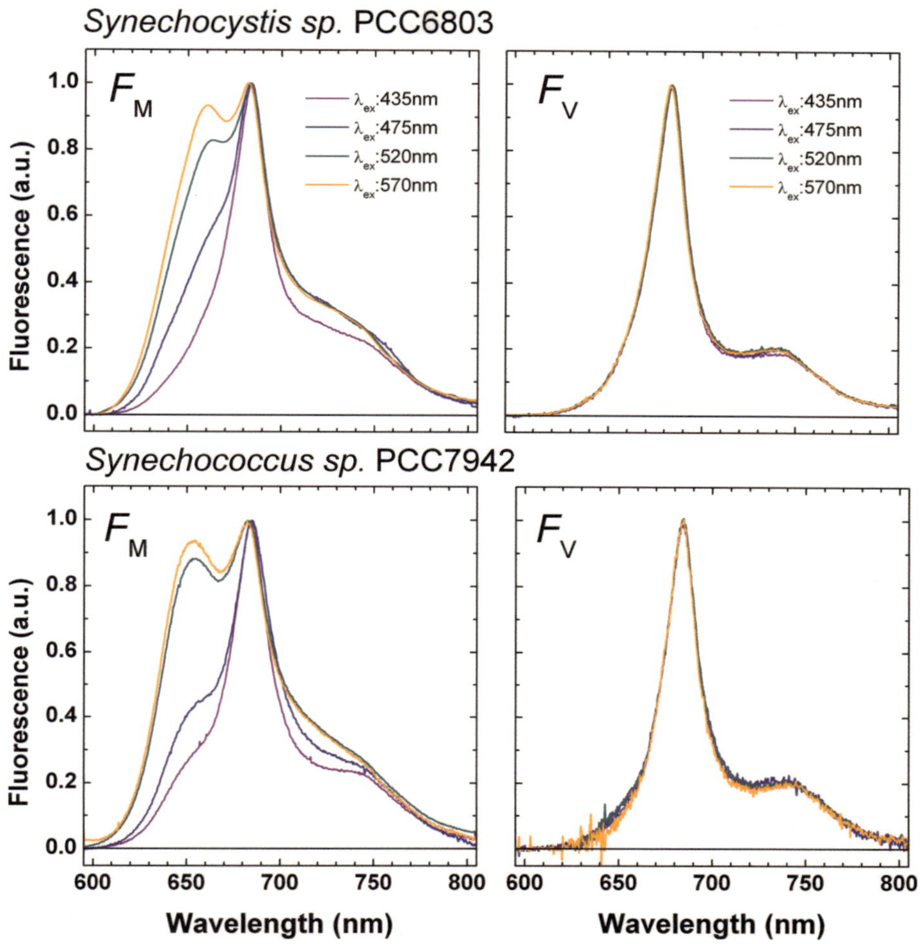

Figure 6.
Comparison of the normalised F_M and calculated $F_V = F_M - F'_0$ emission spectra, in Synechocystis sp. *PCC6803 (top) and* Synechococcus sp. *PCC7942 (bottom), at room temperature, for excitation at 435 nm (violet), 475 nm (blue), 520 nm (green) and 570 nm (orange). Each spectrum is normalised to its maximum value.*

2.3 Contribution of different chromophore-pigment super-complexes to photosynthetic membrane emission

The spectral dependencies of F_V/F_M could be accounted by a relatively simple formalism, based on considering a linear combination of contributions from i) PSII-antenna supercomplexes that give rise to variable fluorescence, F_V, and ii) other supercomplexes, in which the emission yield is independent of PSII redox state. The latter can be identified, in general terms, by the term F_{nV} that collectively accounts for emission of PSI and fractions of decoupled antenna, which are, as shown above, organism-dependent. Eq. (1) can be re-written, explicitly considering the excitation and emission dependency, as follows:

$$\begin{cases} F_0(\lambda_{em}, \lambda_{ex}) = \rho_{II}(\lambda) \cdot \sigma_{II}(\lambda) \cdot \phi_0 + \rho_{nV}(\lambda) \cdot \sigma_{nV}(\lambda) \cdot \phi_{nV} \\ F_M(\lambda_{em}, \lambda_{ex}) = \rho_{II}(\lambda) \cdot \sigma_{II}(\lambda) \cdot \phi_M + \rho_{nV}(\lambda) \cdot \sigma_{nV}(\lambda) \cdot \phi_{nV} \end{cases} \quad (5)$$

where $\rho_{II}(\lambda)$ and $\rho_{nV}(\lambda)$ are the absorption cross-sections of PSII and the pigments showing non-variable fluorescence, respectively; $\sigma_{II}(\lambda)$ and $\sigma_{nV}(\lambda)$ are the

Synechocystis sp. PCC6803								
Excitation	435 nm		475 nm		520 nm		570 nm	
Interferential (Centre, nm)	F'_0	F_V/F_M	F'_0	F_V/F_M	F'_0	F_V/F_M	F'_0	F_V/F_M
650	0.633	0.367	0.802	0.198	0.917	0.083	0.927	0.073
670	0.454	0.546	0.614	0.386	0.729	0.271	0.724	0.276
680	0.419	0.581	0.529	0.471	0.585	0.415	0.568	0.432
690	0.467	0.533	0.552	0.448	0.584	0.416	0.571	0.429
700	0.559	0.441	0.626	0.374	0.684	0.316	0.675	0.325
720	0.613	0.387	0.694	0.306	0.742	0.258	0.729	0.271
Bandpass (cut-on, nm)								
655	0.504	0.496	0.612	0.388	0.688	0.312	0.683	0.317
675	0.508	0.492	0.598	0.402	0.649	0.351	0.636	0.364
695	0.556	0.444	0.635	0.365	0.691	0.309	0.680	0.320
720	0.534	0.466	0.608	0.392	0.675	0.325	0.663	0.337
Synechococcus sp. PCC7942								
Interferential (Centre, nm)	F'_0	F_V/F_M	F'_0	F_V/F_M	F'_0	F_V/F_M	F'_0	F_V/F_M
650	0.750	0.250	0.866	0.134	0.955	0.045	0.972	0.028
670	0.487	0.513	0.678	0.322	0.775	0.225	0.803	0.197
680	0.413	0.587	0.598	0.402	0.629	0.371	0.663	0.337
690	0.446	0.554	0.631	0.369	0.630	0.370	0.666	0.334
700	0.530	0.470	0.710	0.290	0.729	0.271	0.766	0.234
720	0.547	0.453	0.722	0.278	0.774	0.226	0.806	0.194
Bandpass (cut-on, nm)								
655	0.491	0.509	0.677	0.323	0.731	0.269	0.761	0.239
675	0.479	0.521	0.663	0.337	0.691	0.309	0.720	0.280
695	0.516	0.484	0.697	0.303	0.728	0.272	0.756	0.244
720	0.498	0.502	0.677	0.323	0.708	0.292	0.730	0.270

Values of F'_0 and F_V/F_M determined from the F'_0 and F_M emission spectra of model cyanobacteria, after convolution with either ideal interferential filter (Gaussian, FWHM 10 nm) or long-pass filter (cut-on transitions 4 nm at the indicated 50% transmission). Values are normalised to $F_M = 1$ under each excitation/emission condition.

Table 2.
Estimated F'_0 and F_M detected through interferential and band-pass filters in cyanobacteria.

(area normalised) emission spectra of the same components; ϕ_{nV}, ϕ_0 and ϕ_M are the fluorescence yields of the non-variable component, PSII under open and closed centres, respectively, from which stems that

$$\begin{cases} F_V(\lambda_{em},\lambda_{ex}) = F_M(\lambda_{em},\lambda_{ex}) - F_0(\lambda_{em},\lambda_{ex}) = \rho_{II}(\lambda)\cdot\sigma_{II}(\lambda)\cdot(\phi_M-\phi_0) \\ \dfrac{F_V}{F_M}(\lambda_{em},\lambda_{ex}) = \dfrac{\rho_{II}(\lambda)\cdot\sigma_{II}(\lambda)\cdot(\phi_M-\phi_0)}{\rho_{II}(\lambda)\cdot\sigma_{II}(\lambda)\cdot\phi_M+\rho_{nV}(\lambda)\cdot\sigma_{nV}(\lambda)\cdot\phi_{nV}} \end{cases} \quad (6)$$

which satisfies the lack of excitation wavelength dependency of $F_V(\lambda_{em})$ and simply rationalise the differences in the value of F_V/F_M depending on the relative contribution of $F_{nV}(\lambda_{em},\lambda_{ex})$. When this term tends to zero or being negligible with respect to PSII emission at F_M, the expression is reduced to Eq. (2), and the estimation of $\Phi^{Max}_{PC,PSII}$ is relatively unbiased.

Recent reports from our laboratory have demonstrated that $F_{nV}(\lambda)$ at a given excitation wavelength could be extracted from the experimental spectra, simultaneously obtaining an estimation of $\Phi^{Max}_{PC,PSII}$ with reduced bias from spectral contributions other than PSII [26, 27]. In these studies, it was also shown that it was possible to separate spectrally the contributions of PSI and the uncoupled antenna, by considering the explicit contribution of these components to $F_{nV}(\lambda_{em},\lambda_{ex})$:

$$F_{nV}(\lambda_{em},\lambda_{ex}) = \rho_I(\lambda)\cdot\sigma_I(\lambda)\cdot\phi_I + \rho_{ul}(\lambda)\cdot\sigma_{ul}(\lambda)\cdot\phi_{ul} \qquad (7)$$

where the absorption cross-section ($\rho_I(\lambda)$, $\rho_{ul}(\lambda)$), emission bandwidth ($\sigma_I(\lambda)$, $\sigma_{ul}(\lambda)$) and quantum yield (ϕ_I, ϕ_{ul}) of PSI and uncoupled antenna appear as independent terms. This expression is formally more correct because the fluorescence yield of the components not showing variable fluorescence is, in general, not the same. Eq. (7) can be straightforwardly used to replace the $F_{nV}(\lambda_{em},\lambda_{ex})$ term in Eqs. (5) and (6).

In **Figure 7** are shown examples of the decomposition of the emission spectra of the green alga *C. sorokiniana* recorded under F'_0 and F_M conditions, by the independent contributions of PSI, PSII and uncoupled light harvesting antenna. Analogous results were obtained for *C. reinhardtii* [27]. The decomposition of the F'_0 and F_M spectra recorded in Synechocystis is instead shown in **Figure 8**, which, qualitatively, are representative also of the other model cyanobacterium considered, Synechococcus.

In accordance with qualitative data inspection of the spectra, in green algae, the emission spectra are largely dominated by PSII, and, even though the contribution of PSI is not negligible, the one of the uncoupled antenna, fundamentally, is so. The values of $\Phi^{Max}_{PC,PSII}$ derived from the spectral decomposition (0.715 for *C. sorokiniana* and 0.675 for *C. reinhardtii*) are only slightly higher (2–3%) than those obtained directly from the data, at their spectral maxima. This is moreover the case irrespectively of the excitation wavelength, as the same $\Phi^{Max}_{PC,PSII}$ was obtained for the decomposition at all excitation wavelengths [26, 27], and the F_V/F_M maximal values did not change greatly. On the other hand, the decomposition of the emission

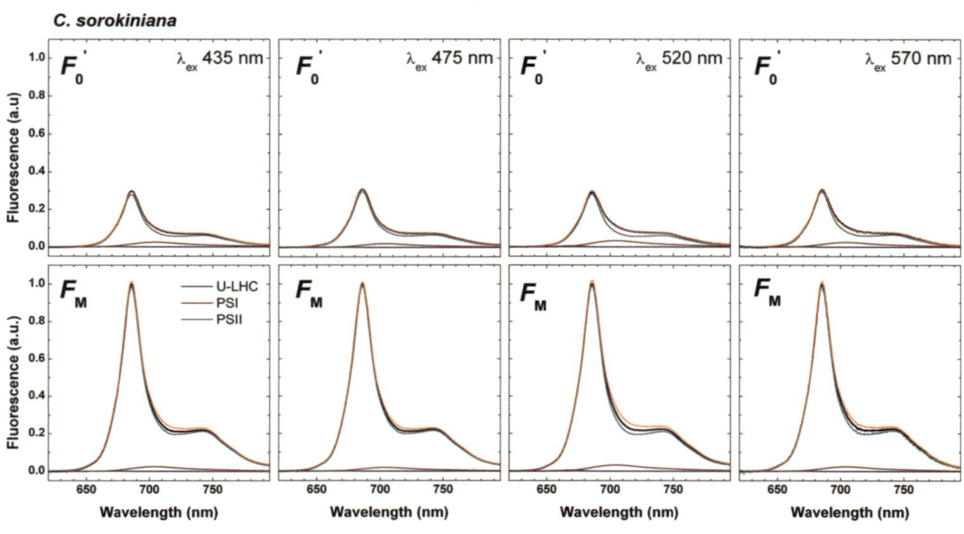

Figure 7.
Decomposition of the F'_0 and F_M emission spectra of C. sorokiniana *in terms of independent contributions of PSI (wine lines), PSII (green lines) and uncoupled light-harvesting component, U-LHC, (blue lines). The black lines are the experimental spectra and the red lines are the results of the decomposition.*

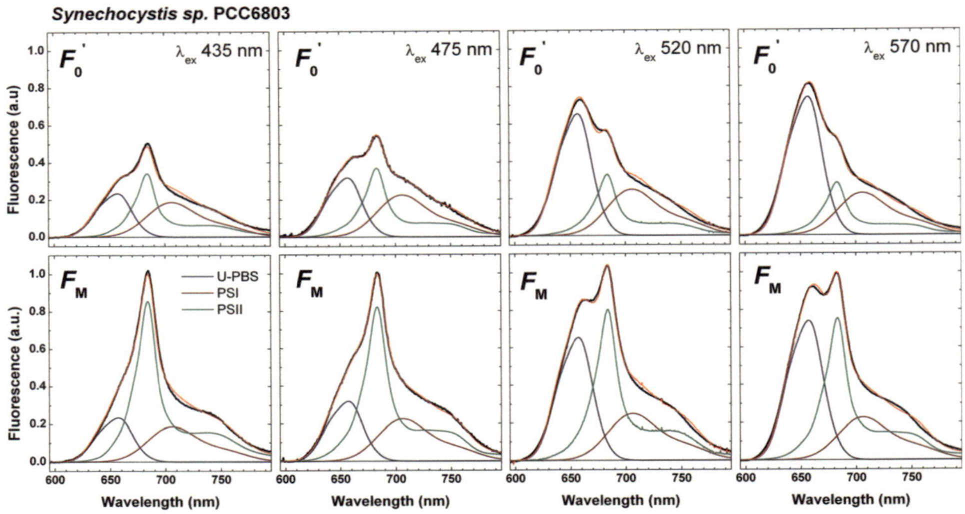

Figure 8.
Decomposition of the F'_0 and F_M emission spectra in Synechocystis *in terms of independent contributions of PSI (wine lines), PSII (green lines) and uncoupled phycobilisomes, U-PBS (blue lines). The black lines are the experimental spectra and the red lines are the results of the decomposition.*

spectra of model cyanobacteria evidences not only a more significant contribution of PSI with respect to PSII, at all excitation wavelengths tested, but also an almost dominant contribution of the uncoupled PBS antenna upon its direct excitation. Thus, the values of $\Phi_{PC,PSII}^{Max}$ (0.625 for Synechocystis and 0.65 for Synechococcus) derived from decomposition, which was imposed to be excitation wavelength-independent, is in general much larger (35–40%) than those obtained directly from experimental data. The only conditions for which deviations are not extremely large are upon detection at maximal emission, 685 nm, and excitation at 435 nm. Under these conditions, the $\Phi_{PC,PSII}^{Max}$ is underestimated only by 5–7.5%. For detection above 700 nm, the underestimation of F_V/F_M is generally larger than 50% and can be up to 75%.

A relatively simple way to compensate for the contribution of the emission from pigment-protein complexes and supercomplexes that do not show variable fluorescence is that of applying a correction to the measured fluorescence levels. This however would in principle require independent evidences that the variable fluorescence spectrum is effectively independent of the excitation wavelength (as shown in **Figures 3** and **6**) and knowledge of the relative emission contribution of F_{nV} at a given pair of excitation/emission wavelengths. If the first condition is met, then F_V shall be considered to be unbiased, and a correction factor, $(\xi_{\lambda_{em},\lambda_{ex}})$, for F_M only would be necessary:

$$\left(\frac{F_V}{F_M}\right)^c_{\lambda_{em},\lambda_{ex}} = \frac{F_{V,\lambda_{em},\lambda_{ex}}}{F_{M,\lambda_{em},\lambda_{ex}}\left(1-\xi_{\lambda_{em},\lambda_{ex}}\right)} \tag{8}$$

A list of $\xi_{\lambda_{em}}$ values, for selected emission wavelengths, and for the excitation wavelengths discussed above, is reported in **Table 3**. Although these values are quite similar between the two green algae, and the two cyanobacteria, considered, it is nonetheless recommendable to use the correction parsimoniously, especially when this yields large increases with respect to the measured F_V/F_M and whenever possible to verify it under both the experimental conditions and the specific growth conditions of the organism. A recommended, empirical, check is represented by acquiring the F_V/F_M ratio at least at two pairs of excitation and emission

	C. sorokiniana				*C. reinhardtii*			
Excitation	435 nm	475 nm	520 nm	570 nm	435 nm	475 nm	520 nm	570 nm
Interferential (Centre, nm)								
650	0.02	0.03	0.05	0.03	0.02	0.03	0.02	0.03
670	0.02	0.02	0.03	0.02	0.02	0.03	0.02	0.03
680	0.02	0.01	0.02	0.02	0.01	0.02	0.02	0.02
690	0.02	0.02	0.03	0.02	0.02	0.03	0.03	0.04
700	0.06	0.04	0.07	0.06	0.04	0.07	0.06	0.08
720	0.09	0.07	0.11	0.09	0.04	0.07	0.06	0.08
Bandpass (cut-on, nm)								
655	0.04	0.03	0.06	0.04	0.03	0.04	0.03	0.05
675	0.05	0.04	0.06	0.05	0.03	0.04	0.04	0.05
695	0.07	0.05	0.09	0.07	0.04	0.06	0.05	0.06
720	0.06	0.05	0.08	0.06	0.03	0.04	0.04	0.05
	Synechocystis sp. PCC6803				*Synechococcus sp.* PCC7942			
Excitation	435 nm	475 nm	520 nm	570 nm	435 nm	475 nm	520 nm	570 nm
Interferential (Centre, nm)								
650	0.10	0.89	0.83	0.83	0.62	0.79	0.93	0.95
670	0.06	0.41	0.47	0.47	0.23	0.45	0.65	0.68
680	0.05	0.23	0.25	0.25	0.13	0.29	0.42	0.45
690	0.08	0.27	0.25	0.25	0.15	0.33	0.42	0.44
700	0.15	0.51	0.43	0.43	0.29	0.52	0.58	0.60
720	0.20	0.64	0.52	0.52	0.37	0.60	0.64	0.66
Bandpass (Cut-on, nm)								
655	0.11	0.46	0.42	0.42	0.24	0.46	0.57	0.59
675	0.12	0.43	0.37	0.37	0.22	0.43	0.50	0.51
695	0.16	0.58	0.44	0.44	0.28	0.51	0.55	0.56
720	0.15	0.57	0.41	0.41	0.24	0.46	0.48	0.49

Correction factors ($\xi_{\lambda_{em},\lambda_{ex}}$) appearing in Eqs. (8), (10) and (11) derived for four excitation wavelengths, and for the expected emissions obtained for convolutions of either ideal interferential filters or long-pass filters (cut-on transitions 4 nm at the indicated 50% transmission).

Table 3.
Correction factors compensating for non-variable fluorescence contributions at F_M.

wavelengths. If the corrected value yields, within errors, the same estimate of $\Phi_{PC,PSII}^{Max}$, then it is likely that the corrected value approaches the un-biased one.

3. Impact of the excitation/emission dependency on the *NPQ* estimations

Non-photochemical quenching of Chl fluorescence is one of the most intensively investigated regulative processes in oxygenic photosynthesis. As stated in the introduction, NPQ being an excited state quenching process, it is natural that fluorescence detection has been by far the most popular instrument to investigate it, particularly *in vivo*. From the mechanistic point of view, NPQ results to be a very complex phenomenon, being controlled more or less synergically by an ensemble of effectors, whose specific action is species-dependent. Moreover, distinct NPQ processes that either develop or relax on different time scales have been reported, with the generally dominant ones being the rapidly forming/relaxing, energy-dependent quenching qE (e.g. [15–17]) and for prolonged illumination, the slowly/irreversible component, often associated with light-induced inhibition, qI component [15–17, 19]. The principal function of NPQ is that of regulating the rate of excitation delivery to the RC of PSII, by increasing the thermal dissipation in the antenna, under conditions in which the photosystem turnover can become limited by the availability of electron acceptors in the thylakoid electron transfer chain.

In land plants, and most green algae, the larger qE component of the NPQ process is dependent on the establishment of large pH gradients across the thylakoid membrane [15–17, 39]. However, whereas in land-plant, the presence of the PsbS subunit is crucial for a rapid and large NPQ establishment [15–19, 39, 40], in green algae, this is less fundamental, and the expression of specific light harvesting complexes (known as lhcs) has a more central role [39, 40]. Carotenoids, particularly those of the so-called xanthophyll cycle, also influence either the extent or the kinetics of formation/relaxation of NPQ [15–17, 19, 39, 40]. In both cases, the main effect is that of promoting the formation of quenching centres within the external antenna matrix of PSII, despite the site of quenching being still a matter of contention. As excited state equilibration within PSII is reached rapidly with respect to the excited state decay at F_M, the non-photochemical quencher could be considered, to a first approximation, to quench homogenously the whole PSII-LHCII supercomplex.

The mechanisms of NPQ in cyanobacteria are understood in less detail. In these organisms, it has been shown that non-photochemical centres are also located in the external PBS antenna (reviewed in [17, 18, 40]) and that NPQ can occur even in mutants lacking the reaction centres [41, 42]. The induction of NPQ has a pronounced actinic light dependency, whose action spectrum closely matches the absorption of the orange carotenoid protein (OCP), which is a key factor in the sensing and possibly the induction of NPQ in this organism. In this scenario, it should then be considered that non-photochemical quenching could occur in the whole PSII-PBS supercomplex as well as in the uncoupled PBS fraction (as well as in both).

For the purposes of this chapter, a detailed discussion of the mechanisms of NPQ is not required, being the focus on the distortion of the measured Chl-based parameters, which for this process is through the "*NPQ* parameter" as defined in Eq. (4).

It is possible to estimate the extent of the bias originating from the emission of PSI and of the energetically uncoupled antenna fraction in a parameter, which shall reflect changes in the fluorescence yield of PSII, by simulating the cellular emission, starting from the decompositions obtained under measured F_M conditions, and by imposing selective non-quenching in PSII (or in PSII and/or the uncoupled antenna), thereby obtaining spectra under F'_M conditions, and successively comparing the *NPQ* which would be retrieved from experimental measurements with the *effective* one used in the simulations.

Using the same formalism adopted so far, when NPQ occurs in PSII only, the non-photochemically quenched emission at F'_M is described by:

$$\begin{cases} F'_{\mathrm{M}}(\lambda_{em}, \lambda_{ex}) = \rho_{II}(\lambda) \cdot \sigma_{II}(\lambda_{ex}) \cdot \phi_{II,QN} + \rho_I(\lambda) \cdot \sigma_I(\lambda) \cdot \phi_I + \rho_{ul}(\lambda) \cdot \sigma_{ul}(\lambda) \cdot \phi_{ul} \\ \phi_{II,QN} = \dfrac{k_f}{\sum_i k_i + [Q_{NP}] \cdot K_{Q,NP}} \end{cases} \tag{9}$$

And when quenching in the uncoupled antenna fraction is necessary, ϕ_{ul} shall be substituted with $\phi_{ul,NQ}$, and, for simplicity we will consider, as a reasonable approximation, that $\phi_{ul} = \phi_M$. For simplicity, the case of cyanobacteria and green algae will be discussed separately.

3.1 Excitation and emission dependency of NPQ estimation in model green algae

The simulated fluorescence emission spectra in *C. sorokiniana* at F'_M conditions, compared with unquenched F_M, are presented in **Figure 9**. Values which would correspond to the *NPQ* parameters in the range of 1–4 for PSII emission were used in the simulations of the F'_M spectra. These values will be hereafter referred to as NPQ_{eff}. In the inset of the figures is shown the correlation between the NPQ_{eff} and the *NPQ*, which will be measured through interference filters centred at 660, 680, 700 and 720 nm, which cover most of the cellular emission bandwidth. For NPQ_{eff} levels below 2, the correspondence between the effective and detected *NPQ* is pretty good, particularly when the emission is monitored at 660 and 680 nm, as the deviation is between 5 and 10% of the imposed value. The correspondence is less satisfactorily when monitoring the emission at 700 and 720 nm, where deviations of

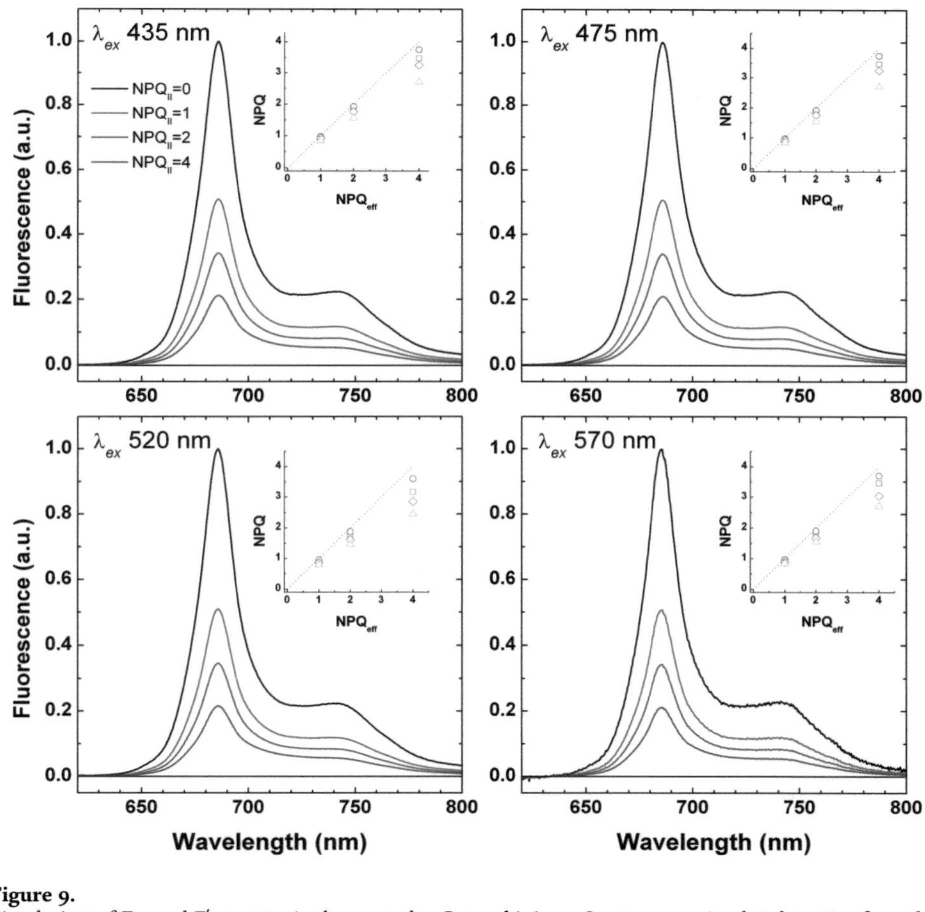

Figure 9.
Simulations of F_M *and* F'_M *spectra in the green alga* C. sorokiniana*. Spectra were simulated starting from the decomposition of* ***Figure 7*** *and by imposing quenching in PSII-LHCII only, corresponding to effective* NPQ_{eff} *values of 1, 2 and 4. The insets show the correlation between* NPQ_{eff} *and the detectable* NPQ *as it would be measured through interferential filters (FWHM 10 nm) centred at 660 (violet squares), 680 (blue circles), 700 (green diamond) and 720 nm (yellow triangles); the dashed lines correspond to un-biased estimations.*

15–20% are estimated already for relatively limited quenching levels, particularly when exciting at 520 and 570 nm. In all cases, deviations are more significant for the largest NPQ_{eff} value tested of 4. This is a rather large value for green algae, which is rarely reported, but is worth testing in perspective. Moreover, an *NPQ* of 4 is almost equivalent to the maximal photochemical quenching at F_0. When monitored at the maximal emission, *NPQ* is only about 10–15% underestimated, with the range depending on the excitation. Yet, for excitation in the yellow-green window and detection at 700 and 720 nm, the underestimation is in the 25–40% range, therefore significant.

The deviations between NPQ_{eff} and *NPQ* show a larger spectral dependence than the F_V/F_M ratio. This is due to the definition of the *NPQ* parameter, in which F'_M rather than F_M appears at the denominator. Since the contribution of any F_{nV} is expected to be the same under this condition, its relative weight would be larger on F'_M which is, by definition, smaller than F_M.

An alternative strategy to quantify non-photochemical quenching is via the non-photochemical quenching yield, (Φ_N, sometimes referred also to as qN), which is defined as $\Phi_N = (F_M - F'_M)/F_M$. In analogy to F_V/F_M, within the large unit connectivity limit, Φ_N can be demonstrated to effectively represent a quenching yield, and therefore, its value falls between 0 and 1. Indeed, when estimating Φ_N, the deviation with respect to its expected value does not exceed 10%.

3.2 Excitation and emission dependency of NPQ estimation in model cyanobacteria

Using the same approach described above, the F'_M spectra were simulated also for the model cyanobacterium *Synechocystis sp.* PCC6803, considering the two scenarios in which NPQ occurs only in the PSII-PBS supercomplex (**Figure 10**) or in it as well as in the uncoupled PBS fraction (**Figure 11**). The same level of NPQ_{eff} (between 1 and 4) is applied to either PSII-PBS or both PSII-PBS and uncoupled PBS, and this value is compared to what is measured through interferential filters, at the same wavelengths already discussed, in the inset of the figures.

It is immediately apparent that, for the simulations of the F'_M spectra shown in **Figure 10**, the discrepancy between the putative measured and the imposed value is extremely large. Underestimations of less than 25% are predicted only for the smaller NPQ_{eff} value considered (1), when exciting preferentially Chl *a* (at 435 nm) and detecting at the maximal emission (680 nm). However, even under these detection conditions, the largest NPQ_{eff} = 4 simulated is predicted to be ~40% underestimated. Furthermore, the underestimations increase progressively with the increasing value of NPQ_{eff} under all measuring conditions simulated. At wavelengths in which PSI emission is significant (700 and 720 nm), *NPQ* underestimations are in the 60–80% range and always exceed 75%, being up to 95% when monitoring on the preferential PBS emission at 660 nm. As discussed for the case of green algae, the discrepancy between the effective and the predicted measured values is larger for *NPQ* than F_V/F_M, that were already significant, due to the largest weight of F_{nV} at F'_M rather than F_M.

The scenario is markedly different when considering quenching in both PSII-PBS and the uncoupled PBS. As shown in **Figure 11**, quenching is more obvious across a larger portion of the emission bandwidth, and this leads to a closer correspondence between NPQ_{eff} and the simulated detectable *NPQ*. In particular, the closest match between effective and measurable *NPQ* is in this scenario attained upon direct monitoring of the PBS emission at 660 nm, and upon their direct excitation at 520 and 570 nm, where deviations are always below 15% of the effective quenching value. For the estimations of *NPQ* when monitoring at the

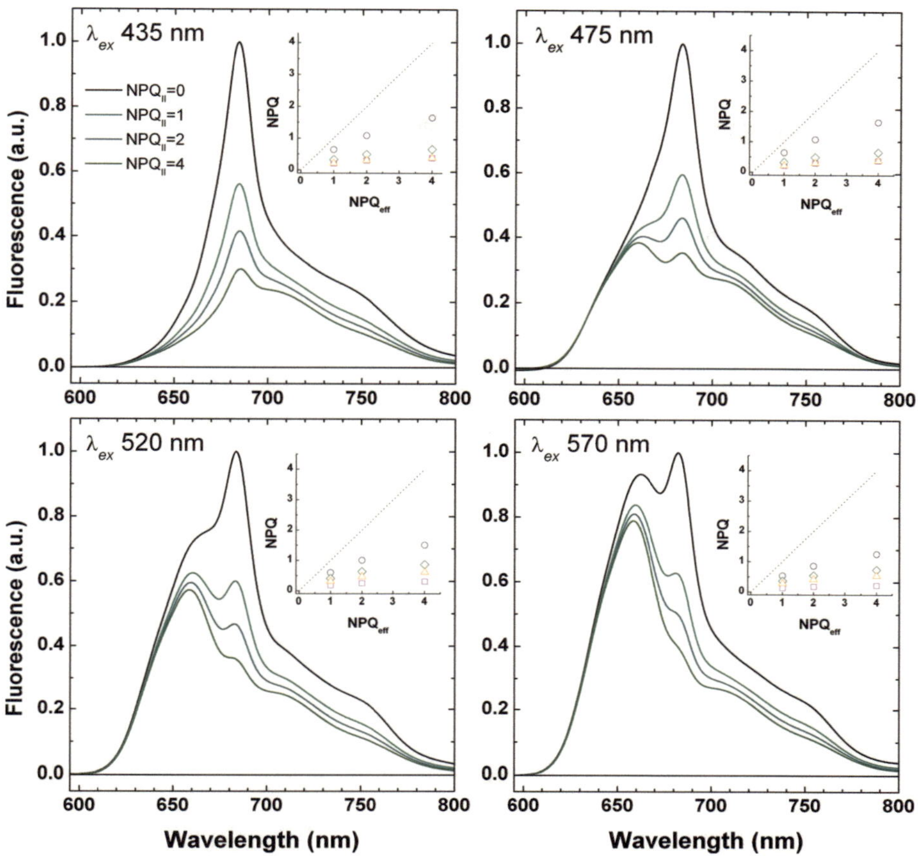

Figure 10.
Simulations of F_M and F'_M spectra in the cyanobacterium Synechocystis. *Spectra were simulated starting from the decomposition of* ***Figure 8*** *and by imposing quenching in PSII-PBS only, corresponding to effective* NPQ_{eff} *values of 1, 2 and 4. The insets show the correlation between* NPQ_{eff} *and the detectable* NPQ *as it would be measured through interferential filters (FWHM 10 nm) centred at 660 (violet squares), 680 (blue circles), 700 (green diamond) and 720 nm (yellow triangles); the dashed lines correspond to un-biased estimations.*

maximal emission, a slightly closer correspondence is predicted when exciting in the blue (435 and 475 nm), but significantly larger values are predicted when exciting preferentially the PBS, despite remaining ~40% underestimated with respect to the effective ones. On the other hand, detection in the red tail (700 and 720 nm) leads to underestimation in the order of 60–80% irrespectively of the excitation wavelength. It is worth mentioning that differences in the spectra of **Figures 10** and **11** (and *NPQ*) might help in discriminating between the two different scenarios, or possible intermediate cases, experimentally.

The non-photochemical quenching yield, Φ_N, results less biased in both cases, however low-bias values (12%) for the scenario of **Figure 10** are predicted only for excitation at 435 nm and detection at 680 nm, whereas for the scenario of **Figure 11**, underestimations between 5 and 15% are predicted for detection at 660 and 680 nm, at all excitation wavelengths considered.

It is in principle possible to correct the measurement parameters, analogously to what is described in Eq. (8), provided the same caveats and even further recommendation of parsimony of use, so that the "corrected" NPQ parameter results:

$$NPQ_c = \frac{F_M - F'_M}{F'_M - \xi_{\lambda_{em},\lambda_{ex}} \cdot F_M} \tag{10}$$

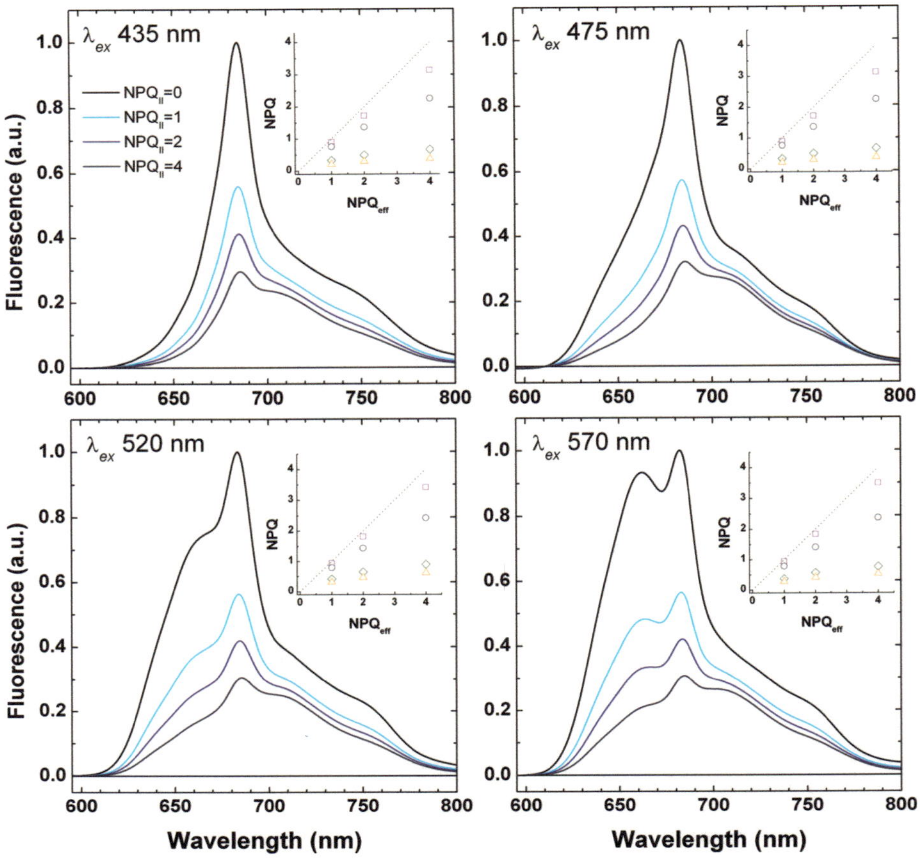

Figure 11.
Simulations of F_M and F'_M spectra in the cyanobacterium Synechocystis. *Spectra were simulated starting from the decomposition of* ***Figure 8*** *and by imposing quenching in PSII-PBS and in the uncoupled PBS population. The same effective* NPQ_{eff} *values of 1, 2 and 4 were imposed in both. The insets show the correlation between* NPQ_{eff} *and the detectable* NPQ *as it would be measured through interferential filters (FWHM 10 nm) centred at 660 (violet squares), 680 (blue circles), 700 (green diamond) and 720 nm (yellow triangles); the dashed lines correspond to un-biased estimations.*

and the photochemical quenching yield:

$$\Phi_{N,c} = \frac{F_M - F'_M}{F_M(1 - \xi_{\lambda_{em},\lambda_{ex}})} \tag{11}$$

4. Conclusions

In this chapter, we discussed the influence of measurement settings on the determination of the two parameters, F_V/F_M and NPQ, which have a very widespread, routine, utilisation in photosynthesis research. A key caveat behind the physical interpretation is that the detected emission stems exclusively or almost exclusively from PSII. By comparing the excitation and emission dependency in two different classes of organisms, it is shown that, particularly in cyanobacteria, the emission from PSI and a fraction of uncoupled antenna, PBS in this case, can lead to large distortions, generally an underestimation, in the determination of F_V/F_M (as a proxy for $\Phi^{Max}_{PC,PSII}$) and even larger ones, of NPQ. Although the underestimation of

$\Phi_{PC,PSII}^{Max}$ in green algae is relatively limited, the one of *NPQ* can also be significant for large non-photochemical quenching.

Simple expressions for empirical correction of the measured F_M level are proposed, and the conditions for their validity and applicability are discussed.

Acknowledgements

This research was supported by: Fondazione Cariplo through the project "Cyanobacterial Platform Optimised for Bioproductions" (CYAO, ref. 2016-0667); sPATIALS3 project, financed by the European Regional Development Fund under the ROP of the Lombardy Region ERDF 2014-2020 - Axis I "Strengthen technological research, development and innovation" - Action 1.b.1.3 "Support for co-operative R&D activities to develop new sustainable technologies, products and services" - Call Hub; CNR project FOE-2019 DBA.AD003.139, BIO-ECO.

Conflict of interest

The authors declare no conflict of interest.

Author details

Stefano Santabarbara[1*], William Remelli[1], Anastasia A. Petrova[1,3] and Anna Paola Casazza[2]

1 Photosynthesis Research Unit, National Research Council, Milano, Italy

2 Istituto di Biologia e Biotecnologia Agraria, National Research Council, Milano, Italy

3 Belozersky Institute of Physico Chemical Biology, Moscow Lomonosov State University, Moscow, Russia

*Address all correspondence to: stefano.santabarbara@cnr.it

References

[1] Papageorgiou GC, Govindjee, editors. Chlorophyll a fluorescence: A signature of photosynthesis. In: Advances in Photosynthesis and Respiration. Vol. 19. Dordrecht, The Nederlands: Springer; 2004

[2] Joliot P, Joliot A. Etudes cinétique de la réaction photochimique libérant l'oxygene au cours de la photosynthese. Comptes rendus de l'Académie des Sciences. 1964;**258**:4622-4625

[3] Butler WL. Energy distribution in the photochemical apparatus of photosynthesis. Annual Review of Plant Physiology. 1978;**29**:345-378

[4] Duysens LNM. Transfer and trapping of excitation energy in photosystem II. Ciba Foundation Symposium. 1978;**61**:323-340

[5] Baker NR. Chlorophyll fluorescence: A probe of photosynthesis *in vivo*. Annual Review of Plant Biology. 2008; **59**:89-113

[6] Searle GF, Tredwell CJ, Barber J, Porter G. Picosecond time-resolved fluorescence study of chlorophyll organisation and excitation energy distribution in chloroplasts from wild-type barley and a mutant lacking chlorophyll b. Biochimica et Biophysica Acta. 1979;**545**:496-507

[7] Moya I, Hodges M, Briantais J-M, Hervo G. Evidence that the variable chlorophyll fluorescence in Chlamydomonas reinhardtii is not recombinant luminescence. Photosynthesis Research. 1986;**10**: 319-325

[8] Holzwarth AR, Wendler J, Haehnel W. Time-resolved picosecond fluorescence spectra of the antenna chlorophylls in *Chlorella vulgaris*. Resolution of photosystem I fluorescence. Biochimica et Biophysica Acta. 1985;**807**:155-167

[9] Hodges M, Moya I. Time-resolved chlorophyll fluorescence studies of photosynthetic membranes: Resolution and characterisation of four kinetic components. Biochimica et Biophysica Acta. 1986;**849**:193-202

[10] Roelofs TA, Lee CH, Holzwarth AR. Global target analysis of picosecond chlorophyll fluorescence kinetics from pea chloroplasts. A new approach to the characterisation of the primary processes in photosystem II alfa- and beta-units. Biophysical Journal. 1992;**61**: 1147-1163

[11] Schansker G, Tóth SZ, Holzwarth AR, Garab G. Chlorophyll a fluorescence: Beyond the limits of the Q_A model. Photosynthesis Research. 2014;**120**:43-58

[12] Magyar M, Sipka G, Kovács L, Ughy B, Zhu Q, Han G, et al. Rate-limiting steps in the dark-to-light transition of photosystem II - revealed by chlorophyll-a fluorescence induction. Scientific Reports. 2018;**8**:2755-2763

[13] Genty B, Briantais J-M, Baker NR. The relationship between the quantum yield of photosynthetic electron transport and quenching of chlorophyll fluorescence. Biochimica et Biophysica Acta. 1989;**990**:87-92

[14] Genty B, Harbinson L, Briantais J-M, Baker NR. The relationship between non-photochemical quenching of chlorophyll fluorescence and the rate of photosystem 2 photochemistry in leaves. Photosynthesis Research. 1990;**25**:249-257

[15] Horton P, Ruban AV, Walters RG. Regulation of light harvesting in green plants. Annual Review of Plant Physiology and Plant Molecular Biology. 1996;**47**:655-684

[16] Ruban AV. Nonphotochemical chlorophyll fluorescence quenching:

Mechanism and effectiveness in protecting plants from photodamage. Plant Physiology. 2016;**170**:1903-1916

[17] Goss R, Lepetit B. Biodiversity of NPQ. Journal of Plant Physiology. 2015; **172**:13-32

[18] Kirilovsky D, Kerfeld CA. The orange carotenoid protein in photoprotection of photosystem II in cyanobacteria. Biochimica et Biophysica Acta. 1817;**2012**:158-166

[19] Caffarri S, Tibiletti T, Jennings RC, Santabarbara S. A comparison between plant photosystem I and photosystem II architecture and functioning. Current Protein & Peptide Science. 2014;**15**: 296-331

[20] Neilson JA, Durnford DG. Structural and functional diversification of the light-harvesting complexes in photosynthetic eukaryotes. Photosynthesis Research. 2010;**106**: 57-57

[21] Grossman AR, Bhaya D, Apt KE, Kehoe DM. Light-harvesting complexes in oxygenic photosynthesis: Diversity, control, and evolution. Annual Review of Genetics. 1995;**29**: 231-288

[22] Croce R, van Amerongen H. Light-harvesting in photosystem I. Photosynthesis Research. 2013;**116**: 153-166

[23] Gobets B, van Grondelle R. Energy transfer and trapping in photosystem I. Biochimica et Biophysica Acta. 2001; **1057**:80-99

[24] Rizzo F, Zucchelli G, Jennings RC, Santabarbara S. Wavelength dependence of the fluorescence emission under conditions of open and closed photosystem II reaction centres in the green alga *Chlorella sorokiniana*. Biochimica et Biophysica Acta. 1837; **2014**:726-733

[25] Byrdin M, Rimke I, Schlodder E, Stehlik D, Roelofs TA. Decay kinetics and quantum yields of fluorescence in photosystem I from *Synechococcus elongatus* with P700 in the reduced and oxidised state: Are the kinetics of excited state decay trap-limited or transfer-limited? Biophysical Journal. 2000;**79**:992-1007

[26] Remelli W, Santabarbara S. Excitation and emission wavelength dependence of fluorescence spectra in whole cells of the cyanobacterium *Synechocystis sp.* PPC6803: Influence on the estimation of photosystem II maximal quantum efficiency. Biochimica et Biophysica Acta. 2018; **1859**:1207-1222

[27] Santabarbara S, Villafiorita Monteleone F, Remelli W, Rizzo F, Menin B, Casazza AP. Comparative excitation-emission dependence of the F_V/F_M ratio in model green algae and cyanobacterial strains. Physiologia Plantarum. 2019;**166**:351-364

[28] Kaňa R, Prásil O, Komárek O, Papageorgiou GC, Govindjee. Spectral characteristic of fluorescence induction in a model cyanobacterium, *Synechococcus sp.* (PCC 7942). Biochimica et Biophysica Acta. 2009; **1787**:1170-1178

[29] Santabarbara S, Jennings RC. The size of the population of weakly coupled chlorophyll pigments involved in thylakoid photoinhibition determined by steady-state fluorescence spectroscopy. Biochimica et Biophysica Acta. 1709;**2005**:138-149

[30] Mullineaux CW, Bittersmann E, Allen JF, Holzwarth AR. Picosecond time-resolved fluorescence emission spectra indicate decreased energy transfer from the phycobilisome to photosystem II in light-state 2 in the cyanobacterium *Synechococcus* 6301. Biochimica et Biophysica Acta. 1990; **1015**:231-242

[31] Mullineaux CW, Holzwarth AR. Kinetics of excitation energy transfer in the cyanobacterial phycobilisome-photosystem II complex. Biochimica et Biophysica Acta. 1991;**1098**:68-78

[32] Campbell D, Hurry V, Clarke AK, Gustafsson P, Öquist G. Chlorophyll fluorescence analysis of cyanobacterial photosynthesis and acclimation. Microbiology and Molecular Biology Reviews. 1998;**62**:667-683

[33] Acuña AM, Snellenburg JJ, Gwizdala M, Kirilovsky D, van Grondelle R, van Stokkum IHM. Resolving the contribution of the uncoupled phycobilisomes to cyanobacterial pulse-amplitude modulated (PAM) fluorometry signals. Photosynthesis Research. 2016;**127**: 91-102

[34] Acuña AM, Kaňa R, Gwizdala M, Snellenburg JJ, van Alphen P, van Oort B, et al. A method to decompose spectral changes in *Synechocystis* PCC 6803 during light-induced state transitions. Photosynthesis Research. 2016;**130**:237-249

[35] Pfündel EE. Deriving room temperature excitation spectra for photosystem I and photosystem II fluorescence in intact leaves from the dependence of F_V/F_M on excitation wavelength. Photosynthesis Research. 2009;**100**:163-167

[36] Pfündel EE, Klughammer C, Meister A, Cerovic ZG. Deriving fluorometer-specific values of relative PSI fluorescence intensity from quenching of F0 fluorescence in leaves of *Arabidopsis thaliana* and *Zea mays*. Photosynthesis Research. 2013;**114**: 189-206

[37] Murata N. Uphill energy transfer from chlorophyll a to phycobilins in the blue-green algae *Anabaena variabilis* and *Anacystis nidulans*. In: Miyachi S, Katoh S, Fujita Y, Shibata K, editors. Special Issue of Plant and Cell Physiology, No. 3. Photosynthetic Organelles. 1977:9-13

[38] Wang RT, Myers J. Reverse energy transfer from chlorophyll to phycobilin in *Anacystis nidulans*. In: Miyachi S, Katoh S, Fujita Y, Shibata K, editors. Special Issue of Plant and Cell Physiology, No. 3. Photosynthetic Organelles. 1977:3-7

[39] Ruban VA. Evolution under the sun: Optimising light harvesting in photosynthesis. Journal of Experimental Botany. 2015;**66**:7-23

[40] Niyogi KK, Truong TB. Evolution of flexible non-photochemical quenching mechanisms that regulate light harvesting in oxygenic photosynthesis. Current Opinion in Plant Biology. 2013; **16**:307-314

[41] Rakhimberdieva MG, Kuzminov FI, Elanskaya IV, Karapetyan NV. Synechocystis sp. PCC 6803 mutant lacking both photosystems exhibits strong carotenoid-induced quenching of phycobilisome fluorescence. FEBS Letters. 2011;**585**:585-589

[42] Kuzminov FI, Karapetyan NV, Rakhimberdieva MG, Elanskaya IV, Gorbunov MY, Fadeev VV. Investigation of OCP-triggered dissipation of excitation energy in PSI/PSII-less *Synechocystis sp*. PCC 6803 mutant using non-linear laser fluorimetry. Biochimica et Biophysica Acta. 2012;**1817**:1012-1021

Chapter 4

The Use of Fluorescence Microscopy to Assess the Suppression of the Development of Cyanobacteria under the Influence of Allelochemicals of Aquatic Macrophytes

Evgeny Kurashov, Larisa Kapustina, Julia Krylova and Galina Mitrukova

Abstract

The luminescent microscopy (LM) is especially convenient for express analysis of the toxicity of various substances or for detecting the degree of inhibition of the physiological state of cyanobacteria populations as a result of the action of certain chemical compounds on them. In natural water bodies, the suppression of the development of the phytoplankton occurs, in particular, under the influence of low-molecular-weight organic compounds (LMWOCs), metabolites-allelochemicals, of aquatic macrophytes. LM, which allows observing the primary or secondary luminescence of microorganisms, was used by us to study changes in the physiological state of the cyanobacteria cultures of *Synechocystis aquatilis* and *Aphanizomenon flos-aquae* under the influence of allelochemicals of water macrophytes in laboratory experiments. We have shown (including using LM) that selected LMWOCs (linoleic, heptanoic, octanoic, tetradecanoic, hexadecanoic, and gallic acids) possess inhibitory allelopathic activity against cyanobacteria. However, their inhibitory effect was different. The highest values of the suppression index (SI > 10) were recorded (in ascending order) for hexadecanoic, linoleic, tetradecanoic, and gallic acids and a mixture of four allelochemicals (heptanoic, octanoic, tetradecanoic, and gallic acids). The creation of a new generation of algaecides/cyanocides based on LMWOCs of aquatic plants is a very promising strategy for combating "algal blooms."

Keywords: luminescent microscopy, cyanobacteria, *Synechocystis aquatilis*, *Aphanizomenon flos-aquae*, allelochemicals, allelopathy, carboxylic acids, gallic acid, algal blooms, aquatic macrophytes

1. Introduction

At present, fluorescence methods play an important role in studies of the physiology of photosynthetic organisms. Luminescent microscopy, which makes it possible to conduct a fairly quick and accurate assessment of the degree of viability of cells of lower and higher plants, manifested in a change in the luminescence of cells, is an integral part of them. This is based on the fact that under microscopy in ultraviolet rays, the cells that differ in their physiological state give different shades of color and brightness of the luminescence.

Luminescent microscopy allows, in particular, to observe the primary fluorescence (autofluorescence) of chlorophyll "a" in the cells of higher and lower plants, as well as in the cells of cyanobacteria, which make it possible to conduct a fairly rapid and accurate assessment of various degrees of cell viability, expressed in changes of their glow [1–3].

In Russia, the basics of applying the method of luminescent microscopy to determine the viability of algae and cyanobacteria cells were developed in the 1960s of the last century [4, 5]. In UV rays, cells with different viabilities were found to yield different colors and brightness of the glow shades. This provides possibilities to differentiate viable and non-viable cells and estimate their number and the degree of degradation of algae and cyanobacteria populations.

The rationale and advantages of this method for fast viability assay for unicellular cyanobacteria, which uses red chlorophyll fluorescence and an unspecific green autofluorescence for the differentiation of viable and non-viable cells without the need of sample preparation, are presented in [6] as well. The method is especially convenient for express analysis of the toxicity of various substances or for detecting the degree of inhibition of the physiological state of cyanobacteria populations as a result of the action of certain chemical compounds on them.

We applied luminescent microscopy to one of the most acute environmental problems of our days, the so-called algal blooms in inland waters, which cause degradation of aquatic ecosystems, deterioration of water quality, and a decrease in the economic potential of reservoirs [7–9]. Among other representatives of photosynthetic plankton, cyanobacteria are the most significant formers of harmful freshwater "algal blooms" (HAB) [10–13]. Inhibiting the growth of undesired cyanobacteria species is crucial for controlling HAB. Moreover, it is most important to prevent the development of HAB, since this will prevent or weaken the tremendous and widespread impact of HAB on human and environmental health, natural and man-made assets, as well as overall ecosystem services [14]. Suppressing of cyanobacteria (and algae of phytoplankton) by allelochemicals released by aquatic macrophytes is reported to be one of the natural mechanisms that maintain a prosperous and clear-water state in aquatic ecosystems (first of all, in shallow lakes and ponds) [15–18].

There are a large number of low-molecular-weight organic compounds (LMWOCs), metabolites of freshwater macrophytes that can perform an allelopathic function in aquatic ecosystems [16, 19–21].

The results of our previous studies using QSAR technology show that carboxylic and gallic acids have a high calculated potential to inhibit the development of cyanobacteria [22].

In order to receive the proofs of the mechanism of suppressing of cyanobacteria by some allelochemicals of aquatic macrophytes, we chose the luminescent microscopy, which allows observing the primary or secondary luminescence of microorganisms. This method was used by us to study changes both in density and in the physiological state of the cyanobacteria cultures of *Synechocystis aquatilis* Sauvageau and *Aphanizomenon flos-aquae* Ralfs ex Bornet and Flahault under the influence of substances-allelochemicals of water macrophytes in a small-scale laboratory microcosm experiments.

2. Material and methods

2.1 Allelochemical identification

In this paper, we use data on the following species of macrophytes: *Ceratophyllum demersum* L., *Potamogeton perfoliatus* L., *Potamogeton natans* L., *Potamogeton obtusifolius* Mert. & W.D.J. Koch, *Potamogeton pectinatus* L., *Nuphar lutea* (L.) Smith., *Nymphaea alba* L., *Myriophyllum spicatum* L., and *Persicaria amphibia* (L.) Gray (**Table 1**). The LMWOCs of these macrophytes were investigated in the plant material, which was collected during the summer of 2013–2017 in the water bodies marked in **Table 1**. The composition and concentrations of LMWOCs in essential oil were determined using a TRACE ISQ gas chromatograph-mass spectrometer (Thermo Electron Corporation) equipped with a quadrupole mass analyzer and Thermo TG-SQC Column (15 m; inner diameter, 0.25 mm; and 0.25 μm film). The technique of sampling macrophytes, sample preparation, and the details of chromatography-mass spectrometric studies are described in detail in [22].

2.2 Assaying allelochemical effects

We used in experiments an axenic strain of *Synechocystis aquatilis* Sauvageau № 1336 from the collection of the living cultures of cyanobacteria, algae and algal parasites (CALU, Collection of Algae of Leningrad University), which was provided by the Centre for Culture Collection of Microorganisms of the Research Park at St. Petersburg University. The strain was isolated from a sample of water taken in the Gulf of Finland near Sosnovy Bor. Another axenic strain of *Aphanizomenon flos-aquae* Ralfs ex Bornet and Flahault was provided by the St. Petersburg Scientific Research Center for Ecological Safety of the Russian Academy of Sciences.

The experiments (three to four replicates) were carried out in microcosms with a volume of 0.5 liters. Cyanobacterial cultures of *S. aquatilis* and *A. flos-aquae* in the exponential phase of growth were introduced into experimental vessels in the form of suspensions.

Cyanobacteria were cultured on medium No. 6 [23]. Cultivation of cyanobacteria and experiments were carried out in a special aquarium using a liquid circulation cryothermostat with cooling and heating {Baths WCR Circulation water bath WCR-MaXircu CR-P8 [Daihan (Witeg)]}. The unit maintained a constant temperature of 25°C during the experiments. The lamp (Lamp Biodesign RIF 80/110/PANORAMA 80/100/DIARAMA 150/200) provided a luminous flux of 1500 lm. The day-night mode (16–8 hours) was set using an adjustable timer (FERON TM50, 3500 W/16A230V).

Due to the promising algicidal effect [24, 25], the following allelochemicals have been used in the present study: linoleic acid, heptanoic acid, octanoic acid, tetradecanoic acid, hexadecanoic acid, and gallic acid. Instead of natural LMWOCs, the purified analogous substances from Acros Organics BVBA and their combination were used in experiments to test their cyanocidal effect on *S. aquatilis* and *A. flos-aquae*.

In experiments, aqueous solutions of these acids were used to test the action of individual compounds. Since some of the allelochemicals used are poorly hydrosoluble, they were dissolved in the organic solvent ethanol in experiments to evaluate their combined allelopathic effects. In experiments, no more than 1 ml of ethanol with the dissolved tested allelochemicals was introduced into the medium in 500 ml vessels to achieve the required concentration in the experimental microcosm. In parallel, experiments with the addition of 1 ml of ethanol without allelochemicals

Species	Water body/coordinates	Number, min-max	(% of total essential oil, min-max)
Ceratophyllum demersum L.	Lakes of Volga-Akhtuba floodplain (Astrakhan Region, Russia) / N 48°29′, E 45°34′	3–5	0.27–14.62
C. demersum	Ponds of the Victory Park, (St. Petersburg, Russia)/N 60° 52.025′, E 30° 19.91′	4–5	4.9–5.7
Potamogeton perfoliatus L.	Lake Ladoga/N 60° 50′, E 31° 33′	1–11	0.10–30.78
P. perfoliatus	Lake Onega/N 61°38′, E 35°31	5–8	1.0–5.28
Potamogeton natans L.	Lakes of Karelian Isthmus, Leningrad Region, Russia/N 61° 07′, E 29° 55′	5–7	8.5–56.39
Potamogeton obtusifolius Mert. & W.D.J. Koch	Ponds of the Victory Park, (St. Petersburg, Russia)/N 60° 52.025′, E 30° 19.91′	3–5	0.3–4.5
Potamogeton pectinatus L.	Lake Ladoga/N 60° 50′, E 31° 33′	2	9.93
P. pectinatus	Lakes of Volga-Akhtuba floodplain (Astrakhan Region, Russia)/N 48°29′, E 45°34'	3–5	29.28–40.0
Nuphar lutea (L.) Smith.	River Ild, Yaroslavl Region, Russia/N 58°0.23′, E 38°13.53′	4–5	19.05–66.03
N. lutea	Lakes of Karelian Isthmus, Leningrad Region, Russia/N 61° 07′, E 29° 55′	4–7	27.43–77.57
N. lutea	The mouth of the Volkhov River, Leningrad Region, Russia/N 60° 07.14′, E 32° 19.57′	1	28.74
Nymphaea alba L.	Lakes of Karelian Isthmus, Leningrad Region, Russia/N 61° 07′, E 29° 55′	5–10	58.06–61.43
N. alba	Lakes of Volga-Akhtuba floodplain (Astrakhan Region, Russia)/N 48°29′, E 45°34′	6–8	49.88–53.05
Myriophyllum spicatum L.	Lakes of Karelian Isthmus, Leningrad Region, Russia/N 61° 07′, E 29° 55′	15–17	49.55–52.93
M. spicatum	Lake Ladoga/N 60° 50′, E 31°33′	9	26.19
M. spicatum	Lake Naroch (Belarus)/N 54°51′, E 26°45′	4	15.41
M. spicatum	Lakes of Volga-Akhtuba floodplain (Astrakhan Region, Russia)/N 48°29′, E 45°34′	5	32.09
M. spicatum	Kazachy River, Astrakhan Region/N 46°13.13′, E 47°52.57′	8	50.22
M. spicatum	Dniester estuary, Black Sea, Ukraine /N 46°05.11′, E 30°27.84′	4–5	43.99–55.39
Persicaria amphibia (L.) Gray	Lake Ladoga/N 60° 50′, E 31° 33′	6–7	50.09–60.63

Table 1.
The number and relative content (% of total essential oil) of the fatty acids in some species of freshwater macrophytes from different water bodies.

of 500 ml of the medium showed that this does not have an inhibiting effect on the development of cyanobacteria.

The effect of allelochemicals in concentrations that may be characteristic of natural bodies of water was tested. The compounds were added into the cyanobacterial cultures in experimental microcosms to achieve their various final concentrations of 0.001, 0.01, 0.1, 1, 10, and 100 mg/l. Different combinations of these concentrations were used in different versions of the experiments.

Suppression index (SI), defined as the cyanobacterial density in control divided by the cyanobacterial density in an experiment with allelochemicals, was used to assess the degree of suppression of cyanobacterial cultural development.

2.3 Assaying cultural density and physiology state of cyanobacteria

The growth of cyanobacterial cultures and their physiological state were monitored using the MIKMED-26 fluorescence microscope with a × 100/1.25 Plan Achromat oil immersion objective at intervals of 2–7 days in different experiments.

Microscopy was performed using ultraviolet radiation (excitation 330–385 nm/emission 400–420 nm). The physiological state of the culture was estimated by the color of the glow of the cells, and the number of viable and dead cells and their ratio (proportion expressed in %) were calculated.

The change in the fluorescence spectrum of cells containing chlorophyll occurs in the following stages [3, 5, 26]:

1. Bright purple-red glow is typical for chlorophyll-containing cells of the highest viability—the cells are in an active state, intensively dividing or preparing for division (logarithmic growth phase).

2. The red glow of lower brightness (dull maroon or pink-red) is inherent in cells in the stationary phase of growth of algae culture.

3. The orange-red hue of the glow is given by cells that are suppressed under the influence of any factor, and the pale orange glow is given by cells with a very low level of vital activity but still alive.

4. Old cells with weakened vital activity glow dimly red.

5. Bluish-green and olive-green glow is typical for dead cells and detritus.

In some experiments, the density of the cyanobacterial culture was estimated by the characteristic of optical density, which was obtained using a photoelectric concentration colorimeter KFC-2MP. Since the optical characteristics of the aquatic environment change with the growth of cyanobacteria in the experimental vessels, this made it possible to correlate the number of cyanobacteria with the optical characteristics of the medium, which can be recorded using KFC-2MP. The optical density of the culture medium was measured in the red region of the spectrum at a wavelength of 670 nm, which is the maximum absorption of chlorophyll-a [27], available in cyanobacteria.

3. Results

3.1 GC-MS studies

The results of our chromatographic-mass-spectrometric studies of the low-molecular-weight metabolome (LMWM) of various aquatic macrophyte species from different types of reservoirs [7, 22, 28–30], and unpublished data show that many carboxylic acids, recognized as active allelochemicals, constitute a significant share of the LMWOC composition of the plant essential oil (**Table 1**).

Plants such as *P. natans, N. lutea, N. alba, M. spicatum,* and *P. amphibia* are the most active producers of fatty acids and therefore can have the greatest allelopathic effect on phytoplankton and cyanobacteria. **Figures 1–3** show sections of the chromatograms of essential oils of some aquatic macrophytes, which present the location of the fatty acids found in them, including those that were used in our experiments to study the suppression of the development of cyanobacteria: linoleic, tetradecanoic, hexadecanoic, and octanoic.

According to our data, the following fatty acids have a high content in the essential oil of aquatic macrophytes studied by us, tetradecanoic acid, hexadecanoic acid, linoleic acid, and linolenic acid, and are part of the major LMWM components of many freshwater macrophytes [20, 22, 24, 31]. Some aquatic macrophytes may contain a particularly significant number of carboxylic acids in their LMWM, for example, *Myriophyllum spicatum* L. (**Figure 3**).

3.2 Small-scale laboratory microcosm experiments

3.2.1 Experiment with S. aquatilis *and linoleic acid*

We added the linoleic acid to the *S. aquatilis* culture medium at concentrations of 0.001, 0.01, and 0.04 mg/l. The initial density of cyanobacteria was 460,000 cells/ml. In the experiment with linoleic acid, significant suppression of the development of *S. aquatilis* culture was observed (**Figure 4**). Moreover, the maximum suppression (SI = 12.5) was noted on the fourth day of the experiment at a concentration of the allelochemical of 0.01 mg/l. At the end of the experiment (on day 12), the greatest inhibition of the development of cyanobacteria (SI = 3) was observed at an allelochemical concentration of 0.04 mg/l (**Table 2**).

Figure 5 presents a visualization of the effect of the allelochemical on cyanobacteria during the experiment. The suppression of the development of culture

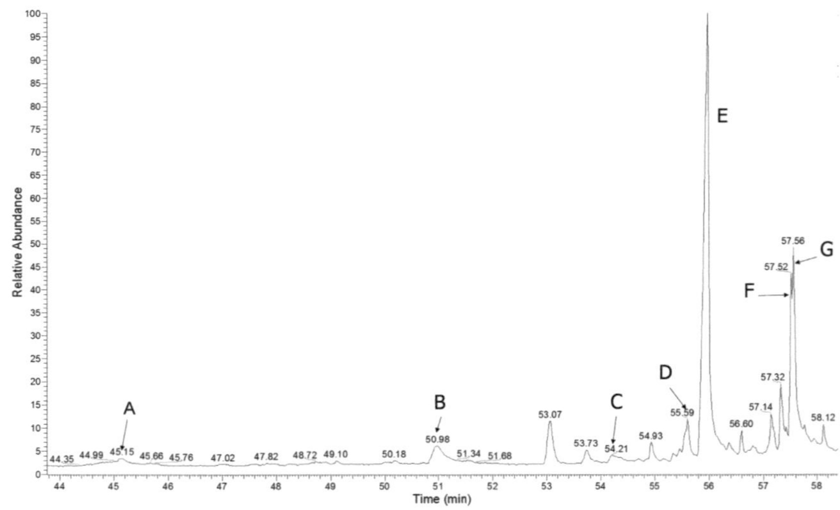

Figure 1.
Chromatogram section of Persicaria amphibia *essential oil (Sortavala skerries, Lake Ladoga) with marked peaks of carboxylic acids: (A) dodecanoic acid, (B) tetradecanoic acid, (C) pentadecanoic acid, (D) palmitoleic acid, (E) hexadecanoic acid, (F) linoleic acid, (G) linolenic acid.*

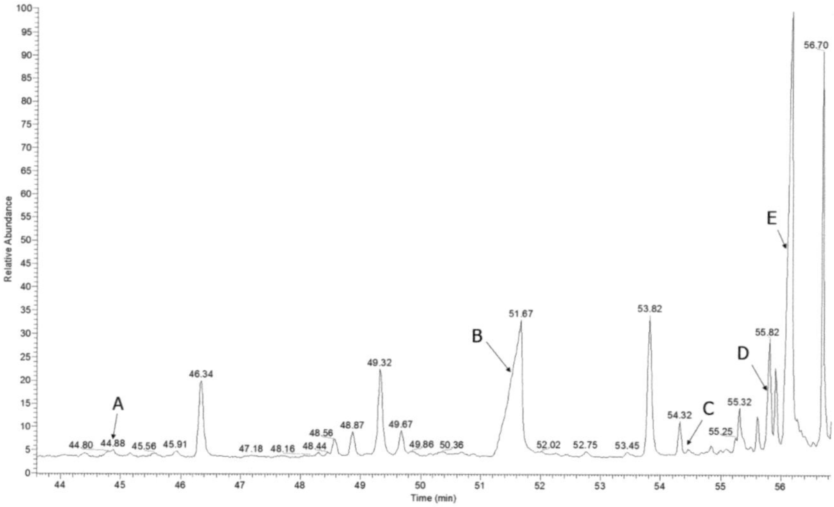

Figure 2.
Chromatogram section of Potamogeton pectinatus *essential oil (Lake Obvalovannoe, Volga-Akhtuba floodplain, Astrakhan Region) with marked peaks of carboxylic acids: (A) dodecanoic acid, (B) tetradecanoic acid, (C) pentadecanoic acid, (D) palmitoleic acid, (E) hexadecanoic acid.*

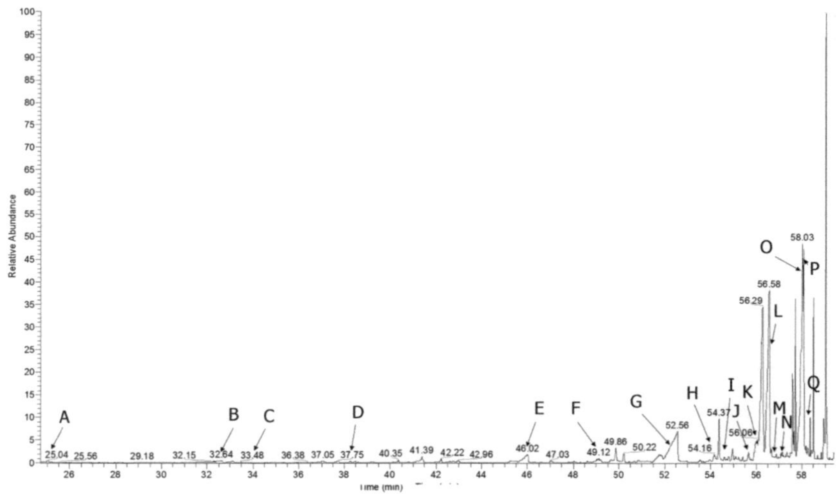

Figure 3.
A chromatogram section of Myriophyllum spicatum *essential oil (Lake Uzkoe, Karelian isthmus, Leningrad region) with marked peaks of carboxylic acids: (A) octanoic acid; (B) nonanoic acid; (C) 2-nonenoic acid; (D) decanoic acid; (E) dodecanoic acid; (F) tridecanoic acid; (G) tetradecanoic acid; (H) pentadecanoic acid; (I) 14-pentadecenoic acid; (J) 9-hexadecenoic acid; (K) hexadecenoic acid, Z-11-; (L) hexadecanoic acid; (M) (E)-heptadec-9-enoic acid; (N) cis-10-heptadecenoic acid; (O) linoleic acid; (P) linolenic acid; (Q) octadecanoic acid.*

and the suppression of its physiological state, which is expressed in a change in the fluorescent glow of cyanobacteria cells from bright red to orange, pink-green, and bluish-green, is clearly seen.

According to fluorescence microscopy data, changes in the proportion of living and dead cyanobacteria cells were observed when acted upon by linoleic acid (**Figure 4b**). Moreover, the maximum percentage of dead cells (36.5%) was seen on day 5 at the maximum of the studied concentrations of linoleic acid (0.04 mg/l)

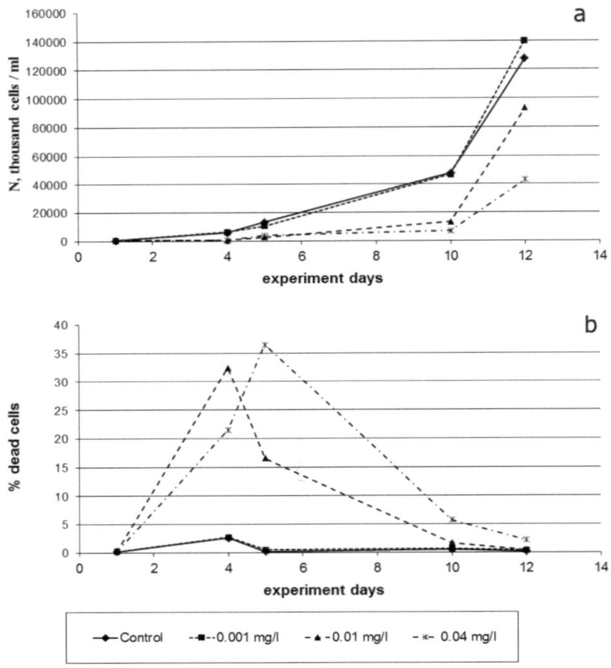

Figure 4.
Development of the culture of S. aquatilis *in the experiment with linoleic acid: (a) the average number (median), (b) % dead cells (median).*

Experiment number/ active allelochemical	Object of impact	Maximum SI/ maximum SI by the end of the experiment	Allelochemical concentration (mg/l) at maximum SI/at the greatest SI at the end of the experiment	Day of experiment with maximum SI/total duration of the experiment
1/linoleic acid	*S. aquatilis*	12.5/3	0.01/0.04	4/12
2/tetradecanoic acid	*S. aquatilis*	7.5/7.5	0.1/0.1	14/14
3/tetradecanoic acid	*S. aquatilis*	14.5/11.9	0.1/0.1	8/10
4/hexadecanoic acid	*S. aquatilis*	10.4/10.4	0.1/0.1	13/13
5/gallic acid	*S. aquatilis*	30/30	100/100	15/15
6/heptanoic acid	*S. aquatilis*	1.9/1.9	1/1	17/17
7/octanoic acid	*S. aquatilis*	3/3	1/1	17/17
8/ mixture of heptanoic, octanoic, tetradecanoic, and gallic acids	*S. aquatilis*	35.3/8.3	10/10	7/13
9/mixture of heptanoic, octanoic, tetradecanoic, and gallic acids	*A. flos-aquae*	17,495/17,495	10/10	23/23

Table 2.
The maximum suppression of the development of cyanobacteria when exposed to various allelochemicals.

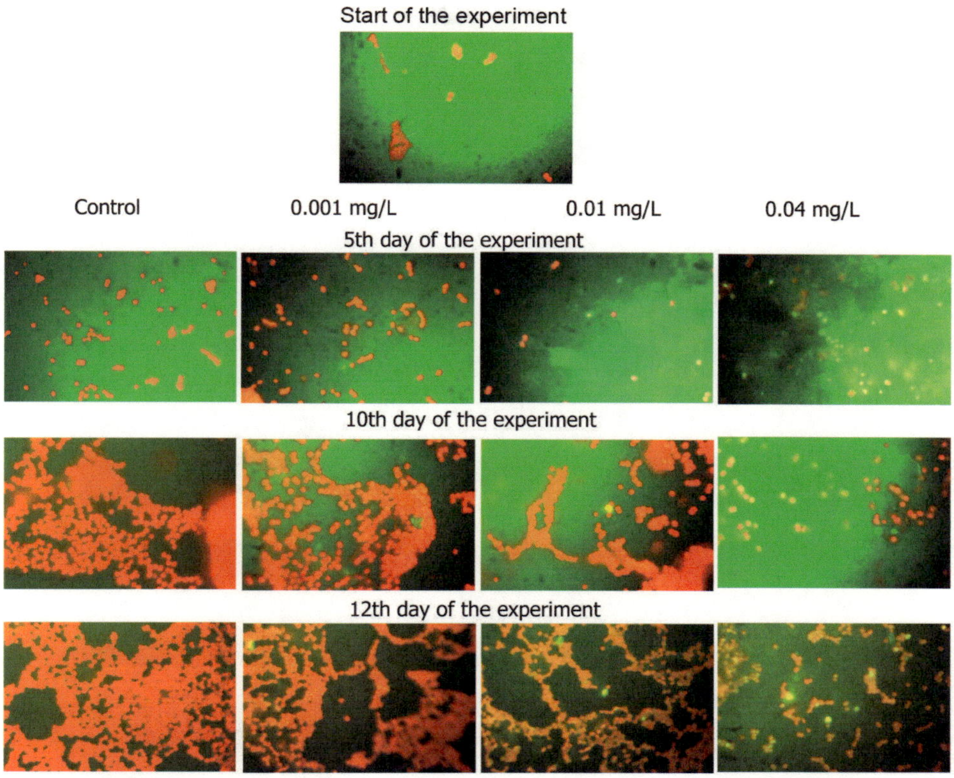

Figure 5.
Visualization of changes in the physiological state of S. aquatilis *culture under the action of linoleic acid.*

(**Table 3**). By the end of the experiment (on the 12th day), the proportion of dead cells was insignificant in all variants of the experiments (**Figure 4b**). The observed decrease in the number of *S. aquatilis* by three times compared to the control indicates a physiological slowdown in the growth of the culture.

3.2.2 Experiments with S. aquatilis *and tetradecanoic acid*

According to some researchers [32, 33], tetradecanoic and hexadecanoic acids can be present in natural waters in concentrations from 0.1 to 1000 μg/l. In this regard, we added tetradecanoic acid to the *S. aquatilis* culture medium at concentrations of 0.001, 0.01, and 0.1 mg/l. The initial density of cyanobacteria was 440,000 cells/ml in the first series and 210,000 cells/ml in the second series.

The difference between the two series of experiments with tetradecanoic acid, the results of which are shown in **Figures 6–9**, consisted in different initial numbers of cyanobacteria, which in the 1st series was two times higher (440,000 vs. 210,000 cells/ml). Also, as in the case of linoleic acid, in two series of experiments with tetradecanoic acid, as a result of microscopic fluorescence analysis, a distinct inhibition of *S. aquatilis* culture was observed (**Figures 6** and 7).

In the first series of experiments, the largest proportion of dead cells (19.6%) was recorded on the second day of the experiment, and at the end of the experiment (on day 14), it was 10.4% (**Table 3; Figure 8**). In the second series, with a lower initial concentration of cyanobacteria, the maximum proportion of dead cells was noted at the end of the experiment (on the 10th day) and amounted to 91.7%.

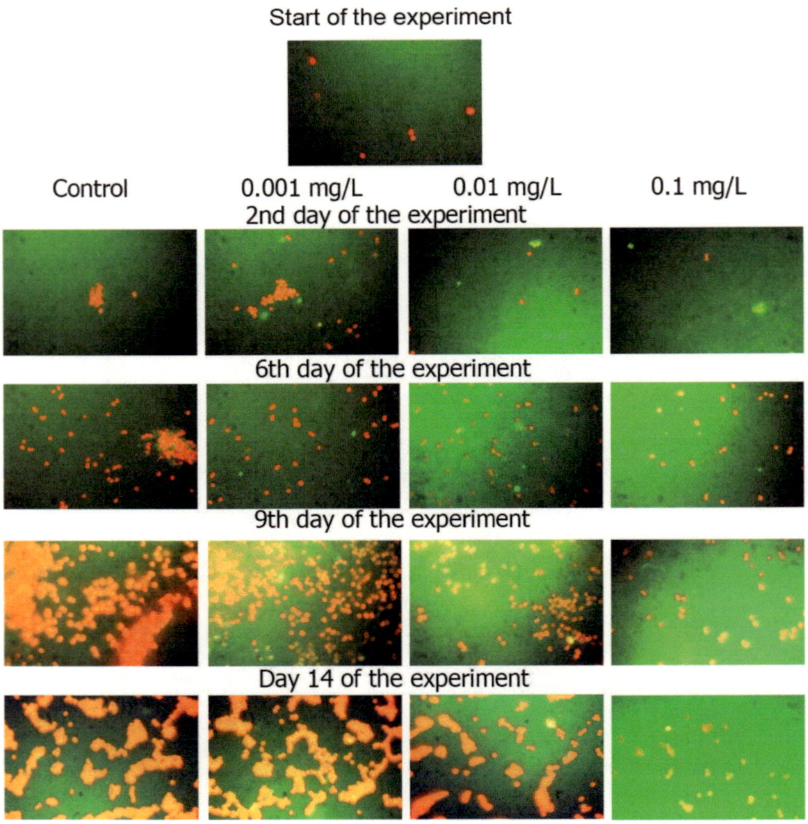

Figure 6.
Visualization of changes in the physiological state of S. aquatilis *culture under the action of tetradecanoic acid (experiment 1).*

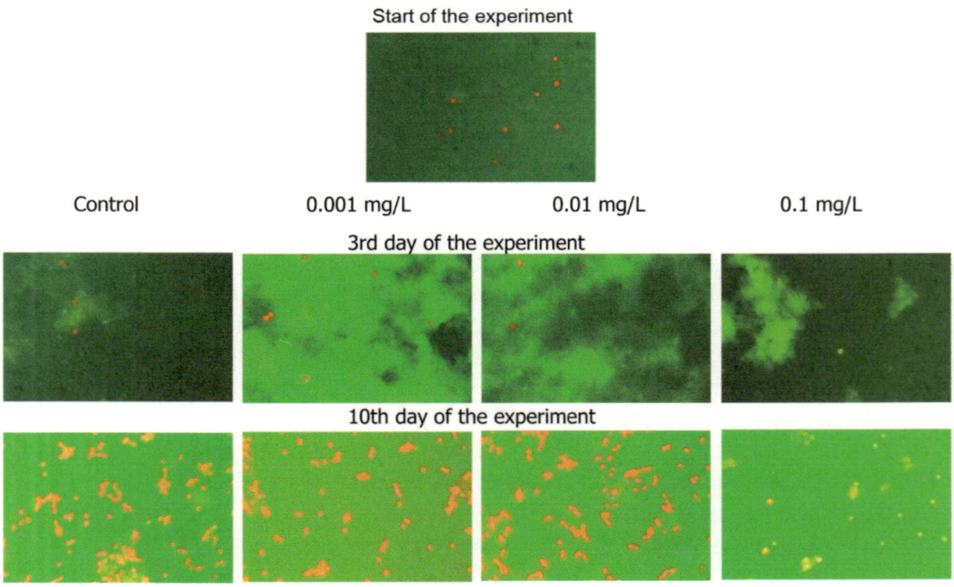

Figure 7.
Visualization of changes in the physiological state of S. aquatilis *culture under the action of tetradecanoic acid (experiment 2).*

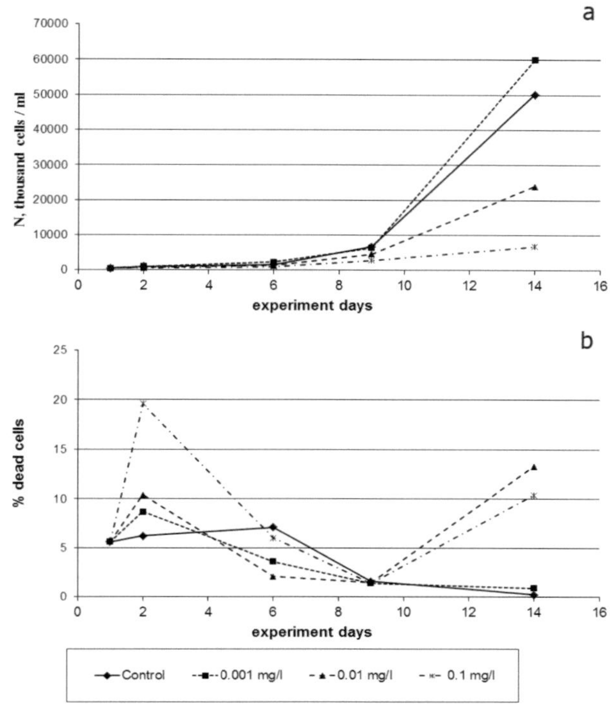

Figure 8.
Development of the culture of S. aquatilis *in experiment No. 1 with tetradecanoic acid: (a) the average number (median), (b) % dead cells (median).*

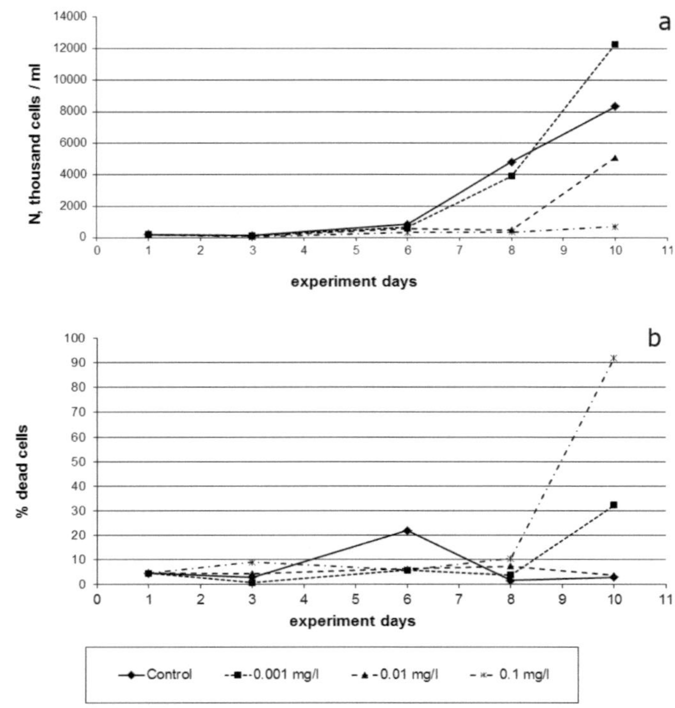

Figure 9.
Development of the culture of S. aquatilis *in experiment No. 2 with tetradecanoic acid: (a) the average number (median), (b) % dead cells (median).*

Experiment number/ active allelochemical	Object of impact	The maximum proportion of dead cells, %/maximum proportion of dead cells (%) at the end of the experiment	Allelochemical concentration (mg/l)/ the maximum proportion of dead cells	Day of the experiment with the maximum proportion of dead cells
1/linoleic acid	*S. aquatilis*	36.5/2.4	0.04/0.04	5/12
2/tetradecanoic acid	*S. aquatilis*	19.6/10.4	0.1/0.1	2/14
3/tetradecanoic acid	*S. aquatilis*	91.7/91.7	0.1/0.1	10/10
4/hexadecanoic acid	*S. aquatilis*	23.4/5.9	0.1/0.1	4/13
5/gallic acid	*S. aquatilis*	22.6/17.6	100/100	13/15
6/heptanoic acid	*S. aquatilis*	3.6/0.8	1/1	10/17
7/octanoic acid	*S. aquatilis*	5.4/1.3	1/1	3/17
8/mixture of heptanoic, octanoic, tetradecanoic, and gallic acids	*S. aquatilis*	4.4/4.4	10/10	13/13
9/mixture of heptanoic, octanoic, tetradecanoic, and gallic acids	*A. flos-aquae*	>95	10/10	23/23

Table 3.
The maximum proportion of dead cells of cyanobacteria when exposed to various allelochemicals.

The color shade of the cells was very different in the control and in the experiments, where the greatest suppression of cyanobacteria under the influence of tetradecanoic acid was noted. Pink-red clusters of cells were observed in the control, while pink-green and orange-green clusters and single cells were seen in the inhibited culture. With a lower initial density of *S. aquatilis* culture, a higher degree of culture inhibition compared to the control (SI = 11.9) was recorded at the end of the experiment than in the first series of the experiment, where this SI value was 7.5 (**Table 2**; **Figures 8** and **9**). The greatest inhibition of the development of cyanobacteria was noted at the highest of the studied concentrations (0.1 mg/l) of tetradecanoic acid in two series of experiments.

The lowest of the studied concentrations (0.001 mg/l) did not exert an adverse effect on cyanobacteria. Moreover, in both series of experiments, its stimulating effect was noted, i.e., the number of *S. aquatilis* in experimental microcosms at this concentration was higher than in the control (**Figures 8** and **9**).

3.2.3 Experiment with S. aquatilis *and hexadecanoic acid*

We added the hexadecanoic acid to the *S. aquatilis* culture medium at concentrations of 0.001, 0.01, and 0.1 mg/l. The initial density of cyanobacteria was 31,000 cells/ml. By the end of the experiment (on day 13), differences in the physiological state of *S. aquatilis* culture in the control and at a concentration of the allelochemical of 0.1 mg/l were clearly visible (**Figure 10**). In the experimental microcosms with allelochemical impact, cyanobacteria cells were present in pink-green clusters, while in the control, orange-red clusters of cells were observed in good physiological condition. Even at a concentration of 0.001 mg/l, hexadecanoic acid led to inhibition of the development of *S. aquatilis* culture (**Figure 11**), which was noticeable on the eighth day of the experiment.

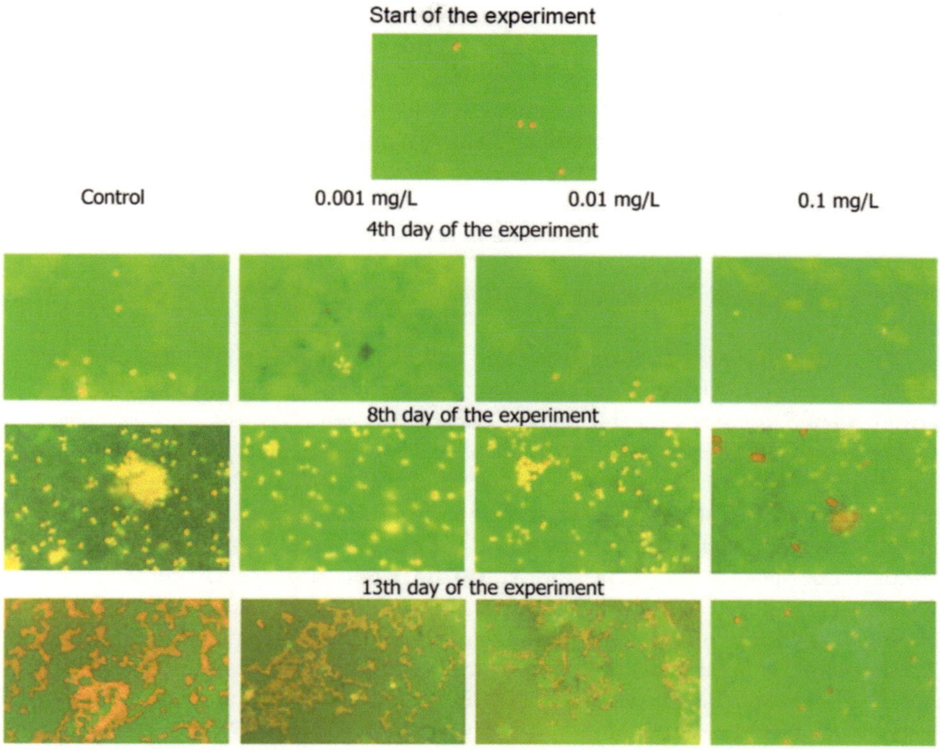

Figure 10.
Visualization of changes in the physiological state of S. aquatilis *culture under the action of hexadecanoic acid.*

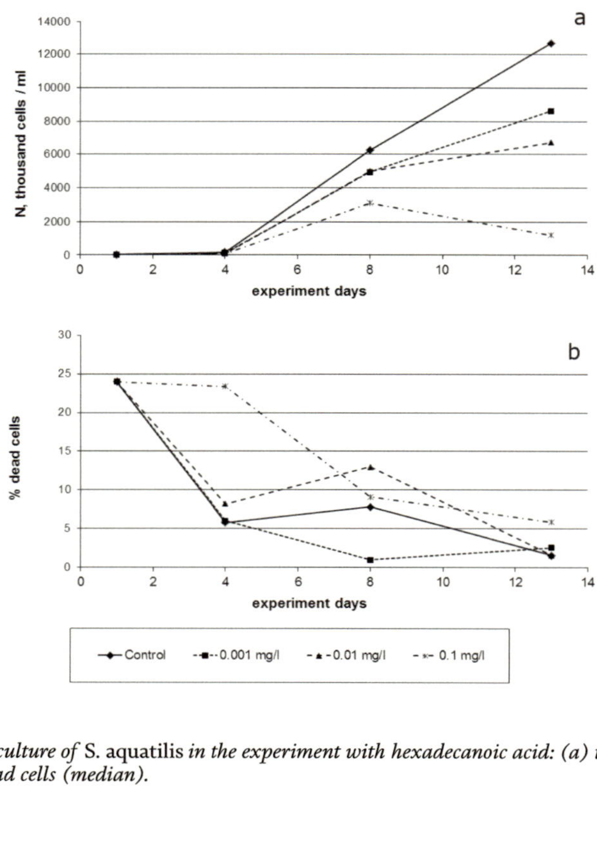

Figure 11.
Development of the culture of S. aquatilis *in the experiment with hexadecanoic acid: (a) the average number (median), (b) % dead cells (median).*

The greatest inhibition (SI = 10.4) was observed at the end of the experiment (on day 13) at a concentration of 0.1 mg/l. At the same concentration, the largest number of dead cells was observed, indicating that the culture was suppressed under the influence of the allelochemical. It was evident already on the fourth day of the experiment (23.4%) (**Figure 11**; **Table 3**). Subsequently, the proportion of dead cells decreased (to 5.9% at the end of the experiment), but a decrease in the number of cyanobacteria was also observed, which indicates the absence of culture growth under the influence of the allelochemical (**Figure 11a**).

3.2.4 Experiment with S. aquatilis *and gallic acid*

In the experiment with *S. aquatilis* and gallic acid, the initial density of cyanobacteria was 40,000 cells/ml. In this experiment, one of the highest values of inhibition of *S. aquatilis* was recorded (**Table 2**; **Figure 12**).

At the end of the experiment (on the 15th day), the density of the *S. aquatilis* culture in the control was 30 times higher than at the maximal studied concentration (100 mg/l) (**Table 2**). However, even at concentrations of 1 and 10 mg/l, inhibition of cyanobacteria development was observed by 3.3–3.5 times when compared to the control. At the same time, the proportion of dead cells (glowing pink-green or greenish-blue) was especially high at a concentration of 100 mg/l, at which the maximum inhibition of culture development was observed (**Tables 2** and **3**; **Figure 12**). It should be noted that, as in the case of tetradecanoic acid, the lowest of the studied concentrations of the allelochemical (0.1 mg/l) led to stimulation of the development of cyanobacteria (**Figure 12**).

3.2.5 Experiments with S. aquatilis *and heptanoic and octanoic acids*

In experiments with heptanoic and octanoic acids, the initial density of cyanobacteria was 73,000 cells/ml. Of all the acids tested, heptanoic acid showed the

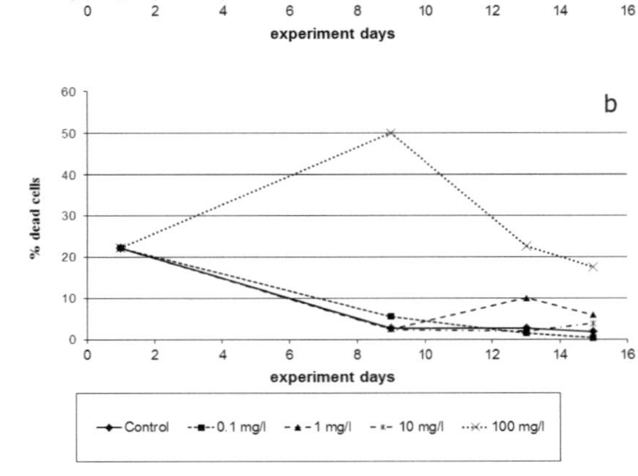

Figure 12.
Development of the culture of S. aquatilis *in the experiment with gallic acid: (a) the average number (median), (b) % dead cells (median).*

least inhibitory effect on *S. aquatilis* culture. So, at the end of the experiment, the density of cyanobacterial cells in the control exceeded that in microcosms with an allelochemical only two times (**Table 2**; **Figure 13a**). The differences in the optical density of the culture were even more minor (**Figure 13b**).

At the same time, it should be noted that the proportion of dead cells in microcosms with the influence of the allelochemical is higher (**Figure 13c**) than the control, although their absolute values were low. The maximum value was only 3.6% on the 10th day of the experiment (**Table 2**).

Octanoic acid showed a relatively low but higher than heptanoic acid inhibitory activity against the cyanobacteria *S. aquatilis*. By the end of the experiment (on day 17), the difference in cyanobacterial population density between the control and the allelochemical exposure was three times (**Table 2**; **Figure 14a**). Throughout the experiment, the level of development of cyanobacteria in the experimental vessels was lower than in the control. This is especially noticeable in the graphs of changes in the optical density of *S. aquatilis* culture and the proportion of dead cells (**Figure 14b** and **c**). The proportion of dead cells when exposed to octanoic acid was always higher than in the control and by the end of the experiment was 1.3%, while the maximum value of this indicator (5.4%) was observed on the third day of the experiment.

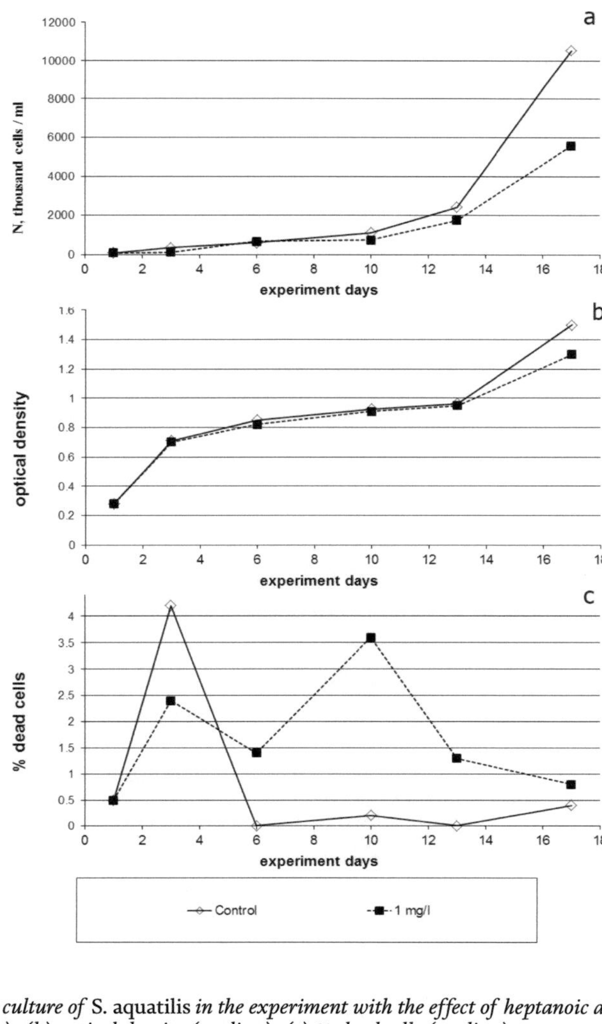

Figure 13.
Development of the culture of S. aquatilis *in the experiment with the effect of heptanoic acid: (a) the average cell density (median), (b) optical density (median), (c) % dead cells (median).*

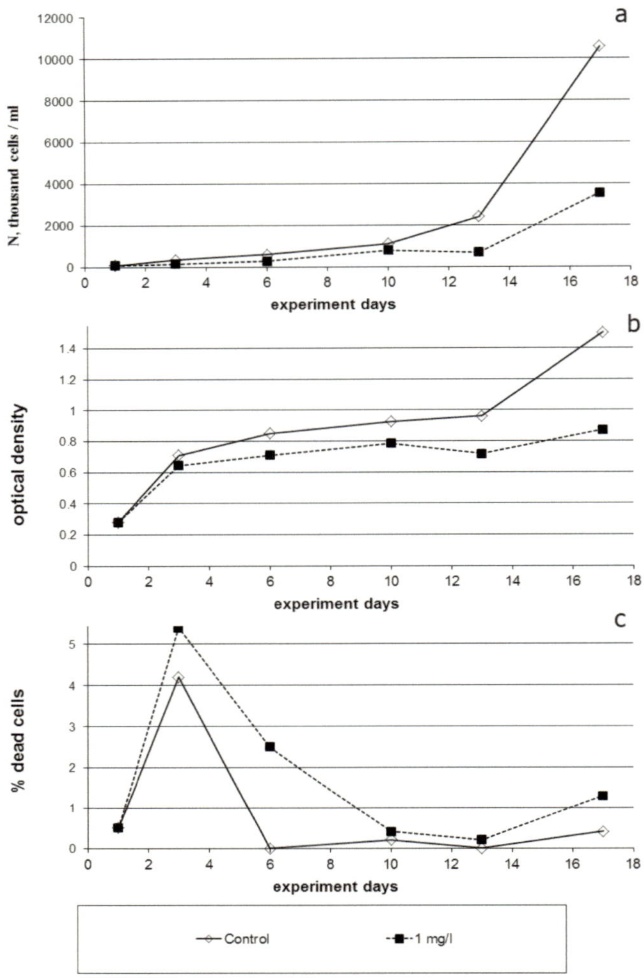

Figure 14.
Development of the culture of S. aquatilis *in the experiment with the effect of octanoic acid: (a) the average cell density (median), (b) optical density (median), (c) % dead cells (median).*

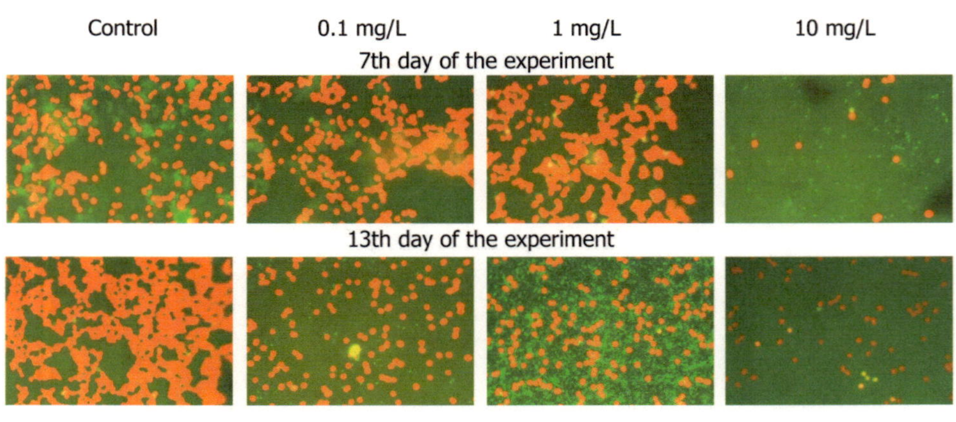

Figure 15.
Visualization of changes in the physiological state the culture of S. aquatilis *in the experiment with the combined effect of heptanoic, octanoic, tetradecanoic, and gallic acids at various concentrations.*

3.2.6 Experiment with S. aquatilis *and the mixture of allelochemicals*

Four of the six allelochemicals (heptanoic, octanoic, tetradecanoic, and gallic acids) were combined to obtain their mixture in organic solvent ethanol. The required amount of allelochemicals in equal amounts was dissolved in 1 ml of ethanol and added to experimental vessels to achieve the final concentrations 0.1, 1, and 10 mg/l. No addition of the mixture of allelochemicals was set as the control 1. No addition of the mixture of allelochemicals but with the addition of 1 ml of ethanol was set as the control 2. The initial density of cyanobacteria was 1014,000 cells/ml.

Visual fixation of the state of *S. aquatilis* culture, shown in **Figure 15**, clearly demonstrates a particularly strong inhibition of the development of *S. aquatilis* at a concentration of a mixture of allelochemicals of 10 mg/l when compared to the control. In experimental vessels with a mixture of four allelochemicals, a lower development of cyanobacteria was observed throughout the experiment and at all used concentrations (0.1, 1, 10 mg/l) according to fluorescence microscopy data when compared to the control (**Figure 16a**). More frequently performed the taking

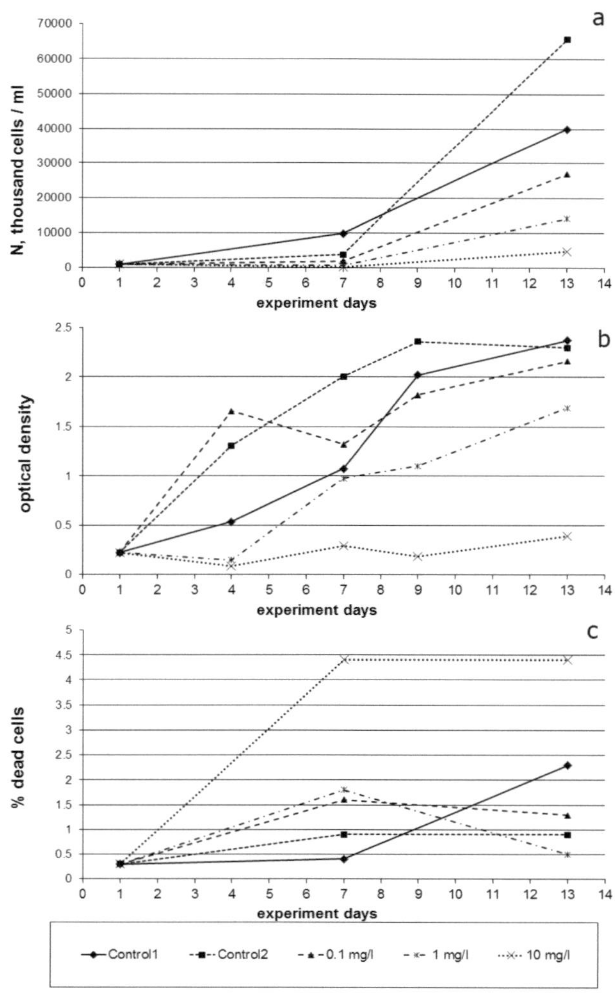

Figure 16.
Development of the culture of S. aquatilis *in the experiment with the combined effect of heptanoic, octanoic, tetradecanoic, and gallic acids at various concentrations: (a) the average number (median), (b) optical density (median), (c) % dead cells (median).*

readings of the optical density of culture development showed its increase at a concentration of 0.1 mg/l on day 4 of the experiment more than in both control variants. Later, by the end of the experiment, the optical density at this concentration was lower than in the control. The cyanobacteria abundance and optical density of the culture at allelochemical concentrations of 1 and 10 mg/l throughout the experiment were lower than Controls 1 and 2 (**Figure 16a** and **b**).

The maximum suppression of the culture of *S. aquatilis* (35.3 times when compared to the Control 1) was noted on the seventh day of the experiment at a concentration of the mixture of allelochemicals of 10 mg/l. By the end of the experiment (on the 13th day) at this concentration, the SI index was 8.3 (**Table 2**; **Figure 16a**). Also, marked suppression of the development of cyanobacterial culture by the end of the experiment was noted at concentrations of 0.1 and 1 mg/l (1.5 and 2.6 times, respectively) (**Figure 16a**).

In our experiments, we detected no negative influence of ethanol on the culture of *S. aquatilis*. Moreover, in the case of the addition of 1 ml of ethanol to experimental microcosm (Control 2), we have seen more intensive growth of cyanobacteria density in Control 2 toward the end of the experiment in comparison with Control 1 without any additions. Fluorescence analysis also showed that the highest proportion of dead cells of *S. aquatilis* (4.4%) was at a concentration of a mixture of allelochemicals of 10 mg/l (**Figure 16c**; **Table 3**).

3.2.7 Experiment with A. flos-aquae *and the mixture of allelochemicals*

Experiments with *A. flos-aquae*, a common bloom-forming cyanobacterium producing a hazardous effect in eutrophic freshwater ecosystems, were carried out only with a mixture of four allelochemicals (heptanoic, octanoic, tetradecanoic, and gallic acids). The initial density of cyanobacteria was 1,064,000 cells/ml.

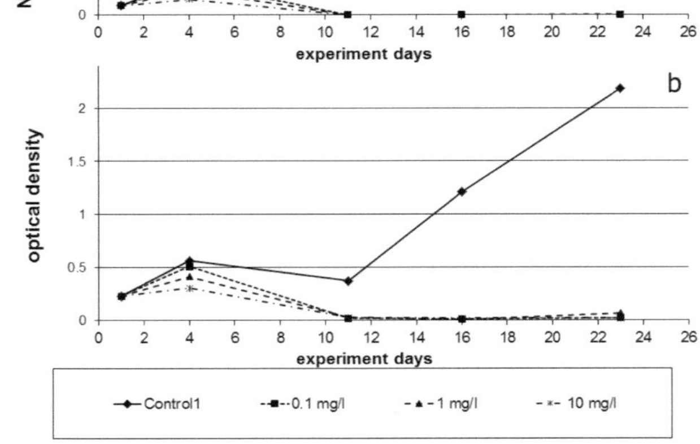

Figure 17.
Development of the culture of A. flos-aquae *in the experiment with the combined effect of heptanoic, octanoic, tetradecanoic, and gallic acids at various concentrations: (a) the average number (median); (b) optical density (median).*

This species of cyanobacteria was especially sensitive to the used mixture of allelochemicals. So, at a concentration of 10 mg/l, a complete suppression of the development of cyanobacteria was observed (**Figure 17**). SI at the end of the experiment (on the 23rd day) was 17,495 (**Table 2**). As this takes place, living cells were practically absent.

The proportion of dead cells was more than 95%. At concentrations of 0.1 and 1 mg/l, a strong suppression of the development of cyanobacteria was also observed. By the end of the experiment, SI values amounted to 14,568 and 2130, respectively. However, the proportion of dead cells at these concentrations was small, no more than 2%. This suggests a strong slowdown in the process of reproduction of cyanobacteria.

At the same time, the proportion of dying cells (orange-green shade) in the control on day 23 of culture development averaged 10.5%. For the most part, the culture contained bright red cells in long chains. This effect may be related to the self-inhibition of the culture, whose density in the control reached 27,236,000 cells/ml by the end of the experiment.

4. Discussion

Ultimately, our results showed that organic acids-allelochemicals are widespread in freshwater macrophytes. Of the macrophytes we studied, the largest number of carboxylic acid-allelochemicals was found in *M. spicatum* (**Table 1**), which is one of the aquatic plants with the highest allelopathic potential [34], especially concerning its effect on cyanobacteria [35]. The information available in the literature [36] and our results show that hexadecanoic acid can be attributed to one of the most common allelochemicals in aquatic macrophytes. In [37] it is noted that hexadecanoic acid is part of the major algal inhibiting allelochemicals in planting water of *Hydrilla verticillata*. Already in early studies of the phytotoxic properties of acids, it was shown that hexadecanoic acid possesses phytotoxic properties [38]. For hexadecanoic acid, our study revealed a rather high cyanocidal activity against *S. aquatilis*. It appeared even at concentrations of 0.01 and 0.1 mg/l. At the same time, for other representatives of photosynthetic plankton, for example, red tide algae, hexadecanoic acid displayed relatively weak activity, with LC_{50} values of 50.6 and 69.5 μg/ml [39]. This may indicate the selectivity of this allelochemical against cyanobacteria.

The studies [33, 38] demonstrated that heptanoic acid also has phytotoxic properties. The work [40] presents data that hexadecanoic acid diminished the seed germination of some epiphytes.

In the samples of aquatic macrophytes we studied, we did not find heptanoic and gallic acids. However, there is ample evidence regarding the presence of heptanoic acid in the composition of metabolites of aquatic macrophytes [41]. In addition, there are results that heptanoic and octanoic acids, being allelochemicals of aquatic macrophytes, can lead to suppressing the development of cyanobacteria [42]. The gallic acid is judged to be the allelochemical compound and phytotoxin [43–45]; it may be found in many aquatic macrophytes [44, 46, 47].

Polyphenols were found to contribute, on average, to 50% of the allelopathic effects of *M. spicatum* on *Microcystis aeruginosa* (Kutzing) Kutzing [48]. As noted earlier [49], the four phenols (pyrogallic acid, gallic acid, (+)-catechin, and ellagic acid) secreted by *M. spicatum* could contribute to the allelopathic effects of *M. spicatum* at 10–100%. Another part of the allelopathic effect of this macrophyte seems to be due to carboxylic acids, which this species synthesizes especially much (**Figure 3**). Gallic acid and carboxylic acids are reported to be active allelochemicals

[16, 50, 51]. The fact that we did not find this compound in the macrophyte samples we examined may be probably related to the method of obtaining the essential oil of these plants by the method of steam hydrodistillation.

It has been shown that allelochemicals such as heptanoic and octanoic acids can be present in natural waters at concentrations of 100–1000 μg/l [32, 33]. There is evidence that heptanoic and octanoic acids, being allelochemicals of aquatic macrophytes, can lead to a degradation of the development of cyanobacteria [42]. It was also shown that perfluoro modifications of hexanoic, heptanoic, octanoic, and nonanoic acids suppressed the development of cyanobacteria and algae (LC_{50}s range: 6.0–24.3 mg/l) [52]. At the same time, cyanobacteria and diatom algae were comparable in sensitivity, but both were more sensitive than green algal species. Fukuda et al. [53] studied the allelopathic potential of heptanoic acid. They reported that a 0.1 M solution of heptanoic acid was able to reduce the viability of crabgrass by 30% compared with the control. Earlier studies with the same compound [54] reported that heptanoic acid was able to significantly reduce the growth of various organelles within the plant Phragmites at a concentration of 1.4 mM. It was shown [33] that the EC50 (half-maximal effective concentration) is 650 μg/l for heptanoic acid.

Linoleic acid, as an allelochemical, is produced not only by aquatic macrophytes but also, for example, by filamentous green algae (*Uronema confervicolum*, a benthic filamentous green alga) [55]. The effect of linoleic acid on cyanobacteria *M. aeruginosa* has been studied in [56]. The results obtained on the inhibition of *M. aeruginosa* (up to 96%) are in good agreement with our data, showing a high ability of linoleic acid to suppress the development of cyanobacteria. As shown in this work, under the influence of linoleic acid algal cell membrane lipid peroxidation occurred, and the membrane permeability increased, accompanied by the damage of cell membranes. It was concluded that the main mechanism of linoleic acid inhibiting cyanobacterial growth is the destruction of the cyanobacterial cell membrane.

At the same time, our observations using fluorescence microscopy showed the presence of a relatively small number of dead cyanobacterial cells, which differed insignificantly (to the large side) from the control. This does not indicate the death of cells of the cyanobacterial culture but their significantly lower reproduction rate than the control due to the general physiological inhibition of the cyanobacterial population. It is reasonable to hypothesize that there may be another mechanism for inhibiting cyanobacteria in addition to what is indicated in [56].

All of the organic acids tested by us showed inhibitory allelopathic activity against cyanobacteria. However, their inhibitory effect was different in relation to the same species (*S. aquatilis*). The highest values of the suppression index (SI > 10) were recorded (in ascending order) for hexadecanoic, linoleic, tetradecanoic, and gallic acids and a mixture of four allelochemicals (heptanoic, octanoic, tetradecanoic, and gallic acids).

Recently, Zhou et al. [57] reported the inhibitory effects of some allelochemicals on cyanobacteria *M. aeruginosa*. Thus, for tetradecanoic acid, the EC_{50} (mg/l) was 15.5 and, for hexadecanoic acid, 18.23 mg/l [57]. The higher level of inhibition concerning *S. aquatilis* for these allelochemicals, noted in our experiments (**Table 2**), seems to indicate a different sensitivity of different cyanobacteria species to the same allelochemicals. This is supported by our results of evaluating the effect of the same combination of allelochemicals on two different species of cyanobacteria, *S. aquatilis*, and *A. flos-aquae*. For the first species, the resulting SI values for concentrations of 0.1, 1, and 10 mg/l on the 13th day of the experiment were 1.5, 2.8, and 8.3, respectively, and for the second species, on the 11th day, 1606, 1165, and 1063 (at the same concentrations), increasing by the end of the experiment (on day 23) to values of 14,568; 2130; and 17,495, respectively. That is, there was a particularly strong suppression of the development of the cyanobacterial culture of *A. flos-aquae* up to complete inhibition, whereas in the case of *S. aquatilis*, the inhibitory effect of the mixture of allelochemicals was not so pronounced. This shows the great sensitivity of *A. flos-aquae* to a selected mixture of allelochemicals.

This is also confirmed by the data of Nakai et al. [48], who did not find a significant effect of suppressing another cyanobacterium *M. aeruginosa* by tetradecanoic and hexadecanoic acids. High effective concentrations (68–78 mg/l) were obtained for tetradecanoic acid against red tide microalgae [58]. Cyanobacterial inhibition ability of the fatty acid may be related to the carbon chain length and the degree of their unsaturation. So, it was shown when studying the inhibitory effect of palmitic acid and stearic acid on the growth of *M. aeruginosa* [36].

The data obtained may indicate a higher specific sensitivity of cyanobacteria to long-chain saturated fatty acids than to algae. The cyanobacterial inhibition mechanism of organic acids might be closely related to their chemical structures [59]. This paper shows that fatty acids changed the permeability of cell membranes, which caused the leakage of cell contents, further damaged the membrane structures, and thus affected the level of development of cyanobacterial population including the synthesis of toxins.

The results received to date suggest that among cyanobacteria we will inevitably be faced with the specificity of the action of various acids-allelochemicals, whose anti-cyanobacterial activities will depend on (i) length of carbon chain, (ii) number of unsaturated linkages, and (iii) positions of any double bonds [48]. To this, we need to add the dependence on the concentration of an allelochemical; when the concentration is insufficient to suppress, the reverse stimulating effect of cyanobacteria development can be observed. In some of our experiments, we observed this stimulating effect, for example, in experiments with tetradecanoic, hexadecanoic, and gallic acids.

Until only recently, little experimental evidence has existed to establish that low concentrations of the allelochemicals can stimulate the development of cyanobacteria. Thus, in experiments with *M. aeruginosa* [42], at the lowest concentration of the two organic acids used (benzoic acid and heptanoic acid), higher development of cyanobacteria was observed than in the control. The effect of the allelochemicals mixture in our experiments was higher than in the case of individual components. Thus, under the action of a mixture of allelochemicals (heptanoic, octanoic, tetradecanoic, and gallic acids) on *S. aquatilis*, the maximum SI value (35.3) was obtained (**Table 2**), while the highest value of SI (30) when on exposure to a single allelochemical was recorded in the case of gallic acid.

A number of studies have shown that a combination of several allelochemicals has a synergistic effect in suppressing the development of cyanobacteria. In [42] it was indicated that benzoic acid and heptanoic acid showed a synergistic effect and that their compound algal inhibition effect was superior to the separate effects of the two organic acids acting independently. It has also been found that when two or three allelochemicals were mixed in specific proportions, the algal inhibition rate increased significantly, thereby indicating allelopathic synergistic interactions [60]. Gniazdowska and Bogatek [61] pointed out that the activity of allelochemicals cannot be explained by just a single mode of action. The synergistic effect of the action of allelochemicals (fatty acids, gallic acids, and pyrogallic acids) in relation to cyanobacteria is indicated in [49, 62–64]. It has also been revealed that allelochemicals from *Chara vulgaris* included three fatty acids: (Z, Z)-9,12-octadecadienoic, tetradecanoic, and hexadecanoic acids. The combined impact of these three fatty acids induced a synergistic inhibitory effect on the growth of toxic cyanobacterium *M. aeruginosa*. Synergistic growth inhibition of these cyanobacteria by a mixture of the polyphenols was reported in [49].

In the experiments with *S. aquatilis, A. flos-aquae*, and the mixture of allelochemicals, the situation of a eutrophic reservoir was actually simulated, i.e., they started at a very high initial density of cyanobacteria, which corresponded to the level of hypertrophic reservoirs [65]. It was very interesting to check whether a combination of substances-allelochemicals of aquatic macrophytes could suppress the intensive development of cyanobacteria. The results obtained gave a positive answer. This suggests that new cyanocides made up of a combination of LMWOCs-allelochemicals have the potential to not only inhibit the development of cyanobacteria populations, preventing them from developing to dangerous levels, but also suppress active HAB.

The method of fluorescence microscopy traces changes that are caused in the culture of cyanobacteria by the action of various LMWOCs-allelochemicals of aquatic macrophytes. It was previously shown that the method of fluorescence microscopy is very useful to distinguish producing and non-producing cells of genetically engineered cyanobacteria *Synechocystis sp.* PCC6803 and *Synechococcus sp.* PCC7002 which are used for the production of biofuels [26].

Several studies have attempted to find compounds with pronounced cyanocidal and algicidal activities among natural plant products [60, 66]. In particular, in [67] it is shown for the first time that there are strong inhibition effects (EC_{50} < 100 μg/l) of the isoquinoline alkaloids on phytoplankton (including cyanobacteria).

5. Conclusion

Recently, scientific interest in cyanobacteria responsible for the HAB in natural reservoirs has increased significantly. Various ways of controlling the number of these microorganisms in aquatic ecosystems are being investigated, and hence, simple and fast methods of evaluating the viability of their cells are required. The majority of common methods (e.g., plating, fluorescent staining) to determine cell viability are preparation demanding, time-consuming, and rather expensive. The advantages of the fluorescence microscopy method, using red chlorophyll autofluorescence and nonspecific green autofluorescence, are high sensitivity, non-destructive evaluation of the vital state of cells, and differentiation of living and dead cells without sample preparation. The method is time-efficient and accurate, and a large number of samples with small amounts of material can be examined in a short time.

In our study, we have shown (including using fluorescence microscopy) that natural compounds, metabolites of aquatic macrophytes, in particular carboxylic acids and gallic acid, can be used to suppress the development of cyanobacteria populations. Their combined effects are likely to be particularly effective. In other words, the creation of a new generation of algaecides/cyanocides based on LMWOCs of aquatic plants is a very promising strategy for combating HAB. In this case, of course, it will be necessary to take into account that the prices of selected natural products for these goals are usually higher than for conventional products (copper sulfate, aluminum salts, simazine, and many others) used to control cyanobacteria, but the new potential cyanocides likely will have important properties, including (1) selective toxicity toward only undesirable cyanobacteria in the phytoplankton and the absence (or minimal toxicity) toward other hydrobionts and (2) rather fast biological degradation, for an optimal state of the aquatic ecosystem [66].

In fairness, it should be noted that not only LMWOCs-allelochemicals can become the basis for creating new effective methods to combat excessive and undesirable development of cyanobacteria but also such methods as the use of new chlorinated nanoproducts, in particular N-halamine derivatized nanoparticles, that, as it turned out, enable selective elimination of blooming cyanobacteria while minimizing the effect on cohabiting phytoplankton species [68].

Judging from the results obtained, we can conclude that the method of fluorescence microscopy can be very useful in revealing the inhibitory mechanisms of cyanobacterial populations in experimental studies on the search for natural compounds (in particular, allelochemical metabolites of aquatic macrophytes), on the basis of which a new generation of cyanocides can be created. This "novel weapon" against blooming cyanobacteria seems to be a potentially more interesting product for further development of technologies on HAB control and management.

Acknowledgements

The work was performed within the framework of the state task of the Russian Academy of Sciences on topic 0154-2019-0002.

We are grateful to the Centre for Culture Collection of Microorganisms of the Research Park at St. Petersburg University as well as St. Petersburg Scientific Research Center for Ecological Safety of the Russian Academy of Sciences for supplying the strains of *S. aquatilis* and *A. flos-aquae* for our experiments.

Conflict of interest

The authors declare that there is no conflict of interest.

Author details

Evgeny Kurashov[1,2]*, Larisa Kapustina[1,2], Julia Krylova[2] and Galina Mitrukova[1]

1 Institute of Limnology, Russian Academy of Sciences, Saint Petersburg, Russia

2 Saint-Petersburg Branch of the Federal State Budgetary Scientific Institution "All-Russian Research Institute of Fisheries and Oceanography" ("GosNiorch" by L.S. Berg), Saint Petersburg, Russia

*Address all correspondence to: evgeny_kurashov@mail.ru

References

[1] Fedorov VD. Methods of studying phytoplankton and its activity. Science. Novosibirsk: Nauka; 1979. 165 p. (In Russian)

[2] Bentley-Mowat JA. Application of fluorescence microscopy to pollution studies on marine phytoplankton. Bottomline. 1982;**28**:203-204

[3] Sadchikov AP. Methods of Studying Freshwater Phytoplankton (Methodological Manual). Moscow: University and School; 2003. p. 156

[4] Goryunova SV. The use of fluorescence microscopy to determine live and dead algae cells. Institute of Microbial Technology. 1952;**2**:64-67

[5] Goryunova SV. The technique of applying the method of fluorescence microscopy for hydrobiological studies. In: Life of the Fresh Waters of the USSR. Vol. 4. Moscow, Leningrad: Publishing House of the USSR Academy of Sciences; 1956. p. 101

[6] Schulze K, López DA, Tillich UM, Frohme M. A simple viability analysis for unicellular cyanobacteria using a new autofluorescence assay, automated microscopy, and ImageJ. BMC Biotechnology. 2011;**11**(1):118. DOI: 10.1186/1472-6750-11-118

[7] Hallegraeff GM. A review of harmful algal blooms and their apparent global increase. Phycologia. 1993;**32**:79-99

[8] Watson SB, Whitton BA, Higgins SN, Paerl HW, Brooks BW, Wehr JD. Harmful algal blooms. In: Wehr JD, Robert GS, Kociolek JP, editors. Freshwater Algae of North America. London: Elsevier Inc.; 2015. pp. 873-920. DOI: 10.1016/B978-0-12-385876-4.00020-7

[9] Lou I, Han B, Zhang W, editors. Advances in monitoring and modelling algal blooms in freshwater reservoirs. In: General Principles and a Case Study of Macau. Dordrecht: Springer Science+Business Media; 2017. p. 154

[10] Paerl HW, Fulton RS, Moisander PH, Dyble J. Harmful freshwater algal blooms, with an emphasis on cyanobacteria. The Scientific World Journal. 2001;**1**:76-113. DOI: 10.1100/tsw.2001.16

[11] Lopez CB, Jewett EB, Dortch Q, Walton BT, Hudnell HK. Scientific Assessment of Freshwater Harmful Algal Blooms. Washington: DC: Interagency Working Group on Harmful Algal Blooms, Hypoxia, and Human Health of the Joint Subcommittee on Ocean Science and Technology; 2008. p. 78

[12] Sharma NK, Choudhary KK, Bajpai R, Rai AK. Freshwater cyanobacterial (blue-green algae) blooms: Causes, consequences and control. In: Nemr AE, editor. Impact, Monitoring and Management of Environmental Pollution. 1st ed. New York: Nova Science Publishers; 2010. pp. 73-95

[13] Gatz L. Freshwater Harmful algal blooms: Causes, challenges and policy considerations. Congressional research service report; 2019. p. 30

[14] Barrington DJ, Xiao X, Coggins LX, Ghadouani A. Control and management of harmful algal blooms. In: Botana LM, Louzao MC, Vilariño N, editors. Climate Change and Marine and Freshwater Toxins. Berlin, Germany: De Gruyter; 2015. pp. 313-358. DOI: 10.1515/9783110333596-012

[15] Erhard D. Allelopathy in aquatic environments. In: Manuel JR, Pedrol N, González L, editors. Allelopathy: A Physiological Process with Ecological Implications. Netherlands: Springer; 2006. pp. 433-450. DOI: 10.1007/1-4020-4280-9_19

[16] Hu H, Hong Y. Algal-bloom control by allelopathy of aquatic macrophytes—A review. Frontiers of Environmental Science & Engineering in China. 2008;**2**(4):421-438

[17] Wang J, Zhu J, Liu S, Liu B, Gao Y, Wu Z. Generation of reactive oxygen species in cyanobacteria and green algae induced by allelochemicals of submerged macrophytes. Chemosphere. 2011;**85**(6):977-982. DOI: 10.1016/j.chemosphere.2011.06.076

[18] Ahluwalia AS. Cyanobacterial and algal allelopathy. In: Cheema Z, Farooq M, Wahid A, editors. Allelopathy. Berlin, Heidelberg: Springer; 2013. pp. 485-509

[19] Macías FA, Galindo JLG, García-Díaz MD, Galindo JCG. Allelopathic agents from aquatic ecosystems: Potential biopesticides models. Phytochemistry Reviews. 2008;**7**:155-178. DOI: 10.1007/s11101-007-9065-1

[20] Kurashov EA, Krylova JV, Mitrukova GG, Chernova AM. Low-molecular-weight metabolites of aquatic macrophytes growing on the territory of Russia and their role in hydroecosystems. Contemporary Problems of Ecology. 2014;**7**(4):433-448. DOI: 10.1134/S1995425514040064

[21] Gomes MP, Garcia QS, Barreto LC, Pimenta LPS, Matheus MT, Figueredo CC. Allelopathy: An overview from micro- to macroscopic organisms, from cells to environments, and the perspectives in a climate-changing world. Biologia. 2017;**72**:113-129. DOI: 10.1515/biolog-2017-0019

[22] Kurashov EA, Mitrukova GG, Krylova YV. Interannual variability of low-molecular metabolite composition in *Ceratophyllum demersum* (Ceratophyllaceae) from a floodplain lake with a changeable trophic status. Contemporary Problems of Ecology. 2018;**11**(2):179-194. DOI: 10.1134/S1995425518020063

[23] Gromov BV, Vepritskiy AA, Titova NN, Mamkayeva KA, Alexandrova OV. Production of the antibiotic cyanobacterin LU-1 by Nostoc linckia CALU 892 (cyanobacterium). Journal of Applied Phycology. 1991;**3**:55-59

[24] Kurashov EA, Fedorova EV, Krylova JV, Mitrukova GG. Assessment of the potential biological activity of low molecular weight metabolites of freshwater macrophytes with QSAR. Scientifica. 2016;**2016**:9. Article ID: 1205680. Available from: https://doi.org/10.1155/2016/1205680

[25] Kurashov EA, Fedorova EV, Krylova JV. The use of the QSAR method to identify the most promising alleochemists against cyanobacteria. Russian Journal of Ecology. 2018;**4**(16):56-61

[26] Schulze K, Lang I, Enke H, Grohme D, Frohme M. The use of fluorescence microscopy and image analysis for rapid detection of non-producing revertant cells of *Synechocystis* sp. PCC6803 and *Synechococcus* sp. PCC7002. BMC Research Notes. 2015;**8**:160. DOI: 10.1186/s13104-015-1112-1

[27] Krasnovsky AA, Kovalev YV. Spectral and kinetic parameters of phosphorescenceof triplet chlorophyll a in the photosynthetic apparatus of plants. Biochemistry (Moscow). 2014;**79**(4):349-361

[28] Krylova JV, Kurashov EA, Mitrukova GG. Component composition of essential oil of *Potamogeton perfoliatus* L. from Lake Ladoga at the beginning of the fructification period. The Chemistry of Plant Raw Materials. 2016;**2**:79-88. DOI: 10.14258/jcprm.2016021189

[29] Kurashov EA, Krylova YV, Egorova AA, Suschenko AS,

Khodonovich VV, Yavid EY. Perspectives for the use of a low molecular weight metabolome of aquatic macrophytes to indicate the ecological state of aquatic ecosystems. Water Chemistry and Ecology. 2018;**1-3**:68-79

[30] Krylova JV, Kurashov EA, Rusanov AG. Comparative analysis of the component composition of the low molecular weight metabolome of water smartweed (*Persicaria amphibia* (L.) Delarbre) from diverse habitats in Lake Ladoga. Transactions of Karelian Research Centre of Russian Academy of Science. Limnology and Oceanography. 2020;**4**:1-21

[31] Kurashov EA, Krylova YV, Chernova AM, Mitrukova GG. Component composition of low molecular volatile organic compounds of Nuphar lutea (Nymphaeaceae) at the beginning of vegetation. Water Chemistry and Ecology. 2013;**5**:67-80

[32] Everall NC, Lees DR. The use of barley straw to control general and blue green algal growth in a Derbyshire reservoir. Water Research. 1996;**30**:269-276

[33] Murray D, Jefferson B, Jarvis P, Parsons SA. Inhibition of three algae species using chemicals released from barley straw. Environmental Technology. 2010;**31**(4):455-466. DOI: 10.1080/09593331003663294

[34] Inderjit M, editor. Chemical Ecology of Plants: Allelopathy in Aquatic and Terrestrial Ecosystems. Basel; Boston; Berlin: Birkhäuser; 2002. p. 272. DOI: 10.1007/978-3-0348-8109-8

[35] Švanys A, Paškauskas R, Hilt S. Effects of the allelopathically active macrophyte Myriophyllum spicatum on a natural phytoplankton community: A mesocosm study. Hydrobiologia. 2014;**737**:57-66. DOI: 10.1007/s10750-013-1782-4

[36] Jiang WX, Jia Y, Wang CY, Wang Q, Tian XJ. The inhibitory effects of palmitic acid and stearic acid on *Microcystis aeruginosa*. Ecology and Environmental Sciences. 2010;**19**(2):291-295

[37] Gao H, Song Y, Lv C, Chen X, Yu H, Peng J, et al. The possible allelopathic effect of Hydrilla verticillata on phytoplankton in nutrient-rich water. Environment and Earth Science. 2015;**73**:5141-5151. DOI: 10.1007/s12665-015-4316-8

[38] Elliott LF, McCalla TM, Waiss A. Crop phytotoxicity. In: Crop Residue Management Systems. Vol. 31. Madison, Wisconsin: ASA Special Publication; 1978. pp. 131-146

[39] Sun X, Jin H, Zhang L, Hu W, Li Y, Xu N. Screening and isolation of the algicidal compounds from marine green alga Ulva intestinalis. Chinese Journal of Oceanology and Limnology. 2016;**34**:781-788. DOI: 10.1007/s00343-016-4383-z

[40] Valencia-Díaz S, Flores-Palacios A, Rodríguez-López V, Ventura-Zapata E, Jiménez-Aparicio A. Effect of host-bark extracts on seed germination in *Tillandsia recurvata*, an epiphytic bromeliad. Journal of Tropical Ecology. 2010;**26**:571-581

[41] Santos LO, Silva FF, Santos LC, Carregosa ISC, Wisniewski AJ. Potential bio-oil production from invasive aquatic plants by microscale pyrolysis studies. Journal of the Brazilian Chemical Society. 2018;**29**(1):151-158. DOI: 10.21577/0103-5053.20170124

[42] Chai T, Zhu HD, Yan HZ, Zhao D, Liu XY, Fu HY. Allelopathic effects of two organic acids on *Microcystis aeruginosa*. IOP Conference Series: Earth and Environmental Science. 2018;**146**:012047. DOI: 10.1088/1755-1315/146/1/012047

[43] Inderjit, Callaway RM. Experimental designs for the study of allelopathy. Plant and Soil. 2003;**256**:1-11

[44] Kumar AS, Bais HP. Allelopathy and exotic plant invasion. In: Baluška F, Ninkovic V, editors. Plant Communication from an Ecological Perspective. Signaling and Communication in Plants. Berlin, Heidelberg: Springer; 2010. pp. 61-74

[45] Favaretto A, Scheffer-Basso SM, Perez NB. Allelopathy in Poaceae species present in Brazil. A review. Agronomy for Sustainable Development. 2018;**38**:22. DOI: 10.1007/s13593-018-0495-5

[46] Rice EL. Allelopathy—An update. The Botanical Review. 1979;**45**:15-109

[47] Gopal B, Goel U. Competition and allelopathy in aquatic plant communities. The Botanical Review. 1993;**59**:155-210

[48] Nakai S, Yoshihara T, Yamada S, Hosomi M. The allelochemicals accounting for the allelopathic effects of *Myriophyllum spicatum* on the cyanobacterium *Microcystis aeruginosa* [Internet]. 2005. Available from: http://www.regional.org.au/au/allelopathy/2005/2/1/2244_nakais.htm#TopOfPage [Accessed: 27 April 2020]

[49] Nakai S, Inoue Y, Hosomi M, Murakami A. *Myriophyllum spicatum*-released allelopathic polyphenols inhibiting growth of blue-green algae *Microcystis aeruginosa*. Water Research. 2000;**34**(11):3026-3032. DOI: 10.1016/s0043-1354(00)00039-7

[50] Li FM, Hu HY. Isolation and characterization of a novel antialgal allelochemical from Phragmites communis. Applied and Environmental Microbiology. 2005;**71**(11):6545-6553

[51] Zhou S, Nakai S, Hosomi M, Sezaki Y, Tominaga M. Allelopathic growth inhibition of cyanobacteria by reed. Allelopathy Journal. 2006;**18**(2): 277-285

[52] Latała A, Nędzi M, Stepnowski P. Acute toxicity assessment of perfluorinated carboxylic acids towards the Baltic microalgae. Environmental Toxicology and Pharmacology. 2009;**28**(2):167-171. DOI: 10.1016/j.etap.2009.03.010

[53] Fukuda M, Tsujino Y, Fujimori T, Wakabayashi K, Böger P. Phytotoxic activity of middle-chain fatty acids I: Effects on cell constituents. Pesticide Biochemistry and Physiology. 2004;**80**(3):143-150. DOI: 10.1016/j.pestbp.2004.06.011

[54] Armstrong J, Armstrong W. Phragmites die-Back: Toxic effects of propionic, butyric and Caproic acids in relation to pH. New Phytologist. 1999;**142**(2):201-217

[55] Allen JL, Ten-Hage L, Leflaive J. Regulation of fatty acid production and release in benthic algae: Could parallel allelopathy be explained with plant defence theories? Microbial Ecology. 2017;**75**(3):609-621. DOI: 10.1007/s00248-017-1082-z

[56] Ni L, Rong S, Gu G, Hu L, Wang P, Li D, et al. Inhibitory effect and mechanism of linoleic acid sustained-release microspheres on *Microcystis aeruginosa* at different growth phases. Chemosphere. 2018;**212**:654-661. DOI: 10.1016/j.chemosphere.2018.08.045

[57] Zhou L, Chen G, Cui N, Pan Q, Song X, Zou G. Allelopathic effects on *Microcystis aeruginosa* and allelochemical identification in the cuture solutions of typical artificial floating-bed plants. Bulletin of Environmental Contamination and Toxicology. 2019;**102**:115-121. DOI: 10.1007/s00128-018-2486-2

[58] Wang R, Wang Y, Tang X. Identification of the toxic compounds produced by *Sargassum thunbergii* to red tide microalgae. Chinese Journal of Oceanology and Limnology. 2012;**30**:778-785. DOI: 10.1007/s00343-012-1294-5

[59] Wu J-T, Chiang Y-R, Huang W-Y, Jane W-N. Cytotoxic effects of free fatty acids on phytoplankton algae and cyanobacteria. Aquatic Toxicology. 2006;**80**(4):338-345. DOI: 10.1016/j.aquatox.2006.09.011

[60] Zuo S, Zhou S, Ye L, Ding Y, Jiang X. Antialgal effects of five individual allelochemicals and their mixtures in low level pollution conditions. Environmental Science and Pollution Research. 2016;**23**:15703-15711. DOI: 10.1007/s11356-016-6770-6

[61] Gniazdowska A, Bogatek R. Allelopathic interactions between plants. Multi site action of allelochemicals. Acta Physiologiae Plantarum. 2005;**27**:395-407. DOI: 2060/10.1007/s11738-005-0017-3

[62] Park MH, Han MS, Ahn CY, Kim HS, Yoon BD, Oh HM. Growth inhibition of bloom-forming cyanobacterium *Microcystis aeruginosa* by rice straw extract. Letters in Applied Microbiology. 2006;**43**:307-312

[63] Zhang TT, He M, Wu AP, Nie LW. Allelopathic effects of submerged macrophyte Chara vulgaris on toxic *Microcystis aeruginosa*. Allelopathy Journal. 2009;**23**:391-402

[64] Zhang TT, Zheng CY, Hu W, Xu WW, Wang HF. The allelopathy and allelopathic mechanism of phenolic acids on toxic *Microcystis aeruginosa*. Journal of Applied Phycology. 2010;**22**:71-77

[65] Dmitriev VV. Diagnostics and Modeling of Aquatic Ecosystems. Saint Petersburg: SPbSU; 1995. p. 215

[66] Matthijs HCP, Jančula D, Visser PM, Maršálek B. Existing and emerging cyanocidal compounds: New perspectives for cyanobacterial bloom mitigation. Aquatic Ecology. 2016;**50**:443-460. DOI: 10.1007/s10452-016-9577-0

[67] Jancĉula D, Gregorová J, Maršálek B. Algicidal and cyanocidal effects of selected isoquinoline alkaloids. Aquaculture Research. 2010;**41**:598-601. DOI: 10.1111/j.1365-2109.2009.02342.x

[68] Sadhasivam G, Gelber C, Zakin V, Margel S, Shapiro OH. N-halamine derivatized nanoparticles with selective cyanocidal activity: Potential for targeted elimination of harmful cyanobacterial blooms. Environmental Science & Technology. 2019;**53**(15):9160-9170. DOI: 10.1021/acs.est.9b01368

Chapter 5

Fluorescence Behavior of Phytoplankton Blooms by Time-Correlated Single-Photon Counting (TCSPC)

Helena C. Vasconcelos, Joao A. Lopes, Maria João Pereira and Afonso Silva Pinto

Abstract

Many aquatic ecosystems, such as lagoons or lakes, are increasingly vulnerable to climate changes and human pressure. The environmental and economic costs of anthropogenic eutrophication are high as well as the applied methods to counteract eutrophication This chapter analyzes the variation in abundance (and biomass) of several phytoplankton families in one of the most well-known volcanic lagoons in the Archipelago of the Azores-Portugal (Furnas Lagoon) and the dynamic correlation between groups of different types of algae that have been established seasonally between 2003 and 2018. For that purpose, the principal component analysis (PCA) technique was used in data series on biomass and abundance of phytoplankton and chlorophyll a, in the time interval considered. The application of PCA techniques in natural phytoplankton populations offers the possibility of making rapid qualitative diagnoses of the trophic state in natural lakes. On the other hand, the fluorescence properties of phytoplankton microorganisms are strongly affected by the physico-chemical properties of natural waters. The fluorescence emission and the lifetime of the different water samples were obtained by photon counting with time correlation (TCSPC), allowing to establish the fluorescent signature of these phytoplankton groups under certain conditions.

Keywords: Furnas Lake, Azores, phytoplankton, component principal analyses, fluorescence excitation, fluorescence lifetime, TCSPC

1. Introduction

The Azores archipelago is located in the North Atlantic Ocean (**Figure 1a**), limited by the parallels 36°55′43″ and 39°43′02″ N and by the meridians 24°46′15″ and 31°16′02″ W. The volcanic genesis of this archipelago explains the origin of many of its lakes and lagoons from volcanoes craters filled with rainwater. Although several of these bodies of water are now protected areas of high environmental value (e.g., Fogo, Sete Cidades, or Furnas Lake), some are undergoing trough an accelerated process of eutrophication (over enrichment with minerals and nutrients inducing an excessively growth of algae). The eutrophication of Furnas Lake (**Figure 1**), due to human agricultural activities, has been known since the 1980s and frequently

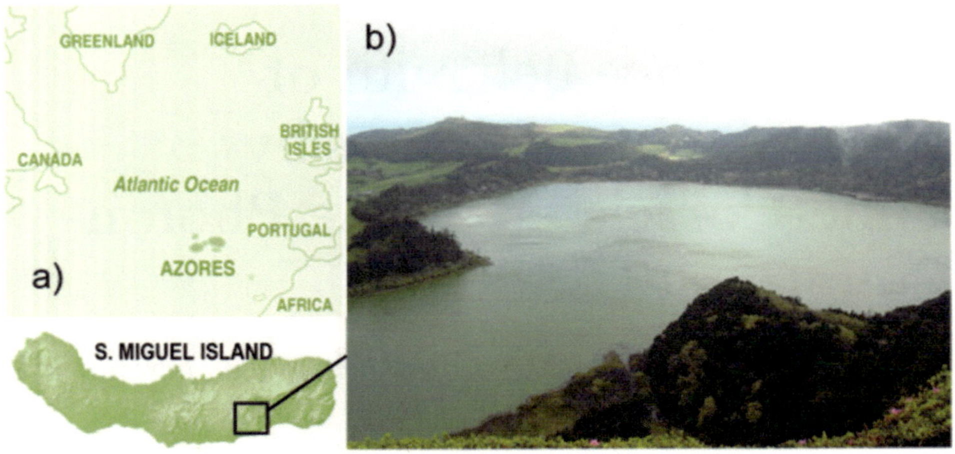

Figure 1.
(a) Azores archipelago localization, adapted from https://sunflowerbooks.co.uk/product/walking-in-the-azores/; (b) Furnas Lake, located in the eastern region of S. Miguel Island.

monitored afterward. The presence and extent of agricultural and forest areas in its basin allowed the drainage of nutrients to Furnas Lake causing algal blooms [1]. Eutrophication due to human activities is considered water pollution, and the phytoplankton community is currently used as a bio-indicator reflecting the lakes ecological state. The growth of phytoplankton often causes increased turbidity of the water, making difficult for sunlight to reach submerged plants. The disappearance of underwater vegetation implies the loss of food, habitats, and dissolved oxygen in water, distressing the trophic state of the ecosystem. Some types of algae can also produce toxins that are harmful vertebrates. Therefore, algae blooms disrupt the normal ecosystem functioning and causing problems along the food chain. The phytoplankton indicators highlight the vulnerability of these ecosystems to human activities since phytoplankton communities present high sensitivity to small changes in environmental conditions. The dynamics of phytoplankton biomass and composition over time reflects changes in environmental conditions. The aquatic ecosystems' degradation is associated with the increase of phytoplankton biomass—the "*blooms*"—a term used to mean a population concentration greater than 20×10^3 ml^{-1} [2]. This phenomenon, in particular in Furnas Lake, has occurred more frequently in the last years due to increase of human pressure [3].

Phytoplankton corresponds to the autotrophic organisms of the plankton community that are suspended in the water column. Phytoplankton includes a set of microalgae and cyanobacteria that have different requirements and responses to environmental variations. They have several photosynthetic pigments, such as chlorophylls (*a*, *b*, *c*, *d* and *f*), carotenoids (carotenes and xanthophylls), and phycobiliproteins. Chlorophyll *a* is, however, the key photosynthetic component, that is necessary in order for chloroplasts to convert light energy into chemical energy. The remaining chlorophylls, carotenoids, and phycobiliproteins (phycoerythrin and phycocyanin) act as accessory pigments in photosynthesis, expanding the range of light wavelengths that can be used in photosynthesis (e.g., chlorophylls *d* and *f*) [4] or protecting the photosystems against excessive light (e.g., carotenoids) [5]. Therefore, one can take advantage of the spectral optical properties inherent to the photosynthetic pigments of phytoplankton species, such as absorption and fluorescence, to obtain information about their characteristics by "spectral sets," that can be used to estimate, for instance, the phytoplankton biomass or composition in lakes, reservoirs, or aquaculture [6, 7]. With the exception of chlorophylls *d* and *f* that absorb in far-red light [4], the chlorophylls absorb wavelengths in the visible blue and red spectrum, but not in the green spectrum. Chlorophyll makes plants and algae appear green because it reflects the green wavelengths, while absorbing all other colors. The different forms (*a*, *b*, and *c*) each reflects slightly different ranges of green wavelengths. The remaining pigments absorb different wavelengths (**Figure 2**) and reflect accordingly, but they all act as accessory pigments for chlorophyll *a* in photosynthesis. Carotenoids, for instance, absorb in the blue/green wavelength region and reflect yellow, red, and orange wavelengths.

Each type of pigment or sets of accessory pigments can be identified by the specific wavelength pattern they absorb from visible light [8, 9]. This allows discriminate optical signals between species of phytoplankton. In non-stressful conditions, when accessory pigments absorb photons, they become excited and transfer their absorbed energy between them, establishing a sequence of biophysical events until

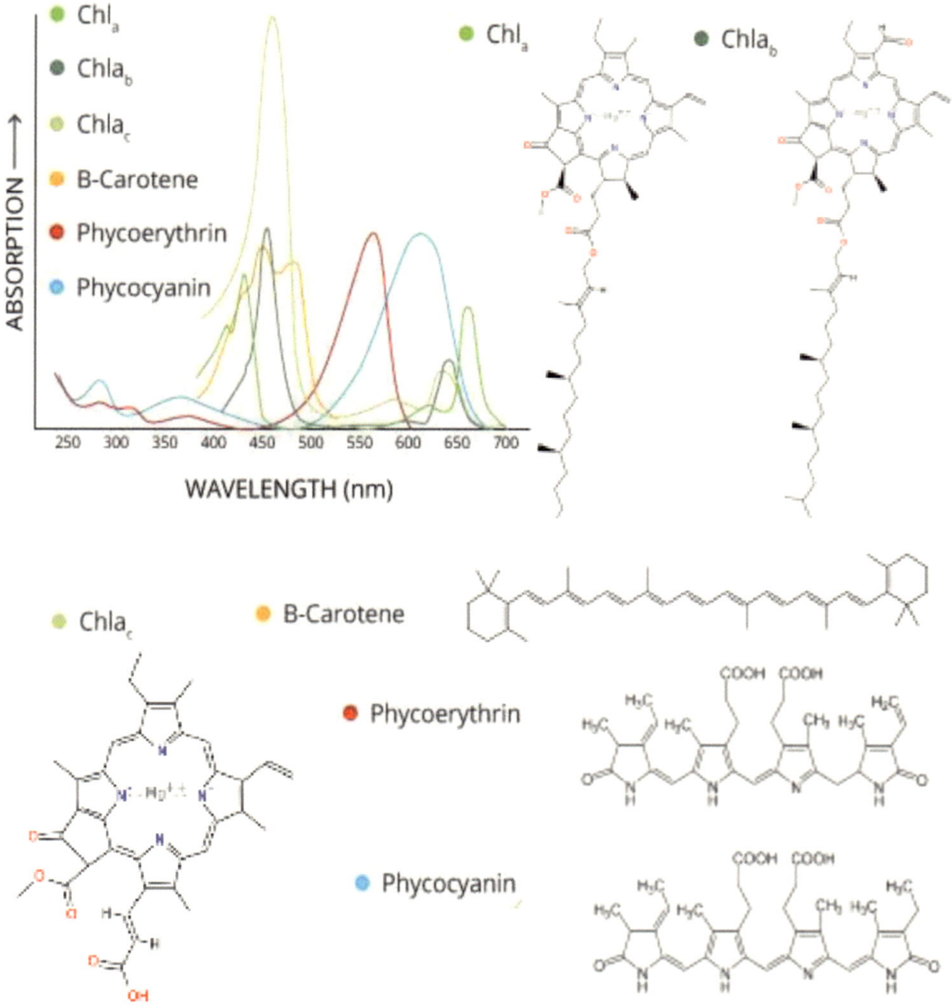

Figure 2.
Optical absorption spectra of photosynthetic pigments. Adapted from https://www.fondriest.com/environmental-measurements/parameters/water-quality/algae-phytoplankton-chlorophyll/.

reaching the chlorophyll a pigment (terminal of the excitation transfer channel), which emit fluorescence [10]. Chlorophylls will cause two main peaks in absorption spectra, a primary peak at blue wavelength (~440 nm) and a secondary peak in red part of the spectra (~675 nm) (**Figure 3**). The existence of other pigments (dependent on species and taxa) will cause the broadening of blue peak and appearance of additional absorption maxima. These taxa absorption peaks have been used as an optical spectroscopy tool in situ [11].

All photosynthetic plants, algae, and cyanobacteria contain chlorophyll *a*, while only green plants and green algae phytoplankton contain chlorophyll b, along with a few types of cyanobacteria in fresh water. The predominant forms of phytoplankton are diatoms, golden brown algae, green algae, blue green algae, and dinoflagellates. The taxonomy of phytoplankton is complex, and there are differences in classification systems. However, the classification currently followed in the study of phytoplankton in the Azores lakes comprises seven groups of phytoplankton *taxa* [14]: *Cyanophyta* (blue-green algae or cyanobacteria), *Chlorophyta* (chlorophytes or green algae), *Euglenophyta* (euglenoids), *Dinophyta* (dinoflagellates), *Cryptophyta* (cryptomonads), *Chrysophyta* (golden algae), and *Bacillariophyta* (diatoms). The blue-green algae, despite not having chloroplasts, have photosynthetic pigments, chlorophyll a and phycobiliproteins, which give them a predominantly blue color.

Eutrophication was recognized as a water pollution problem in European and North American lakes and reservoirs in the mid-twentieth century [15]. Since then, many efforts have been done to develop and apply appropriate mitigation methods and strategies [16]. The most usual procedure to study phytoplankton in water samples involves microscopic observation of the samples collected, thereby generating variable data that need to be careful processed (taxonomic technique).

A spectroscopic technique proposed in the 1980s [16] uses the excitation and emission characteristics associated with the fluorescence of the various photosynthetic pigments, to produce the fluorescence spectral signatures of specific water bodies. This allows the monitorization and detection of possible changes, due to environmental characteristics variation that can have impact on the relative

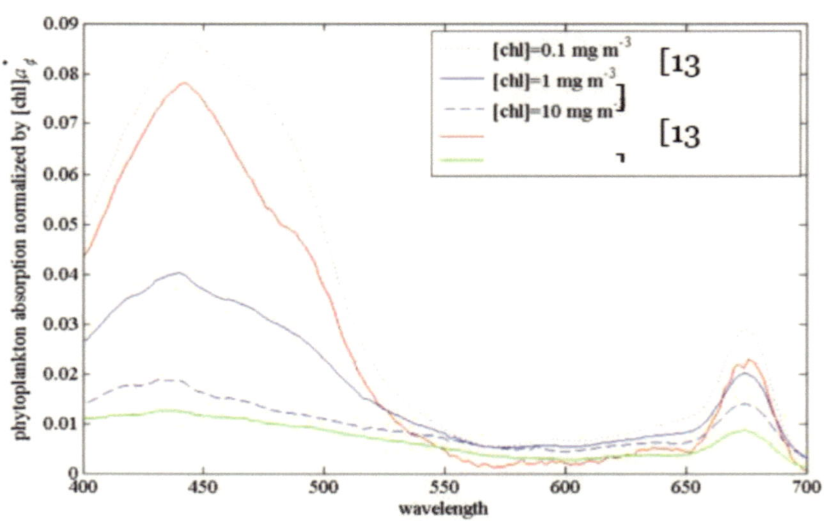

Figure 3.
Chlorophyll normalized phytoplankton absorption [12, 13]. Adapted from https://oceanopticsbook.info/index.php/view/optical-constituents-of-the-ocean/phytoplankton.

growth of phytoplankton populations. Since chlorophyll *a* and various pigments exhibit fluorescence in vivo and the qualitative and quantitative composition of phytoplankton pigments varies between taxonomic groups, its fluorescence properties also vary between taxonomic groups. Therefore, this tool can be used to discriminate changes in different *taxa* abundance [17, 18]. For example, diatoms and dinoflagellates exhibit typically broader excitation spectra than green algae and euglenoids. In cyanobacteria, the emission due to phycoerythrin is superior to the emission due to chlorophyll *a*, which is very much reduced. This main difference in the emission spectra usually makes cryptomonads and cyanobacteria easily separable from other types of phytoplankton, while differences in the excitation spectra can be used to separate green algae from diatoms and dinoflagellates [16]. Previous work has shown the effectiveness of fluorescence measurements in the characterization and discrimination of "spectral sets" of phytoplankton [8].

Differences in the spectral signature of some algal groups which differed in their pigment types are shown in **Figure 4** [19].

The measurements and interpretation of in vivo fluorescence signals from phytoplankton pigments can be affected by several sources, e.g., the algal physiology, the light conditions (quenching of fluorescence at high light intensity), the biomass concentration, and the differences in pigment composition between and within species [20]. Generally, factors such as environmental pollution and excess of nutrients alter the phytoplankton dynamics and consequently the rate of photosynthesis, which is revealed in the amount of fluorescence.

In this work, a rapid qualitative diagnosis of the trophic state of Furnas Lake was carried out. Associations established between the various phytoplankton communities (groups of taxa) over time have been identified using principal component analysis (PCA) [21]. In particular, the period from 2010 to 2011 was evaluated, looking for the most frequent taxa correlations in the different seasons. Additionally, the fluorescence spectra of some of these phytoplankton communities were obtained, as well as the lifetime of the fluorescence decays, using the time-correlated single-photon counting technique (TCSPC) [22].

The aim of this study was to use the potential of TCSPC for the rapid characterization of associated taxa in phytoplankton communities in vivo: (1) reveal the spectral signature and excitation/emission fluorescence spectrum associated with these communities and (2) their fluorescence lifetimes.

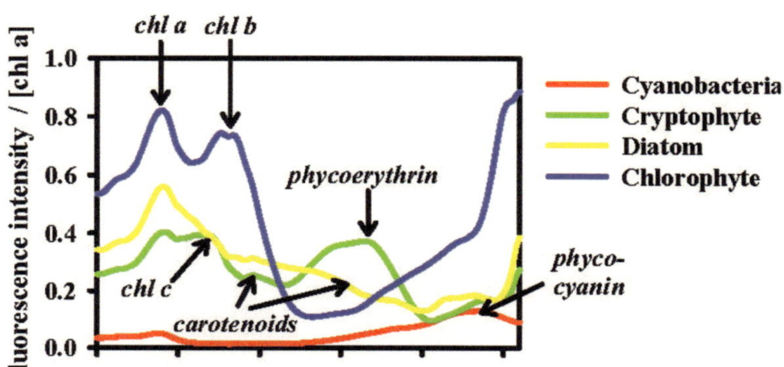

Figure 4.
Excitation spectra of some phytoplankton species measured at the maximum emission of chlorophyll a (682 nm) [19].

2. General overview about spectroscopy and fluorescence

Spectroscopy is the study of the interaction of electromagnetic radiation with matter. The natural world is filled with several colors that derive from the interaction process between matter and the visible radiation. The origins of spectroscopy date back to the seventeenth century, to the Newton's sunlight prisms work and observations of the multicolored band of light that he called spectrum. Since then, the terms "spectrum" and "spectroscopy" have been generalized in experiments of light scattered and decomposed into its different components.

Towards the end of the 19th century, J. Balmer discovery the spectral series of the hydrogen atom (1885), which allowed to determine the length of the hydrogen spectral lines. It was the discrete nature of atomic spectra that led to the formulation of some of the fundamental principles of quantum mechanics and, subsequently, their ability to explain spectroscopic observations.

The interaction of radiation with the matter can be manifested through absorption, emission, or scattering phenomena and be evaluated by spectroscopic methods. These are framed in the physical techniques of structural characterization of materials and are typically based on the absorption or emission of radiation with certain wavelengths (and photon energies), originating transitions between quantized levels. The results (optical spectra) are a representation of intensities as a function of wavelengths. Spectroscopy techniques are therefore methods that use radiation to analyze material properties and can be classified according to the type of transition they produce (**Table 1**). The response of matter to changes of incident radiation that falls within the range of 200 nm (UV) to 3000 nm (NIR) is generally evaluated by optical spectroscopy (UV-VIS-NIR regions). Photons emitted or absorbed when such changes occur are responsible for a large number of physical and chemical properties of matter. Electromagnetic radiation, commonly called light, is represented by electromagnetic waves characterized by a frequency (ν) and a wavelength (λ), related by the equation $\lambda\nu = c$, where c is the speed of light for a given medium. However, light also has particle properties (photons) with well-defined energy $E = h\nu$, where h is the Planck constant. Therefore, the light of wavelength λ is made up of photons with energy $E = hc/\lambda$.

Many of the physical-chemical processes that take place at the microscopic level are only explained on the basis of quantum mechanics. In particular, the concepts of quantified states of matter (Planck, 1900) and the nature of light (Einstein, 1905) describe how atoms interact with light and how electrons and energy levels within atoms are responsible for the absorption and light-emitting properties.

Transitions types	Techniques
Nuclear	Mössbauer spectroscopy
Electronic	Ultraviolet-Visible Spectroscopy (UV-Vis-NIR), ultraviolet photoelectron spectroscopy (UPS), and X-ray photoelectron spectroscopy (XPS)
Vibrational	Fourier-transform infrared spectroscopy (FTIR) and Raman
Rotational	Microwave rotational spectroscopy
Magnetic	Nuclear magnetic resonance (NMR) and electron paramagnetic resonance (EPR)

Table 1.
Spectroscopy techniques.

The main energy levels can only assume discrete values and is allowed only specific spatial distributions of electrons (called electronic energy levels); these levels are broken down into vibrational levels, which indicate the possible modes of vibration of the molecules. The first energy level (closest to the nucleus) is called the fundamental state (or ground state), and in it, the energies (electronic and vibrational) exhibit the minimum values. The other energy states of the molecule are called excited states.

It is easily to shown experimentally that a beam of electromagnetic radiation undergoes attenuation (intensity change, dI) when passes through an infinitely small part of matter (dx), following the Beer-Lambert law, which relates the absorption of light with the properties of the material crossed:

$$dI = -\alpha I dx \tag{1}$$

where I represents the light intensity and x the beam displacement.

$$\frac{dI}{I} = -\alpha dx \rightarrow \int_{I_o}^{I} \frac{dI}{I} = -\int_{o}^{x} dx \tag{2}$$

$$\ln \frac{I}{I_o} = -\alpha x \rightarrow I(x) = I_o e^{-\alpha x} \tag{3}$$

where I_0 is the radiation intensity and α is the absorption coefficient (cm^{-1}), which is a characteristic of each material. The transmittance (T) of a sample with a thickness d is defined as:

$$T = \frac{I}{I_o} = e^{-\alpha d} \tag{4}$$

The absorption cross section σ (cm^2) is defined as follows:

$$\sigma = \frac{\alpha}{N} \tag{5}$$

where N represents the concentration of absorbent centers (per cm^3), and σ is related to the capacity of the material to absorb the radiation at a specific wavelength λ.

Physically the photons absorbed by the electronic structure raise the energy of the molecule's electrons to quantum states less stable (excited).

2.1 Fluorescence emission

The luminescence can be view like the reverse of the absorption process. If the material absorbs radiation and, consequently, is excited, it can then return to its ground state by spontaneous emission of light; this process of de-excitation is known as luminescence. The phenomenon is called fluorescence if this emission of light (emission of a photon) occurs quickly from a singlet excited state, in which the excited electron does not change the spin orientation and remains unpaired, which allows its return to the fundamental state. The typical fluorescence lifetime is in the order of 10^{-9} s, which corresponds to a fluorescence emission rate ~10^{8} s^{-1} [23, 24]. A fluorescence emission spectrum is a curve of fluorescence intensity (I) versus wavelength (λ).

Basically, a photon of appropriate energy is able to excite an electronic transition in which an electron from the singlet state S_0 passes to a higher energy level of S_1 (**Figure 5a**). This excitation (absorption of a photon of light) occurs instantly (femtoseconds). Once the material is excited at a wavelength where it absorbs, the system can return to its ground state. Fluorescence is a radiative mechanism by which excited electrons can relax in a fast process (nanoseconds), by emitting light from the lowest excited state (S1) to the ground state (S0). The energy difference between the two states is dissipated by emitting a photon.

An interesting detail is that the emitted light has a longer wavelength than that of absorption radiation (excitation) (effect named Stokes Shift), as shown in **Figure 5b**. This is due mostly because the energy loss during the electronic excitation is also due to transition of electrons to lower vibrational levels, caused by release of thermal energy due to internal conversions, named as non-radiative decay. The fluorescence is therefore independent of the emission wavelength from the excitation wavelength (Kasha's rule), i.e., the fluorescence emission will always originate from the vibrational ground state of the lowest excited singlet level, S1. Even if the electron is excited to higher vibrational levels of S1, it mainly undergoes relaxation from S1 to S0. **Figure 5c** shows a simplified Jablonski diagram, which illustrates the energy states of a molecule and the transitions between them. Transitions between states are represented by vertical lines in order to show the instantaneous nature of light absorption. The times for the occurrence of the transitions are in fact very short, in the order of 10^{-15} s, which does not allow displacements of the nucleus (mass greater than that of electron), a fact known as the Franck-Condon principle [23]. To achieve stability, the molecule dissipates the

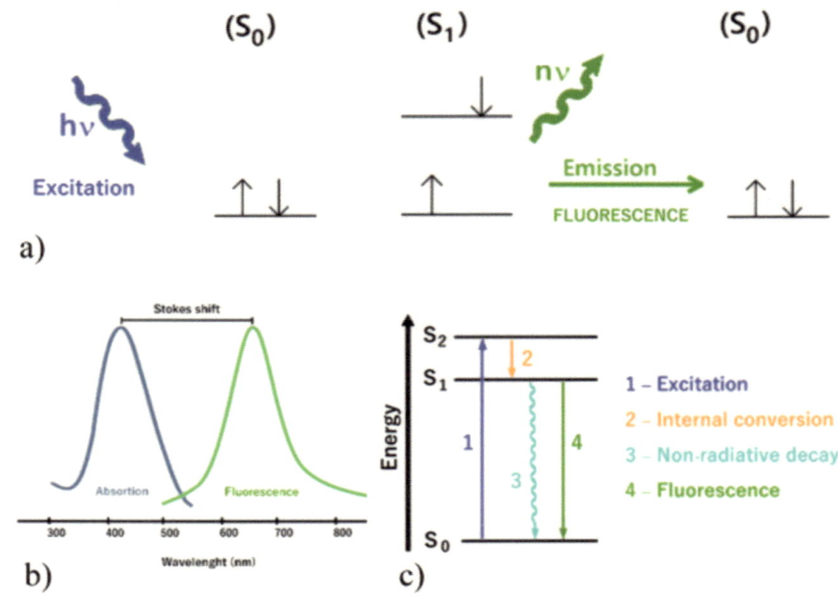

Figure 5.
(a) Excitation and fluorescence emission after absorption of light; (b) Stokes shift; and (c) Jablonski diagram (simplified).

energy absorbed by radiative and/or non-radiative decay processes (**Figure 1c**). The radiative decay processes involve the emission of photons, and this emission can occur by fluorescence in a time range of the order of nanoseconds of lifetime τ, typically less than 10^{-8} s [23]. The non-radiative processes occur by internal conversion, due to vibrations and rotations of the molecule or, in some cases, by energy transfer to neighboring molecules through collisions.

Transitions between states are represented by vertical lines in order to show the instantaneous nature of light absorption. The times for the occurrence of the transitions are in fact very short, in the order of 10^{-15} s, which does not allow displacements of the nucleus (mass greater than that of electron), a fact known as the Franck-Condon principle. At room temperature, thermal energy (E ~ KT) is not sufficient to cause a significant population of excited vibrational states. Photon absorption occurs in functional groups of molecules with lower vibrational energy, and the fluorophore is excited to some higher vibrational level of its S1 singlet states.

2.2 Fluorescence lifetime and quantum yield

The fluorescence lifetime of a substance generally represents an average value of time that a molecule remains in the excited state before returning to the ground state. The characteristics of the fluorescent decay can reveal details about the interactions of fluorophores with its neighborhood. The fluorescence decay time is sensitive and can be affected by local molecular surroundings, which may cause changes in decay time [23, 25–27].

This technique reveals the change of fluorescence intensity as a function of time after the establishment of the excited state. It displays the evolution of the excited state reactions and the mechanism by which they return back to the ground state. Measurements of fluorescent lifetime are difficult as these times are typically in the nanosecond range, making it necessary to use high-speed electronic devices and suitable detectors. The excitation of a fluorophore with an ultra-short pulse of light, at time $t = 0$, results in an initial population (N_0) of fluorophores in the excited S1 singlet state through the absorption of photons. The excited molecules can decay to ground state by radiative or non-radiative processes giving a signal of intensity decay. The rate of change of the population of excited states, $N(t)$, after instantaneous pulse excitation, follows the equation:

$$\frac{dN(t)}{dt} = -W_T N(t) \tag{6}$$

where the total decay rate constant, W_T, is the sum of the rate constants for all radiative (R) and non-radiative (NR) decay processes ($W_T = W_R + W_{NR}$). Following integration of Eq. (6), the concentration of the excited states $N(t)$ is given by the following equation:

$$N(t) = N_o e^{-W_T t} \tag{7}$$

where N_0 is the density of excited electrons at t = 0 just after the pulse excitation be absorbed. The de-excitation process can be observed experimentally by analyzing the lifetime decay of the emitted light (fluorescence decay) (**Figure 6**). The parameter monitored during fluorescence lifetime experiments is the fluorescence

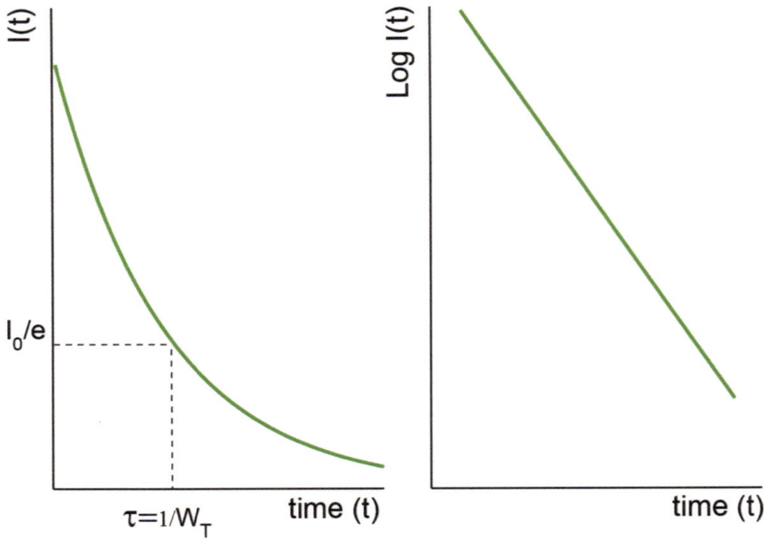

Figure 6.
Scheme of a fluorescence decay curve using pulse method. Adapted from Ref. [23].

intensity, $I(t)$, which is the rate of the emission of photons and is related to the excited state concentration by the following equation:

$$I(t) = W_R N(t) = I_o e^{-W_T t} \tag{8}$$

where $I_0 = W_R N_0$ is the intensity at $t = 0$.

Therefore, the intensity decays exponentially after the initial excitation pulse. The fluorescence lifetime of the excited state, τ, can be represented as $\tau = (1/W_T)$. The fluorescence lifetime represents the time in which the emitted intensity decays to 1/e of that initially excited and gives us the total rate of decay (radiative + non-radiative); therefore, τ is given by the following equation:

$$\frac{1}{\tau} = \frac{1}{\tau_o} + W_{NR} \text{ and } \frac{1}{\tau_o} = W_R \tag{9}$$

where τ_0 is the radiative lifetime. Fluorescence quantum yield, η, can be obtained through the Eq. (10), in terms of τ_0 and τ, as follows:

$$\eta = \frac{W_R}{W_R + W_{NR}} = \frac{\tau}{\tau_o} \tag{10}$$

Briefly, η is defined as the ratio between the radiative decay rate and the total decay rate. The total decay rate can be obtained experimentally as the inverse of the lifetime decay of the emitted light (obtained in the laboratory), while the rate of radiative decay, usually more difficult to obtain, is the inverse of radiative lifetime. The lifetime can be influenced by quenchers, i.e., the process that leads to a decrease in fluorescence (quenching is a form of internal conversion) [23].

Quantum yield is a dimensionless gauge to assess the efficiency of fluorescence emission relative to all of the possible ways for relaxation and is expressed as the ratio of photons emitted to the number of photons absorbed. This means that the quantum yield corresponds to the probability that an excited molecule will produce an emitted photon (fluorescence). Quantum yields range between a value of 0 and 1.

The emission spectrum can be obtained at different times after the absorption of the excitation pulse; this experimental procedure is called luminescence resolved in time. The basic idea of this technique is to obtain the emission spectrum with a certain delay t, with respect to the excitation pulse within a time window Δt.

The lifetime can also be considered as the average value of the time that a fluorophore remains in the excited state. However, for very complex decay laws, such as multi-exponential or non-exponential decays, the adjustment is made by a sum of exponentials of the type [23]:

$$I(t) = \sum_i B_i e^{-\frac{t}{\tau_i}} \tag{11}$$

where τ_i is the lifetime of component i of the decay and B_i is the corresponding pre-exponential factor. Thus, the average lifetime is given by the following equation:

$$\tau_{av} = \frac{\sum_i B_i \tau_i^2}{\sum_i B_i \tau_i} \tag{12}$$

3. Methods

3.1 Study area

The Azores are an Archipelago of nine islands in the North Atlantic Ocean, from 37° to 40 N latitude and 25 to 31°W 125 longitude. Its location coincides with the confluence of three tectonic plates which makes this archipelago a complex geodynamic system with a volcanic genesis [28].

The Furnas Lake is located on the island of São Miguel in a volcanic crater. The lake has an area of 1.87 km^2 and a maximum length and width equal to 2025 and 1600 m, respectively [29]. Agriculture and fertilized pastures are the most common activities in its hydrographic basin, contributing to the lake's nutrient load and causing increasing eutrophication. Over the past 20 years, a local meteorological station recorded average air temperatures in this region ranging from 11°C in winter to 20°C in summer (data provided by Azores Regional Environmental Office). The climatic characteristics in the region favor the occurrence of severe rains, and as a result, large amounts of nutrients are easily transported to the lake. Because the temperature in the S. Miguel island is adequate for phytoplankton growth throughout the year and the nutritional conditions favor cyanobacteria development [2], the lake has exhibited several episodes of algal blooms over time [30]. Although the lake is part of a protected landscape area, the economic activities that contribute to local development often override environmental authorities, and the use of fertilization around the lake has facilitated the entry of nutrients. In 1988/89, the lake was classified as eutrophic (average Carlson's TSI for chlorophyll *a* was 66) [30]. Carlson's Trophic State Index (TSI) is a common method for characterizing a lake's trophic state. This method uses Secchi disc transparency, chlorophyll *a*, and phosphorus measurements.

Previously studies showed that the phytoplankton populations in Furnas Lake exhibited a high abundance divided into two main groups, diatoms and cyanobacteria. In spring, there is a very high relative abundance of Bacillariophyta (diatoms). In summer and autumn, there is a presence of persistent blooms of cyanobacteria, indicators of reduced biodiversity in the lake [3]. The bloom-forming cyanobacteria dominate the phytoplankton of the eutrophic Furnas Lake, as expected during the summer thermal stratification period [31].

In order to mitigate this problem, the Azores government has initiated several actions to counteract the Furnas Lake eutrophication [32–35].

3.2 Sampling and available data

The sampling collection took place between January and December 2011. **Table 2** exhibits the sampling conditions, and **Table 3** summarizes the labeling of samples and date of collection.

The used biological data [chlorophyll a and phaeopigments (μg/L), abundance (cell/L), and phytoplankton biomass (10^{-9} mg/L)] were obtained from the Azores Regional Government—Water Monitoring website [36] in the periods 2003–2018. The selected phytoplankton data totaled seven phyla (**Table 4**) and corresponded to the analysis of samples collected according to **Table 2**. For the assessment of trophic status, information on chlorophyll *a* concentration (an indicator of phytoplanktonic biomass) and phaeopigments were used.

The Matlab program, version R2016b (MathWorks, Natick, MA) was used to perform PLS Toolbox version 8.2.1 for Matlab (Eigenvector Research, Manson, WA) for the PCA analysis.

3.3 Time-correlated single-photon counting (TCSPC)

Fluorescence techniques are methods that exploit the phenomenon of fluorescence. They have been widely used in the characterization, in vivo, of phytoplankton communities of diverse aquatic ecosystems [6, 7]. These methods allow estimate the concentration of chlorophyll a (or other pigments) and the electron transfer rates during photosynthesis. The pigments whose function is to collect light and transfer the absorbed energy to the reaction centers (antenna pigments)

Sampling type	Sampling equipment	Sampling technique	Laboratory method
Collection of water samples	"Van Dorn" water collection bottle	Each water sample was collected at various depth levels (surface, middle, and bottom) and the ultimate sample consisting of homogenization of identical volumes of water collected at each of the sampling depths	Phytoplankton samples were kept fresh and in the dark until analysis, which started in 24–48 h after collection, to minimize pigment degradation.

Table 2.
Sampling conditions.

Sample labeling	Date of collection
A1	2011, later spring (June)
A2	2011, winter (February)
A3	2011, early autumn (December)
A4	2011, spring (May)
A5	2011, summer (August)

Table 3.
Labeling of samples and date of collection.

Scope	Descriptors (variables)
Biological (phytoplankton phyla)	Cyanophyta (aka. Cyanobacteria), Chlorophyta, Euglenophyta, Dinophyta, Chrysophyta, Cryptophyta, and Bacillariophyta
Biological (trophic assessment)	Chlorophyll *a* and phaeopigments

Table 4.
Data groups and variables in each group.

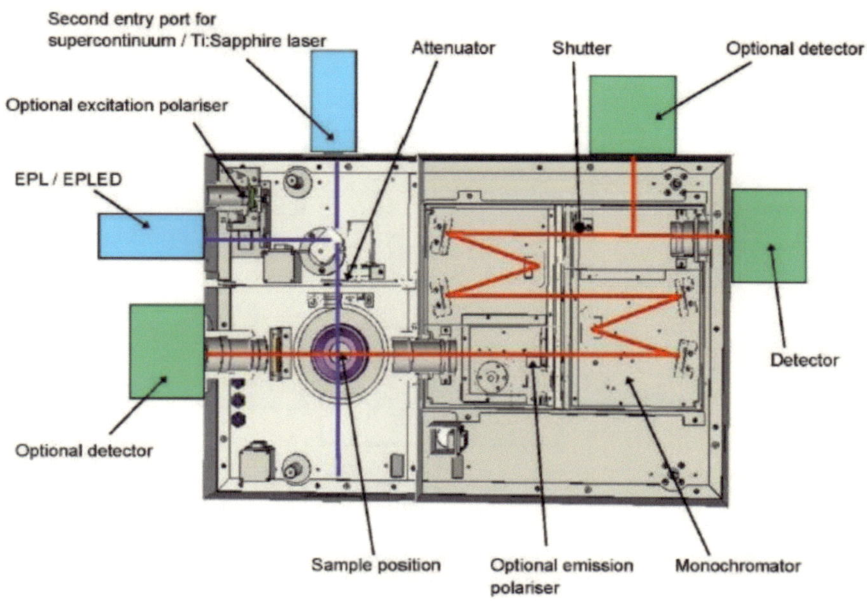

Figure 7.
Layout of the Spectrofluorometer LifeSpec II [37] (optional and main components).

vary between the phytoplankton taxa. Therefore, the characteristics of the phytoplankton in vivo fluorescence, including its fluorescence efficiency, depend on the taxonomy. Chlorophyll *a* fluorescence reflects the concentration of this endogenous pigment.

Fluorescence is a phenomenon that competes with the use of photons absorbed by the antenna pigments in photosynthesis. However, if the photon capture mechanism is necessary to perform photosynthesis changes, increasing its efficiency or decreasing it, the fluorescence intensity $[I(t)]$ will also be changed. Thereby, the main goal of measuring fluorescence decay profiles is to determine the $I(t)$ function. Typically, the decays observed are set to a sum of exponentials [23].

There are several commercial processes and instruments to induce the fluorescence of phytoplankton photosynthetic pigments in vivo. Since the fluorescence lifetime of the organic molecules is in the ns or ps order, it requires the use of high temporal resolution techniques and making use of excitation pulses to follow the pattern of fluorescent emissions over time. Time-correlated single-photon counting (TCSPC) spectroscopy is one of the most suitable techniques for allowing the measuring of shorter times and multi-exponential decay with great accuracy.

A spectrofluorometer of the type LifeSpec II from Edinburgh Instruments [37] was the instrument used in this work. The measurements were performed with a pulsed EPL-406 nm diode laser with 80 ps of pulse duration as the excitation source. All measurements were performed at room temperature. **Figure** 7 shows the layout of the measuring device.

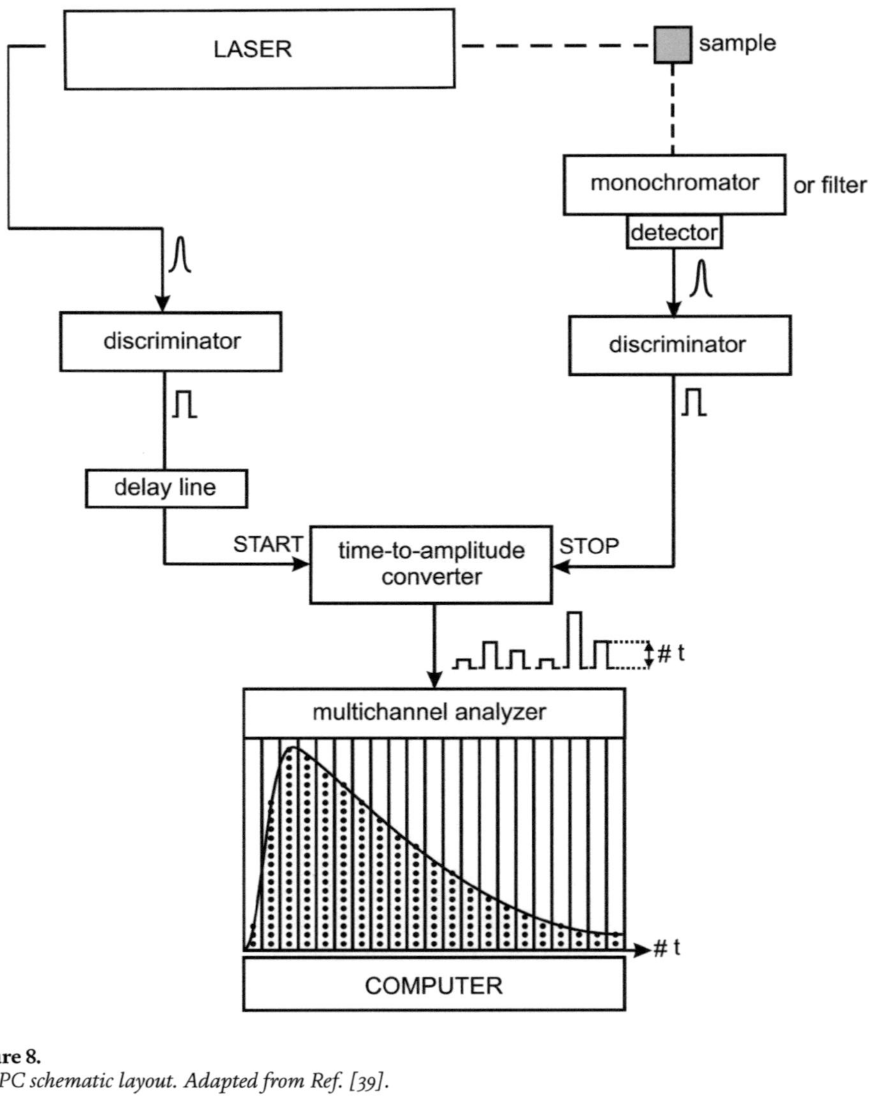

Figure 8.
TCSPC schematic layout. Adapted from Ref. [39].

The principle of TCSPC is the detection of single photons and the measurements of their arrival times with respect to the excitation light pulse [38]. TCSPC is a statistical method, and a repetitive light source is used to accumulate a sufficient number of photon events for a required statistical data precision.

Briefly, TCSPC measures the time delay between two signals, the signal of fluorescence that comes from the sample and the signal of an electrical pulse generator which triggers the laser source. After being triggered by the electrical pulse generator, the excitation laser produces pulsed light with the same time profile as the electrical pulse. These electrical pulses are guided, at the same time, to the laser and to the detector and that time is set as 0. Then, the detector compares the time delay between both signals, that which triggered the laser and that emitted from the sample, and then it sets a count corresponding to the exact time delay of the photon relative to the electrical pulse. TCSPC can be compared to a fast stopwatch with two inputs (**Figure 8**) [39]. The clock is started by the START signal pulse and stopped by the STOP signal pulse. The process is repeated several times and so more and more photons are added on the histogram. The resulting histogram counts versus channels will represent the fluorescence intensity versus time [38].

The term single-photon counting comes from the fact that in the pause between two consecutive excitations (or pulses), only one photon, emitted by the sample, can be detected. When this photon is detected, it added to the previous photons which had the same time delay on the histogram.

The emitted photons were collected at the maximum emission length of the studied samples. Before the sample measurements, the equipment's instrument response function (IRF) was measured. The IRF represents the shortest decay curve that can be measured with that equipment.

4. Results

4.1 Analysis of principal components (PCA)

Phytoplankton populations in the Azores lakes are represented by their most common phyla, namely, Bacillariophyta, Chlorophyte, Cryptophyta, Chrysophyta, Cyanophyta [aka cyanobacteria (Cyanoprokaryota)], Dinophyta (Pyrrophyta), and Euglenophyta [3]. The trophic state and composition of the phytoplankton community can be made through temporal (abundance and/or biomass) records and further analysis of patterns of variation of these communities by statistical techniques. A PCA of the phytoplankton data was performed, and two models were created considering the abundance data (measured in cells/L) and biomass data (measured in 10^{-9} mg/L).

4.1.1 Multivariate analysis of phytoplankton abundance

Figure 9 shows the correlations between the seven abundant phytoplankton phyla in Furnas Lake and identifies algal assemblages showing how these associations correlate with each other in the period 2003–2017.

The graph is analyzed as follows: the variables (descriptors) that are close to each other have a high positive correlation and the variables that are diagonally opposite are negatively correlated. For example, it seems that the Chlorophyta and Chrysophyta phyla appear very close, meaning that there is a high correlation between the abundance of these two phytoplankton phyla. On the fourth quadrant, we can see that Cyanophyta (Cyanoprokaryota) is inversely correlated with Bacillariophyta, while it exhibits more affinity with Pyrrophyta. On the other

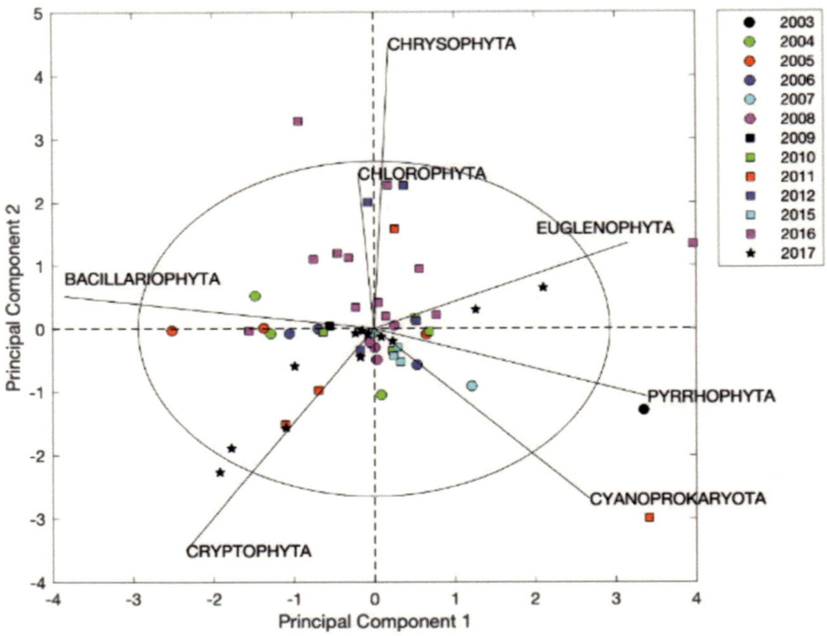

Figure 9.
Multivariate analysis (PCA) of abundance (cells/L) for phytoplankton descriptors (colored by year).

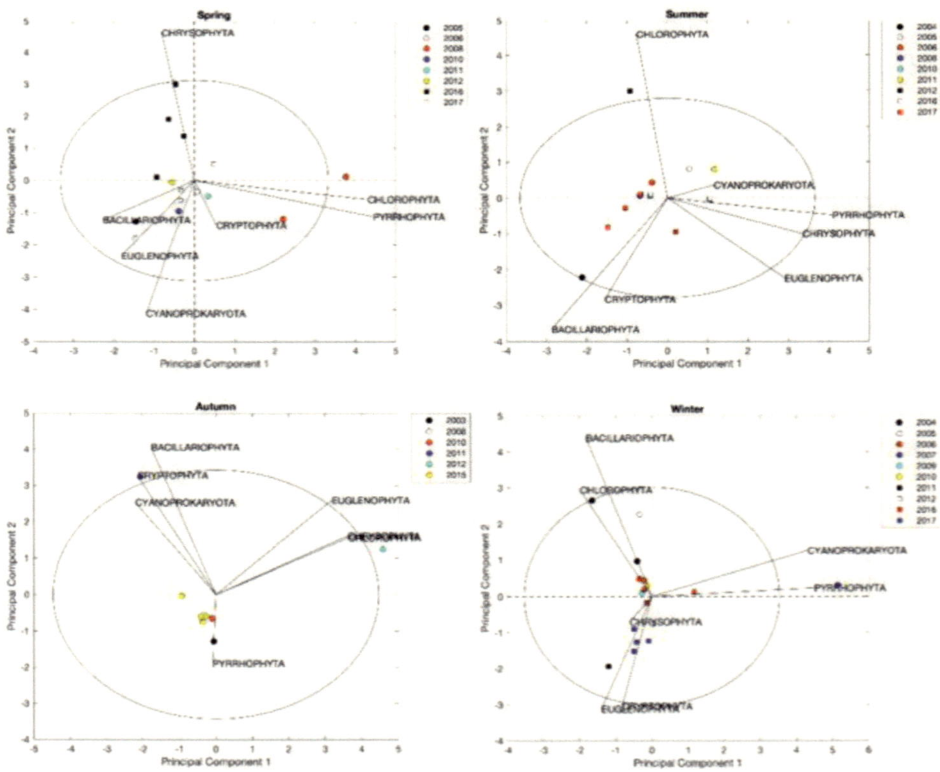

Figure 10.
Multivariate analysis (PCA) of abundance (cells/L) for phytoplankton descriptors, by seasons (colored by year).

hand, although most of the points are found to be located very close to the center without much year variability, some are distant, corresponding to situations where the abundance of phytoplankton phyla increases far beyond the limit of the corresponding circle at the confidence interval. For example, in 2011, it seems that a bloom of Cyanoprokaryota occurred, during which the abundance of these organisms reached maximum values. Records show that blooms have not occurred as pronounced since 2006–2007. Records also show that the abundance of Chlorophyta had a very significant increase in 2016 when it was practically non-existent in previous years. However, this situation can be considered sporadic, since in 2017 there was again a decrease.

Figure 10 exhibits the phytoplankton correlations throughout the seasons, where the high abundance is divided into two main groups, diatoms and cyanobacteria. In spring, there is a very high relative abundance along years of Bacillariophyta (diatoms). In summer, there is a presence of persistent blooms of cyanobacteria, indicators of reduced biodiversity in the lagoon [3]. The bloom-forming cyanobacteria dominate the phytoplankton of the eutrophic Furnas Lake, as expected during the summer thermal stratification period [31]. In particular, in 2011, the record shows that bloom-forming cyanobacteria reached a maximum value. In Furnas Lake, cyanobacteria often dominate the phytoplankton population in summer, early autumn and later spring, whereas during the winter and early spring, they are replaced mainly by Bacillariophyta and Chlorophyta. The data set considers summer from July to September and autumn from October to December.

4.1.2 Multivariate analysis of phytoplankton biomass

For biomass data (**Figure 11**), the analysis is relatively similar to that for abundance data, except that Chlorophyta appears to have a negative correlation with Cyanoprokaryota. Warming temperatures of the lakes since there is also

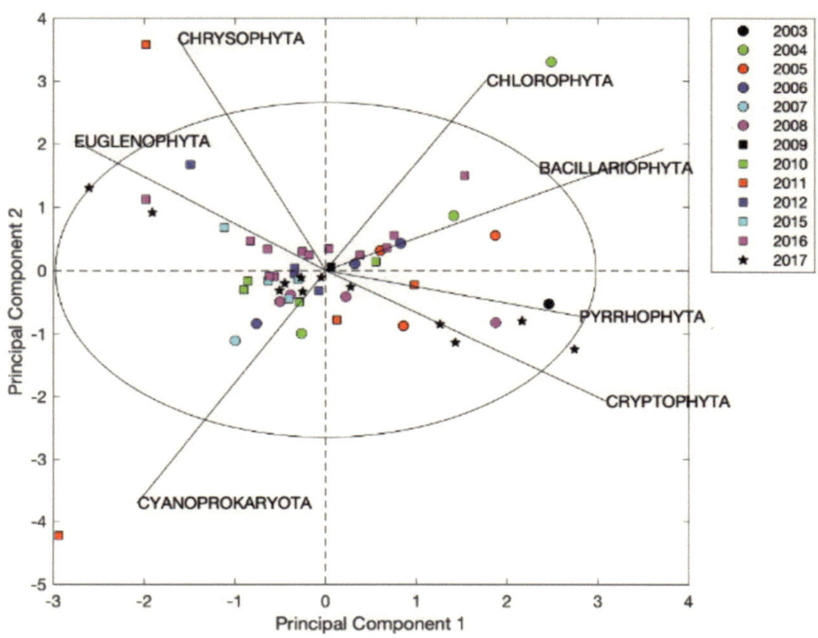

Figure 11.
Multivariate analysis (PCA) of biomass (10^{-9} mg/L) for phytoplankton descriptors (colored by year).

a correlation between the green algae and cyanobacteria with warmer seasons (summer), which leads to limited transport between layers causing a lake stratification.

The analysis of the results, although still preliminary, allows to some extent confirm that the phytoplankton composition in the lake of Furnas reflects its trophic status, which is obviously higher (i.e., of low water quality) when is found a predominance of Cyanoprokaryota, while a lower trophic state (oligotrophic or mesotrophic) seems to be related to the presence of Chlorophyta and also Bacillariophyta. Moreover, Chlorophyta is negatively correlated with Cyanoprokaryota, especially in summer, but correlated positively with Bacillariophyta, suggesting competition between these two groups (**Figure 12**) [40].

Chlorophyta can also be predominant in lakes with high concentrations of phosphorus [41], such as the lake of Furnas, and may change with the dynamics of the eutrophication state [42]. It is therefore probable that competitive interactions between Chlorophyta, Cyanoprokaryota, and Bacillariophyta can occur under nitrogen and phosphorous gradients [40].

Since 2002, Lake Furnas shows a constant state of eutrophication [3], with average annual values of the trophic state index based on Secchi disk between 57 and 70, characteristic of eutrophic media, and average annual values of the trophic state index based on phosphorus total between 45 and 67, characteristic of mesotrophic and eutrophic media. The value 45, obtained from this last index, allowed a mesotrophic classification in 2014. Between 2013 and 2016, the annual values of the TSI corresponding to chlorophyll varied between 54 and 55, characteristic of eutrophication [3].

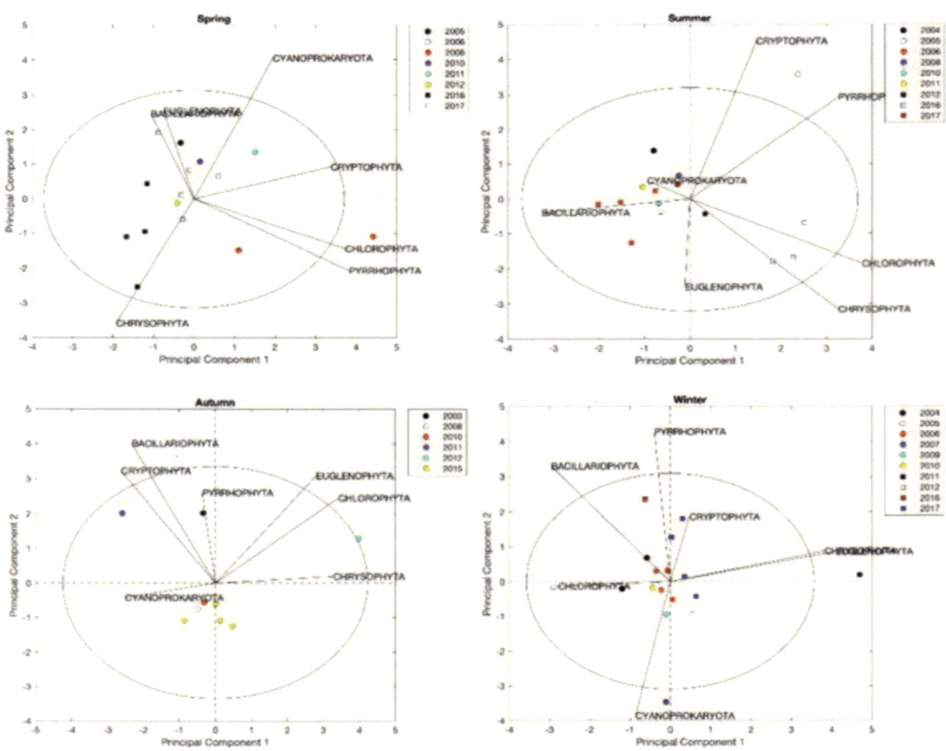

Figure 12.
Multivariate analysis (PCA) of biomass (10^{-9} mg/L) for phytoplankton descriptors, by seasons (colored by year).

4.1.3 Multivariate analysis of chlorophyll a and phaeopigments

Concerning the analysis of chlorophyll *a* and phaeopigments, the results are recorded in the scores of **Figure 13**. The chlorophyll *a* degradation products can often form a significant fraction of phaeopigments, which correspond to byproducts formed by the loss of Mg atom (**Figure 2**), generally under acidic conditions [43]. As phaeopigments may originate from chlorophyll *a*, some correlation between these two descriptors was expected. Taking into account the apparent occurrence of a bloom of Cyanoprokaryota in 2011, a more detailed analysis of these two descriptors was conducted between 2010 and 2011.

The multivariate analysis of the concentrations of chlorophyll a and phaeopigments reveal an expected positive correlation and some patterns that are worth to discuss. This correlation is stable over time. The analysis segregated by year reveals different tendencies of abundance. While in 2010 and 2012, the abundance of both indicators is higher in autumn and winter, especially in winter. This trend was not observed in 2011 where the concentrations were higher in spring and summer. It is noteworthy to say that this analysis was based on a relatively short dataset encompassing approximately three observations per season.

A global analysis to these data considering years from 2003 to 2018 reveal some prevalence patterns (**Figure 14**). While the correlation between the two indicators stays relatively stable over time, it can be observed a high concentration of chlorophyll *a* in water in 2004 and 2009 (eventually one can include here also 2005 and 2006 at least in some periods of the year). Concerning the levels of phaeopigments, there is a clear evidence of the gradual increase starting in 2011. It is clear the increasing occurrence of phaeopigments after 2011 that was still observable in 2018. For some years (e.g., 2016), it was found that simultaneously with a high occurrence of phaeopigments in some periods of the year, there is also an observation of very low amounts in other periods of the year, widening the range of concentration observed in a single year.

The seasonal variation is controlled mainly by solar radiation and temperature, which will determine the stratification of the water column and the availability of nutrients.

4.2 Fluorescence behavior

Phytoplankton organisms have a wide range of photosynthetic pigments that are of great interest in identifying species in freshwater samples representing different phytoplankton communities [11]. Although microscopy is a method widely used to determine the structure of the phytoplankton community, it is an exhausting and time-consuming process. Hence, the great interest is in fluorescence studies of these communities, using the knowledge of the specific excitation bands of the photosynthetic pigments, in order to have access to the probable types of phytoplankton communities. But for this task, it is necessary to know that there are two main regions of excitation of phytoplankton pigments: the blue (400–520 nm) and red (630–700 nm) regions, which can be seen in **Figure 4**. In the green (515–600 nm) region, there are some absorption bands attributed to phycobilins. Fluorescent spectroscopy began to be used extensively in the characterization of phytoplankton taxa from the 1980s, particularly after the development of an “ataxonomic tool” [16]. In vivo fluorescence methods are widely used to characterize phytoplankton communities, but the taxonomic identification of algae taxa is still a challenge. Measurements of the fluorescence intensity of phytoplankton communities are affected by varying concentrations of biomass, and therefore, a variable contribution to the fluorescence signal can be expected in different subpopulations, as well

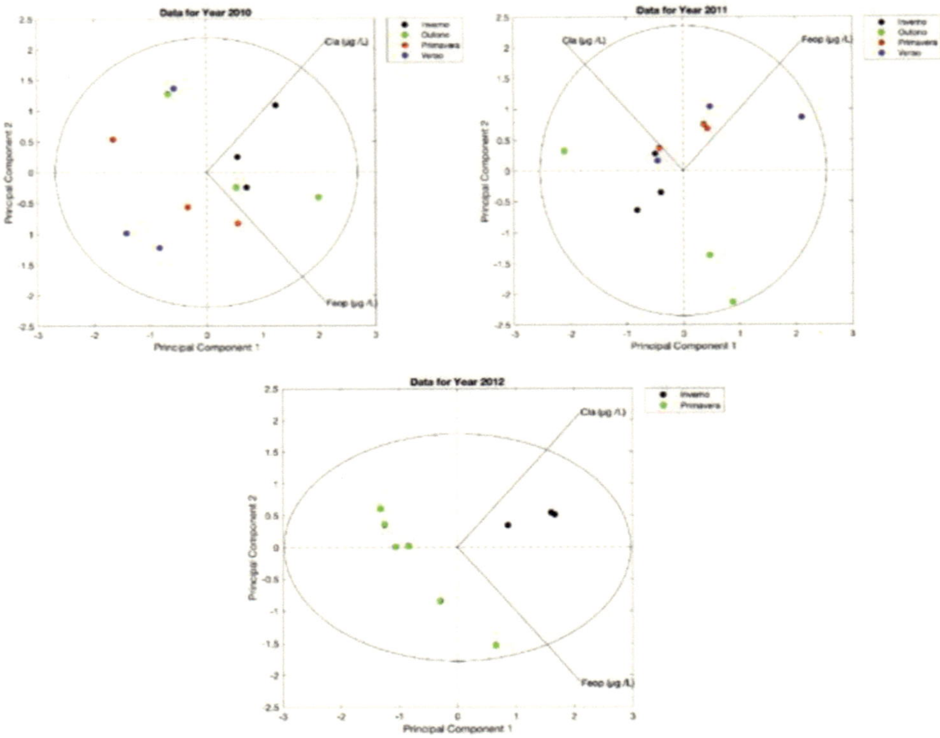

Figure 13.
Multivariate analysis (PCA) of chlorophyll a and phaeopigments (2010–2012, colored according to season).

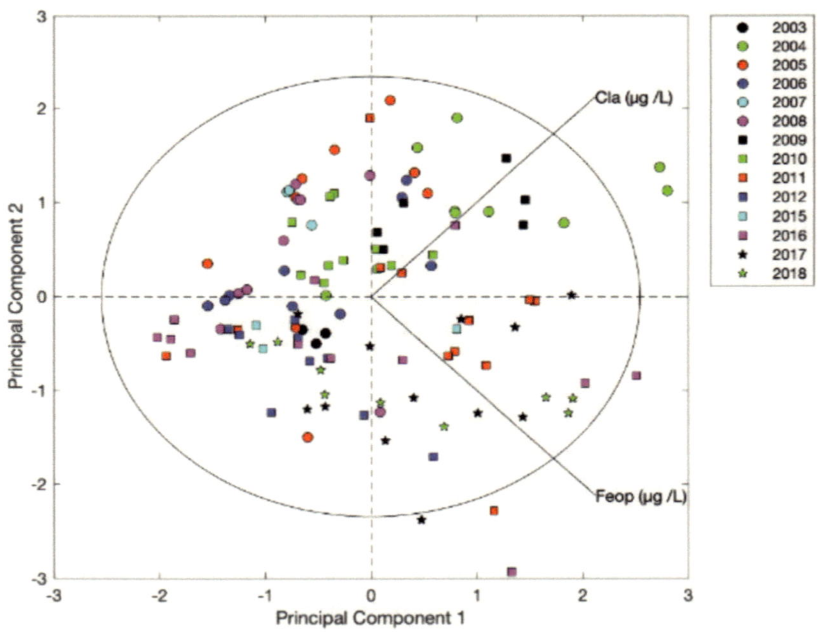

Figure 14.
Multivariate analysis (PCA) of chlorophyll a and phaeopigments (colored according by year).

as in the variability of pigment composition between species. The relationships between fluorescence, chlorophyll concentration, and phytoplankton biomass are functions of light (by extinction) and also functions of the physiological status and composition of the phytoplankton community [44].

Until now, any spectral approach based on spectral data does not allow for a real separation of the phytoplankton taxa but can provide information on specific taxa clusters. The in vivo fluorescence emission of chlorophyll *a* (680 nm) arises from PSII. The light harvesting pigments in PSII are taxa specific, and thus, there are changes in the shapes of excitation spectra of the main taxa phytoplankton [17].

The in vivo excitation peak of chlorophyll *a* at 440 nm and its emission at 680 nm played an important role in the fluorescence spectra of natural phytoplankton, while the excitation wavelengths of phycocyanin in the range 550–650 nm and the emission peak around 645 nm played a major role in the fluorescence spectra of cyanobacteria [45]. The analysis of the shape of the spectrum, amplitude, and spectral position (spectral signature) can be associated with a certain phytoplankton community. The algae divisions, Dinophyta, Bacillariophyta, Chrysophyta, Cyanophyta, Cryptophyta, and Chlorophyta, are characterized by similar fluorescence excitation spectra resulting from the similar composition of their antenna pigments [46].

4.2.1 Fluorescence excitation, emission spectra, and lifetime

For all taxa, light can be absorbed by chlorophyll *a* itself (440 nm); for chlorophytes also by chlorophyll *b* (480 nm) and for chromophytes by chlorophyll c (460–470 nm) and carotenoids (490–560 nm), affecting the fluorescence excitation specimen. The pigment system of cyanobacteria is however different because exhibit weak fluorescence from chlorophyll *a* but contain phycobilins that fluoresce at lower wavelengths. The cryptophytes also exhibit phycobilins and chlorophyll *a* fluorescence.

Based on the spectral behavior, it is possible to identify four major taxa groups which can be fitted with distinguishing spectra [16]: the first group is the Bacillariophyta (diatoms and dinoflagellates) which displayed in the excitation spectra a broad band with a maximum near 470 nm and a shoulder near 500 nm; the second group, chlorophyte, exhibits a spectrum similar in shape to that of the previous group but narrower. The third and fourth groups are the cryptomonads and cyanobacteria, respectively, which have similar emission due to phycobilins.

The excitation/emission spectra and lifetimes of samples A1 to A5 have been measured with a Spectrophotometer LifeSpec II. A 406 nm pulsed diode laser with excitation pulse of 80 ps was used to pump the sample contained inside a spectroscopic cuvette. The detection has been achieved with a time-correlated single-photon counting (TCSPC) detector.

In our results of fluorescence excitation spectra (**Figure 15**), we observed two emission bands for samples A1, A2, and A4, one caused by chlorophyll fluorescence with a pronounced peak centered at 475 nm and the other (a shoulder) centered at 500 nm. The ratios of fluorescence intensity (FI): FI 475/FI500 are >1 except for sample A3, which is <1, since it exhibits a different spectrum shape. Sample A5 exhibits only a large band, centered at 525 nm, identified as phycoerythrin. The emission fluorescence spectra of samples A1 and A6 are shown in **Figure 16**.

Part of the energy absorbed by the antenna pigments that is not used in the process of photosynthesis is that it can turn into fluorescent radiation.

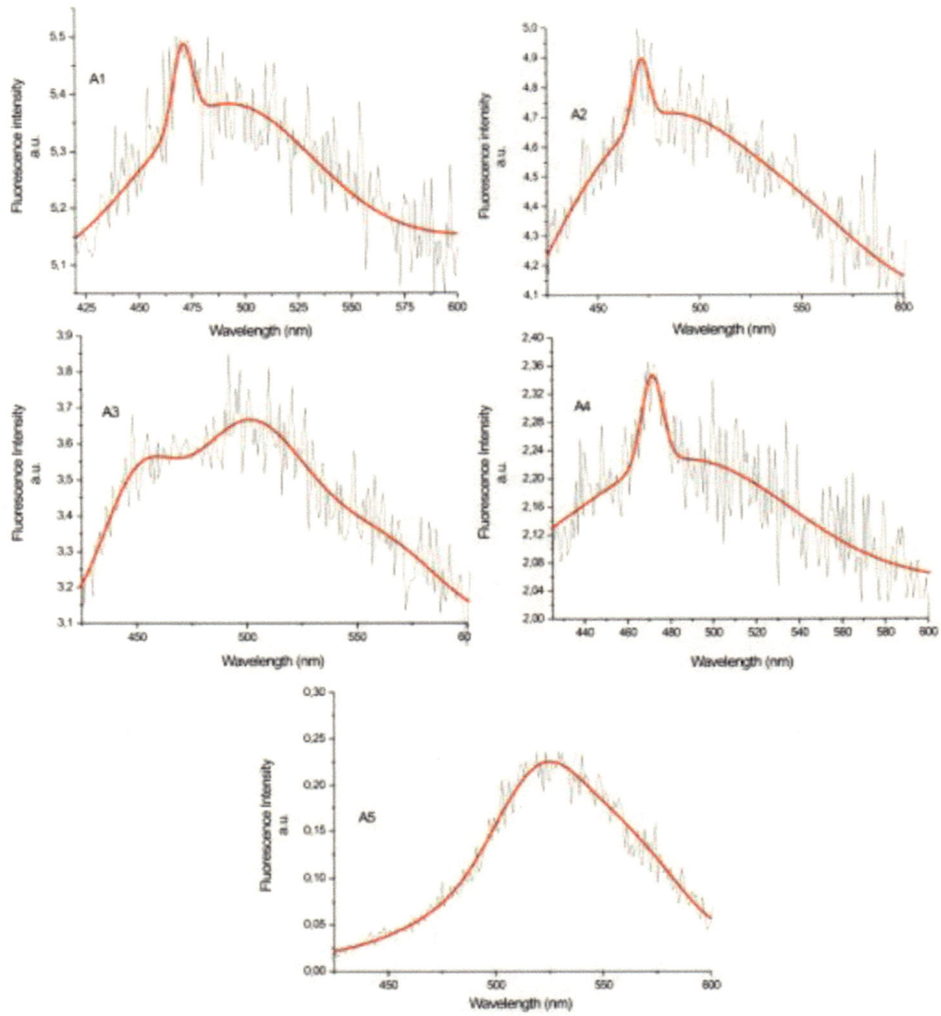

Figure 15.
Fluorescence excitation spectra of samples A1 to A5.

Fluorescence is an important feature for identifying a specific phytoplankton taxa group. However, any change in the photosynthetic mechanism can either increase or decrease the fluorescence intensity. For example, if the antenna pigment set is affected, preventing radiation from reaching the reaction center, then fluorescence will decrease. On the other hand, the fluorescence emitted at 685 nm may increase if the antenna pigments redirect the absorbed energy to other reaction centers that will saturate, favoring fluorescence [47].

The lifetime decays of each sample (**Figure 17**) have been measured at the peak of each emission spectrum. Also, the instrumental response of the equipment (IRF) has been measured and is shown together with the decays in **Figure 17**.

The fluorescence decays are consistent with a multi-exponential equation, which can be useful to estimate an average lifetime that represents the decay rate of the system.

In order to quantify the lifetime of each sample, it has been calculated a three-exponential fitting of the decay curve (Eq. 13) that gives an estimation about the average lifetime as defined in Eq. 14:

$$I(t) = B_1 \cdot e^{(t/\tau_1)} + B_2 \cdot e^{(t/\tau_2)} + B_3 \cdot e^{(t/\tau_3)} \quad (13)$$

$$\tau_{av} = \frac{B_1\tau_1^2 + B_2\tau_2^2 + B_3\tau_3^2}{B_1\tau_1 + B_2\tau_2 + B_3\tau_3} \quad (14)$$

The fitting has been managed to do after a measurement data and IRF reconvolution analysis. The average lifetimes and fitting parameters are given in **Table 5**.

The measured fluorescence decay is not a simple exponential decay. It is instead a quasi-continuous exponential distribution or various distributions. This is because the pigments (for example, chlorophyll) are located in different organisms, and there is more than a single rate in the sample contributing to the fluorescence signal.

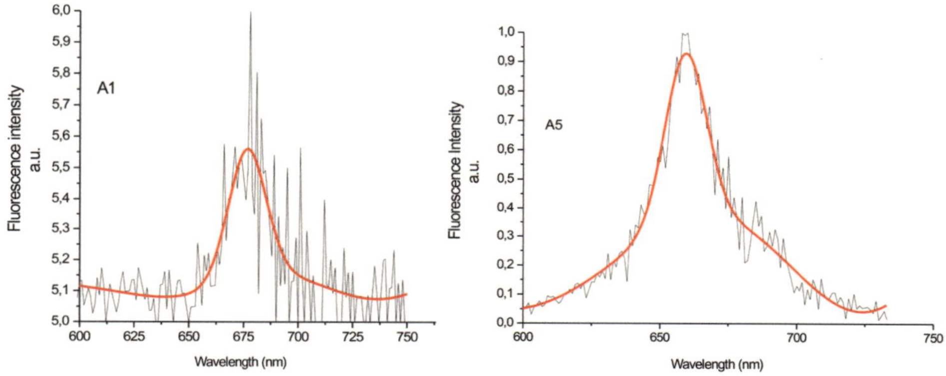

Figure 16.
Fluorescence emission spectra of samples A1 and A5.

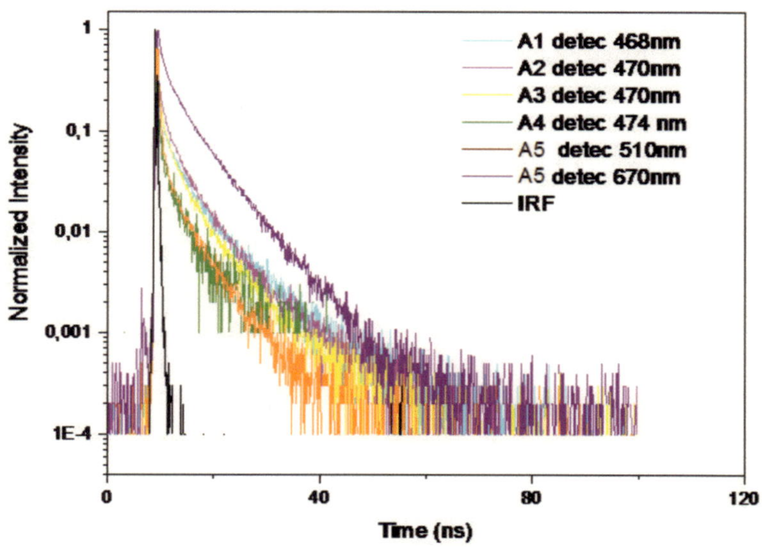

Figure 17.
Decay curves at different water samples represented distinct phytoplankton groups. IRF: instrumental response function.

Samples	λ_{detec} (nm)	τ_1 (ns)	τ_2 (ns)	τ_3 (ns)	B_1	B_2	B_3	τ_{av} (ns)
A1	468	0.0590	1.5440	6.5940	4.9635	0.0617	0.0253	2.2785
A2	470	0.0780	1.3160	5.3850	3.0872	0.0938	0.0401	2.3166
A3	470	0.0690	1.2410	5.0870	3.7274	0.0753	0.0308	1.8346
A4	474	0.0390	0.0650	2.2820	0.0111	4.4557	0.0351	0.5447
A5	510	0.0800	0.5340	4.2390	3.4867	0.1410	0.0201	0.9642
	670	0.1990	1.5170	6.1320	0.9297	0.1910	0.1343	4.2566

Table 5.
Fluorescence lifetime values for the different water samples represented distinct phytoplankton groups.

Acknowledgements

The authors wish to thank Prof. Fernando Lahoz (University of La Laguna, Tenerife-Spain) for the fluorescence measurements. The contribution of Dr. Carla Pérez-Rodríguez is also gratefully recognized.

Conflict of interest

The authors declare no conflict of interest.

Author details

Helena C. Vasconcelos[1,2*], Joao A. Lopes[3], Maria João Pereira[1,4] and Afonso Silva Pinto[5]

1 Science and Technology Faculty (FCT), University of the Azores, Ponta Delgada, Azores, Portugal

2 CEFITEC, Center for Physics and Technological Research, Faculty of Science and Technology (FCT), New University of Lisbon, Lisbon, Portugal

3 Research Institute for Medicines (iMed.ULisboa), Faculty of Pharmacy, University of Lisbon, Lisbon, Portugal

4 Biotechnology Centre of Azores, Ponta Delgada, Azores, Portugal

5 Innovation Green Azores (IGA), Ponta Delgada, Azores, Portugal

*Address all correspondence to: helena.cs.vasconcelos@uac.pt

References

[1] Cruz J, Pacheco D, Porteiro J, Cymbron R, Mendes S, Malcata A, et al. Analysis of long-term monitoring data and remediation measures. Science of The Total Environment. São Miguel, Azores: Sete Cidades and Furnas lake eutrophication; 2015:520. DOI: 10.1016/j.scitotenv.2015.03.052

[2] Oliver R, Ganf G. Freshwater blooms. In: Whitton B, Potts M, editors. The Ecology of Cyanobacteria: Their Diversity in Time and Space. The Netherlands: Kluwer Academic Publishers; 2000. pp. 149-194

[3] LL/SMG, 2013-2016. Book of the Ponds. Available from: https://servicos-sraa.azores.gov.pt/grastore/DRA/RH/Book-Lagoas-SMG-2013-2016.pdf [Accessed: 25 March 2020]

[4] Allakhverdiev SI, Kreslavski VD, Zharmukhamedov SK, et al. Chlorophylls d and f and their role in primary photosynthetic processes of cyanobacteria. Biochemistry (Moscow). 2016;**81**:201-212. DOI: 10.1134/S0006297916030020

[5] Telfer A. Too much light? How beta-carotene protects the photosystem II reaction centre. Photochemical & Photobiological Sciences: Official Journal of the European Photochemistry Association and the European Society for Photobiology. 2006;**4**:950-956. DOI: 10.1039/b507888c

[6] Santos-Ballardo D, Rossi S, Hernández V, Vázquez R, Unceta C, Caro-Corrales J, et al. A simple spectrophotometric method for biomass measurement of important microalgae species in aquaculture. Aquaculture. 2015;**448**:87-92. DOI: 10.1016/j.aquaculture.2015.05.044

[7] Escoffier N, Bernard C, Hamlaoui S, Groleau A, Catherine A. Quantifying phytoplankton communities using spectral fluorescence: The effects of species composition and physiological state. Journal of Plankton Research. 2015;**37**(1):233-247. DOI: 10.1093/plankt/fbu085

[8] Beutler M, Wiltshire KH, Meyer B, Moldaenke C, Luring C, Meyerhofer M, et al. A fluorometric method for the differentiation of algal populations in vivo and in situ. Photosynthesis Research. 2002;**72**:39-53

[9] Millie DF, Schofield OME, Kirkpatrick GJ, Johnsen G, Evens TJ. Using absorbance and fluorescence spectra to discriminate microalgae. European Journal of Phycology. 2002;**37**:313-322

[10] Krause GH, Weis E. Chlorophyll fluorescence and photosynthesis: The basics. Annual Review of Plant Physiology and Plant Molecular Biology. 1991;**42**:313-349

[11] Kirkpatrick G, Millie DF, Moline MA, Schofield O. Absorption-based discrimination of phytoplankton species in naturally mixed populations. Limnology and Oceanography. 2000;**42**:467-471

[12] Bricaud A, Babin M, Morel A, Claustre H. Variability in the chlorophyll-specific absorption coefficients of natural phytoplankton: Analysis and parameterization. Journal of Geophysical Research. 1995;**100**(C7):13321-13332

[13] Ciotti AM, Lewis MR, Cullen JJ. Assessment of the relationships between dominant cell size in natural phytoplankton communities and the spectral shape of the absorption coefficient. Limnology and Oceanography. 2002;**47**(2):404-417

[14] Available from: http://algaevision.myspecies.info/node/3510

[15] Rodhe W. Crystallization of eutrophication concepts in North Europe. In: Eutrophication, Causes, Consequences, Correctives. Washington D.C.: National Academy of Sciences; 1969. pp. 50-64

[16] Yentsch CS, Phinney DA. Spectral fluorescence: An ataxonomic tool for studying the structure of phytoplankton populations. Journal of Plankton Research. 1985;**7**(5):617-632. DOI: 10.1093/plankt/7.5.617

[17] Yentsch CS, Yentsch CM. Fluorescence spectral signatures: The characterization of phytoplankton populations by the use of excitation and emission spectra. Journal of Marine Research. 1979;**37**:471-483

[18] MacIntyre HL, Lawrenz E, Richardson TL, Suggett DJ, et al. Taxonomic discrimination of phytoplankton by spectral fluorescence. In: Chlorophyll a Fluorescence in Aquatic Sciences: Methods and Applications, Developments in Applied Phycology. Vol. 4. Dordrecht: Springer; 2010. pp. 129-169

[19] Seppälä J. Spectral absorption and fluorescence characteristics of the Baltic Sea phytoplankton. ICES CM. 2003;**50**:2-6

[20] Sathyendranath S, Lazzara L, Prieur L. Variations in the spectral values of specific absorption of phytoplankton. Limnology and Oceanography. 1987;**32**:403-415

[21] Jolliffe I. Principal component analysis. In: Lovric M, editor. International Encyclopedia of Statistical Science. Berlin, Heidelberg: Springer; 2011

[22] Erdmann R, Enderlein J, Wahl M. Time Correlated Single-Photon Counting and fluorescence Spectroscopy. 2005

[23] Lakowicz JR. Principles of Fluorescence Spectroscopy. 3rd ed. New York: Springer; 2006

[24] van Holde KE, Johnson WC, Ho PS. Principles of Physical Biochemistry. 2nd ed. Upper Saddle River, NJ: Prentice Hall; 1998

[25] Demas JN. Excited State Lifetime Measurements. New York: Academic Press; 1983

[26] Connor DVO, Philips D. Time Correlated Single Photon Counting. London: Academic Press; 1984

[27] Birch DJS, Imhof R. In: Lakowicz JR, editor. Topics in Fluorescence Spectroscopy. Vol. 1. New York: Plenum Press; 1991

[28] Antunes P, Rodrigues FC. Azores volcanic lakes: Factors affecting water quality. Water Quality: Current Trends and Expected Climate Change Impacts. 2011;**348**:106-114

[29] Andrade C, Viveiros F, Cruz JV, Coutinho R, Silva C. Estimation of the CO_2 flux from Furnas volcanic Lake (São Miguel, Azores). Journal of Volcanology and Geothermal Research. 2016;**315**:51-64. DOI: 10.1016/j.jvolgeores.2016.02.005

[30] Santos M, Muelle H, Pacheco D. Cyanobacteria and microcystins in lake Furnas (S. Miguel Island-Azores). Limnetica. 2012;**29**:107-118

[31] Wetzel RG. Limnologia. Lisboa: Fundação Calouste Gulbenkian; 1993. p. 919

[32] Available from: http://www.azores.gov.pt/Portal/pt/entidades/sraf/noticias/Governo+dos+A%C3%A7ores+anuncia+medidas+de+combate+%C3%A0+eutrofiza%C3%A7%C3%A3o+na+Lagoa+das+Furnas.htm

[33] Available from: https://www.publico.pt/2016/03/23/ecosfera/

noticia/acores-lancam-obra-para-combater-a-eutrofizacao-da-lagoa-das-furnas-1726994

[34] Available from: https://www.aprh.pt/congressoagua2002/pdf/p20.pdf

[35] Available from: https://www.acorianooriental.pt/noticia/regiao-destaca-esforco-no-combate-a-eutrofizacao-nas-lagoas-294433

[36] Available from: https://www.azores.gov.pt/Gra/srrn-drotrh/menus/principal/Monitorização/

[37] Available from: https://www.edinst.com/products/lifespec-ii-dedicated-lifetime-spectrometer/

[38] Available from: https://www.edinst.com/blog/what-is-tcspc/

[39] Valeur B, Berberan-Santos MN. In: Valeur B, editor. Molecular Fluorescence: Principles and Applications. Weinheim, Germany: Wiley-VCH Verlag GmbH; 2001. ISBNs: 3-527-29919-X (Hardcover); 3-527-60024-8 (Electronic)

[40] Paul WJ, Hamilton DP, Ostrovsky I, Miller SD, Zhang A, Muraoka K. Catchment land use and trophic state impacts on phytoplankton composition: A case study from the Rotorua lakes' district, New Zealand. In: Salmaso N, Naselli-Flores L, Cerasino L, Flaim G, Tolotti M, Padisák J, editors. Phytoplankton Responses to Human Impacts at Different Scales. Developments in Hydrobiology. Vol. 221. Dordrecht: Springer; 2012

[41] Jensen JP, Jeppesen E, Kristensen P, Christensen PB, Søndergaard M. Nitrogen loss and denitrification as studied in relation to reductions in nitrogen loading in a shallow, hypertrophic lake (Lake Søbygård, Denmark). Internationale Revue der Gesamten Hydrobiologie. 1992;77:29-42

[42] Jeppesen E, Søndergaard M, Jensen JP, Havens KE, Anneville O, Carvalho L, et al. Lake responses to reduced nutrient loading – An analysis of contemporary long-term data from 35 case studies. Freshwater Biology. 2005;**50**:1747-1771. DOI: 10.1111/j.1365-2427.2005.01415.x

[43] Moed J, Hallegraeff G. Some problems in the estimation of chlorophyll-a and phaeopigments from pre and post-acidification spectrophotometrie measurements. International Review of Hydrobiology. 1978;**63**:787-800. DOI: 10.1002/iroh.19780630610

[44] Dickey TD. Recent advances and future directions in multi-disciplinary in situ oceanographic measurement systems. In: Rotllsoiild BJ, editor. Toward a Theory of Biological Physical Interactions in the World Ocean. Dordrecht, The Netherlands: Kluwer Academic; 1988. pp. 555-598

[45] Campbell D, Hurry V, Clarke A, Gustafsson P, Oquist G. Chlorophyll fluorescence analysis of cyanobacterial photosynthesis and acclimation. Microbiology and Molecular Biology Reviews: MMBR. 1998;**62**:667-683. DOI: 10.1128/MMBR.62.3.667-683.1998

[46] Roy S, Llewellyn C, Egeland E, Johnsen G, editors. Phytoplankton Pigments: Characterization, Chemotaxonomy and Applications in Oceanography (Cambridge Environmental Chemistry Series). Cambridge: Cambridge University Press; 2011. DOI: 10.1017/CBO9780511732263

[47] Falkowski PG, Kolber Z. Variations in chlorophyll fluorescence yields in phytoplankton in the world oceans. Australian Journal of Plant Physiology. 1995;**22**:341-355

Chapter 6

Investigation on the Chlorophyll-*a* Content of Phytoplankton in the Sea of Azov and the Don River by the Fluorescence Method

Pavel Plyaka, Galina Glushchenko, Viktoriya Gerasyuk, Alexey Kleshchenkov, Klim Grigorenko, Margarita Shevchenko, Yuriy Yurasov, Georgy Valov, Igor Tron, Gayane Popovyan and Sergey Berdnikov

Abstract

Results of in situ fluorescence investigations on chlorophyll-*a* (chl-*a*) of phytoplankton in the Sea of Azov (May to June 2018) and in the Don River estuary (September to November 2019) are presented. Continuous data series of chl-*a* fluorescence were obtained with flow-through fluorometers. Discrete reference measurements of chl-*a* concentration and phytoplankton biomass were performed by standard methods for the sake of comparison and eventual corrections. The fluorescence intensity values measured in a lateral surface salinity gradient of the Sea of Azov were found correlating with the obtained data on the chl-*a* concentration (R^2 = 0.88, n = 27) and phytoplankton biomass (R^2 = 0.90, n = 11). Instead, there was a weak correlation (R^2 = 0.40, n = 33) between the fluorescence and phytoplankton biomass found in the estuary. This disparity in correlations was explained by the difference in conditions during measurements, which affected the fluorescence. There were no significant changes in both temperature and coenotic composition of phytoplankton in the course of the marine expedition. The measurements on the river covered a period characterized by the seasonal variations in the phytoplankton composition and by noticeable temperature fluctuations.

Keywords: chlorophyll-*a*, intravital fluorescence, chlorophyll-*a* concentration, biomass and coenotic composition of phytoplankton, the sea of Azov, the Don River

1. Introduction

Fluorescence methods for chlorophyll-*a* (chl-*a*) estimation are widely used in modern natural sciences, including hydrobiology. Despite the rapid development of satellite methods for the assessment of phytoplankton concentration in natural water, including case II waters (i.e., turbid and productive) in the Sea of Azov [1], fluorescence methods are used for the correction of satellite imagery [2] and as substantive assessment methods for phytoplankton distribution [3, 4].

Standard methods for determining chl-*a* concentration in water are based on light absorption [5, 6]; they are time-consuming and require samples in large quantities. Initially, fluorescence was only used to determine the chl-*a* concentration in extracts [7]. Later on, the technology advancement has made modern fluorometers became not inferior to spectrophotometers in the accuracy of the pigment content determination [8–10]. Nowadays, the high speed and ease of measurements, good sensitivity, and a variety of instrumentations are the undoubted advantages of fluorescence methods [8, 11].

Recently, scientists have been expressing a growing interest in the in situ fluorescence studies of chl-*a*, which is the main plant pigment. In contrast to solutions, the fluorescence intensity of living chlorophylls depends not only on their concentration in microalgae but also on the state of the photosynthesis system [12]. The specific yield of fluorescence depends on the state of the photosynthesis electron transport chain and can be significantly reduced by spending a fraction of the absorbed energy on photochemical reactions. In living phytoplankton, not all absorbed light energy goes into optical emission. In all situations, living organisms containing chlorophyll strive to maximize the use of absorbed light energy for synthesis and return only the excess of it in the form of radiation. Photosynthetic activity depends on the condition of living cells, which is determined by the parameters of the environment at a given moment and by the previous state [13]. The higher is the photosynthetic capacity of microalgae, the lower is the specific yield of fluorescence.

The fluorescence of living chlorophylls depends on temperature [14] and illumination [15]. Both have to be controlled for eventual corrections of the measurements considering the inertia the photosynthetic system demonstrates at rearrangement. The dissolved oxygen, acidity [16], dissolved salts [16, 17], and other abiotic factors of the aquatic environment all have an impact on fluorescence [16]. Ultimately, researchers are mostly interested in the quantity of biomass rather than in the intensity of fluorescence [18, 19]. To get reliable estimations of biomass, it is necessary to consider not only the environmental parameters mentioned above but also changes in microalgae species composition [19].

Despite the mentioned difficulties in quantitative analysis of phytoplankton distribution, fluorescence methods are widely used for its assessment in either fresh or salt waters. Among monitoring methods a special place is occupied by vessel surveys using a continuous vessel-based horizontal and vertical profiling [20]. Vessels allow to collect large datasets with a precise time and geo-referencing. While the volume of measured data is comparable to that provided by satellites, it significantly exceeds in accuracy and detail [20]. Fluorometers have a clear advantage over satellites allowing for continuous measurements of chl-*a* concentration during day-and-night cycles. Measurements can be automated, and data can be transmitted in real time, enabling quick adjustments of the research program according to findings. A regular sampling on-board for subsequent measurements in an on-ground laboratory is a common practice during vessel expeditions [2, 20, 21]. Collected samples allow for further in-depth analysis with identification of the composition of phytoplankton species [19].

A number of studies carried out in the most developed areas of the World Ocean by using continuous series of fluorescence data enabled a deeper inquiry into the seasonal and spatial dynamics of phytoplankton in aquatic ecosystems. The findings were used for the prediction of the productivity of waters and their conditions under climate change and anthropogenic pressure [22]. Monitoring of phytoplankton is essential for planning actions against the growth of harmful algal blooms. It is possible to determine the bio-productivity of water based on chl-*a* concentration [23]. In addition, fluorescence-based monitoring of water quality in rivers permits a rapid response to bio-induced deviations of its parameters at water-treatment sites for tap-water distribution networks [24].

In the Taganrog Bay of the Sea of Azov, the phytoplankton concentration was estimated from the in situ chl-*a* fluorescence measurements for the first time in June 2008 [25]. The studies continued then in 2014–2016, thanks to the grant (RFMEFI60714X0059) from the Russian Ministry of Science and Education. In July 2016, continuous measurements of chl-*a* concentration and of dissolved oxygen in the surface layer of the Sea of Azov were performed on the research vessel (R/V) "Deneb" using a custom designed fluorometer [26] in order to justify the new method for calculation of gross primary production based on satellite data [27]. Simultaneously, the flow-through fluorometer, which later has been employed for the present study, was developed and installed into the vessel's seawater system for the tests [28]. They concluded successfully and the device is still in operation. The experience gained about the development and operation of flow-through fluorometers in case II waters with complex optical properties determined the goals and objectives of this work.

In this contribution we report on the complex studies of spatial distribution and temporal dynamics of phytoplankton in the Sea of Azov and the Don River, where significant variations have been noted previously. The main task was to establish the relationship between the parameters of the aquatic environment, the total biomass, the phytoplankton species composition, and chl-*a* fluorescence measurements. It was required to determine the range of applicability of flow-through fluorometers operating on vessel s and at permanent stations. To foresee the changes in water quality in the studied waters, it is necessary to understand the main trends that affect the state of microalgae.

2. Materials and methods

The empirical data for analysis were collected during the Southern Scientific Center of the Russian Academy of Sciences (SSC RAS) expedition on R/V "Deneb" to the Sea of Azov in May to June 2018 and directly from the natural water intake point of JSC "Rostovvodokanal" within the confines of Rostov-on-Don City in September to November 2019 (**Figure 1**).

The in situ fluorescence of chl-*a* in phytoplankton was measured using flow-through fluorescence sensors developed in the SSC RAS. The sensors of a similar design [29] were installed on R/V "Deneb" and on the intake of the water supply system of JSC "Rostovvodokanal." The installations were different in water pumping systems and in customized adaptation to the location and operation conditions. Similar to other studies, fluorescence of chl-*a*, which is the most abundant pigment in microalgae, was registered. The excitation light had a maximum intensity at a wavelength of 447.5 nm with a bandwidth of 20 nm [30]. The source was a set of powerful LEDs equipped with the SZS-20 stained-glass filters. The fluorescence emission was cut out in the 650–790 nm range by a combined filter and was detected by a silicon photodiode. The signal was amplified, digitized, and averaged over a set of pulses. Taking into account the well-known temporal behavior of fluorescence [16, 31], the excitation was performed by 10 μs pulses with a current-stabilized power.

The intake for seawater submitted to the flow system of the fluorometer on-bard of R/V "Deneb" was situated at a depth of 0.5 m, while that of JSC "Rostovvodokanal" for the water supply system was located at the river bottom horizon at least 1 m from the bottom.

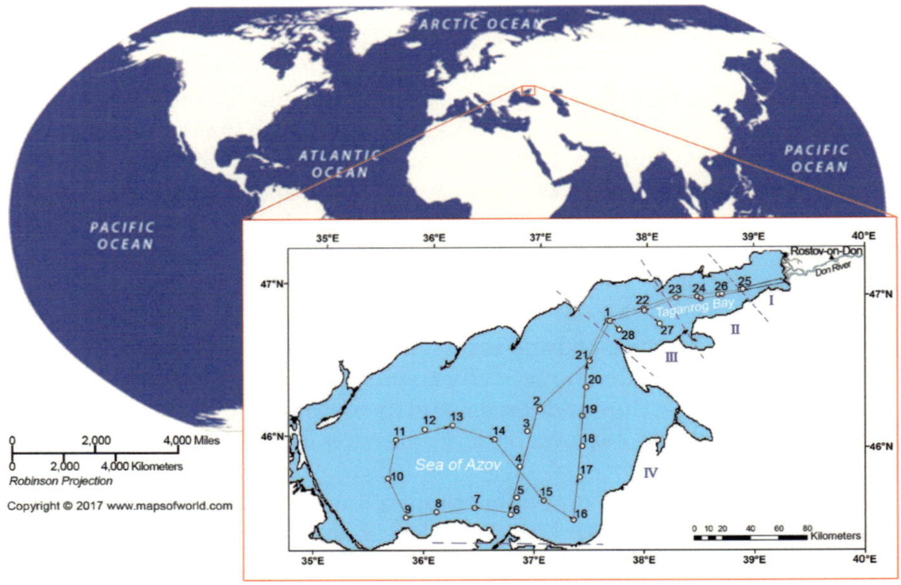

Figure 1.
Schematic map of sampling stations in May to June 2018 in the sea of Azov. Designations: 1–26, station numbers (arrows indicate the route of the R/V "Deneb"); I–IV, conditional division of the sea into four areas, eastern (I), central (II), western (III) areas of the Taganrog Bay, and central part of the sea (IV).

The continuous series of fluorescence measurements on living chl-*a* were transmitted via satellite and cellular channels to the server of SSC RAS and then presented on the website http://meteo.ssc-ras.ru. For the sake of calibration, a set of samples was collected at the stations, and the chl-*a* content was determined by the standard method [6] using an Evolution 3000 double-beam spectrophotometer. Another set of samples from the stations was used for the quantitative and qualitative analysis of phytoplankton composition according to the previously reported method [32]. The vertical profile of temperature and electrical conductivity of water was determined with the SBE 19 oceanographic probe and the horizontal distribution of thermohaline characteristics along the vessel's route using SBE 21 thermosalinograph.

3. Results and discussion

3.1 Chlorophyll-*a* concentration and phytoplankton development in the Sea of Azov under a salinity gradient condition research result

The fluorescence values of phytoplankton chl-*a* measured in situ during the expedition to the Sea of Azov were confined in the range from 4.0 to 10.8 a.u. The corresponding concentrations of chl-*a* were estimated by spectrophotometry of the extracts and ranged from 1.2 to 30.8 μg/l, respectively. A continuous series of fluorescence values and spatial changes in the concentration of chl-*a* measured in situ during the expedition are presented in **Figure 2**.

Figure 2A shows that fluorescence decreases from the level of 5.0–10.8 a.u. in the Taganrog Bay to the level of 4.0–8.0 a.u. in the central part of the Sea of Azov. The values of chl-*a* concentration at complex stations have a similar distribution pattern across the water as presented in **Figure 2B**. The fluorescence intensity and

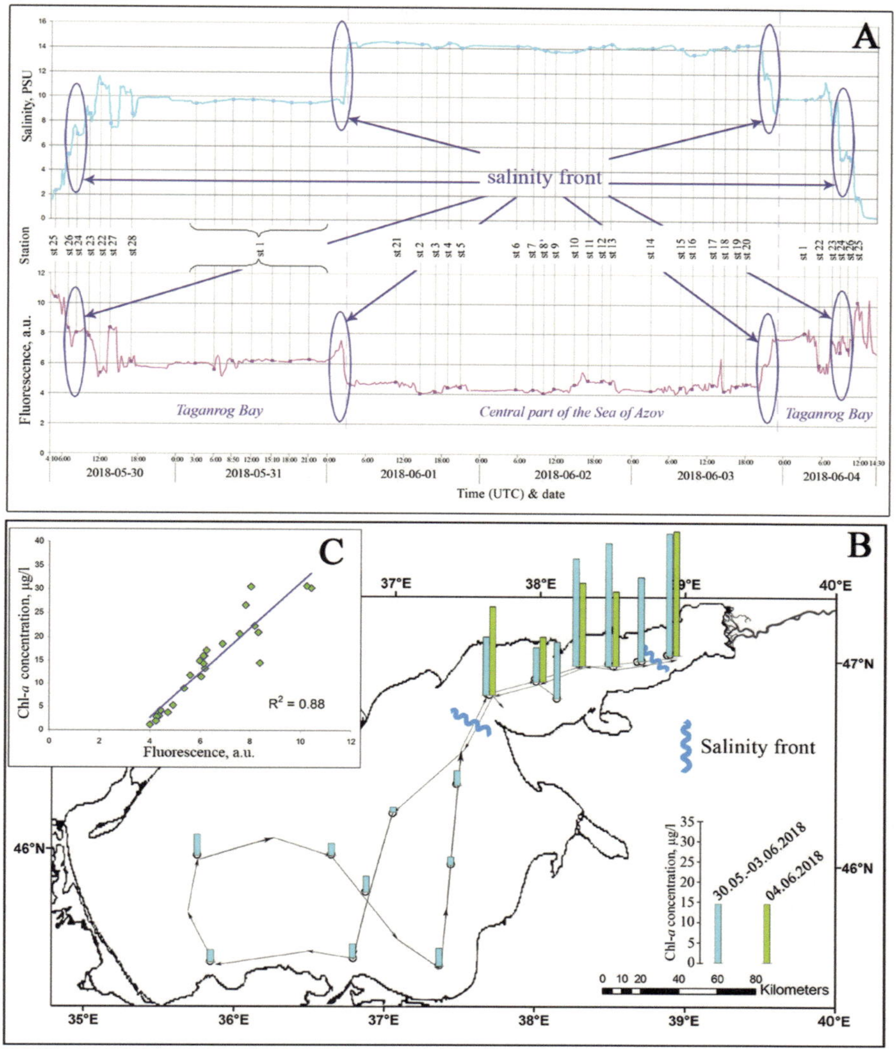

Figure 2.
*(A) Fluorescence of chl-*a *in phytoplankton (a.u.) measured in situ during the R/V "Deneb" survey (arrows indicate the salinity fronts); (B) the spatial distribution of chl-*a *concentration (µg/l) across the sea of Azov; (C) the strength of relationship between these quantities.*

the concentration of chl-*a* demonstrate a tight relationship (R^2 = 0.88, n = 27) over the entire set of examined data (**Figure 2C**).

Such a rather wide range of fluorescence and chl-*a* concentration values can be explained by differences in quantitative indicators as well as by a variety in the phytoplankton species composition in the areas of different seawater salinity. The measured water salinity in the Sea of Azov ranges from 1 to 14 PSU. In particular, there was a sharp increase in salinity (from 1 to 6 PSU) identified in the eastern part of the Taganrog Bay when moving West from the mouth of the Don to the Sea of Azov. Similar increments were also observed in the central part (from 7 to 11 PSU) and in the western one (from 11 to 14 PSU). The water salinity in the central part of the Sea of Azov was not less than 14 PSU [33, 34]. The phytoplankton community was characterized by the rich diversity in species, a large quantity and biomass particularly in the Taganrog Bay, where the salinity gradient is the most pronounced.

Variety is dwindling toward the central part of the Sea of Azov, where there were no sharp fluctuations in salinity detected.

In the studied waters, some specific features in the structure of phytoplankton were noted (**Figure 3**). The Taganrog Bay is rich in species compared to the central part of the Sea of Azov. There were 141 taxa of eukaryotic algae and cyanobacteria belonging to 8 divisions identified in the algal plankton from the bay. In the waters of the central part of the Sea of Azov, the number of revealed species decreased to 48 and they represent 6 divisions.

Such a distribution of phytoplankton in the studied waters is typical not only for the Sea of Azov but also for all estuary-type waters. The fact is that the Taganrog Bay is the zone where waters of various origins mix, leading to the intensification of production and destruction processes [35]. Both marine and freshwater algae can develop here, as it has been observed in the present study. There were 106 species with a rank below the genus (154), constituting 69% of the total number of identified algal taxa with known ecological characteristics. Among them, 21% of species were marine (diatoms and dinophytes), while the remaining 79% belonged to freshwater and brackish-water algae (mainly green algae, cyanobacteria, and diatoms as well).

Since 2007, the composition and spatial structure of phytoplankton in the Taganrog Bay and the central part of the Sea of Azov have been transformed, due to a decrease in feeding from rivers and the associated increase in water salinity. This transformation has led to the situation when habitats of freshwater-brackish-water species are reducing while habitats of marine species are expanding [36]. The tendency of periodic expansions of the marine phytoplankton habitat toward the delta of the Don River was noted in 2017, when the average biomass of phytoplankton had reached its highest values in the eastern and central regions of the Taganrog Bay [37]. In the recent years (comparing to 2002–2012), the maximum concentration of chl-*a* in the western part of the bay has decreased and shifted eastern [38]. In 2018, the Don River experienced the most severe high water over the past 15 years, reaching the maximum in May to June. It also had some impact on the phytoplankton species composition and spatial distribution in the Taganrog Bay.

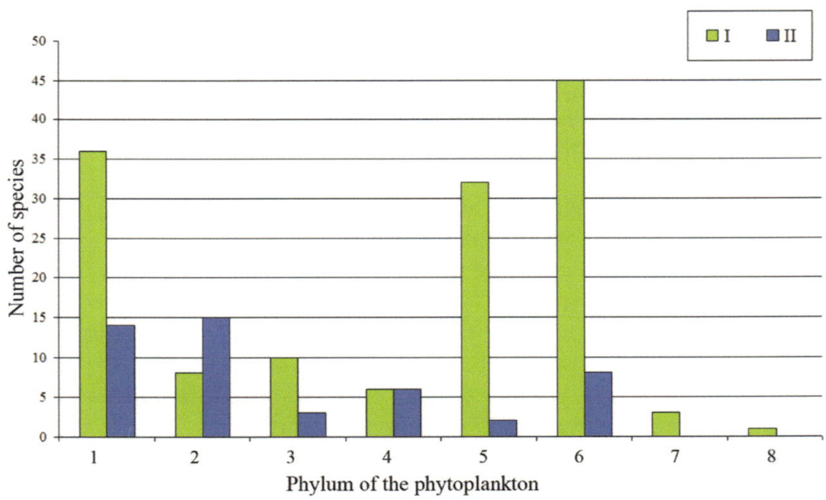

Figure 3.
The number of identified species of different phyla of phytoplankton (1–8) in the Taganrog Bay (I) and in the central part of the Sea of Azov (II): 1, Bacillariophyta; 2, Miozoa; 3, Euglenozoa; 4, Cryptophyta; 5, Cyanobacteria; 6, Chlorophyta; 7, Charophyta; 8, Cercozoa.

The data obtained during the expedition concerning the fluorescence and concentration of chl-*a* as well as the abundance and biomass of microalgae demonstrate the intensive development of phytoplankton throughout the Taganrog Bay water. Also, according to the continuous data series from the flow-through fluorometer, two additional sharply defined zones with high values of chl-*a* concentration were found in the frontal salinity zones (**Figure 2A**), which were formed due to the high water in 2018. The qualitative analysis of microalgae has shown that the distribution of freshwater-brackish-water species is not limited by the eastern and central parts of the bay but extends to the western part too.

In the almost entire water area under study, which was evenly warm (t = 19.5 ± 0.5°C), the dominance of *Coscinodiscus granii* Gough and *Actinoptychus senarius* (Ehrenberg) Ehrenberg from the diatom plankton complex was noted. In the Taganrog Bay, they were subdominated by cyanobacteria *Woronichinia* cf. *compacta* (Lemmermann) Komárek and Hindák and green alga species of the *Oocystis* genus. This happened due to the beginning of summer succession.

A graphical representation (**Figure 4**) of the obtained microalgae biomass shows that the total biomass significantly decreases with increasing salinity and the phytoplankton biomass is significantly lower in the central part of the Sea of Azov than in the Taganrog Bay. The total phytoplankton biomass also led to high concentrations of chl-*a*, because these parameters have a tight relationship (R^2 = 0.90, n = 11). Also, a reliable relationship (R^2 = 0.88, n = 27) was obtained for the concentration of chl-*a* measured in laboratory conditions in connection to the in situ fluorescence data from the flow-through fluorometer operating continuously on a vessel. Results of the analysis prove the efficiency of the method used in the present study of the phytoplankton in the Sea of Azov under conditions of varying salinity. The principles of operation and the design of the flow-through fluorometer allow for a larger number of measurements, in order to obtain data in conditions difficult to access for sampling (e.g., when the vessel cannot be stopped, anchored, etc.).

3.2 Chlorophyll-*a* concentration in the Don River delta under temperature variations

In contrast to the studies in the Sea of Azov, observations in the Don River are held in September to November 2019. During that period, the mineralization of water in the river was changing in a narrow range of 0.6–0.8 g/l, the temperature of water lowered from 23.4 to 7.4°C [39], and the completion of the summer stage with transition to autumnal phytoplankton succession was observed (**Figure 5**). A seasonal change in dominant species has occurred accompanied by the reduction of the total biomass of phytoplankton. The predominant cyanobacteria in September and partly in October (*Microcystis* spp., *Aphanizomenon flosaquae* Ralfs ex Bornet & Flahault) were gradually replaced by diatoms (*Nitzschia* spp., *Gyrosigma* sp., *Ulnaria ulna* (Nitzsch) Compère). The total phytoplankton biomass decreased from 0.94 to 0.17 mg/l. The fluorescence values were decreasing from 14.95 to 4.84 c.u. over the period of studies, which corresponded to the range of chl-*a* concentrations from 35.7 to 5.36 μg/l.

The observed fluctuations in the measured fluorescence have an implicit correlation with the phytoplankton biomass variations (R^2 = 0.40, n = 33). It might be attributed to the influence of water temperature [40]. However, from the available dataset, it was not possible to derive a reliable relationship between the fluorescence and water temperature (R^2 = 0.20, n = 7), which can be explained that rather a small dataset was collected under conditions of abrupt changes in temperature. It has been clearly seen (**Figure 5**) that the values of fluorescence and of biomass demonstrate

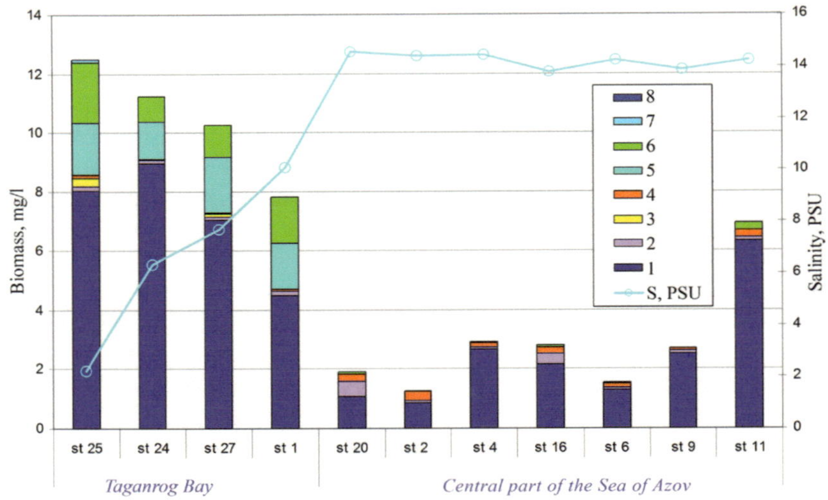

Figure 4.
Variations in water salinity (S, PSU) and phytoplankton (1–8) biomass (B, mg/l) in the Taganrog Bay and in the central part of the sea of Azov: 1, Bacillariophyta; 2, Miozoa; 3, Euglenozoa; 4, Cryptophyta; 5, Cyanobacteria; 6, Chlorophyta; 7, Charophyta; 8, Cercozoa.

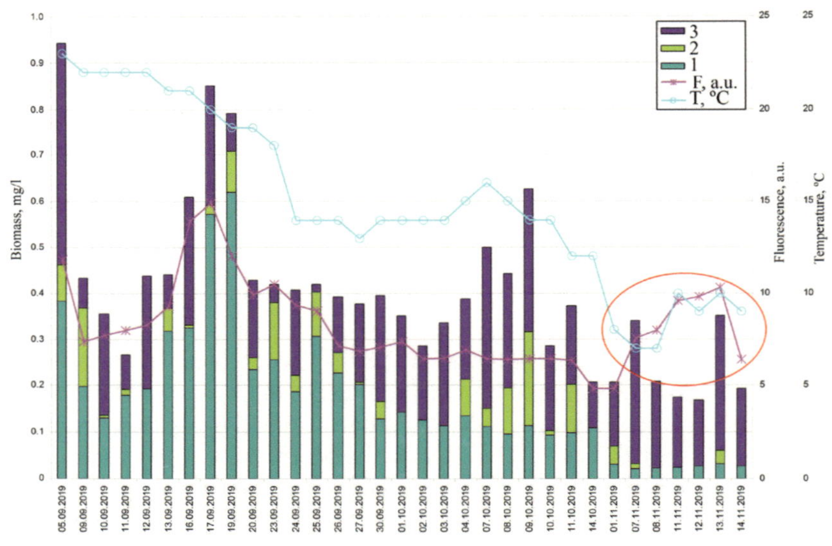

Figure 5.
Variations in the biomass (B, mg/l) of the major divisions of phytoplankton (1–3), water temperature (T, °C), and chl-a fluorescence (F, a.u.) in September to November 2019: 1, Cyanobacteria; 2, Chlorophyta; and 3, Bacillariophyta. Red oval denotes a region of decrease in water temperature ≤10°C and coenotic change of phytoplankton.

the same trend in the temperature range of 13–20°C. When the temperature decreases below 8°C, a coenotic change in phytoplankton occurs, apparently, causing a change in the chl-*a* fluorescence (highlighted by a red oval in **Figure 5**).

In the future, it is necessary to continue collecting empirical data, especially under the changes in the dominant phytoplankton groups and significant variations of water temperature, in order to reveal the influence of environmental parameters on the chl-*a* fluorescence and the correct interpretation of the results. Nevertheless, the fluorescence method of rapid assessment of water entering the water supply network is very useful and allows for a round-the-clock remote monitoring. It is a good alternative to satellite methods, the use of which is very limited for rivers with a narrow channel and does not provide information about water at the depth of installation of water intake systems.

4. Conclusions

As a result of the studies, a reliable relationship was established between the fluorescence intensity of chl-*a* measured in the Sea of Azov and the Don River by flow-through fluorometer and the concentration of chl-*a* determined by standard methods in a wide range of environmental conditions and species compositions of phytoplankton.

Results of continuous vessel-based chl-*a* fluorescence measurements, total biomass values, and species composition of microalgae in the selected probes evidently demonstrate that salinity is one of the main factors regulating the distribution of phytoplankton in the Sea of Azov.

It was shown that the high water of 2018 is temporarily affected by the spatial distribution of freshwater and brackish-water species of phytoplankton in the waters of the Taganrog Bay. But, the general trend of marine microalgae species expansion from the Sea of Azov to the Taganrog Bay under conditions of weakening river flow and growing salinity remains unchanged for over 15 years.

Long-term observations in the lower reaches of the Don have demonstrated that the development of phytoplankton and variations in its species composition reflected in the changes of chl-*a* fluorescence are strongly influenced by temperature.

Fluorescence measurements have demonstrated to be an effective method for studying water productivity and rapid diagnostics of phytoplankton status. Despite the relative simplicity in design and operation, the developed flow-through fluorometers allowing for a round-the-clock remote monitoring of the environment have become an effective research tool, providing sufficient data for revealing spatial and temporal dynamics.

The fluorescent method of monitoring the quality of water entering the urban water supply system is promising for taking preventive measures to combat water blooming and rising population of cyanobacteria.

Acknowledgements

The authors are grateful to the staff of the Southern Scientific Center of the RAS, E.P. Oleinikov for kindly provided data on abiotic parameters of water, and V.V. Sayapin and G.K. Tymoshenko for their help in sampling.

Conflict of interest

The authors declare no conflict of interest. This article does not contain any research using animals as objects.

Thanks

The work was carried out within the State assignments of SSC RAS for 2020, G.Yu. Glushchenko, M.S. Shevchenko, K.S. Grigorenko, and

A.V. Kleshchenkov were supported by the project No. AAAA-A18-118122790121-5, while P.S. Plyaka and V.S. Gerasyuk were supported by the project No. AAAA-A19-119040390084-3.

Author details

Pavel Plyaka[1], Galina Glushchenko[1*], Viktoriya Gerasyuk[1], Alexey Kleshchenkov[1], Klim Grigorenko[1], Margarita Shevchenko[1], Yuriy Yurasov[1], Georgy Valov[1], Igor Tron[2], Gayane Popovyan[2] and Sergey Berdnikov[1]

1 Federal Research Centre, The Southern Scientific Centre of the Russian Academy of Sciences (SSC RAS), Rostov-on-Don, Russia

2 Joint-Stock Company, Rostovvodokanal, Rostov-on-Don, Russia

*Address all correspondence to: gluschenko_gala@mail.ru

References

[1] Moses WJ, Gitelson AA, Berdnikov S, Povazhnyy V. Estimation of chlorophyll-*a* concentration in case II waters using MODIS and MERIS data—Successes and challenges. Environmental Research Letters. 2009;**4**(4):045005. DOI: 10.1088/1748-9326/4/4/045005

[2] Moses WJ, Saprygin V, Gerasyuk V, Povazhnyy V, Berdnikov S, Gitelson AA. OLCI-based NIR-red models for estimating chlorophyll-*a* concentration in productive coastal waters—A preliminary evaluation. Environ. Res. Commun. 2019;**1**(1):011002. DOI: 10.1088/2515-7620/aaf53c

[3] Karabashev GS. Flyuorescenciya v Okeane [Fluorescence in the Ocean]. Leningrad; 1987. pp. 199 (in Russian)

[4] Antal TK, Venediktov PS, Matorin DN, Ostrowska M, Woźniak B. Measurement of phytoplankton photosynthesis rate using a pump-and-probe fluorometer. Oceanologia. 2001;**43**(3):291-313

[5] Lutz VA, Sathyendranath S, Head EJH. Absorption coefficient of phytoplankton: Regional variations in the North Atlantic. Marine Ecology Progress Series. 1996;**135**:197-213

[6] State standard 17.1.4.02-90 Voda. Metodika spektrofotometricheskogo opredeleniya xlorofilla-a. Gosudarstvenny'j kontrol 'kachestva vody' [Water. Method for spectrophotometric determination of chlorophyll-*a*. State control of water quality]. Moscow; 2001. p. 688 (In Russian)

[7] Yentsch CS, Menzel DW. A method for the determination of phytoplankton chlorophyll and pheophytin by fluorescence. Deep Sea Research. 1963;**10**:221-231. DOI: 10.1016/0011-7471(63)90358-9

[8] DosSantos ACA, Calijuri MC, Moraes EM, Adorno MAT, Falco PB, Carvalho DP, et al. Comparison of three methods for chlorophyll determination: Spectrophotometry and fluorimetry in samples containing pigment mixtures and spectrophotometry in samples with separate pigments through high performance liquid chromatography. Acta Limnologica Brasiliensia. 2003;**15**(3):7-18

[9] Scheiber U, Schliwa U, Bilger W. Continuous recording of photochemical and non-photochemical chlorophyll fluorescence quenching with a new type of modulation fluorometer. Photosynthesis Research. 1986;**10**:51-62. DOI: 10.1007/BF00024185

[10] Kolber ZS, Prasil O, Falkowski PG. Measurements of variable chlorophyll fluorescence using fast repetition rate techniques: Defining methodology and experimental protocols. Biochimica et Biophysica Acta. 1998;**1367**:88-106. DOI: 10.1016/s0005-2728(98)00135-2

[11] Zeng L, Li D. Review article development of in situ sensors for chlorophyll concentration measurement. Journal of Sensors. 2015;**903509**:16. DOI: 10.1155/2015/903509

[12] Juneau P, Harrison PJ. Comparison by PAM fluorometry of photosynthetic activity of nine marine phytoplankton grown under identical conditions. Photochemistry and Photobiology. 2005;**81**(3):649-653. DOI: 10.1562/2005-01-13-RA-414.1

[13] Baker NR. Chlorophyll fluorescence: A probe of photosynthesis *in vivo*. Annual Review of Plant Biology. 2008;**59**:89-113. DOI: 10.1146/annurev.arplant.59.032607.092759

[14] Lorenzen CJ. A method for the continuous measurement of in vivo

chlorophyll concentration. Deep Sea Research. 1966;**13**:223-227. DOI: 10.1016/0011-7471(66)91102-8

[15] Ostrowska M. Model of the dependence of the sun-induced chlorophyll *a* fluorescence quantum yield on the environmental factors in the sea. Optics Express. 2012;**20**(21):23300-23317. DOI: 10.1364/OE.20.023300

[16] Krause GH, Weis E. Chlorophyll fluorescence as a tool in plant physiology. Photosynthesis Research. 1984;**5**:139-157. DOI: 10.1007/BF00028527

[17] Belkhodja R, Morales F, Abadia A, Cómez-Aparisi J, Abadia J. Chlorophyll fluorescence as a possible tool for salinity tolerance screening in barley. Plant Physiology. 1994;**104**:667-673. DOI: 10.1104/pp.104.2.667

[18] Huot Y, Babin M, Bruyant F, Grob C, Twardowski MS, Claustre H. Does chlorophyll *a* provide the best index of phytoplankton biomass for primary productivity studies? Biogeosciences Discussions. 2007;**4**: 707-745. DOI: 10.5194/bgd-4-707-2007

[19] Catherine A, Escoffier N, Belhocine A, Nasri AB, Hamlaoui S, Yéprémian C, et al. On the use of the FluoroProbe phytoplankton quantification method based on fluorescence excitation spectra for large-scale surveys of lakes and reservoirs. Water Research. 2012;**46**:1771-1784. DOI: 10.1016/j.watres.2011.12.056

[20] Borstad GA, Gower JFR. Phytoplankton chlorophyll distribution in the Eastern Canadian Arctic. Arctic. 1984;**37**(3):226-233. DOI: 10.14430/arctic2195

[21] Demers S, Therriault J-C, Legendre L, Neveux J. An in vivo fluorescence method for the continuous in situ estimation of phytoplankton photosynthetic characteristics. Marine Ecology: Progress Series. 1985;**1**:21-27. DOI: 10.3354/meps027021

[22] Sorokina V, Berdnikov S. Nutrient loading of the Don and Kuban on the ecosystem of the Sea of Azov. Water Resources. 2018;**45**(6):920-934. DOI: 10.1134/S0097807818060143

[23] Berdnikov S, Kulygin V, Dashkevich L. Sezonnaya i mnogoletnyaya dinamika pervichnoj produkcii Azovskogo morya po danny'm spektrometra MERIS [Seasonal and long-term dynamics of the primary production of the Sea of Azov according to the MERIS spectrometer]. Sovremenny'e problemy' distancionnogo zondirovaniya Zemli iz kosmosa [Modern Problems of Remote Sensing of the Earth from Space]. 2019;**16**(2):173-185. DOI: 10.21046/2070-7401-2019-16-2-173-185 (in Russian)

[24] Zamyadi A, McQuaid N, Prevost M, Dorner S. Monitoring of potentially toxic cyanobacteria using an online multi-probe in drinking water sources. Journal of Environmental Monitoring. 2011;**14**:579-588. DOI: 10.1039/c1em10819k

[25] Shavykin A, Berdnikov S, Saprygin V, Verbitsky R. Neprery'vny'e izmereniya okeanologicheskix parametrov v pripoverxnostnom sloe Taganrogskogo zaliva. 1. Opredelenie xlorofilla a fluorimetricheskim metodom [Continuous measurements of oceanological parameters in the surface layer of the Taganrog Bay. 1. Determination of chlorophyll *a* by fluorimetric method]. Vestnik Yuzhnogo nauchnogo centra RAN [Bulletin of the Southern Scientific Center of RAS]. 2010;**6**(3):39-48 (in Russian)

[26] Povazhnyy V, Povazhnyy A, Zolotareva A. Measurements of chlorophyll-A in vivo fluorescence and water reflectance by consumer digital cameras for remote sensing

data verification. In: 16th International Multidisciplinary Scientific GeoConference SGEM 2016; June 28–July 6, 2016; Book 2. 2016. pp. 963-970. DOI: 10.5593/SGEM2016/B22/S10.123. Available from: www.sgem.org

[27] Berdnikov S, Dashkevich L, Kulygin V, Povazhny V. Ocenka pervichnoj produkcii Azovskogo morya po danny'm distancionnogo zondirovaniya [Evaluation of the primary production of the Sea of Azov according to remote sensing]. Nauka Yuga Rossii [Science of the South of Russia]. 2018;**14**(2):55-65. DOI: 10.23885/2500-0640-2018-14-2-55-65 (in Russian)

[28] Plyaka P, Povazhny V. Avtonomny'j fluorimetricheskij kompleks dlya opredeleniya soderzhaniya xlorofilla fitoplanktona i obshhego vzveshennogo veshhestva v vodnoj srede [Autonomous fluorimetric complex for determining the chlorophyll content of phytoplankton and total suspended matter in the aquatic environment]. Utility Model Patent RU 177930 U1, 03.16.2018. Application № 2016149468 from 15.12.2016 (in Russian)

[29] Budnik A, Plyaka P. Opy't sozdaniya fluorimetricheskix kompleksov dlya ispol'zovaniya na stacionarny'x i mobil'ny'x stanciyax kontrolya vodnoj sredy' [Experience in creating fluorimetric complexes for use in stationary and mobile stations for monitoring the aquatic environment]. Inzhenerny'j vestnik Dona [Engineering Herald of Don] [Internet]. 2019. Available from: http://ivdon.ru/ru/magazine/archive/n8y2019/6124 [Accessed: 20 March 2020] (in Russian)

[30] Povazhnyi V. A fluorometer on the basis of powerful light emitting diodes for determination of the chlorophyll "a" concentration. Oceanology. 2014;**54**(3):387-391. DOI: 10.1134/S0001437014030102

[31] Misra AN, Misra M, Singh R. Chlorophyll fluorescence in plant biology [Internet]. Biophysics. Croatia, Rijeka: IntechOpen. 2012. pp. 171-192. Available from: https://www.intechopen.com/books/biophysics/chlorophyll-fluorescence-in-plant-biology [Accessed: 20 March 2020]

[32] Sadchikov A. Metody Izucheniya Presnovodnogo Fitoplanktona: Metodicheskoe Rukovodstvo [Methods for the Study of Freshwater Phytoplankton: A Methodological Guide]. Moscow: University and school; 2003. pp. 157 (In Russian)

[33] Berdnikov S, Kleschenkov A, Grigorenko K, Oleinikov E, Moskovets A, Dashkevich L, et al. Rezul'taty' morskix nauchny'x issledovanij Yuzhnogo nauchnogo centra Rossijskoj akademii nauk (YuNCz RAN) v Azovskom more v 2003-2018. Chast' 1: Gidrologiya i gidroximiya. [The results of marine scientific research of the Southern Scientific Center of the Russian Academy of Sciences (SSC RAS) in the sea of Azov in 2003-2018. Part 1: Hydrology and hydrochemistry]. Vodny'e bioresursy' i sreda obitaniya [Aquatic Bioresources and Habitat]. 2019;**2**(3):7-19 (in Russian)

[34] Grigorenko K, Oleynikov E, Grigorenko E. Vliyanie polovod'ya Dona 2018 g. na termoxalinnuyu strukturu Azovskogo moray [The effect of the Don flood in 2018 on the thermohaline structure of the Sea of Azov]. Nauka Yuga Rossii [Science of the South of Russia]. 2019;**15**(3):63-70. DOI: 10.7868/S25000640190307 (in Russian)

[35] Kovaleva G. Fitoplankton Azovskogo morya i prilegayushchih vodoemov [Phytoplankton of the Azov Sea and adjacent basins]. In: Matishov GG, editor. Azovskoe more v konce XX-nachale XXI vekov: geomorfologiya, osadkonakoplenie, pelagicheskie soobshchestva [The Sea of Azov in the Late XX-Early

XXI Centuries: Geomorphology, Sedimentation, Pelagic Communities]: Apatites. 2008. pp. 134-223 (in Russian)

[36] Safronova L, Luzhnjak O. Transformaciya fitoplanktona Azovskogo morya v usloviyax sovremennogo osoloneniya [Transformation of the phytoplankton of the Sea of Azov in conditions of modern salinization]. In: Morskie biologicheskie issledovaniya: dostizheniya i perspektivy' [Marine Biological Research: Achievements and Prospects]; 19-24 September 2016; Sevastopol: ECOSI-Hydrophysics. Vol. 2. 2016. pp. 417-420 (in Russian)

[37] Korneva LG, Glushchenko G. Composition and seasonal succession of phytoplankton of Taganrog Bay in the sea of Azov and the downstream reaches of the Don River under a changing climate. Inland Water Biology. 2020;**13**(1):23-30. DOI: 10.1134/S1995082920010071

[38] Saprygin V, Berdnikov S, Kulygin V, Dashkevich L, Mestetskiy LM. Spatial distribution and seasonal dynamics of the chlorophyll *a* concentration in the Sea of Azov based on MERIS images. Oceanology. 2018;**58**(5):689-699. DOI: 10.1134/S0001437018050132

[39] Sea temperature. [Internet]. 2020. Available from: https://seatemperature.ru/monthly/don-rostov-sea-temperature-in-september-7007 [Accessed: 04 April 2020]

[40] Rice E, Stewart G. Analysis of interdecadal trends in chlorophyll and temperature in the Central Basin of Long Island sound. Estuarine, Coastal and Shelf Science. 2013;**128**:64-75. DOI: 10.1016/j.ecss.2013.05.002

Section 2

Fluorescence Methods in Medicine

Chapter 7

Fluorescence Fluctuation Techniques for the Investigation of Structure-Function Relationships of G-Protein-Coupled Receptors

Robert T. Youker and Danielle Voet

Abstract

G-protein-coupled receptors (GPCRs) are seven transmembrane receptors that form the largest superfamily of signaling proteins, and the family members function in a diverse array of metabolic pathways including cardiac function, immune response, neurotransmission, smell, taste, cell differentiation and growth, and vision. It is becoming clear that alteration in the quaternary structure of the GPCR receptor can have a profound impact on signaling capabilities. Biochemical, biophysical, physiological, x-ray crystallographic, and computational methods have been used over the last 40–50 years to study the structure and function of GPCRs. Evidence from these studies confirm that GPCRs can be organized as monomers, dimers, and higher-order oligomers. However, many times, these methods require extraction of the receptor from its native environment and high levels of expression and only provide a snapshot of information. A need arose for techniques that could measure the assembly and disassembly of receptors at few-to-single molecule resolution in their native environment at fast time scales. In the last 20 years, fluorescence fluctuation techniques have filled this need and provided new insight into the dynamics of GPCR organization in the absence and presence of ligands, agonists, and antagonists. In this book chapter, we provide a brief introduction to GPCR structure and function [Section 1]. An overview of the theoretical basis for fluorescence fluctuations techniques (FFTs) and how FFTs can be used to study the oligomeric structure of GPCRs in live and fixed cells is explained [Section 2]. We discuss the advantages and limitations of FFTs [Section 3], and finally, we summarize select case studies on GPCR structure and function revealed by FFT experiments [Section 4].

Keywords: fluorescence spectroscopies, molecular brightness, GPCR, oligomer, dimer, SpIDA, protein dynamics

1. Introduction

G-protein-coupled receptors (GPCRs), also known as hepta-helical or seven transmembrane receptors, form the largest superfamily of signaling proteins that comprise 3–5% of genes in animals with approximately 1000 and 800 genes in the human and mice genomes, respectively [1–4]. GPCRs are evolutionarily conserved from single cell excavates – asymmetric protists that usually contain an "excavated" feeding groove – to multicellular animals. Recent phylogenetic analysis of 75 genomes suggest evolution of these ancient GPCRs and associated cytoplasmic signaling machineries occurred through lineage-specific diversification with repeated domain shuffling (**Figure 1**) [5]. Many of the gene families needed for GPCR signaling were present in the last common ancestor of eukaryotes [6–9]. Interestingly, architectural diversification of the N-terminal protein domain of GPCRs was concomitant with the diversification of these ancient protein families [10]. These observations support the theory that the GPCR core signaling machinery evolved first followed by extensive receptor radiation and selective pressures (positive and negative) may have affected evolution of GPCR subgroups differently [5, 11].

The extensive diversification of GPCRs was most likely necessary as eukaryotes evolved from unicellular to multicellular organisms and to exploit new environmental niches, thus sensing and responding to novel stimuli [12]. GPCR family members can be activated by an extraordinary array of molecules that include amines, amino acids, cytokines, glycoproteins, hormones, ions, nucleotides, peptides, and even photons of light, in the case of the Rhodopsin family. Multiple classification systems have been developed for GPCRs (e.g. A-F system and GRAFS system) using a variety of approaches that include motif-based, pharmacological-based, machine learning, or phylogenetic tools [3, 13–15]. The GRAFS system relies on structural and phylogenetic information to classify Metazoan GPCRs [2, 3, 10]. The GRAFS system divides GPCRs into the <u>G</u>lutamate (Class C of A-F system), <u>R</u>hodopsin (Class A), <u>A</u>dhesion (Class B), <u>F</u>rizzled (Class F), and <u>S</u>ecretin (Class B) families. This classification system has been extended to several lineage-specific GPCR families (insect odorant receptors, nematode chemoreceptors, vertebrate vomeronasal receptors) and non-metazoan members such as cAMP (Class E), GPR-108-like families and ITR-like families [16]. The Fungi Ste2 (Class D), Ste3 (Class D), and Git3 receptors and the plant abscisic receptor are also categorized as GPCRs [17, 18]. An important difference between the two widely used classification systems are that GRAFS further divides class B into the Secretin and Adhesion receptor subfamilies, while the A-F system does not.

A majority of physiological functions, including cell differentiation and growth, immune response, metabolism, neurotransmission, smell, taste, and vision are regulated by GPCR signaling. Not surprisingly, disruption of GPCR signaling can cause a wide variety of disease such as cardiac dysfunction, diabetes, nervous system disorders, obesity, inflammation, and even cancer. Estimates range from 30 to 40% of all FDA approved drugs target GPCRs given the central roles of these receptors in human physiology [19, 20]. There are greater than 300 protein structures representing 64 unique GPCRs and the vast majority of solved structures belong to the class A subfamily (GPCR database, accessed May 7, 2020, [21]). Structurally, GPCRs all contain an extracellular domain, a counter-clockwise transmembrane helical bundle, and an intracellular domain. The extracellular domains have the least degree of conservation and vary in size and shape. For example, most class A members have small extracellular domains compared to class B (secretion family) and class C (glutamate family) members that possess long N-terminal domains of approximately 120 and 600 amino acids, respectively (**Figure 1**).

Most GPCRs bind ligands directly to the extracellular N-terminus and/or a binding pocket composed of the extracellular region and transmembrane helices. Typically, GPCRs signal through a heterotrimeric G-protein complex whereby ligand binding to the receptor causes a conformational change and the receptor acts as a guanine nucleotide exchange factor (GEF) to activate the alpha-subunit (Gα) by facilitating GTP binding [22]. Activated Gα dissociates from the G$\beta\gamma$ subunits to allow interaction with downstream effectors such adenylate cyclase,

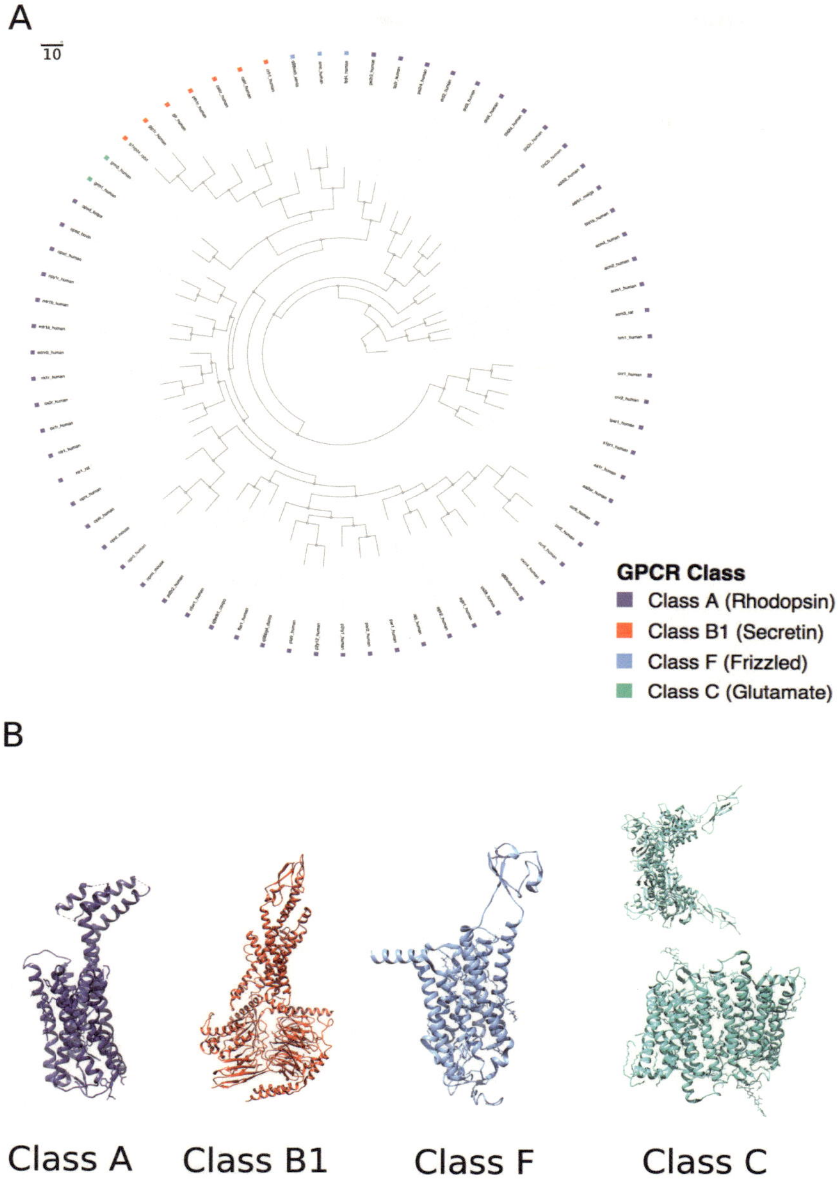

Figure 1.
Neighbor-joining phylogenetic tree of select human GPCRs and representative crystal structures. (A) Information for GPCR phylogenetic tree (steps used): Seventy human GPCRs with crystal structures deposited in GPCRdb were used to construct neighbor-joining distance calculation on conserved segments of sequence/structure with bootstrapping (10 replicates) with regular branch lengths depicted. GPCR classes A (purple), B1 (red), F (light blue), and C (green) are represented. Scale bar = 10. Tree constructed using PHYLIP and jsPhyloSVG packages implemented through GPCRdb website. (B) Crystal structures of class A (serotonin receptor, PBD: 6bqh), class B1 (glucagon receptor, PDB: 6lmk), class F (frizzled 4 receptor, PDB: 6bd4), and class C (extracellular domain from metabotropic glutamate receptor 3 (PDB: 2e4w) and 7TM from bovine rhodopsin (PDB: 1gzm) as a structural model to represent class because no full-length structures exist, https://doi.org/10.1038/aps.2011.186). Structures are colored to match key in part a.

or phospholipase C (PLC), leading to a second messenger signaling cascade [22]. The Gβγ dimeric complex recruits G-protein-coupled receptor kinase (GRK) to the activated receptor and the GRK phosphorylates the GPCR on serine or threonine residues residing in the intracellular C-terminal tail, or the third intracellular loop [23–25]. Phosphorylation of the receptor leads to β-arrestin binding followed by adaptor and clathrin recruitment that triggers endocytosis of the receptor to attenuate signaling. Concomitantly, the Gα subunit and regulators of G-protein signaling (RGS) facilitate re-association of the heterotrimeric complex through GTP hydrolysis, thus resetting the switch for another round of signaling. The GPCR is then either recycled back to the cell surface or degraded in the lysosome and this completes the receptor life cycle (for review see [22]). Recent evidence suggests that in some cases, GPCRs can signal from early endosomes, trans-Golgi network (TGN), mitochondria, and nuclear membranes (for review see [1, 26, 27]). This observed intracellular GPCR signaling is dynamic and is thought to play an important role in synaptic plasticity, learning and memory [26].

In addition, G-protein independent signaling through arrestins, arrestin domain containing proteins (ARDCs) and GRKs has been reported (for review see [4, 28–30]). For example, arrestins can bind to mitogen-activated protein kinase (MAPK) pathway members both at the plasma membrane and from endosomal membranes to trigger G-protein independent MAPK signaling [31, 32]. Interestingly, arrestin activation of extracellular-signal-regulated kinase (ERK) appears to bias cytoplasmic instead of nuclear MAPK signaling possibly through inhibition of ERK translocation to the nucleus [32].

The ability of GPCRs to activate G-protein-mediated, arrestin-mediated, or another potential signaling pathway is called biased signaling [33]. This signaling complexity is due to the ability of GPCRs to bind to multiple ligands, interact with an array of regulators (as mentioned above) and cross-regulate each other leading to a balanced cellular/functional response [34, 35]. The signaling complexity of GPCRs is also due to the dynamic quaternary state of the receptors. The quaternary state of a protein describes whether it is composed of one, or more, folded chains of amino acids. GPCRs are well known to function as monomers and dimers but can also form larger quaternary structures, termed higher-order oligomers that contain three, or more, molecules in a complex [36]. The quaternary state of the receptor can affect its ability to be activated and may influence the type of downstream signaling pathway that is engaged, such as plasma membrane versus intracellular membrane signaling. Therefore, it is vitally important to quantify temporal and spatial GPCR conformational changes in order to gain insight into GPCR signaling [37].

There are a plethora of experimental methods to measure GPCR oligomeric structure that can be classified as biochemical, biophysical, physiological, x-ray crystallographic, and computational [38]. Biochemical methods include co-immunoprecipitation (co-IP) and blue native PAGE (BN-PAGE) that require extraction of the protein from their native cellular membranes. Förster resonance energy transfer (FRET), fluorescence complementation, and fluorescence recovery after photobleaching (FRAP) are examples of biophysical methods where receptors are labeled and measured in their native environment [37, 39]. Total internal reflection fluorescence microscopy (TIRFM), single-molecule imaging, and spatial intensity distribution analysis (SpIDA) are considered physiological methods for studying GPCRs in live and fixed cells. Methods such as x-ray crystallography, NMR spectroscopy and cryo-electron microscopy (cryo-EM) are capable of determining structures at atomic resolution, and only NMR can determine three-dimensional protein structures in solution [40]. Finally, computational methods, such as molecular dynamic (MD) simulations, provide insight into dynamic behavior of GPCRs at the atomic level based on x-ray crystallographic structures. In the last two decades, fluorescence fluctuation techniques have become a vital tool for measuring the dynamic oligomeric changes of proteins, including GPCRs, because of their single molecule sensitivity, high spatial resolution, and ability to measure physiologically-relevant concentrations in live cells [41–43]. We will focus on these existing and newly developed fluorescence techniques and no other well established and traditionally used biophysical/biochemical techniques to study GPCRs that are mentioned above (for reviews see [37–40]).

In this book chapter, we provide an overview of the theoretical basis for fluorescence fluctuations techniques (FFTs) and how FFTs can be used to study the oligomeric structure of GPCRs in live and fixed cells [Section 2]. We discuss the advantages and limitations of FFTs [Section 3]. Finally, we summarize select case studies on GPCR structure and function revealed by FFT experiments [Section 4].

2. Overview of fluorescence fluctuation techniques (FFTs)

Fluorescence fluctuation spectroscopy (FFS) techniques, a.k.a. FFTs, refers to a suite of spectroscopic/microscopic techniques that exploit the fluctuations detected in the signal emitted from a fluorescent molecule to extract molecular dynamic information, such as diffusion rate and oligomeric size of the molecule. Measurements can be made on purified molecules *in vitro* in a cuvette, or on molecules inside live cells with the aid of a confocal microscope. FFTs can be classified based on method of collection of the fluorescence fluctuation information, such as a spot method versus scanning method, or by how the information is analyzed, such as space, time, or both (**Figure 2**). The following sections describe several of the most common FFTs that have been developed, especially for GPCR studies, but are not an exhaustive list of all developed applications (for in-depth review see [41, 44–46]).

2.1 Fluorescence correlation spectroscopy (FCS)

Fluorescence correlation spectroscopy (FCS) was developed in the 1970s at the Karolinska Institute in Stockholm and Cornell University in the U.S. [47, 48]. Today, it is a common spot method implementation, whereby the signal fluctuations from a small volume (>1 femtoliter) are observed, recorded, and an autocorrelation function (ACF) is calculated to determine molecular processes. The main process causing the fluctuations are the diffusion of the fluorescent molecules through the detection volume, but other processes that create fluctuations can be measured, including active transport, flow, rotational diffusion, excited state transitions, chemical reactions, and bleaching [49, 50]. FCS was used in the seminal experiments by Magde, Elson, and Webb to measure diffusion of DNA and kinetics of ethidium bromide binding [47]. FCS measurements were confined to *in vitro* (cuvette) setups because cell experiments using widefield microscopes had "large" excitation and detection volumes that prevented observation of fluorescence fluctuations. Invention of confocal and two-photon excitation microscopies, with greatly reduced volumes, expanded the technique to living systems in the 1990s.

The primary parameters determined from FCS measurements are the average diffusion rate and concentration of observed molecules (**Table 1**). A variation of FCS called fluorescence cross-correlation spectroscopy (FCCS) requires labeling of two molecules with different fluorescent dyes and can be used to study molecular interactions [41]. The amplitude of the ACF plot is inversely proportional to the average number (N) of fluorescent molecules traversing the detection volume and gives FCS a large dynamic range with the ability to measure pico- to nanomolar concentrations of molecules with background fluorescence and detector technology limiting cellular measurements to ~1–400 nM [45, 51]. Concentrations higher than a couple hundred nanomolars cause a reduction in fluctuations and inability to calculate the ACF. Diffusion of fluorescent molecules can be measured over six

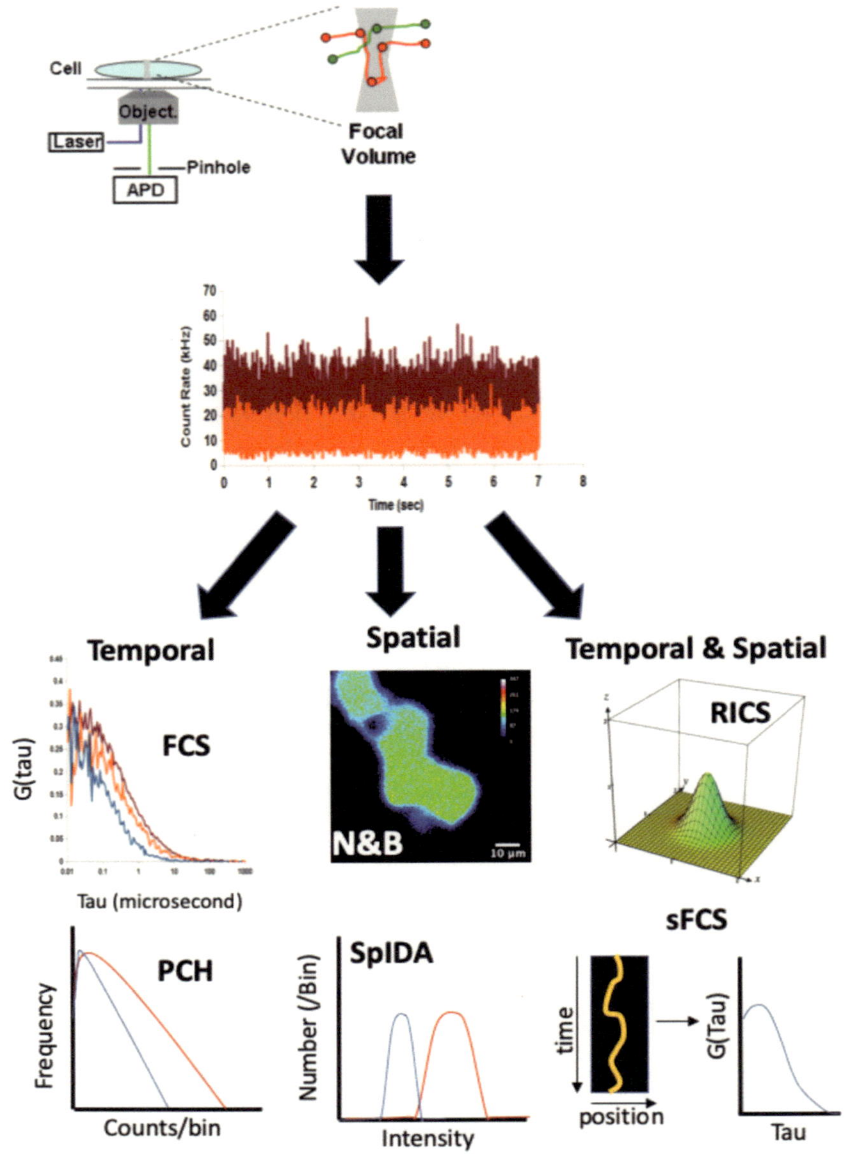

Figure 2.
Categories of fluorescence fluctuation spectroscopies (FFS). Fluorescence signal is recorded from molecules moving through focal volume (~1 femtoliter) of a confocal or two-photon microscope. The fluctuations in the signal can be analyzed to extract molecular parameters (e.g. diffusion and brightness) in the temporal, spatial, or temporal and spatial domains. Representative techniques that analyze in the temporal domain are FCS and PCH. Techniques that analyze in the spatial domain include N&B and SpIDA. The RICS and sFCS analyses exploit information in the temporal and spatial domains. For FCS, blue curve represents fast diffusing species, and dark red curve represents slower species. For PCH and SpIDA, the blue curves represent small species (e.g. monomers) and the dark red curves represent larger species (e.g. oligomers).

orders of magnitude using FCS (microsecond to millisecond) with the limitations at the upper limit being photobleaching for slowly moving particles/complexes and inability to detect fluctuations, as mentioned above. FCS has been used to measure the diffusion and size of nucleic acids, lipids, individual proteins, and protein complexes in live cells. Recent examples of processes investigated with FCS/FCCS include viral capsid assembly, mRNA diffusion, enzyme subunit dissociation, system wide measurements of protein diffusion in yeast, gene expression

Technique name	Temporal resolution	Spatial resolution	Parameters extracted
FCS/FCCS	Nanosecond to microsecond	–[b]	Mobility, concentration, interaction
PCH/moment	–[a]	–[b]	Brightness, oligomerization
sFCS/sFCCS	Microsecond to millisecond	Diffraction limited	Mobility, concentration, interaction
RICS/cc-RICS	Microsecond to millisecond	0.8 μm	Mobility, concentration, interaction maps
ICS/ICCS	—	0.5 μm	Concentration, oligomerization
N&B/cc-N&B	—	Diffraction limited	Mobility, concentration, interaction maps
SpIDA	—	<30 nm[c]	Concentration, oligomerization
FIF	—	–[d]	Abundance and stability of oligomers

[a]*Depends on binning and detector characteristics.*
[b]*Spot measurement.*
[c]*Assuming sufficient signal-to-noise ratio (SNR).*
[d]*Depends on size and shape of ROIs and SNR.*

Table 1.
Comparison of select fluorescence fluctuation spectroscopies.

profiles, protein-lipid interactions, and receptor-ligand interactions [52–60]. FCCS can be used to quantify and distinguish different populations of receptors, protein complexes, protein-lipid interactions, and protein-DNA interactions [61–69]. FCS/FCCS are spot measurements where the excitation laser is kept immobile leading to high temporal resolution but measurements in several locations are required if a larger area of the specimen is to be analyzed.

Importantly, for multicomponent systems — where there are two different dye labeled molecules or different states of the same species — the diffusion rates for the species must differ by ~1.6-fold or greater in order for FCS to resolve the multiple species [70]. For this reason and others listed below (see Section 2.2), a large body of additional FFTs have been developed based on photophysical properties of molecules. Techniques such as photon counting histogram (PCH) analysis and fluorescence intensity distribution analysis (FIDA) exploit moments of the fluorescent signal (i.e. mean, variance, higher order) to determine molecular brightness (ε) and resolve multicomponent systems [71–74]. Methods centered around moment analysis and related FFTs are discussed in Section 2.2 to below.

2.2 Photon counting histogram (PCH) and higher-order analyses

2.2.1 PCH and FIDA analyses

Intensity (1st moment) and variance (2nd moment) information collected during FCS experiments can be used to determine protein size and *N* provided the concentrations of investigated molecules are constant or homogenous during measurement (e.g. amount of receptor monomer and dimer). Importantly, biological systems are complex and the fluorescently labeled molecules/structures usually exist in multiple states whose concentrations are changing temporally. Furthermore, the molecule's brightness contributes nonlinearly to the ACF and, therefore, molecular concentrations cannot be unambiguously determined without knowing the molecular brightness (ε) [70, 75]. Thus, it can be difficult to extract protein size and N accurately using FCS with complex systems [70, 75].

Both PCH and fluorescent intensity distribution (FIDA) analyses do not suffer from the same limitations as FCS. These analyses measure the ε and associated fluorescence intensity distributions to resolve species in multicomponent systems and are amenable to systems where the species have similar diffusion rates [71–74]. The two techniques are considered "mathematically equivalent" and only differ in their treatment of the microscope point spread function (for explanation see, [76]). To implement these techniques, photon counts per sampling time are plotted versus frequency and then fitted using theoretical models to determine N and ε, expressed as counts per second per molecule, of the different species. Importantly, the same fluorescent traces used to calculate ACF for FCS analysis can also be used to calculate histograms for PCH analysis. In fact, several commercial hardware/software implementations provide FCS and PCH analysis simultaneously during data acquisition (e.g. Zeiss, Picoquant). Both PCH and FIDA techniques are routinely used to measure protein oligomerization in cells and cellular organelles [67, 77–80]. Homodimers were detected for serotonin 5-hydroxytryptamine 2C (5-HT2C) receptor using FCS and PCH analyses [81–83]. In one study, PCH analysis of freely diffusing class A GPCRs (5-HT2A, α1b-AR, β2-AR, M1, M2, and D1) were observed to be predominantly homodimers in the plasma membrane of HEK293 cells [82]. The stoichiometric binding of ligands to several GPCRs (chemokine, peptide, and small-molecule ligands) was detected using FIDA in a high-throughput assay and similar results were obtained compared to traditional radioligand binding experiments [84].

A mathematically equivalent approach to PCH/FIDA is moment analysis, based on Mandel's Q-parameter, and provides the same information [85–87]. Moment analysis can be difficult in small cells, such as bacteria, because the cell volume can be easily bleached and the cell only partially overlaps with the microscope point spread function (PSF) that can negatively influence intensity fluctuations. These two problems can cause error in brightness calculations [88, 89]. A modified Q-parameter measurement called mean Q-value of segmented photon count data (MSQ) combined with axial scans was developed to overcome these unique challenges faced in cells that are small [89].

2.2.2 Fluorescence cumulant analysis (FCA)

An approach similar to PCH is fluorescence cumulant analysis (FCA). This method employs cumulant functions in place of probability distribution functions and the same photon-counting data used for PCH can be analyzed by FCA [90]. Muller found that FCA gave identical ε and N values as PCH and both methods are limited to data sampling times shorter than the diffusion times of the molecules. In contrast to PCH, the FCA theory can be expanded through analytical expressions (e.g. 2nd factorial cumulant) to overcome this limitation and be extended to arbitrary sampling times [90]. FCA is also suited for data containing random independent events because of the additive properties of cumulants that simplify the analysis.

2.2.3 Temporal extension of PCH, FIDA, and FCA

PCH, FIDA, and FCA provide a snapshot of N and ε of molecules but not dynamic, or temporal information. These techniques can be modified to extract temporal information through recording multiple counting intervals simultaneously. These time-binned versions of PCH, FIDA, and FCA are called photon counting multiple histogram (PCMH), fluorescence intensity multiple distribution analysis (FIMDA), and time-integrated fluorescence cumulant analysis (TIFCA), respectively [91–95]. These techniques apply a global analysis, consideration of many dwell-times, to a set of photon count histograms, thus improving signal-to-noise (S/N) ratio of the data and allowing resolution of fluorescent species by both time-dependent (e.g. translational diffusion) and time-independent (e.g. N and ε) parameters.

Recently, a modified version of PCH was developed called correlated PCH (cPCH) that is a generalization of dual-color PCH analysis and combines aspects of FCS, PCH and moment analysis [96, 97]. This inherently dual-channel technique can also be applied to single channel measurements and histograms from the same channel, or different channels are acquired and analyzed from different timepoints. The application of cPCH involves calculating photon counting histograms where the time bins occur at time-differences (τ) between the two detection channels. This allows the analysis of correlation and amplitude of the signal simultaneously. The cPCH method is described to be less computationally intensive and to provide a better approximation for the diffusion dependence of the molecular species compared to PCMH, FIDA, and TIFCA. As proof of concept, Scales and Swain present simulated data for ligand binding to a receptor that has two-binding sites [96]. Scales and Swain lay the theoretical foundation for cPCH but only provide simulations to provide proof of concept and it remains to be seen if the full potential of this technique is realized using real experimental data. In fluorescence intensity and lifetime distribution analysis, lifetime data of the fluorescent species can also be included to provide an additional parameter to resolve the investigated molecules [98]. Photon arrival time distributions can also be used to measure ε and correlation functions simultaneously for measuring macromolecular interactions through the analysis of two-dimensional photon pair histograms [99].

The development of these moment/cumulant techniques extended the sensitivity and resolution capabilities of measuring complex mixtures of fluorescent macromolecules compared to FCS/FCCS. The subsequent extension of these techniques to a global analysis that incorporates temporal information allowed measurement of molecular dynamics. Dual-color versions of these techniques allow the observation of two different labeled fluorescent macromolecules [73, 97]. However, these spot techniques — at the time — were cumbersome for measurements involving multiple, or large region of interests (ROIs). This necessitated the development of scanning FFTs and multiplexing approaches to acquire spatial information across large regions of samples, such as entire cells or tissues (Section 2.3).

2.3 Scanning FFTs (single and multiplexing)

2.3.1 Scanning and multiplexing FCS

In the 1990s, Koppel et al. linearly scanned the laser across samples to measure, with high spatial resolution, the concentration dynamics of fluorescently-labeled DNA [100]. This technique is called scanning mode FCS (sFCS) and allows the measurement of a larger area of the specimen without the hassle of multiple time-consuming single point measurements (for review see [46]). In sFCS, fluctuations are collected in adjacent volumes, at a fast rate, and in the same volume during different orbits, at a slower rate, thus spatial and temporal information are encoded in the data. This allows for correction of slow sample movements that can interfere with analysis in spot measurements like FCS. The increased spatial resolution achieved with sFCS is at the expense of decreased temporal resolution but for most measurements in live cells this is not a major issue because molecules are diffusing at rates of 0.1 μm^2/sec (membrane proteins) to 25 μm^2/sec (soluble proteins) that is within the temporal resolution of sFCS (microsecond to millisecond) (**Table 1**). The sFCS approach provides better signal-to-noise ratio, greatly reduced chance of photobleaching, and better accuracy for measuring macromolecules with slow dynamics. Initial implementations of sFCS kept the laser stationary and moved the sample but recent setups move the laser beam instead and linear and raster-scanned patterns can be used, leading to different spatial and temporal resolutions [46].

One motivation for multiplexing FCS measurements was due to the need to have single molecule sensitivity and high-spatial resolution to measure a large number of samples in biochip microarray analysis, both during fabrication and implementation of the chip.

Multiplexing of a spot method such as FCS requires the acquisition of fluorescence fluctuations from multiple ROIs in parallel without sacrificing temporal resolution (micro-to-nanoscale). Many strategies have been used to multiplex FCS for confocal microscopes and most techniques employ an array of detectors and custom optical elements to create multifocal excitation spots and detection [101–103].

Strategies used to create multifocal spots include, the use of a 2 × 2 fan-out diffractive optical element, parallel fiber optics, and spatial light modulators [101, 103–105]. These early attempts at multiplexing relied on single photon avalanche photodiodes (SPAD) and photomultiplier tubes (PMTs) due to their high sensitivity and response times (for review of detectors see [43]). However, only small numbers of excitation volumes (4–7) could be generated ranging from PMT, SPAD, and complementary metal oxide semiconductor (CMOS) detectors [101, 103, 105]. Many groups directed their efforts toward implementing electron multiplying charged coupled devices (EMCCDs) because the detectors have high quantum efficiencies, single molecule sensitives, and the flexible detection area with microsecond resolution [106]. A custom fabricated CMOS single photon detector has been shown to collect 64 channels, an 8-fold improvement on earlier studies [107]. A multiplexed setup measuring five ROIs measured the dynamics of the heat shock transcription factor in the nucleus of heat stressed mammalian cells [108]. Confocal volumes need to be sparsely separated (10 or more pixels) in order to avoid cross-talk between signals, limiting confocal multiplexing approaches. One way to circumvent this limitation is to use a spinning disk confocal microscope that can scan a large number (1000s) of widely spaced pinholes temporally across the sample [109]. In this approach, each pixel is only excited for a brief period of time, leading to reduced signal-to-noise ratios.

Both confocal and spinning disk multiplexing approaches suffer from issues with interference from out-of-focus light restricting these techniques to thin samples. Employing total internal reflection (TIR) microscopy, or single-plane illumination microscopy (SPIM) can eliminate this problem and extend multiplexed FCS to thick biological samples [44]. Most SPIM-FCS microscopy have a larger volume compared to confocal because two medium-NA (0.8) objectives are needed to create the plane for illumination and detection [102, 110, 111]. This type of FCS imaging has been used to record ~1 million ACFs simultaneously at 25 frames per second using next generation scientific CMOS detectors [102]. These custom-built systems have mostly been used for detecting dyes and fluorescent beads in solution for proof-of-concept and only a handful of studies performed in cells and organisms have been reported to date [110]. SPIM-FCS and TIR-FCS have reduced bleaching and photodamage compared to other multiplexed FCS approaches because of the 100 nm illumination plane for TIR and ~ 1.0-µm light beam width for SPIM-FCS. Recently, SPIM-FCS has also been extended to a two-color version to investigate molecular interactions [112]. Combining single spot FCS measurements with automation to create a high-throughput workflow is an alternative to multiplexing and can be easier to implement with biological specimens [113, 114].

2.3.2 Raster image correlation spectroscopy (RICS)

Raster image correlation spectroscopy (RICS) is an imaging analog of FCS that extracts an averaged (2D) spatial ACF from a series of images acquired by a laser scanning confocal microscope [115–117]. The inherent time information encoded in the scanned image(s) can be used to measure molecular dynamics at microsecond, millisecond, and second timescales depending on if the correlation is calculated from pixels that are adjacent, in successive lines, or successive frames, respectively. The ACFs for immobile species have an isotropic Gaussian function, while species with diffusive behavior have a stretched Gaussian in the direction of the fast scanning. Therefore, the ACF in RICS is a product of the correlations due to diffusion and scanning. This aspect of the analysis makes RICS better suited for measuring anisotropic behavior (diffusion rate varying in space and time) compared to traditional single spot FCS. Processes such as auxin transporter diffusion in plant cell plasma membrane, dynamics of adhesion complex protein paxillin, thyrotropin-releasing hormone receptor diffusion under cholesterol depletion, and Ras nanoclustering in plasma membrane have been studied using RICS [117, 117–120]. RICS was combined with 3D-orbital tracking and general polarization analysis to measure fast and slow fluctuations of labeled lysosomes and gain insight into lysosomal trafficking and metabolism [121].

A useful feature of RICS is the ability to create a high-resolution spatial diffusion map of the investigated macromolecule. The RICS technique can be implemented with cross-correlation (ccRICS) between two molecular species to investigate clustering of a kinase at the plasma membrane, degradation of exogenous DNA in cells, measure remodeling of adhesion complexes, and regulation of sub-apical trafficking of a phosphate cotransporter [122–129]. A modified version of RICS has been developed to minimize spectral cross-talk using statistical weighting of the signals [130].

The pixel size and dwell time must be compatible in order to implement RICS successfully [117, 124]. For example, scanning faster than the molecule's diffusion rate makes the species appear to be immobile. In contrast, the particle will diffuse away before being detected in subsequent pixels if scanning is performed at a rate slower than the characteristic diffusion of the molecule. Practically, pixel dwell times of 2–100 μsec and pixel sizes of 0.025–0.2 μm should be sufficient to measure the dynamics of a range of macromolecules. For example, a dwell time of 25 μs and pixel size ≤0.05 μm can measure the diffusion of a 25 kDa cytoplasmic protein by the second or third pixel scanned [115, 124]. Longfils et al. observed a 3–5-fold decrease in error on the diffusion constant by decreasing pixel size from 100 to 50 nm and they also observed consistent high quality RICS data over a board concentration range from nM to μM [131]. Minimum ROI of 2 μm × 2 μm is recommended by Brown et al. to prevent under-sampling. Overall, RICS is a powerful technique for measuring diffusion and complex dynamics across a range of timescales at the expense of a 2–3-fold lower resolution compared to FCS [124].

2.3.3 Image correlation spectroscopies

Image correlation spectroscopy (ICS), imaging cross-correlation spectroscopy (ICCS), spatiotemporal imaging correlation spectroscopy (STICS), and k-space imaging correlation spectroscopy (kICS) are additional methods that are imaging versions of FCS (for review see [132, 133]). All of these methods rely on the acquisition of a time series of images from a confocal, two-photon, or TIR microscopy. The series of images amount to fluorescence intensity recorded as a function of time and space, $i(x,y,t)$, and the generalized spatiotemporal function is defined as

$$r_{ab}(\xi,,\eta,,\tau)=\frac{\delta i_a(x,,y,,t)\delta i_b(x+\xi,,y+\eta,,t+\tau)}{i_a(x,,y,,t)_t\, i_b(x,,y,,t+\tau)_{t+\tau}}$$

where the fluorescence fluctuation, $\delta i(x,,y,,t)$, is defined as $\delta i(x,,y,,t)=i(x,,y,,t)-i(x,,y,,t)_t$ [132]. This equation is for two different emission wavelength detection channels (a, b) and in the case of autocorrelation of one channel ($a = b$) the subscripts are dropped. The different ICS variants are just simplified versions of this equation for the specified limits. Importantly, all images acquired on the fluorescence microscope are a convolution of the emission of the fluorophore with the microscope's PSF. This convolution causes spreading of the diffraction limited fluorophore signal over several pixels in the image and this signal is correlated over space and time. This is a powerful technique because large numbers of stochastic fluctuations can be reduced to a small number of physically important parameters due to spatiotemporal averaging. An overriding assumption of ICS and related techniques are that the biological system being investigated is stationary in space, or time. This condition is not necessarily met in living systems such as cells that can move in space and time and are heterogeneous. In addition, conventional applications of these spatial imaging techniques are insensitive to interactions and molecular movements below the diffraction limit of ~200 nm. In the original ICS, the correlation function is calculated for each pixel from the raw data (images) and fitted to an analytical model to extract parameters such as cluster density and degree of aggregation of the molecular species is determined [134]. The amplitude peak of the calculated correlation is proportional to the number of particles. Broadening of the correlation peak can be caused by movement of the investigated particle and this initially limited ICS to the study of large clusters on the cell surface [135].

Srivastava et al. were able to enhance the ICS technique by extending into the time domain with temporal imaging correlation spectroscopy (TICS) [136]. The correlation function is calculated within and between images but only a single correlation function was calculated per image frame meaning there was no spatial, or vectoral information. Wiseman et al. combined ICS and TICS approaches to create spatiotemporal image correlation spectroscopy (STICS) [137, 138]. In this technique, there is dynamic and vectoral information encoded in the correlation peaks. In addition, sub-regions of samples could be quantified and velocity information determined. A two-color version of STICS was developed and given the name spatiotemporal image cross-correlation spectroscopy (STICCS) [139]. This technique was used to measure the cotransport of several integrin subtypes with paxillin and distinguish diffusion versus active transport [139]. A drawback to STICS and TICS is the potential photobleaching and dye photophysics (e.g. blinking) that can interfere with analysis [140].

A more recent variant of ICS is labeled k-space image correlation spectroscopy (kICS) and this technique utilizes 2D fast Fourier transformation to analyze the image data in the frequency domain (k-space ACF) [141]. Analyzing in k-space makes it easy to separate molecular dynamics from the artifactual photophysics of the dye. An additional advantage of kICS is the dimensions of the PSF microscopy system are not needed for the analysis. In addition, complex mixtures of proteins undergoing directed transport and diffusion can be measured with kICS. The kICS technique has been used to study the transport dynamics of integrins and agonist induced clustering of CFTR in cholesterol microdomains [141, 142]. ICS and its modifications have been used with ~30–150 ms frame rates restricting these techniques to the study of large and slow-moving cellular structures on the cell surface and some intracellular organelles.

2.3.4 Number and brightness (N&B) analysis

Number and brightness (N&B) analysis, also known as scanning N&B, is a scanning extension to PCH, FIDA, and moment analyses developed by Digman et al. [143]. In this technique, the mean intensity and variance (i.e., 1st and 2nd moments) are calculated from each pixel of a series of images. The apparent brightness (B) of the molecule at each pixel can be calculated as the ratio of the variance to the average intensity (σ^2 / k). The number of mobile molecules (N) can be calculated as the ratio of total intensity to B (k^2 / σ^2) [143]. The parameter N is directly related to *n* particles. N&B analysis can be implemented using a confocal or widefield fluorescent microscope equipped with either point, or array detectors [143, 144]. It is important to correct for detector noise, background signal, and artifactual motion such as specimen drift or movement that can contribute to variance and lead to error [124, 145]. N&B analysis has been successfully implemented to measure oligomeric complexes in live cells such as remodeling of adhesion complexes, activation of ErbB receptors, Ebola viral capsid assembly, DNA and protein aggregation or oligomerization [79, 143, 146–151]. The cross-correlation N&B (ccN&B) approach extends N&B to two-colors and has been used to measure stoichiometric binding of focal adhesion and integrin protein complexes [116, 127, 152, 153].

Similar to RICS, it is important to select the proper pixel dwell time in order to capture the fluctuations. For example, a pixel dwell time of 25 μs is reasonable for a protein diffusing at ~20 $\mu m^2/s$ [143]. Autofluorescence, light scattering, and detector noise can all contribute to the signal variance, potentially contributing error to N&B analysis. Many times, these variances are independent and thus the "noise" is equal to the sum of the total variance [143]. Importantly, the variance due to the mobile molecules/species fluctuations will vary with the square of the brightness ($\sigma^2_n = \varepsilon^2 n$) resulting in a B value >1 and immobile species will have a B value of 1. A simple plot of variance as a function of intensity (increasing illumination) should lead to a quadratic relationship if the signal variance arises from the true molecular species and not "noise". Therefore, subtraction of 1 from the B value will give a true molecular brightness.

A limitation of N&B analysis, in its original form, was the inability to detect mixtures of several species in the same pixel and only weighted average brightness could be obtained.

Hortiguela et al. developed a statistical resampling of raw fluctuation data to measure coexisting oligomer species in each pixel and this improved N&B methodology is called enhanced N&B analysis (eN&B) [154, 155]. One drawback to eN&B is analysis requires 200 frames and can take from seconds to minutes to complete one timepoint. Overall, N&B analysis can robustly identify spatially heterogenous clusters of macromolecules in an image and generate a *spatiotemporal map of said clusters, but these are averaged values, unless eN&B is employed.*

2.3.5 Spatial intensity distribution analysis (SpIDA) and fluorescence intensity fluctuation (FIF) spectroscopy

Spatial intensity distribution analysis (SpIDA) was inspired by PCH analysis and creates fluorescence intensity histograms that are subsequently analyzed using super-Poissonian distribution functions [36]. In SpIDA, the density and quantal brightness (QB) of the fluorescently-labeled protein are calculated and can be used to determine expression level and oligomerization state. A given protein's QB can be compared to the QB of control proteins that are known to be monomer, or larger oligomers, therefore determining the oligomeric state of the protein of interest [36]. As a result, information regarding the densities of fluorescent molecules and their QB are obtained. While PCH is focused on the time domain, SpIDA is centered around the spatial domain of acquired images. The intensity histograms distributed from SpIDA analyses can elucidate the presence of dimers from monomers in a population based on intensity fluctuations, making this method of analysis beneficial in comparison to previous techniques discussed above. Importantly, the SpIDA technique is compatible with conventional immunostaining methods used on fixed cells and tissues that is a distinct advantage over PCH and N&B analyses [156]. SpIDA analysis software is freely available from the De Koninck lab (https://ydklab.org/tools). SpIDA analysis was employed to measure the QB and surface density of GPCR proteins in live and fixed cells. GPCRs or associated signaling proteins measured using the SpIDA approach include GABA(B) in spinal cord tissue, tyrosine kinase, secretin receptor, protein aggregation, and kinetics of fluorophore uptake in tissue culture cells [156–161].

Both N&B and eN&B fail to separate distinct oligomeric species based on brightness distribution. SpIDA can resolve multiple oligomeric species but at the expense of reduced resolution [44, 154, 162]. One and two-dimensional fluorescence intensity fluctuation (FIF) spectroscopy was developed to measure the effects of agonists and antagonists on the concentration, size, and stability of membrane receptor oligomers [162, 163]. Stoneman et al. were motivated to develop a rapid method that could accurately identify and measure mixtures of receptor oligomer sizes from individual pixels of images. Their technique requires the segmentation of a large ROI into hundreds of smaller ROIs (200–500 pixels) and thousands of concentration and brightness values are calculated from these ROIs. The brightness distributions calculated from the intensity fluctuations spread across the image pixels are used to generate spectrograms. The spectrograms, created for each segmented ROI, can then be used to measure the size and abundance of oligomeric species. The authors used FIF spectroscopy to probe the dynamic change in receptor oligomer size for the secretin (Class B GPCR) and epidermal growth factor receptors [162].

3. Advantages and disadvantages of FFTs

As mentioned above, clear advantages for FFTs include single molecule sensitivity, measurements at physiological concentrations, and high spatial, and temporal resolution. Protein complex stoichiometries and kinetics can also be measured with dual-color versions of FFTs. FFTs rely on fluorescence fluctuation measurements in space, time, or both to extract the molecular parameters. There are many factors that can affect the quality of the fluctuation data collected. Factors that can negatively impact FFT analysis include fluctuations not from the molecule of interest (i.e. cell movement, stage drift), fluorophore photobleaching, immobile species, illumination geometry artifacts, undesired photophysics of fluorophore, and detector characteristics [45, 164, 165].

The measurement time, or scan speed, must be appropriately matched with the macromolecule's diffusion rate in order to acquire sufficient fluctuations for analysis. It is also important to distinguish the molecule's fluctuation from other factors that can contribute to the signal variance. Slow cell or organelle movement can add additional fluctuations that create artifacts during analysis [88, 89, 164]. For example, Hennen et al. observed that a nuclear localized EGFP protein exhibited a concentration dependent increase of brightness [164]. The cytosolic version did not exhibit this unexpected behavior. After further analysis, they concluded that the anomalous result was due to movement of the nuclear membrane and employing MSQ corrected the problem. MSQ analysis was also used to correct for incomplete PSF overlap and photobleaching for measurements in bacteria [89]. Undulations affecting FFT measurements have been observed for other membranes and several strategies, including lowering the measurement temperature (22°C or 30°C instead of 37°C), algorithmic filtering of data, multiple short measurements, and removal of distorted correlation curves were employed to improve the analysis [166–169]. These extra-fluctuation effects can be problematic especially when sample thickness does not exceed the size of the observation volume. Finite sample thickness can lead to a bias in the brightness calculation but this can be corrected by performing an z-scan and using this data to better model the PSF shape when analyzing the FFT data [145, 170].

Photobleaching is more problematic for spot methods and less an issue for scanning FFTs and can lead to incorrect determination of molecular parameters. The two main approaches for correcting photobleaching are mathematical fitting of the experimental signal decay or moving average filters that do not rely on fitting [171]. Both mono- and bi-exponential functions have been applied to correct for the signal decay in FFT data with varying success [88, 168]. It can be difficult to correct using a fitting method because the decay rates can vary substantially from cell to cell [172]. Moving average filters, for imaging data, work by calculating the local average of each pixel in a time series (all images) and the average of the neighboring time-points (global) in the series. Next, the fluctuation at the local pixel is defined as the deviation from the local average and then the local pixel value is changed to the global average value plus the addition of the fluctuation value. The two methods for calculating the local average are called the box-car method, a.k.a. sliding window, and the exponential filtering method [143]. Filtering algorithms were applied to imaging FFT data to remove contributions from cellular movement or photobleaching and thus enhance the accuracy of the calculated molecular parameters [140, 143, 155, 173]. Lange et al. used a wavelet shrinking algorithm to reduce the false positive signal from 69% to 4% for the negative control in FCCS measurements [174]. Recently, Nolan et al. introduced the Robin Hood algorithm as an improved filtering method for live cell images with limited photon budget (i.e. low intensity signals) [175]. In this algorithm, counts are subtracted from high intensity images and then added to low intensity images in the series. The overall counts are preserved in the image series in contrast to the box-car or exponential methods where the filtering process can lead to loss of information because of the numerical processing (natural to real number).

FFT analysis can be negatively influenced by dead-time and afterpulse effects. The dead-time and afterpulse effects are intrinsic characteristics of detectors and can cause underestimation and overestimation of molecular brightness, as well as other molecular parameters (e.g. rotational diffusion), respectively [176, 177]. The dead-time of a detector is the fixed time after photon registration (e.g. 50 nanoseconds for actively quenched APD) where another event cannot be detected. If the signal intensity being measured is too great, or the temporal period is too short then a certain number of photons will not be registered leading to underestimation of ε.

The afterpulse effect is when there are additional spurious events registered after the real photon detection event and this leads to overestimation. For example, an APD can generate a nano- to microsecond delay that leads to a ~1% probability of detecting spurious events [176]. These two effects need to be corrected especially at higher concentrations [95, 177]. Hillesheim et al. conducted PCH measurements on a fluorescent dye in solution and calculated N and ε with or without correction for dead-time and afterpulse effects [177]. They found relative error >10% when there was more than one molecule present in the observation volume if corrections were not employed. One practical method to remove afterpulsing effects from FCS measurements is to split the signal being measured between two detectors and then calculate the cross-correlation (signal to self). Correlating the signal to itself can remove the spurious events and compensate for the afterpulsing effect. Importantly, individual detectors of the same brand, type, and make require fundamentally different mathematical models to correct for these effects and thus universal models cannot be generated and each detector must be characterized individually [178]. Fukushima et al. applied a two-detector approach to N&B analysis called TD-N&B to correct for detector artifacts when they measured the dynamics of glucocorticoid receptor dimerization [179].

The photophysics of the fluorescent dye, or protein can negatively influence the FFT analysis. For example, it is well known that a significant portion of certain fluorescent protein molecules do not properly mature and/or have long-lived triplet "dark" states (~10–70%) [180, 180–184]. These two processes can lead to a large number of molecules that do not fluorescence. In FCCS, if 100% of two labeled species are interacting then the CCF should be ideally 1.0, but in many experimental setups the value is ~0.5 which can lead to incorrect K_d determinations. Maturation problems, triplet state fraction, and photobleaching contribute to this discrepancy. The impact on FCCS analysis due to these effects can be corrected to provide label-independent K_d values [66, 180]. Correction factors can be employed to account for incomplete maturation of fluorescent proteins and this approach was used successfully to measure the oligomeric status of the homo-trimeric influenza A Hemagglutinin (HA) glycoprotein in live cells [183].

Overall, many of the disadvantages, or limitations of FFT measurements can be overcome with proper experimental design, and correction factors in the analytical models used to fit the data. It is important to thoroughly characterize your experimental setup, especially the detectors. Unfortunately, studying fast chemical processes necessitate the above-mentioned correction factors. Ideally, it is better to design experiments where the fluctuations being measured are not on the same time scale as the detector traits (i.e. dead-time, afterpulse) and that photobleaching, and other photophysical processes that have a negative impact, be minimized.

4. FFTs reveal new information on GPCR structure and function

Different GPCRs classes are known to vary in their oligomeric structure due to the varying propensity for dimerization and avidity of their interactions. Furthermore, agonist/antagonist ligands are known to alter GPCR quaternary structure. Generally speaking, the ability of GPCRs to associate with self, or other family members aligns with the receptor family. For example, class A GPCRs tend to transiently associate and multiple surfaces are required for the interaction [185]. In contrast, class C GPCRs form obligate dimers and higher-order oligomers [186]. While class B GPCRs appear to have an intermediate behavior to A and C receptors [187]. The class A family represent the largest class of GPCRs and it is heavily debated whether these receptors can form dimers, or larger structures due to transient interactions [188]. The significance and molecular basis for these receptor complexes are poorly understood.

Historically, these GPCR oligomeric changes were observed using biochemical approaches such as Co-IP, FRET, BiFC, and BN-PAGE experiments. Often, different experimental approaches have yielded opposite conclusions concerning the oligomeric state of the GPCR. In the past decade, FFTs have been employed to measure the dynamics of GPCRs in live cells and these methods are revealing new insights into GPCR structure/function and may shed light onto the disparate results (**Table 2**).

Receptor name*	Technique	Species observed (in the absence or presence of ligand)	References
M1-muscarinic	FCS/PCH	Dimer	[82]
M2-muscarinic	FCS/PCH	Dimer	[82]
Serotonin (5-HT_{2A})	FCS/PCH	Dimer	[82]
Adrenergic (α_{1b}-AR)	FCS/PCH	Dimer	[82]
Adrenergic (β_2-AR)	FCS/PCH	Dimer	[82]
D1-Dopamine	FCS/PCH	Dimer	[82]
Serotonin (5-HT_{2C})	FCS/PCH	Dimer	[81, 83]
$GABA_{B1}$ & $GABA_{B2}$	SpIDA	Monomer/dimer	[157]
Secretin	SpIDA	Monomer/dimer	[160]
Secretin	2D-FIF	Monomer/dimer/higher-order oligomers	[162]
Neuropeptide Y	FCS	Clustering	[168]
D3-Dopamine	SpIDA	Monomer/dimer/higher-order oligomers	[193]
Serotonin (5-HT_{2C})	SpIDA	Monomer/dimer/higher-order oligomers	[194]
M1-Muscarinic	SpIDA	Monomer/dimer/higher-order oligomers	[197]
M3-Muscarinic	SpIDA	Monomer	[197]
μ-opioid	FCS/PCH/N&B	Tetramers/oligomers	[204]
δ-opioid	FCS/PCH	Tetramers/oligomers	[204]
A1-Adenosine & A2a-Adenosine	FCS	heterodimers	[205, 206]**
B2-Bradykinin	FCS/sFCS	Small/large aggregates	[207]
Complement (C5a)	FCS	Clusters	[208]

**GFP or YFP-tagged.*
***Fluorescent ligand used to label receptor.*

Table 2.
Summary of fluorescence fluctuation spectroscopy studies on GPCRs.

4.1 Muscarinic acetylcholine M_2 receptor: monomer, dimer, or tetramer?

One prominent example of controversy in the field is the quaternary state of the muscarinic acetylcholine M_2 receptor (M_2R). The M_2R was reported to be predominantly a monomer with capability to form dimers based on TIRF microscopy experiments in CHO cells, primary cardiac cells, and cardiac tissues [189]. In contrast, Herrick-Davis et al. reported M_2R was predominantly a dimer based on FCS/PCH analysis in transfected cell lines [82], while FRET, single molecule photobleaching (smPB), and ligand binding assays yielded results consistent with a tetrameric arrangement for M_2R [190, 191]. What factors could account for these seemingly contradictory results?

Recent biochemical studies on three distinct class A GPCRs using proteoliposomes reported that environmental factors, including membrane curvature and receptor density, affect degree of homo-oligomerization [192]. A concentration dependent increase in the number and complexity of oligomeric species has been observed for both the 5-HT2C and dopamine D3 receptors (D_3R) based on SpIDA analyses [193, 194]. In addition, the experimental temperature used can have a profound impact on the GPCR being investigated. Nenasheva et al. observed a two-fold change in mobility for every 10°C change in temperature, suggesting alterations in protein-protein interactions or protein-lipid interactions or both [189].

Expression level differences between the smPB and FCS experiments for M_2R seem unlikely given there were ~880,000 receptors per cell and ~52,000–620,000 receptors per cell, respectively [82, 195]. These levels of expression are within an order of magnitude for M_2R in rat cardiomyocytes that are estimated at 1.7×10^6 copies per cell [196]. The discrepancies in the M_2R results could be explained by photobleaching in the FCS/PCH measurements. It is well known that slow moving membrane proteins are prone to photobleaching during FCS measurements and photobleaching "shortens" the apparent diffusion time of the species [45, 79, 165]. Photobleaching can be difficult to detect because fluorescent molecules replenish the bleached molecules leading to a steady intensity recording that does not decrease with time as expected. Stavesich et al. refer to this scenario as "cryptic" bleaching [165]. Therefore, it is possible that slower moving, and/or immobile tetrameric species were not "visible" in the FCS experiments but were in the FRET and single-particle experiments. The authors acknowledge that less mobile GPCR clusters in membrane microdomains would not be detectable in their FCS/PCH analysis [82].

4.2 SpIDA analyses of GPCR organization and dynamics

Many FFT studies performed to date have utilized the SpIDA technique to measure mEGFP-tagged GPCR proteins [36, 160, 188, 193, 194]. The original studies that employed SpIDA measured the organization of $GABA_{B1}$ and $GABA_{B2}$ receptors alone and in combination through antibody staining in rat spinal cords [157]. Godin et al. demonstrated through single- and double-labeling of B_1 and B_2 receptors that SpIDA returned the expected oligomerization state of monomer (single label) and dimer (double label). The authors also measured the dynamics of EGFR-GFP in absence and presence of 20 nM EGF [157]. Short-term incubation with EGF increased the amount of dimeric EGFR-GFP without any change in total protein on the surface suggesting that changes in monomer levels were not due to receptor internalization. These measurements were performed on a range of expression levels from 100,000 to 600,000 receptors/cell (32–220 EGFR per μm^2) suggesting the oligomeric changes seen are not concentration dependent.

Milligan et al. measured 5-HT2C in the absence and presence of several distinct chemical ligands. They observed mostly monomeric 5-HT2C at low expression levels and a concentration dependent increase to larger oligomeric sizes. Several antagonists, including SB221284, disrupted the larger oligomeric species into monomers and this was a time-dependent effect that was reversible by washing out the chemical [194]. Similar observations were seen for D_3R where increased receptor expression in the plasma membrane shifted the receptor organization to dimer and higher-order oligomers [193]. Interestingly, several antagonists, including spiperone and haloperidol, promoted monomerization of the D_3R while others, such as eticlopride, nemonapide, and clozapine, had no effect. Importantly, Marsango et al. created a mutated version of D_3R ($Asp_{110}Ala$) that reduced its affinity for spiperone and this mutant receptor had less conversion to monomer species in the presence of ligand. This result suggests that the ligand affects the receptor-receptor interactions and not the lipid membrane, or another associated protein. Molecular dynamic simulations suggested that that spiperone binding increased the distance between transmembrane IV-V and haloperidol binding increased the distance between transmembrane I-II, which could disrupt interactions [193].

Measurements of the muscarinic M_1 receptor using SpIDA revealed mostly monomer organization in the basal state with some dimers/larger species (75:25 ratio) and addition of the antagonist pirenzepine caused a large shift to larger species (73.5%) [197]. Importantly, sustained treatment with pirenzepine also caused increase receptor expression from ~50 to ~120 receptor copies per μm^2. Unfortunately, the authors were not able to determine if the shift to larger species upon antagonist incubation was due to binding, or increased expression. Similar to D_3R, other antagonists had no effect on receptor association, such as the non-selective agents' atropine and *N*-methylscopolamine [197]. Disruption of the actin cytoskeleton with cytochalasin D enhanced M_1 receptor oligomerization and this is not surprising given, cytochalasin D is known to redistribute caveolar/lipid raft-resident proteins, including GPCRs (β_1-adrenergic receptors (β_1-AR, β_2-AR)), and increase cAMP levels [198]. The enhanced M_1 receptor oligomerization was most likely caused by altered interactions of the GPCR with adenylate cyclase and $G\alpha_s$ due to disruption of lipid rafts/caveolae [198]. In contrast, the M_3 receptor quaternary organization was not affected upon incubation with either pirenzepine or telenzepine antagonists [197]. Interestingly, the M_3 receptor was measured as mostly monomers (~80%) based on SpiDA analysis but FRET analysis suggested the receptor existed as a mixture of dimers and tetramers [197, 199]. It is important to note that for the FRET study M_3 receptor-Citrine and M_3-receptor-Y149C/A239-Cerulean were co-expressed and used for measurements. The Y149C/A239 mutation alters receptor ligand selectivity to allow activation by a synthetic chemical instead of acetylcholine [200]. The FRET study does not appear to have used monomeric versions of Citrine and Cerulean unlike the SpIDA study (mEGFP) and this could have led to oligomerization artifacts [201].

Most FFT studies on GPCRs have focused on class A receptors but two recent studies were performed on the peptide hormone secretin receptor (class B). Ward et al. observed a mixed population of monomers (33.8%) and dimers (59.2%) with only a small portion of receptors forming higher-order oligomers (7.1%) based on SpIDA analysis [160]. Addition of sodium butyrate, known to increase expression of GPCRs from viral promoters, led to increased expression and formation of higher-order oligomers (22% versus 7.1%) [160, 202]. Mutation of residues Gly^{243} and Ile^{247} to Ala in the lipid-facing side of transmembrane IV disrupted secretin receptor oligomerization, reduced cAMP production but did not change ligand binding affinity to the receptor [203]. SpIDA analysis indicated this mutant receptor was 90% monomeric and there was no change in receptor association upon sodium butyrate addition. Interestingly, cholera toxin inhibition of $G\alpha_s$ expression in cells did not affect secretin oligomerization, suggesting receptor protein-protein interactions mediate dimer formation [160]. Short-term incubation of the mutant secretin receptor with secretin caused an increase in dimeric population but the wildtype receptor organization remained unchanged. It is known that the receptor can be internalized as a monomer, and the authors hypothesize that the increase in dimer formation could be due to clathrin-mediated clustering instead of genuine receptor-receptor interaction, but this remains to be determined.

4.3 FCS, PCH, and N&B analyses of GPCR organization and dynamics

In addition to SpIDA studies, FCS, PCH and N&B analyses have been used to probe the dynamic assembly/disassembly of GPCRs [81–83, 168, 204–208]. Golebiewska et al. measured the oligomeric assembly of opioid receptor (OR) complexes in Neuro-2a cells [204]. They observed eYFP-μOR diffused in the plasma membrane at a rate of 7.3×10^{-9} cm^2/sec suggesting the receptor is organized as dimers and tetramers based on FCS measurements. This diffusion rate was not affected upon incubation with 1 μM morphine ligand for 24 hours. However, co-expression of eYFP-μOR or eGFP-μOR with unlabeled μOR led to a decrease in the diffusion rate suggesting the formation of higher order oligomers. Moreover, incubation with a range of morphine concentrations over 24 hours significantly increased diffusion in a dose-dependent manner and this effect was reversible upon co-incubation with the antagonist naloxone. These results suggested that morphine disrupted the OR complexes into smaller species. In contrast to the FCS results, the eGFP-μOR molecular brightness increased (dimers to tetramers) in the presence of morphine based on N&B analysis. Furthermore, co-expression of eGFP-μOR with unlabeled μOR had a molecular brightness the same as eGFP-μOR alone but with a greatly reduced diffusion rate. The simplest model to explain this result was if two eGFP-μOR are in complex with two μOR forming a mixed tetramer of labeled and unlabeled receptors. Importantly, the brightness values for these complexes was broad which could mean there was anywhere from one to three eGFP-μOR. Their FFT measurements coincided with the results they obtained from FRET.

Pharmacological modulation of adenosine receptor function can be used to treat heart arrhythmias, asthma, stroke, and renal failure [209]. Adenosine A_1 and A_{2a} heterodimers displayed a faster diffusion rate in the plasma membrane of CHO cells compared to homodimers (5.6×10^{-9} versus 4.0×10^{-9} cm^2/sec) based on FCS experiments [205]. Briddon et al. used a fluorescently labeled A_1 agonist to quantify binding to the adenosine A_1 receptor in single cells [206]. They identified at least two populations of receptors with varying diffusion times (tens to hundreds of millisecond) upon incubation with the agonist [206, 210]. Furthermore, there was a reduction in the amount of receptor complexes in the presence of an antagonist suggesting disruption of oligomer formation.

Bradykinin receptors (subtype B1 and B2) are GPCRs that mediate vascular contribution to inflammation and edema [211]. Single point and scanning FCS were used to measure the mobility of bradykinin subtype B2 receptor-GFP (B_2R-GFP) expressed in HEK293 cells [207]. The B_2R-GFP diffusion rate was measured to be 3.5×10^{-9} cm^2/sec in the plasma membrane by single point FCS and a similar rate was observed whether one- or two-photon microscopy was employed. Scanning FCS identified three populations of receptor with a large portion (~72%) of receptors located in slow-moving aggregates. The FCS results combined with binding studies and FRET suggested that B_2R complexes are dynamic and that pre-formed receptor-G protein complexes initiate rapid signaling through a limited number of molecules.

BiFC was combined with FCS analysis to identify and measure complex dynamics between the neuropeptide Y (NPY) receptor and β-arrestin [168]. The NPY receptors play critical roles in food intake, bone metabolism, and cardiovascular regulation [212]. Y1 and Y2 receptors were measured to have similar diffusion rates ~2.0×10^{-9} cm^2/sec and treatment with NPY to cause internalization decreased the diffusion rate of Y1-GFP receptor but not Y2-GFP receptor. Mutations in β-arrestin that prevented recruitment and internalization blocked the agonist induced diffusion changes observed for Y1-receptor. NPY treatment of Y1-receptor also caused an increase in brightness (~1.5 fold) based on PCH measurements suggesting receptor clustering. Overall, Kilpatrick et al. interpreted their results to mean they were observing early events in arrestin dependent clustering and endocytosis of Y1-receptor. An important observation in this study was in relation to one of their control proteins used for PCH analysis. The authors used a membrane anchored GAP-GFP and were surprised to find the brightness of it was the same as Y1-GFP which they measured as a dimer based on FRET. They noted that the concentration level for GAP-GFP in the plasma membrane was 0.05 mM based on FCS. This concentration is very close to the K_d (0.1 mM) measured for GFP *in vitro* [213]. Their conclusion was that their membrane embedded monomeric control was probably clustering as a dimer possibly due to lipid microdomains.

4.4 Experimental caveats of FFT measurements on GPCRs

Most FFTs are ensemble measurements that report an average of the population compared to single particle experiments that focus on a single molecule and this should be kept in mind when trying to compare FFTs to single particle tracking experiments. It is very important to confirm that monomeric and dimeric controls for FFT measurements are appropriate for the analysis. Controls used in the above discussed studies included roundabout receptor 1 (dimeric axonal guidance receptor), EGFR (dimeric), CD-86 (monomeric), CD-28 (dimeric), GABA heterodimer, palmitoylated-mEGFP (PM1) and palmitoylated-myristoylated-mEGFP-mEGFP (PM2) [82, 160, 162, 214]. For fixed cells/tissues, samples stained with secondary antibody only or background staining outside cells (primary and secondary) were assumed to have a nonspecific and monomeric distribution for the fluorescence signal [157]. Confirmation using a complimentary biochemical (e.g. blue native gel, column chromatography) approach to confirm oligomeric status of the control would be helpful. In addition, maintaining reasonable expression levels (lower than K_d, if known) to prevent non-specific interaction of control proteins is critical for proper calibration and determination of number of receptor subunits.

Many FFT studies chose to measure the receptor at lower temperatures (22–23°C) to minimize cellular movements and this should be kept in mind when interpreting dynamics that are not at a physiologically relevant temperature. SpIDA measurements for 5-HT2C, D_3R, M_1 receptor, M_3 receptor and secretin receptor were performed on the basolateral/ventral surface of cells [83, 193, 194, 197]. Receptor organization appears to vary depending on the membrane region measured. For example, 5-HT2C receptor measured at the apical surface of choroid plexus epithelial tissue (32 receptors/μm^2) and dorsal surface of HEK293 cells are dimeric (~30 receptors/μm^2) but at the basolateral/ventral surface is mostly monomeric (50 receptors/μm^2) when measured at similar receptor densities [83, 194].

What factors besides technical shortcomings/artifacts could account for the observed differences in the oligomeric structure of 5-HT2C receptor? Oligomeric differences could be due to organizational differences in cytoskeletal and associated membrane proteins and/or differences in membrane tension [215, 216]. Clathrin forms membrane invaginations (pits) during endocytosis but can also form flat lattice sheets (FLCs) and the amount of lattice sheets can vary up to two-fold higher on the ventral versus dorsal membrane for HeLa and HEK293 cells [215]. Quantitative super-resolution imaging suggests that FLCs can functionally compartmentalize the plasma membrane to regulate signaling and FLCs can recruit or exclude the GCPR chemokine receptor 5 (CCR5). The interactions between CCR5 and FLCs can be long-lived (minutes up to an hour) upon agonist stimulation independent of endocytosis [215]. Furthermore, lipid composition, curvature and membrane tension can affect GPCR oligomerization partly driven by receptor-membrane hydrophobic matching [217, 218]. Importantly, membrane tension is

~4 times greater in regions that are connected to the cytoskeleton versus those that are not and affects clathrin assembly [219, 220]. The proportional contributions of these various processes could vary in the different membrane domains (dorsal versus ventral) of the cell leading to alteration in GPCR oligomerization.

Recently, there is a debate over the size and location of ROI to be taken for certain image FFTs such as SpIDA and 2D-FIF [163, 221, 222]. This debate stems from what one considers to be a biologically-relevant length scale. Annibale and Lohse suggested the implementation of an additional step in 2D-FIF protocol that includes case-by-case selection of homogenous regions of the membrane in order to exclude fluorescence "hotspots" [221]. It is not surprising to observe heterogeneities of image data as one "zooms" in given the compartmentalization of the plasma membrane due to cortical actin and clathrin structures [215, 223]. Their reasoning for this modification is that hotspots occur due to sub-cellular structures and/or artifacts unrelated to receptor oligomerization, adding to the heterogeneity of data, thus confounding the analysis. Stoneman et al. responded to Annibale and Lohse's concerns by pointing out that the brightness heterogeneity (~1.5) observed in their ROI sampling falls within the distribution of their monomeric control. This result implies two symmetrical data points exist 1.5 above and below the center of the Gaussian peak for the distribution that would represent 0.5 monomer [163]. Clearly, 0.5 of a monomer is not possible and this variation can be partly explained by micro-environment of fluorophore, fluorophore orientation with respect to polarization of excitation light and three dimensional folds of membrane [163]. Stoneman et al. do acknowledge that brightness fluctuations of 2-fold or greater are needed for the brightness spectrums to identify larger oligomeric species. They also emphasize that 2D-FIF inherently employs a rigorous low-pass filter during generation of spectrogram to remove the contribution from "hotpsots" and this is less subjective then manual selection. The Annibale's brightness values (supplemental Table 1, [163]) using Stoneman's data do not agree with Stoneman's original published work and current re-analyses [162, 163, 221]. Stoneman suggests that this could be due to a flaw in their analysis procedure (i.e. correction for detector variance and intensity background). There is also disagreement concerning whether SpIDA and 2D-FIF fail to discriminate large oligomers from large immobile background spots. Finally, both Stoneman and Annibale agree that combining spatial and temporal information (e.g. N&B approach with FIF) would be a robust approach. Overall, FFTs are powerful tools to measure dynamic changes in GPCRs organization in their native membrane environment provided the microscopic system is thoroughly characterized, appropriate controls are employed, sufficient number/size of ROI is measured and careful analysis performed.

5. Conclusions

GPCRs are dynamic receptors that can form transient homo- and hetero-oligomers, and alteration in receptor quaternary states can impact signaling ability. Recent FFT experiments have shed new light on the affect agonists, antagonists, and ligands have on the oligomeric status of several class A and one class B receptor. These fluorescence fluctuation analyses provide complimentary information to the more traditional biochemical, FRET, and radioligand studies performed on GPCRs. The power of these analyses is the ability to measure receptor dynamics at microsecond time scales on a few molecules with micrometer spatial resolution. Unfortunately, there is still controversy and disagreement on the oligomeric status of many GPCRs when comparing previous results to the more recent SpIDA, FCS, PCH, and N&B data. Some of the discrepancies could be contributed to measurement location in cell membrane, or experimental conditions (e.g. different temperature, photobleaching). Other likely contributors to these differences are the limitations of measuring ensembles of molecules versus single particle tracking and the spatial/temporal scales of the different techniques. Researchers are beginning to combine multiple techniques such as FFTs with super-resolution or other biochemical approaches (BiFC-FCS) to compensate for individual shortcomings. Application of these combined approaches to study GPCR dynamics not just in the plasma membrane but other internal organelles in cells and live animal tissues will hopefully help to resolve conflicting experimental results.

Acknowledgements

Our sincere thanks to Dr. Amanda Storm for proofreading our manuscript, and our apologies to the authors whose research we were unable to cite due to page constraints.

Author details

Robert T. Youker* and Danielle Voet
Department of Biology, Western Carolina University, Cullowhee, North Carolina, USA

*Address all correspondence to: rtyouker@wcu.edu

References

[1] Calebiro D, Godbole A. Internalization of G-protein-coupled receptors: Implication in receptor function, physiology and diseases. Best Practice & Research Clinical Endocrinology & Metabolism. 2018;**32**:83-91

[2] Fredriksson R, Schiöth HB. The repertoire of G-protein–coupled receptors in fully sequenced genomes. Molecular Pharmacology. 2005;**67**:1414-1425

[3] Fredriksson R, Lagerström MC, Lundin L-G, et al. The G-protein-coupled receptors in the human genome form five main families. Phylogenetic analysis, paralogon groups, and fingerprints. Molecular Pharmacology. 2003;**63**:1256-1272

[4] Gurevich VV, Gurevich EV. The structural basis of arrestin-mediated regulation of G-protein-coupled receptors. Pharmacology & Therapeutics. 2006;**110**:465-502

[5] de Mendoza A, Sebé-Pedrós A, Ruiz-Trillo I. The evolution of the GPCR signaling system in eukaryotes: Modularity, conservation, and the transition to metazoan multicellularity. Genome Biology and Evolution. 2014;**6**:606-619

[6] Derelle R, Lopez P, Guyader HL, et al. Homeodomain proteins belong to the ancestral molecular toolkit of eukaryotes: Evolution of homeodomains. Evolution & Development. 2007;**9**:212-219

[7] Fritz-Laylin LK, Prochnik SE, Ginger ML, et al. The genome of *Naegleria gruberi* illuminates early eukaryotic versatility. Cell. 2010;**140**:631-642

[8] Wickstead B, Gull K, Richards TA. Patterns of kinesin evolution reveal a complex ancestral eukaryote with a multifunctional cytoskeleton. BMC Evolutionary Biology. 2010;**10**:110

[9] Grau-Bové X, Sebé-Pedrós A, Ruiz-Trillo I. A genomic survey of HECT ubiquitin ligases in eukaryotes reveals independent expansions of the HECT system in several lineages. Genome Biology and Evolution. 2013;**5**:833-847

[10] Lagerström MC, Schiöth HB. Structural diversity of G protein-coupled receptors and significance for drug discovery. Nature Reviews Drug Discovery. 2008;**7**:339-357

[11] Hu G-M, Mai T-L, Chen C-M. Visualizing the GPCR network: Classification and evolution. Scientific Reports. December 2017;**7**:1-15, 15495. DOI: 10.1038/s41598-017-15707-9. [Epub ahead of print]

[12] Strotmann R, Schröck K, Böselt I, et al. Evolution of GPCR: Change and continuity. Molecular and Cellular Endocrinology. 2011;**331**:170-178

[13] Attwood TK. A compendium of specific motifs for diagnosing GPCR subtypes. Trends in Pharmacological Sciences. 2001;**22**:162-165

[14] Horn F. GPCRDB information system for G protein-coupled receptors. Nucleic Acids Research. 2003;**31**:294-297

[15] Sgourakis NG, Bagos PG, Papasaikas PK, et al. A method for the prediction of GPCRs coupling specificity to G-proteins using refined profile hidden Markov models. BMC Bioinformatics. 2005;**6**:104

[16] Nordstrom KJV, Sallman Almen M, Edstam MM, et al. Independent HHsearch, Needleman-Wunsch-based, and motif analyses reveal the overall hierarchy for most of the G protein-coupled receptor families.

Molecular Biology and Evolution. 2011;**28**:2471-2480

[17] Pandey S, Assmann SM. The Arabidopsis putative G protein–coupled receptor GCR1 interacts with the G protein α subunit GPA1 and regulates abscisic acid signaling. The Plant Cell. 2004;**16**:1616-1632

[18] Krishnan A, Almén MS, Fredriksson R, et al. The origin of GPCRs: Identification of mammalian like rhodopsin, adhesion, glutamate and frizzled GPCRs in fungi. PLoS One. 2012;**7**:e29817

[19] Sriram K, Insel PA. G protein-coupled receptors as targets for approved drugs: How many targets and how many drugs? Molecular Pharmacology. 2018;**93**:251-258

[20] Hauser AS, Attwood MM, Rask-Andersen M, et al. Trends in GPCR drug discovery: New agents, targets and indications. Nature Reviews Drug Discovery. 2017;**16**:829-842

[21] Pándy-Szekeres G, Munk C, Tsonkov TM, et al. GPCRdb in 2018: Adding GPCR structure models and ligands. Nucleic Acids Research. 2018;**46**:D440-D446

[22] Hanlon CD, Andrew DJ. Outside-in signaling—A brief review of GPCR signaling with a focus on the Drosophila GPCR family. Journal of Cell Science. 2015;**128**:3533-3542

[23] Pitcher JA, Freedman NJ, Lefkowitz RJ. G protein–coupled receptor kinases. Annual Review of Biochemistry. 1998;**67**:653-692

[24] Luttrell LM, Ferguson SSG, Daaka Y, et al. β-Arrestin-dependent formation of β_2 adrenergic receptor-Src protein kinase complexes. Science. 1999;**283**:655-661

[25] Palczewski K, Buczyłko J, Kaplan MW, et al. Mechanism of rhodopsin kinase activation. The Journal of Biological Chemistry. 1991;**266**:12949-12955

[26] Jong Y-JI, Harmon SK, O'Malley KL. Intracellular GPCRs play key roles in synaptic plasticity. ACS Chemical Neuroscience. 2018;**9**:2162-2172

[27] Eichel K, von Zastrow M. Subcellular organization of GPCR signaling. Trends in Pharmacological Sciences. 2018;**39**:200-208

[28] Liggett SB. Phosphorylation barcoding as a mechanism of directing GPCR signaling. Science Signaling. 2011;**4**:pe36-pe36

[29] Shenoy SK, Lefkowitz RJ. β-Arrestin-mediated receptor trafficking and signal transduction. Trends in Pharmacological Sciences. 2011;**32**:521-533

[30] Charest PG, Oligny-Longpré G, Bonin H, et al. The V2 vasopressin receptor stimulates ERK1/2 activity independently of heterotrimeric G protein signalling. Cellular Signalling. 2007;**19**:32-41

[31] McDonald PH, Chow CW, Miller WE, et al. Beta-arrestin 2: A receptor-regulated MAPK scaffold for the activation of JNK3. Science. 2000;**290**:1574-1577

[32] DeFea KA, Zalevsky J, Thoma MS, et al. β-Arrestin–dependent endocytosis of proteinase-activated receptor 2 is required for intracellular targeting of activated Erk1/2. Journal of Cell Biology. 2000;**148**:1267-1282

[33] Steen A, Larsen O, Thiele S, et al. Biased and G protein-independent signaling of chemokine receptors. Frontiers in Immunology. 23 June 2014;**5**(277):1-13. Article ID: 277. DOI: 10.3389/fimmu.2014.00277. [Epub ahead of print]

[34] Vischer HF, Watts AO, Nijmeijer S, et al. G protein-coupled receptors:

Walking hand-in-hand, talking hand-in-hand?: GPCR oligomerization and crosstalk. British Journal of Pharmacology. 2011;**163**:246-260

[35] Inoue A, Raimondi F, Kadji FMN, et al. Illuminating G-protein-coupling selectivity of GPCRs. Cell. 2019;**177**:1933-1947.e25

[36] Pediani JD, Ward RJ, Marsango S, et al. Spatial intensity distribution analysis: Studies of G protein-coupled receptor oligomerisation. Trends in Pharmacological Sciences. 2018;**39**:175-186

[37] Ward RJ, Milligan G. Structural and biophysical characterisation of G protein-coupled receptor ligand binding using resonance energy transfer and fluorescent labelling techniques. Biochimica et Biophysica Acta (BBA) - Biomembranes. 2014;**1838**:3-14

[38] Guo H, An S, Ward R, et al. Methods used to study the oligomeric structure of G-protein-coupled receptors. Bioscience Reports. 30 April 2017;**37**:1-19. DOI: 10.1042/BSR20160547. [Epub ahead of print]

[39] Cottet M, Faklaris O, Maurel D, et al. BRET and time-resolved FRET strategy to study GPCR oligomerization: From cell lines toward native tissues. Frontiers in Endocrinology. 2012;**3**(92):1-14. Article ID: 92. DOI: 10.3389/fendo.2012.00092. [Epub ahead of print]

[40] Shimada I, Ueda T, Kofuku Y, et al. GPCR drug discovery: Integrating solution NMR data with crystal and cryo-EM structures. Nature Reviews Drug Discovery. 2019;**18**:59-82

[41] Weidemann T, Mücksch J, Schwille P. Fluorescence fluctuation microscopy: A diversified arsenal of methods to investigate molecular dynamics inside cells. Current Opinion in Structural Biology. 2014;**28**:69-76

[42] González Bardeci N, Angiolini JF, De Rossi MC, et al. Dynamics of intracellular processes in live-cell systems unveiled by fluorescence correlation microscopy: FCS in biological studies. IUBMB Life. 2017;**69**:8-15

[43] Youker RT. Detectors for super-resolution & single-molecule fluorescence microscopies. In: Britun N, Nikiforov A, editors. Photon Counting—Fundamentals and Applications. Rijkea, Croatia: InTech; 21 March 2018. DOI: 10.5772/intechopen.71943. [Epub ahead of print]

[44] Bag N, Wohland T. Imaging fluorescence fluctuation spectroscopy: New tools for quantitative bioimaging. Annual Review of Physical Chemistry. 2014;**65**:225-248

[45] Youker RT, Teng H. Measuring protein dynamics in live cells: Protocols and practical considerations for fluorescence fluctuation microscopy. Journal of Biomedical Optics. 2014;**19**:090801

[46] Gunther G, Jameson DM, Aguilar J, et al. Scanning fluorescence correlation spectroscopy comes full circle. Methods. 2018;**140-141**:52-61

[47] Magde D, Elson E, Webb WW. Thermodynamic fluctuations in a reacting system—Measurement by fluorescence correlation spectroscopy. Physical Review Letters. 1972;**29**:705-708

[48] Ehrenberg M, Rigler R. Rotational brownian motion and fluorescence intensify fluctuations. Chemical Physics. 1974;**4**:390-401

[49] Schwille P, Kummer S, Heikal AA, et al. Fluorescence correlation spectroscopy reveals fast optical excitation-driven intramolecular

dynamics of yellow fluorescent proteins. Proceedings of the National Academy of Sciences. 2000;**97**:151-156

[50] Schwille P, Haupts U, Maiti S, et al. Molecular dynamics in living cells observed by fluorescence correlation spectroscopy with one- and two-photon excitation. Biophysical Journal. 1999;**77**:2251-2265

[51] Briddon SJ, Kilpatrick LE, Hill SJ. Studying GPCR pharmacology in membrane microdomains: Fluorescence correlation spectroscopy comes of age. Trends in Pharmacological Sciences. 2018;**39**:158-174

[52] Rat V, Pinson X, Seigneuret F, et al. Hepatitis B virus core protein domains essential for viral capsid assembly in a cellular context. Journal of Molecular Biology. May 2020;**432**(13):3802-3819. DOI: 10.1016/j.jmb.2020.04.026. [Epub ahead of print]

[53] Fujita H, Oikawa R, Hayakawa M, et al. Quantification of native mRNA dynamics in living neurons using fluorescence correlation spectroscopy and reduction-triggered fluorescent probes. Journal of Biological Chemistry. 2020;**295**(23):7923-7940

[54] Fukuda T, Kawai-Noma S, Pack C-G, et al. Large-scale analysis of diffusional dynamics of proteins in living yeast cells using fluorescence correlation spectroscopy. Biochemical and Biophysical Research Communications. 2019;**520**:237-242

[55] Nomura Y, Nakamura T, Feng Z, et al. Direct quantification of gene expression using fluorescence correlation spectroscopy. Current Pharmaceutical Biotechnology. 2007;**8**:286-290

[56] Betaneli V, Mücksch J, Schwille P. Fluorescence correlation spectroscopy to examine protein–lipid interactions in membranes. In: Kleinschmidt JH, editor. Lipid-Protein Interactions. New York, NY: Springer New York. 2019. pp. 415-447

[57] Vasconcelos L, Lehto T, Madani F, et al. Simultaneous membrane interaction of amphipathic peptide monomers, self-aggregates and cargo complexes detected by fluorescence correlation spectroscopy. Biochimica et Biophysica Acta (BBA) - Biomembranes. 2018;**1860**:491-504

[58] Wohland T, Friedrich K, Hovius R, et al. Study of ligand-receptor interactions by fluorescence correlation spectroscopy with different fluorophores: Evidence that the homopentameric 5-hydroxytryptamine type 3 $_{As}$ receptor binds only one ligand †. Biochemistry. 1999;**38**:8671-8681

[59] Jakobs D, Sorkalla T, Haberlein H. Ligands for fluorescence correlation spectroscopy on G protein-coupled receptors. Current Medicinal Chemistry. 2012;**19**:4722-4730

[60] Zemanová L, Schenk A, Hunt N, et al. Endothelin receptor in virus-like particles: Ligand binding observed by fluorescence fluctuation spectroscopy. Biochemistry. 2004;**43**:9021-9028

[61] Maeder CI, Hink MA, Kinkhabwala A, et al. Spatial regulation of Fus3 MAP kinase activity through a reaction-diffusion mechanism in yeast pheromone signalling. Nature Cell Biology. 2007;**9**:1319-1326

[62] Slaughter BD, Schwartz JW, Li R. Mapping dynamic protein interactions in MAP kinase signaling using live-cell fluorescence fluctuation spectroscopy and imaging. Proceedings of the National Academy of Sciences. 2007;**104**:20320-20325

[63] Tiwari M, Oasa S, Yamamoto J, et al. A quantitative study of internal and

external interactions of homodimeric glucocorticoid receptor using fluorescence cross-correlation spectroscopy in a live cell. Scientific Reports. December 2017;7:1-16. Article ID: 4336. DOI: 10.1038/s41598-017-04499-7. [Epub ahead of print]

[64] Jastrzebska B, Comar WD, Kaliszewski MJ, et al. A G protein-coupled receptor dimerization Interface in human cone opsins. Biochemistry. 2017;**56**:61-72

[65] Teichmann A, Gibert A, Lampe A, et al. The specific monomer/dimer equilibrium of the corticotropin-releasing factor receptor type 1 is established in the endoplasmic reticulum. Journal of Biological Chemistry. 2014;**289**:24250-24262

[66] Kaliszewski MJ, Shi X, Hou Y, et al. Quantifying membrane protein oligomerization with fluorescence cross-correlation spectroscopy. Methods. 2018;**140-141**:40-51

[67] Chen J, Nag S, Vidi P-A, et al. Single molecule in vivo analysis of toll-like receptor 9 and CpG DNA interaction. PLoS One. 2011;**6**:e17991

[68] Li Y, Shivnaraine RV, Huang F, et al. Ligand-induced coupling between oligomers of the M2 receptor and the Gi1 protein in live cells. Biophysical Journal. 2018;**115**:881-895

[69] Rogacki MK, Golfetto O, Tobin SJ, et al. Dynamic lateral organization of opioid receptors (kappa, mu_{wt} and mu_{N40D}) in the plasma membrane at the nanoscale level. Traffic. 2018;**19**:690-709

[70] Meseth U, Wohland T, Rigler R, et al. Resolution of fluorescence correlation measurements. Biophysical Journal. 1999;**76**:1619-1631

[71] Chen Y, Müller JD, So PTC, et al. The photon counting histogram in fluorescence fluctuation spectroscopy. Biophysical Journal. 1999;**77**:553-567

[72] Kask P, Palo K, Ullmann D, et al. Fluorescence-intensity distribution analysis and its application in biomolecular detection technology. Proceedings of the National Academy of Sciences. 1999;**96**:13756-13761

[73] Kask P, Palo K, Fay N, et al. Two-dimensional fluorescence intensity distribution analysis: Theory and applications. Biophysical Journal. 2000;**78**:1703-1713

[74] Müller JD, Chen Y, Gratton E. Resolving heterogeneity on the single molecular level with the photon-counting histogram. Biophysical Journal. 2000;**78**:474-486

[75] Yu L, Tan M, Ho B, et al. Determination of critical micelle concentrations and aggregation numbers by fluorescence correlation spectroscopy: Aggregation of a lipopolysaccharide. Analytica Chimica Acta. 2006;**556**:216-225

[76] Meng F, Ma H. A comparison between photon counting histogram and fluorescence intensity distribution analysis. The Journal of Physical Chemistry B. 2006;**110**:25716-25720

[77] Saffarian S, Li Y, Elson EL, et al. Oligomerization of the EGF receptor investigated by live cell fluorescence intensity distribution analysis. Biophysical Journal. 2007;**93**:1021-1031

[78] Slaughter BD, Huff JM, Wiegraebe W, et al. SAM domain-based protein oligomerization observed by live-cell fluorescence fluctuation spectroscopy. PLoS One. 2008;**3**:e1931

[79] Youker RT, Bruns JR, Costa SA, et al. Multiple motifs regulate apical sorting of p75 via a mechanism that involves dimerization and higher-order oligomerization. Molecular Biology of the Cell. 2013;**24**:1996-2007

[80] Malengo G, Andolfo A, Sidenius N, et al. Fluorescence correlation spectroscopy and photon counting histogram on membrane proteins: Functional dynamics of the glycosylphosphatidylinositol-anchored urokinase plasminogen activator receptor. Journal of Biomedical Optics. 2008;**13**:031215

[81] Herrick-Davis K, Grinde E, Lindsley T, et al. Oligomer size of the serotonin 5-Hydroxytryptamine 2C (5-HT $_{2C}$) receptor revealed by fluorescence correlation spectroscopy with photon counting histogram analysis: Evidence for homodimers without monomers or tetramers. Journal of Biological Chemistry. 2012;**287**:23604-23614

[82] Herrick-Davis K, Grinde E, Cowan A, et al. Fluorescence correlation spectroscopy analysis of serotonin, adrenergic, muscarinic, and dopamine receptor dimerization: The oligomer number puzzle. Molecular Pharmacology. 2013;**84**:630-642

[83] Herrick-Davis K, Grinde E, Lindsley T, et al. Native serotonin 5-HT2C receptors are expressed as homodimers on the apical surface of choroid plexus epithelial cells. Molecular Pharmacology. 2015;**87**:660-673

[84] Rudiger M, Haupts U, Moore KJ, et al. Single-molecule detection technologies in miniaturized high throughput screening: Binding assays for G protein-coupled receptors using fluorescence intensity distribution analysis and fluorescence anisotropy. Journal of Biomolecular Screening. 2001;**6**:29-37

[85] Sanchez-Andres A, Chen Y, Müller JD. Molecular brightness determined from a generalized form of Mandel's Q-parameter. Biophysical Journal. 2005;**89**:3531-3547

[86] Qian H, Elson EL. On the analysis of high order moments of fluorescence fluctuations. Biophysical Journal. 1990;**57**:375-380

[87] Mandel L. Sub-Poissonian photon statistics in resonance fluorescence. Optics Letters. 1979;**4**:205

[88] Hur K-H, Macdonald PJ, Berk S, et al. Quantitative measurement of brightness from living cells in the presence of photodepletion. PLoS One. 2014;**9**:e97440

[89] Hur K-H, Mueller JD. Quantitative brightness analysis of fluorescence intensity fluctuations in *E. coli*. Plos One. 2015;**10**:e0130063

[90] Müller JD. Cumulant analysis in fluorescence fluctuation spectroscopy. Biophysical Journal. 2004;**86**:3981-3992

[91] Perroud TD, Huang B, Zare RN. Effect of bin time on the photon counting histogram for one-photon excitation. ChemPhysChem. 2005;**6**:905-912

[92] Palo K, Mets Ü, Jäger S, et al. Fluorescence intensity multiple distributions analysis: Concurrent determination of diffusion times and molecular brightness. Biophysical Journal. 2000;**79**:2858-2866

[93] Wu B, Müller JD. Time-integrated fluorescence cumulant analysis in fluorescence fluctuation spectroscopy. Biophysical Journal. 2005;**89**:2721-2735

[94] Wu B, Singer RH, Mueller JD. Time-integrated fluorescence cumulant analysis and its application in living cells. In: Methods in Enzymology. Amsterdam, Netherlands: Elsevier; 2013;**518**:99-119

[95] Melnykov AV, Hall KB. Revival of high-order fluorescence correlation analysis: Generalized theory and biochemical applications. The

Journal of Physical Chemistry B. 2009;**113**:15629-15638

[96] Scales N, Swain PS. Resolving fluorescent species by their brightness and diffusion using correlated photon-counting histograms. PLoS One. 2019;**14**:e0226063

[97] Chen Y, Tekmen M, Hillesheim L, et al. Dual-color photon-counting histogram. Biophysical Journal. 2005;**88**:2177-2192

[98] Palo K, Brand L, Eggeling C, et al. Fluorescence intensity and lifetime distribution analysis: Toward higher accuracy in fluorescence fluctuation spectroscopy. Biophysical Journal. 2002;**83**:605-618

[99] Laurence TA, Kapanidis AN, Kong X, et al. Photon arrival-time interval distribution (PAID): A novel tool for analyzing molecular interactions. The Journal of Physical Chemistry B. 2004;**108**:3051-3067

[100] Koppel DE, Morgan F, Cowan AE, et al. Scanning concentration correlation spectroscopy using the confocal laser microscope. Biophysical Journal. 1994;**66**:502-507

[101] Ohsugi Y, Kinjo M. Multipoint fluorescence correlation spectroscopy with total internal reflection fluorescence microscope. Journal of Biomedical Optics. 2009;**14**:014030

[102] Singh AP, Krieger JW, Buchholz J, et al. The performance of 2D array detectors for light sheet based fluorescence correlation spectroscopy. Optics Express. 2013;**21**:8652

[103] Gösch M, Serov A, Anhut T, et al. Parallel single molecule detection with a fully integrated single-photon 2×2 CMOS detector array. Journal of Biomedical Optics. 2004;**9**:913

[104] Blom H, Johansson M, Hedman A-S, et al. Parallel fluorescence detection of single biomolecules in microarrays by a diffractive-optical-designed 2×2 fan-out element. Applied Optics. 2002;**41**:3336

[105] Colyer RA, Scalia G, Rech I, et al. High-throughput FCS using an LCOS spatial light modulator and an 8×1 SPAD array. Biomedical Optics Express. 2010;**1**:1408

[106] Burkhardt M, Schwille P. Electron multiplying CCD based detection for spatially resolved fluorescence correlation spectroscopy. Optics Express. 2006;**14**:5013

[107] Colyer RA, Scalia G, Villa FA, et al. Ultra high-throughput single molecule spectroscopy with a 1024 pixel SPAD. In: Enderlein J, Gryczynski ZK, Erdmann R, editors. Single Molecule Spectroscopy and Imaging IV. Vol. 7905. San Francisco, California, USA: SPIE Proceedings; 2011. p. 790503

[108] Kloster-Landsberg M, Herbomel G, Wang I, et al. Cellular response to heat shock studied by multiconfocal fluorescence correlation spectroscopy. Biophysical Journal. 2012;**103**:1110-1119

[109] Sisan DR, Arevalo R, Graves C, et al. Spatially resolved fluorescence correlation spectroscopy using a spinning disk confocal microscope. Biophysical Journal. 2006;**91**:4241-4252

[110] Wohland T, Shi X, Sankaran J, et al. Single plane illumination fluorescence correlation spectroscopy (SPIM-FCS) probes inhomogeneous three-dimensional environments. Optics Express. 2010;**18**:10627

[111] Buchholz J, Krieger J, Bruschini C, et al. Widefield high frame rate single-photon SPAD imagers for SPIM-FCS. Biophysical Journal. 2018;**114**:2455-2464

[112] Krieger JW, Singh AP, Garbe CS, et al. Dual-color fluorescence cross-correlation spectroscopy on a single plane illumination microscope (SPIM-FCCS). Optics Express. 2014;**22**:2358-2375

[113] Wachsmuth M, Conrad C, Bulkescher J, et al. High-throughput fluorescence correlation spectroscopy enables analysis of proteome dynamics in living cells. Nature Biotechnology. 2015;**33**:384-389

[114] Fu X, Song Y, Masud A, et al. High-throughput fluorescence correlation spectroscopy enables analysis of surface components of cell-derived vesicles. Analytical and Bioanalytical Chemistry. 2020;**412**:2589-2597

[115] Digman MA, Brown CM, Sengupta P, et al. Measuring fast dynamics in solutions and cells with a laser scanning microscope. Biophysical Journal. 2005;**89**:1317-1327

[116] Digman MA, Wiseman PW, Choi C, et al. Stoichiometry of molecular complexes at adhesions in living cells. Proceedings of the National Academy of Sciences. 2009;**106**:2170-2175

[117] Brown CM, Dalal RB, Hebert B, et al. Raster image correlation spectroscopy (RICS) for measuring fast protein dynamics and concentrations with a commercial laser scanning confocal microscope. Journal of Microscopy. 2008;**229**:78-91

[118] Suzuki T, Ainai A, Nagata N, et al. A novel function of the N-terminal domain of PA in assembly of influenza a virus RNA polymerase. Biochemical and Biophysical Research Communications. 2011;**414**:719-726

[119] Brejchová J, Sýkora J, Ostašov P, et al. TRH-receptor mobility and function in intact and cholesterol-depleted plasma membrane of HEK293 cells stably expressing TRH-R-eGFP. Biochimica et Biophysica Acta (BBA) - Biomembranes. 2015;**1848**:781-796

[120] Laňková M, Humpolíčková J, Vosolsobě S, et al. Determination of dynamics of plant plasma membrane proteins with fluorescence recovery and raster image correlation spectroscopy. Microscopy and Microanalysis. 2016;**22**:290-299

[121] Begarani F, D'Autilia F, Signore G, et al. Capturing metabolism-dependent solvent dynamics in the lumen of a trafficking lysosome. ACS Nano. 23 January 2019. DOI: 10.1021/acsnano.8b07682. [Epub ahead of print]

[122] Norris SCP, Humpolíčková J, Amler E, et al. Raster image correlation spectroscopy as a novel tool to study interactions of macromolecules with nanofiber scaffolds. Acta Biomaterialia. 2011;**7**:4195-4203

[123] Digman MA, Wiseman PW, Horwitz AR, et al. Detecting protein complexes in living cells from laser scanning confocal image sequences by the cross correlation raster image spectroscopy method. Biophysical Journal. 2009;**96**:707-716

[124] Digman MA, Gratton E. Scanning image correlation spectroscopy. BioEssays. 2012;**34**:377-385

[125] Sasaki A, Yamamoto J, Jin T, et al. Raster image cross-correlation analysis for spatiotemporal visualization of intracellular degradation activities against exogenous DNAs. Scientific Reports. November 2015;**5**:1-9. Article ID: 14428. DOI: 10.1038/srep14428. [Epub ahead of print]

[126] Hendrix J, Lamb DC. Implementation and application of pulsed interleaved excitation for dual-color FCS and RICS. Methods in Molecular Biology. 2014;**1076**:653-682

[127] Choi CK, Zareno J, Digman MA, et al. Cross-correlated fluctuation

analysis reveals phosphorylation-regulated paxillin-FAK complexes in nascent adhesions. Biophysical Journal. 2011;**100**:583-592

[128] Kapoor-Kaushik N, Hinde E, Compeer EB, et al. Distinct mechanisms regulate Lck spatial organization in activated T cells. Frontiers in Immunology. 2016;7:83

[129] Dobrinskikh E, Lanzano L, Rachelson J, et al. Shank2 contributes to the apical retention and intracellular redistribution of NaPiIIa in OK cells. American Journal of Physiology: Cell Physiology. 2013;**304**:C561-C573

[130] Schrimpf W, Lemmens V, Smisdom N, et al. Crosstalk-free multicolor RICS using spectral weighting. Methods. 2018;**140-141**:97-111

[131] Longfils M, Smisdom N, Ameloot M, et al. Raster image correlation spectroscopy performance evaluation. Biophysical Journal. 2019;**117**:1900-1914

[132] Kolin DL, Wiseman PW. Advances in image correlation spectroscopy: Measuring number densities, aggregation states, and dynamics of fluorescently labeled macromolecules in cells. Cell Biochemistry and Biophysics. 2007;**49**:141-164

[133] Wiseman PW. Image correlation spectroscopy: Principles and applications. Cold Spring Harbor Protocols. 2015;**2015**:336-348

[134] Petersen NO, Höddelius PL, Wiseman PW, et al. Quantitation of membrane receptor distributions by image correlation spectroscopy: Concept and application. Biophysical Journal. 1993;**65**:1135-1146

[135] Keating E, Nohe A, Petersen NO. Studies of distribution, location and dynamic properties of EGFR on the cell surface measured by image correlation spectroscopy. European Biophysics Journal. 2008;**37**:469-481

[136] Srivastava M, Petersen NO. Diffusion of transferrin receptor clusters. Biophysical Chemistry. 1998;**75**:201-211

[137] Hebert B, Costantino S, Wiseman PW. Spatiotemporal image correlation spectroscopy (STICS) theory, verification, and application to protein velocity mapping in living CHO cells. Biophysical Journal. 2005;**88**:3601-3614

[138] Wiseman PW. Spatial mapping of integrin interactions and dynamics during cell migration by image correlation microscopy. Journal of Cell Science. 2004;**117**:5521-5534

[139] Toplak T, Pandzic E, Chen L, et al. STICCS reveals matrix-dependent adhesion slipping and gripping in migrating cells. Biophysical Journal. 2012;**103**:1672-1682

[140] Kolin DL, Costantino S, Wiseman PW. Sampling effects, noise, and photobleaching in temporal image correlation spectroscopy. Biophysical Journal. 2006;**90**:628-639

[141] Kolin DL, Ronis D, Wiseman PW. k-space image correlation spectroscopy: A method for accurate transport measurements independent of fluorophore photophysics. Biophysical Journal. 2006;**91**:3061-3075

[142] Abu-Arish A, Pandžić E, Kim D, et al. Agonists that stimulate secretion promote the recruitment of CFTR into membrane lipid microdomains. Journal of General Physiology. 2019;**151**:834-849

[143] Digman MA, Dalal R, Horwitz AF, et al. Mapping the number of molecules and brightness in the laser scanning microscope. Biophysical Journal. 2008;**94**:2320-2332

[144] Unruh JR, Gratton E. Analysis of molecular concentration and brightness from fluorescence fluctuation data with an electron multiplied CCD camera. Biophysical Journal. 2008;**95**:5385-5398

[145] Macdonald P, Johnson J, Smith E, et al. Brightness analysis. In: Methods in Enzymology. Vol. 518. Academic Press (imprint of Elsevier). pp. 71-98

[146] Adu-Gyamfi E, Digman MA, Gratton E, et al. Investigation of Ebola VP40 assembly and oligomerization in live cells using number and brightness analysis. Biophysical Journal. 2012;**102**:2517-2525

[147] James NG, Digman MA, Gratton E, et al. Number and brightness analysis of LRRK2 oligomerization in live cells. Biophysical Journal. 2012;**102**:L41-L43

[148] Ossato G, Digman MA, Aiken C, et al. A two-step path to inclusion formation of Huntingtin peptides revealed by number and brightness analysis. Biophysical Journal. 2010;**98**:3078-3085

[149] Mieruszynski S, Briggs C, Digman MA, et al. Live cell characterization of DNA aggregation delivered through lipofection. Scientific Reports. September 2015;**5**:1-9. Article ID: 10528. DOI: 10.1038/srep10528. [Epub ahead of print]

[150] Sameni S, Zhang R, Digman MA. Number and molecular brightness analysis reveals Htt25Q protein aggregation upon the uptake of Htt97Q aggregates. Biochemical and Biophysical Research Communications. 2020;**522**:133-137

[151] Adu-Gyamfi E, Soni S, Jee C, et al. A loop region in the N-terminal domain of Ebola virus VP40 is important in viral assembly, budding, and egress. Viruses. 2014;**6**:3837-3854

[152] Bachir AI, Zareno J, Moissoglu K, et al. Integrin-associated complexes form hierarchically with variable stoichiometry in nascent adhesions. Current Biology. 2014;**24**:1845-1853

[153] Lagarrigue F, Vikas Anekal P, Lee H-S, et al. A RIAM/lamellipodin–talin–integrin complex forms the tip of sticky fingers that guide cell migration. Nature Communications. December 2015;**6**:8492. DOI: 10.1038/ncomms9492. [Epub ahead of print]

[154] Hortigüela V, Larrañaga E, Cutrale F, et al. Nanopatterns of surface-bound EphrinB1 produce multivalent ligand–receptor interactions that tune EphB2 receptor clustering. Nano Letters. 2018;**18**:629-637

[155] Cutrale F, Rodriguez D, Hortigüela V, et al. Using enhanced number and brightness to measure protein oligomerization dynamics in live cells. Nature Protocols. 2019;**14**:616-638

[156] Barbeau A, Swift JL, Godin AG, et al. Spatial intensity distribution analysis (SpIDA). In: Methods in Cell Biology. Amsterdam, Netherlands: Elsevier. 2013;**117**:1-19

[157] Godin AG, Costantino S, Lorenzo L-E, et al. Revealing protein oligomerization and densities in situ using spatial intensity distribution analysis. Proceedings of the National Academy of Sciences. 2011;**108**:7010-7015

[158] Godin AG, Rappaz B, Potvin-Trottier L, et al. Spatial intensity distribution analysis reveals abnormal oligomerization of proteins in single cells. Biophysical Journal. 2015;**109**:710-721

[159] Hamrang Z, McGlynn HJ, Clarke D, et al. Monitoring the kinetics of CellTrace™ calcein red-orange AM intracellular accumulation with spatial intensity distribution analysis.

Biochimica et Biophysica Acta (BBA) - General Subjects. 2014;**1840**:2914-2923

[160] Ward RJ, Pediani JD, Harikumar KG, et al. Spatial intensity distribution analysis quantifies the extent and regulation of homodimerization of the secretin receptor. The Biochemical Journal. 2017;**474**:1879-1895

[161] Rattray Z, Zindy E, Buzza KM, et al. Evaluation of temporal aggregation processes using spatial intensity distribution analysis. In: McManus JJ, editor. Protein Self-Assembly. New York, NY: Springer New York. 2019. pp. 141-155

[162] Stoneman MR, Biener G, Ward RJ, et al. A general method to quantify ligand-driven oligomerization from fluorescence-based images. Nature Methods. 2019;**16**:493-496

[163] Stoneman MR, Biener G, Raicu V. Reply to: Spatial Heterogeneity Molecular brightness. Biophysics. 11 November 2019. DOI: 10.1101/822296. Preprint. [Epub ahead of print]

[164] Hennen J, Hur K-H, Saunders CA, et al. Quantitative brightness analysis of protein oligomerization in the nuclear envelope. Biophysical Journal. 2017;**113**:138-147

[165] Stasevich TJ, Mueller F, Michelman-Ribeiro A, et al. Cross-validating FRAP and FCS to quantify the impact of photobleaching on in vivo binding estimates. Biophysical Journal. 2010;**99**:3093-3101

[166] Ries J, Schwille P. New concepts for fluorescence correlation spectroscopy on membranes. Physical Chemistry Chemical Physics. 2008;**10**:3487

[167] Ries J, Bayer M, Csúcs G, et al. Automated suppression of sample-related artifacts in fluorescence correlation spectroscopy. Optics Express. 2010;**18**:11073

[168] Kilpatrick LE, Briddon SJ, Holliday ND. Fluorescence correlation spectroscopy, combined with bimolecular fluorescence complementation, reveals the effects of β-arrestin complexes and endocytic targeting on the membrane mobility of neuropeptide Y receptors. Biochimica et Biophysica Acta (BBA) - Molecular Cell Research. 2012;**1823**:1068-1081

[169] Khmelinskaia A, Mücksch J, Conci F, et al. FCS analysis of protein mobility on lipid monolayers. Biophysical Journal. 2018;**114**:2444-2454

[170] Ozgen H, Schrimpf W, Hendrix J, et al. The lateral membrane organization and dynamics of myelin proteins PLP and MBP are dictated by distinct galactolipids and the extracellular matrix. PLoS One. 2014;**9**:e101834

[171] Nolan R, Iliopoulou M, Alvarez L, et al. Detecting protein aggregation and interaction in live cells: A guide to number and brightness. Methods. 2018;**140-141**:172-177

[172] Bakker E, Swain PS. Estimating numbers of intracellular molecules through analysing fluctuations in photobleaching. Scientific Reports. December 2019;**9**. DOI: 10.1038/s41598-019-50921-7. [Epub ahead of print]

[173] Trullo A, Corti V, Arza E, et al. Application limits and data correction in number of molecules and brightness analysis: N&B-limits and corrections. Microscopy Research and Technique. 2013;**76**:1135-1146

[174] Lange JJ, Wood CJ, Marshall WA, et al. Correction of bleaching artifacts in high content fluorescence correlation spectroscopy (HCS-FCS) data. In: Enderlein J, Gregor I, Gryczynski ZK, et al., editors. Single Molecule Spectroscopy and Superresolution Imaging VI. San Francisco, California, USA: SPIE Proceedings; 2013. p. 859006

[175] Nolan R, Iliopoulou M, Siebold C, et al. Robin Hood: Non-fitting, non-smoothing image detrending for bleaching correction. bioRxiv. 14 June 2019:1-13. DOI: 10.1101/667824. Preprint, [Epub ahead of print]

[176] Ishii K, Tahara T. Correction of the afterpulsing effect in fluorescence correlation spectroscopy using time symmetry analysis. Optics Express. 2015;**23**:32387

[177] Hillesheim LN, Müller JD. The dual-color photon counting histogram with non-ideal photodetectors. Biophysical Journal. 2005;**89**:3491-3507

[178] Ziarkash AW, Joshi SK, Stipčević M, et al. Comparative study of afterpulsing behavior and models in single photon counting avalanche photo diode detectors. Scientific Reports. December 2018;**8**(5076):1-8. DOI: 10.1038/s41598-018-23398-z. [Epub ahead of print]

[179] Fukushima R, Yamamoto J, Ishikawa H, et al. Two-detector number and brightness analysis reveals spatio-temporal oligomerization of proteins in living cells. Methods. 2018;**140-141**:161-171

[180] Foo YH, Naredi-Rainer N, Lamb DC, et al. Factors affecting the quantification of biomolecular interactions by fluorescence cross-correlation spectroscopy. Biophysical Journal. 2012;**102**:1174-1183

[181] Widengren J, Mets U, Rigler R. Fluorescence correlation spectroscopy of triplet states in solution: A theoretical and experimental study. The Journal of Physical Chemistry. 1995;**99**:13368-13379

[182] Balleza E, Kim JM, Cluzel P. Systematic characterization of maturation time of fluorescent proteins in living cells. Nature Methods. 2018;**15**:47-51

[183] Dunsing V, Luckner M, Zühlke B, et al. Optimal fluorescent protein tags for quantifying protein oligomerization in living cells. Scientific Reports. December 2018;**8**. DOI: 10.1038/s41598-018-28858-0. [Epub ahead of print]

[184] Hendrix J, Flors C, Dedecker P, et al. Dark states in monomeric red fluorescent proteins studied by fluorescence correlation and single molecule spectroscopy. Biophysical Journal. 2008;**94**:4103-4113

[185] Hern JA, Baig AH, Mashanov GI, et al. Formation and dissociation of M1 muscarinic receptor dimers seen by total internal reflection fluorescence imaging of single molecules. Proceedings of the National Academy of Sciences. 2010;**107**:2693-2698

[186] Chun L, Zhang W, Liu J. Structure and ligand recognition of class C GPCRs. Acta Pharmacologica Sinica. 2012;**33**:312-323

[187] Ferré S, Casadó V, Devi LA, et al. G protein–coupled receptor oligomerization revisited: Functional and pharmacological perspectives. Pharmacological Reviews. 2014;**66**:413-434

[188] Milligan G, Ward RJ, Marsango S. GPCR homo-oligomerization. Current Opinion in Cell Biology. 2019;**57**:40-47

[189] Nenasheva TA, Neary M, Mashanov GI, et al. Abundance, distribution, mobility and oligomeric state of M2 muscarinic acetylcholine receptors in live cardiac muscle. Journal of Molecular and Cellular Cardiology. 2013;**57**:129-136

[190] Redka DS, Morizumi T, Elmslie G, et al. Coupling of G proteins to reconstituted monomers and tetramers of the M_2 muscarinic receptor. Journal of Biological Chemistry. 2014;**289**:24347-24365

[191] Shivnaraine RV, Kelly B, Sankar KS, et al. Allosteric modulation in monomers and oligomers of a G protein-coupled receptor. eLife. 6 May 2016;**5**:1-26, e11685. DOI: 10.7554/eLife.11685. [Epub ahead of print]

[192] Walsh SM, Mathiasen S, Christensen SM, et al. Single proteoliposome high-content analysis reveals differences in the homo-oligomerization of GPCRs. Biophysical Journal. 2018;**115**:300-312

[193] Marsango S, Caltabiano G, Jiménez-Rosés M, et al. A molecular basis for selective antagonist destabilization of dopamine D3 receptor quaternary organization. Scientific Reports. May 2017;7:1-17, 2134. DOI: 10.1038/s41598-017-02249-3. [Epub ahead of print]

[194] Ward RJ, Pediani JD, Godin AG, et al. Regulation of oligomeric organization of the serotonin 5-hydroxytryptamine 2C (5-HT_{2C}) receptor observed by spatial intensity distribution analysis. Journal of Biological Chemistry. 2015;**290**:12844-12857

[195] Shivnaraine RV, Fernandes DD, Ji H, et al. Single-molecule analysis of the supramolecular organization of the M_2 muscarinic receptor and the $G\alpha_{i1}$ protein. Journal of the American Chemical Society. 2016;**138**:11583-11598

[196] Pathak A, Smih F, Galinier M, et al. Insulin downregulates M2-muscarinic receptors in adult rat atrial cardiomyocytes: A link between obesity and cardiovascular complications. International Journal of Obesity. 2005;**29**:176-182

[197] Pediani JD, Ward RJ, Godin AG, et al. Dynamic regulation of quaternary organization of the M_1 muscarinic receptor by subtype-selective antagonist drugs. Journal of Biological Chemistry. 2016;**291**:13132-13146

[198] Head BP, Patel HH, Roth DM, et al. Microtubules and actin microfilaments regulate lipid raft/caveolae localization of adenylyl cyclase signaling components. Journal of Biological Chemistry. 2006;**281**:26391-26399

[199] Patowary S, Alvarez-Curto E, Xu T-R, et al. The muscarinic M3 acetylcholine receptor exists as two differently sized complexes at the plasma membrane. Biochemical Journal. 2013;**452**:303-312

[200] Armbruster BN, Li X, Pausch MH, et al. Evolving the lock to fit the key to create a family of G protein-coupled receptors potently activated by an inert ligand. Proceedings of the National Academy of Sciences. 2007;**104**:5163-5168

[201] Snapp EL. When is a monomer not a monomer? The top three ways your favorite fluorescent protein oligomerizes in cells. 2016. Available from: https://blog.addgene.org/when-is-a-monomer-not-a-monomer-the-top-three-ways-your-favorite-fluorescent-protein-oligomerizes-in-cells [Accessed: 04 June 2020]

[202] Mueller A, Mahmoud NG, Goedecke MC, et al. Pharmacological characterization of the chemokine receptor, CCR5. British Journal of Pharmacology. 2002;**135**:1033-1043

[203] Harikumar KG, Pinon DI, Miller LJ. Transmembrane segment IV contributes a functionally important interface for oligomerization of the class II G protein-coupled secretin receptor. Journal of Biological Chemistry. 2007;**282**:30363-30372

[204] Golebiewska U, Johnston JM, Devi L, et al. Differential response to morphine of the oligomeric state of μ-opioid in the presence of δ-opioid receptors. Biochemistry. 2011;**50**:2829-2837

[205] Briddon SJ, Gandía J, Amaral OB, et al. Plasma membrane diffusion of g protein-coupled receptor oligomers. Biochimica et Biophysica Acta (BBA) - Molecular Cell Research. 2008;**1783**:2262-2268

[206] Briddon SJ, Middleton RJ, Cordeaux Y, et al. Quantitative analysis of the formation and diffusion of A1-adenosine receptor-antagonist complexes in single living cells. Proceedings of the National Academy of Sciences. 2004;**101**:4673-4678

[207] Philip F, Sengupta P, Scarlata S. Signaling through a G protein-coupled receptor and its corresponding G protein follows a stoichiometrically limited model. Journal of Biological Chemistry. 2007;**282**:19203-19216

[208] Licht SS, Sonnleitner A, Weiss S, et al. A rugged energy landscape mechanism for trapping of transmembrane receptors during endocytosis †. Biochemistry. 2003;**42**:2916-2925

[209] Kaiser SM, Quinn RJ. Adenosine receptors as potential therapeutic targets. Drug Discovery Today. 1999;**4**:542-551

[210] Briddon SJ, Middleton RJ, Yates AS, et al. Application of fluorescence correlation spectroscopy to the measurement of agonist binding to a G-protein coupled receptor at the single cell level. Faraday Discussions. 2004;**126**:197

[211] Marceau F, Bachelard H, Bouthillier J, et al. Bradykinin receptors: Agonists, antagonists, expression, signaling, and adaptation to sustained stimulation. International Immunopharmacology. 2020;**82**:106305

[212] Brothers SP, Wahlestedt C. Therapeutic potential of neuropeptide Y (NPY) receptor ligands. EMBO Molecular Medicine. 2010;**2**(11):429-439

[213] Zacharias DA. Partitioning of lipid-modified monomeric GFPs into membrane microdomains of live cells. Science. 2002;**296**:913-916

[214] Zakrys L, Ward RJ, Pediani JD, et al. Roundabout 1 exists predominantly as a basal dimeric complex and this is unaffected by binding of the ligand Slit2. Biochemical Journal. 2014;**461**:61-73

[215] Grove J, Metcalf DJ, Knight AE, et al. Flat clathrin lattices: Stable features of the plasma membrane. Molecular Biology of the Cell. 2014;**25**:3581-3594

[216] Pontes B, Monzo P, Gauthier NC. Membrane tension: A challenging but universal physical parameter in cell biology. Seminars in Cell & Developmental Biology. 2017;**71**:30-41

[217] Mondal S, Johnston JM, Wang H, et al. Membrane driven spatial organization of GPCRs. Scientific Reports. December 2013;**3**. DOI: 10.1038/srep02909. [Epub ahead of print]

[218] Botelho AV, Huber T, Sakmar TP, et al. Curvature and hydrophobic forces drive oligomerization and modulate activity of rhodopsin in membranes. Biophysical Journal. 2006;**91**:4464-4477

[219] Dai J, Sheetz MP. Membrane tether formation from blebbing cells. Biophysical Journal. 1999;**77**:3363-3370

[220] Boulant S, Kural C, Zeeh J-C, et al. Actin dynamics counteract membrane tension during clathrin-mediated endocytosis. Nature Cell Biology. 2011;**13**:1124-1131

[221] Annibale P, Lohse MJ. Spatial heterogeneity in molecular brightness. Nature Methods. 2020;**17**:273-275

[222] Annibale P, Lohse MJ. Molecular Brightness analysis of GPCR oligomerization in the presence of spatial heterogeneity. 2019. arXiv:190710841 [physics, q-bio]. Available from: http://arxiv.org/abs/1907.10841. [Accessed: 05 June 2020]

[223] Sadegh S, Higgins JL, Mannion PC, et al. Plasma membrane is compartmentalized by a self-similar cortical actin meshwork. Physical Review X. 9 March 2017;7:1-10, 011031. DOI: 10.1103/PhysRevX.7.011031. [Epub ahead of print]

Chapter 8

Study of the Fluorescence Intensity of Histones HP1gamma, H3K9me3, H3K27me3, and Histone Deacetylases SIRT1 and SIRT6 in Dermal Fibroblasts of Patients with Breast Cancer and Patients at Risk of Cancer

Mirya Kuranova, Aleksandra Nozdracheva, Nadezhda Pleskach, Daria Pasekova and Aleksey Manikhas

Abstract

Breast cancer is by far the most common cancer among women. The heterogeneous nature of the onset of this disease has already been proven and a search is underway for possible epigenetic mechanisms that affect the course of treatment. The indirect immunofluorescence method was used to study the fluorescence intensity of histones HP1γ, H3K9me3, H3K27me3, and histone deacetylases SIRT1 and SIRT6 in dermal fibroblasts of patients with breast cancer and patients at risk of developing cancer. The results of the study showed a significant difference in the level of fluorescence intensity of histones H3K9me3 and H3K27me3 in the cells of patients with breast cancer from the values in the cells of healthy donors. Also, the results of the study suggest that the heterochromatin protein HP1γ is a possible marker of the heterogeneity of breast cancer.

Keywords: breast cancer, immunofluorescence, HP1γ, H3K9me3, H3K27me3, SIRT1, SIRT6

1. Introduction

Despite significant advances in cancer research, breast cancer remains a serious problem and, according to WHO, is the most common cancer worldwide affecting women. It accounts for 16% of all oncological diseases in women with incidence rates of 99.4 per 100,000 women aged 13–90 years. According to forecasts in the coming years, its frequency and mortality will increase significantly.

Modern medicine shows that progress in the fight against breast cancer over the past decade is slow. This is not surprising given the important limitations of currently available targeted treatments. The reasons for the high internal and acquired resistance to available target drugs include their temporary antitumor activity and the lack of consideration for origins of heterogeneity of tumors. Understanding this extremely complex heterogeneity is crucial in the "war" against breast cancer and metastasis.

Oncogene activation in mammalian cells leads to proliferative stress and the induction of aging, which limits tumor growth. Thus, aging is the physiological mechanism of tumor suppression, which prevents progression from benign tumor foci to malignant tumors. Induction of aging by oncogene activation is called oncogene-induced senescence, OIS. Oncogene inactivation, on the other hand, can also cause aging. MYC inactivation causes cell aging, and regression in various tumor samples, such as lymphoma, osteosarcoma, and hepatocellular carcinoma (HCC) [1].

Recently, researchers have been interested in breast cancer in young women. Current evidence suggests that in women under the age of 45, breast cancer is undoubtedly the main cause of cancer mortality. This type of cancer is most likely very heterogeneous and has potentially aggressive and complex biological features. However, management strategies, recommendations, and options are not based on age, and the "complex" biology of this type of cancer remains uncertain and unexplored [2].

Recent studies suggest a significant association between biological aging and objective cognitive performance in breast cancer survivors. Future prospective studies are needed to confirm the causative role of biological aging as a driver of cognitive decline after cancer treatment [3].

To study the processes of activation of cell aging in tumor cells, it is important to understand the effect of chromatin structure on cell proliferation, aging of cells and tissues, and the treatment of cancer. The foci of heterochromatin associated with aging (SAHF) are specialized domains of facultative heterochromatin that form in aging human cells. Although SAHFs are highly densified domains of heterochromatin, they largely exclude other chromatin domains in telomeres and pericentromeres, which are themselves believed to be heterochromatic.

HP1 proteins are important components of a dynamic nuclear response that detects and eliminates defects in the epigenetic information encoded in chromatin through histone modifications and DNA methylation. Defects in this "chromatin repair" response in transformed cells can contribute to the preferential killing of cancer cells using cancer epigenetic therapy, which is still under clinical development [4].

In breast cancer, the level of the H3K27me3 protein rises to maintain the LOX gene in a repressed state [5]. The global level of H3K9me3 protein, on the contrary, decreases during transformation, and its distribution over chromatin is reprogrammed. Perhaps, this is due to the mechanism of the occurrence of breast cancer by regulating gene expression and chromatin stability during cell transformation [6]. Epigenetic modifications of the H3K9me3 and H3K27me3 histones may be useful as diagnostic/prognostic biomarkers and/or therapeutic targets for breast cancer, since DNA and histone modifications are involved in the activation of PD-1, CTLA-4, TIM-3, and LAG-3 in tissue of breast tumors [7, 8].

The intratumoral heterogeneity due to the presence of cancer stem cells (cancer stem cell [CSC]) is the main cause of drug resistance and distant metastases of breast cancer. Histone deacetylase SIRT1 outlines a central role in the regulation of CSC and the SIRT1 pathway can be a promising therapeutic strategy [9, 10]. Overexpression of SIRT1 significantly stimulates breast cancer growth both in vitro and in vivo, while SIRT1 knockdown inhibits cancer phenotypes [11]. Also, histone deacetylase SIRT6 may be a therapeutic marker. Sirtuin 6 (SIRT6) is associated with longevity and is also a tumor suppressor. The survival of patients with breast cancer was positively correlated with the abundance of SIRT6 and inversely correlated with phosphorylation of SIRT6 at Ser (338) [12]. SIRT1 and SIRT6 play a role in fine-tuning the acetylation of the antitumor suppressor FOXO3 (Forkhead Box O3) and sensitivity to lapatinib. The results of immunohistochemistry for different subtypes of breast cancer showed that the high levels of SIRT6/1 are associated with constant high expression of FOXO3, which in turn is associated with deregulation/inactivation of FOXO3 and a poor prognosis of the course of the disease [13].

2. Material and methods

2.1 Study group selection

We studied three groups:

1. Patients with breast cancer: two patients 30 (BC30SP1) and 55 (BC55SP1) years old.

2. Breast cancer risk group: a 30-year-old person with 5382insC mutation in the BRCA1 gene (BRCA1SP) and two mothers of patients with ataxia-telangiectasia—47 (AT8SP) and 45 (AT9SP) years old. Heterozygous carriers of mutations in the ATM gene are at risk of tumors, including breast cancer [14].

3. Healthy donors: two donors without breast cancer or a history of oncological diseases of 30 (N9SP) and 55 (FK19) years old.

2.2 Cell line production

Cell lines of dermal fibroblasts were obtained by the migration method from post-operative material. Lines of dermal fibroblasts BRCA1SP, BC30SP1, and BC55SP1 were obtained from a skin biopsy, which was performed under local anesthesia. Skin biopsies were placed in Petri dishes, filled with complete MEM growth medium (Gibco, USA) containing 15% bovine fetal serum (Gibco, USA), 100 IU/ml penicillin, 100 μg/ml streptomycin (Gibco, USA), 25 μg/ml fungicon, 0.3 mg/ml L-glutamine, and cultured in a CO_2 incubator at +37°C in an atmosphere with 5% CO_2 until a subconfluent monolayer was obtained. Next, the cells were rinsed with a solution of phosphate-buffered saline (PBS), removed from Petri dishes using 0.25% Trypsin-Versene solution (Gibco, USA) and scattered on coverslips (Carl Zeiss, Germany).

2.3 Immunocytochemical analysis

By indirect immunofluorescence method, the cell lines were stained with antibodies to chromatin proteins HP1γ, H3K9me3, H3K27me3, and histone deacetylases SIRT1 and SIRT6.

According to the standard protocol and usage annotations of antibodies, the cells were fixed on ice with a 4% formaldehyde solution for 10 minutes at +40°C followed by fixation with ice-cold 70% ethanol solution in PBS (15 min at room temperature), washed with a solution of PBS, permeabilized with a 1.5% X-100 triton solution in PBS (15 min at room temperature), washed with a PBS solution 3 times for 5 minutes, and incubated in a blocking solution of 5% BSA in PBS (1 h at room temperature), followed by washing with PBS. As the first antibodies, mouse IgG to H3K9me3, H3K27me3, and SIRT1 (Cell Signaling, UK) and rabbit IgG to the HP1γ protein (Abcam, UK), H3K27me3, SIRT6, and SIRT7 (Cell Signaling, UK) were used. Goat antibodies against mouse or rabbit IgG conjugated with Alexa Fluor 488 and 546 (Invitrogen, USA) at a 1: 500 dilution were used as the second antibodies. To prevent the rapid burnout of immunofluorescence dyes, the preparations were enclosed in antifading using SlowFade Gold Antifade reagent with DAPI (Invitrogen, USA).

2.4 Microscopy

Images from stained preparations were obtained using an AXIOVERT200-DFC420 fluorescent inverted microscope (Carl Zeiss AG, Germany) equipped with a Leica DFC 420 camera and an E×40/0.75 lens. In each line, using the Fiji program, fluorescence intensity in 100 cells was determined.

2.5 Statistical analysis

For the parameter of statistical analysis, the standard deviation of the mean values was chosen. Statistical analysis was performed in Google Sheets and Excel.

3. Results

The fluorescence intensity of chromatin proteins HP1γ, H3K9me3, H3K27me3, and histone deacetylases SIRT1 and SIRT6 was measured in the studied lines (**Figure 1**).

During the fluorescence intensity analysis, we studied both average intensity and intensity distribution. Intensity distribution in all cell lines was visualized with histograms plotted in Google Sheets in two ways:

- with automatic bin size selection
- with manual bin size selection—5 RFU, chosen empirically.

Histograms of the distribution of fluorescence intensities make it possible to assess how heterogeneous the cell line is for the studied protein and also to detect the presence of several subpopulations in cell lines.

Overlapping histograms of the fluorescence intensities of different cell lines with the same bin size allows comparing the distribution of the same protein in different cell lines.

3.1 HP1γ fluorescence intensity

The highest value of the average HP1γ fluorescence intensity was detected in dermal fibroblasts of the 30-year-old patient with breast cancer, and it was 8.25 times higher than that of the 55-year-old patient with breast cancer, 3.5

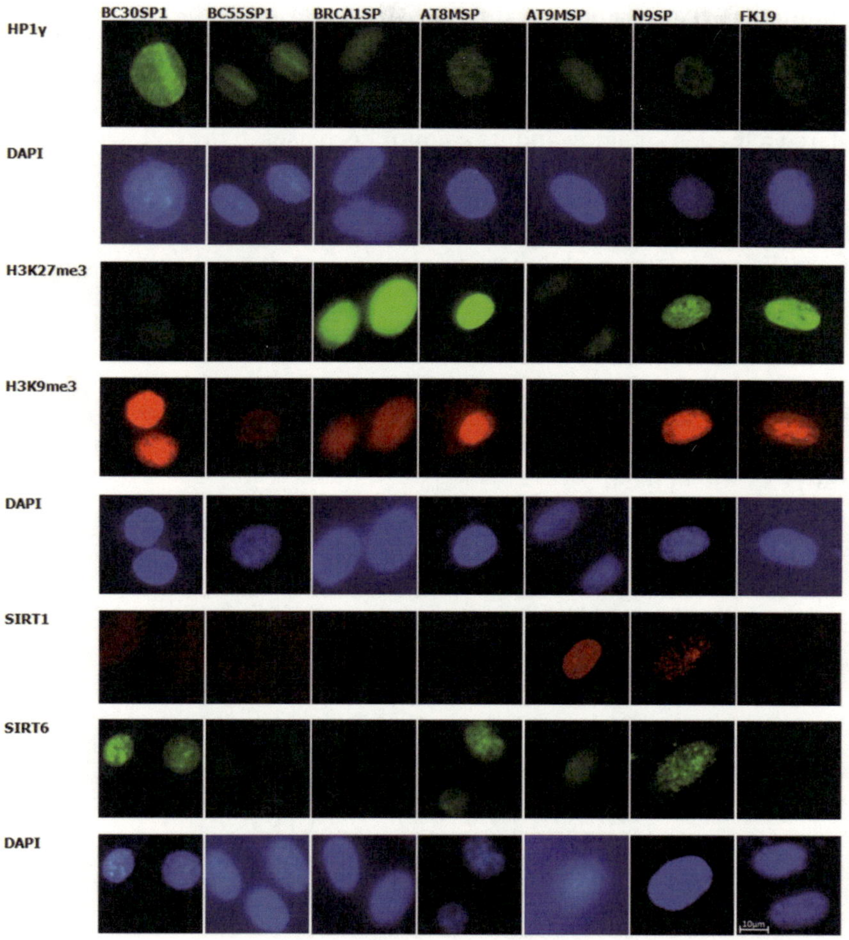

Figure 1.
Immunocytochemical analysis of the studied lines.

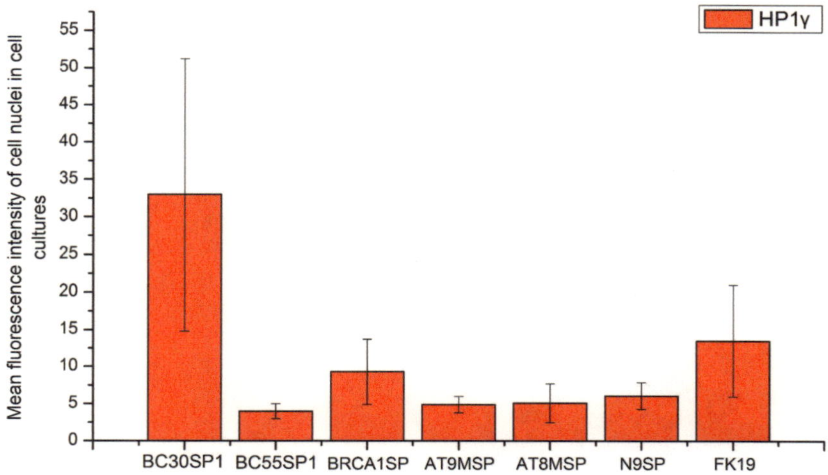

Figure 2.
Average fluorescence intensity of HP1γ in the nuclei of dermal fibroblasts of the studied lines.

times more than that of a patient at risk with a mutation in the *BRCA1* gene—BRCA1SP; 6.7 and 6.3 times more than mothers of patients with AT (AT9SP and AT8SP) (**Figure 2**). Such difference in the values of HP1γ fluorescence intensity can be explained by the fact that the histone heterochromatin protein HP1γ is considered a marker of aging and its amount decreases in the cell with age. In addition, the nature of the cancer of a 30-year-old patient, most probably, is different than that of a 55-year-old patient.

The average fluorescence intensity levels of HP1γ in healthy donors did not significantly differ from each other, despite the fact that the data differed by a factor of 6.1 and 13.5 RFU in lines N9SP and FK19, respectively. A significant difference in the average HP1γ fluorescence intensity of the BS30SP1 line was observed only in relation to the line of one healthy N9SP donor and was 5.4 times greater.

The absence of a significant difference between the cells of healthy donors is explained by the large value of the standard deviation in the population of lines due to the large spread of data.

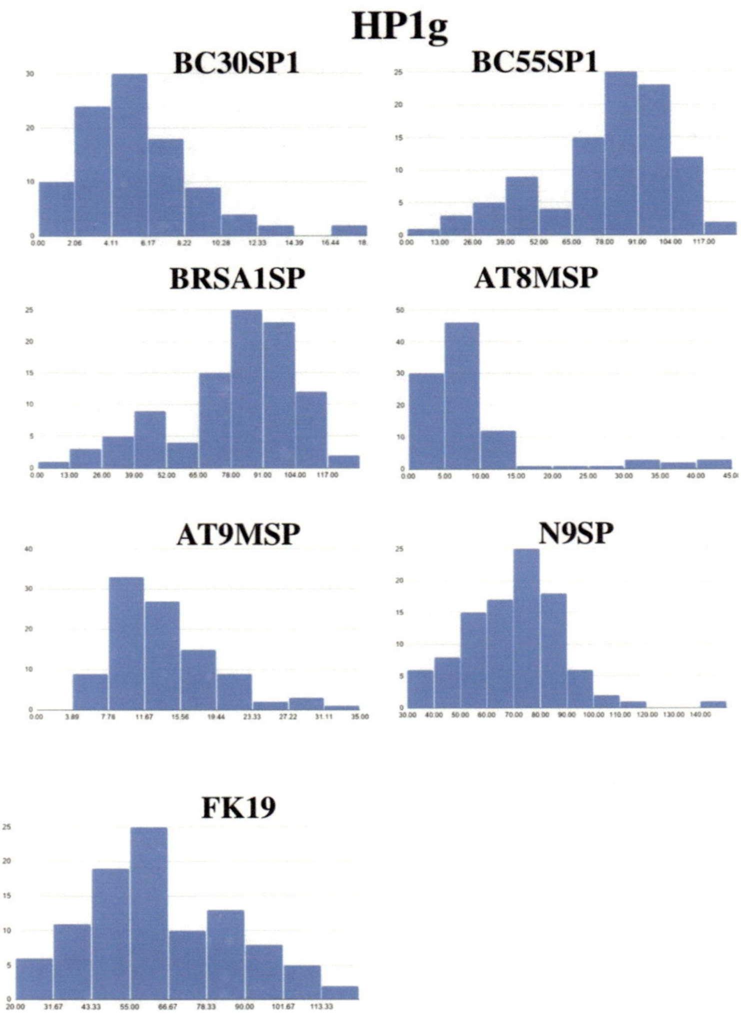

Figure 3.
Histograms of fluorescence intensity distribution of HP1γ.

The histogram of the distribution of HP1γ fluorescence intensities (**Figure 3**) in the cell lines showed the presence of two subpopulations in the BC55SP1 and BRCA1SP lines, each with its own distribution. In the 55-year-old healthy donor line, the boundary between the two subpopulations was less pronounced. Perhaps, this is due to age-related epigenetic changes in the HP1γ protein. In the remaining lines, insignificant second intensity peaks in some cells were observed, which may be associated with some rare cell changes. Interestingly, the picture of the distribution of HP1γ fluorescence intensity in patients with breast cancer differed from each other, as did the values of the average intensity.

When overlapping the histograms of HP1γ fluorescence intensity distribution in different cell lines with the same bin size, one can see how widely the HP1γ

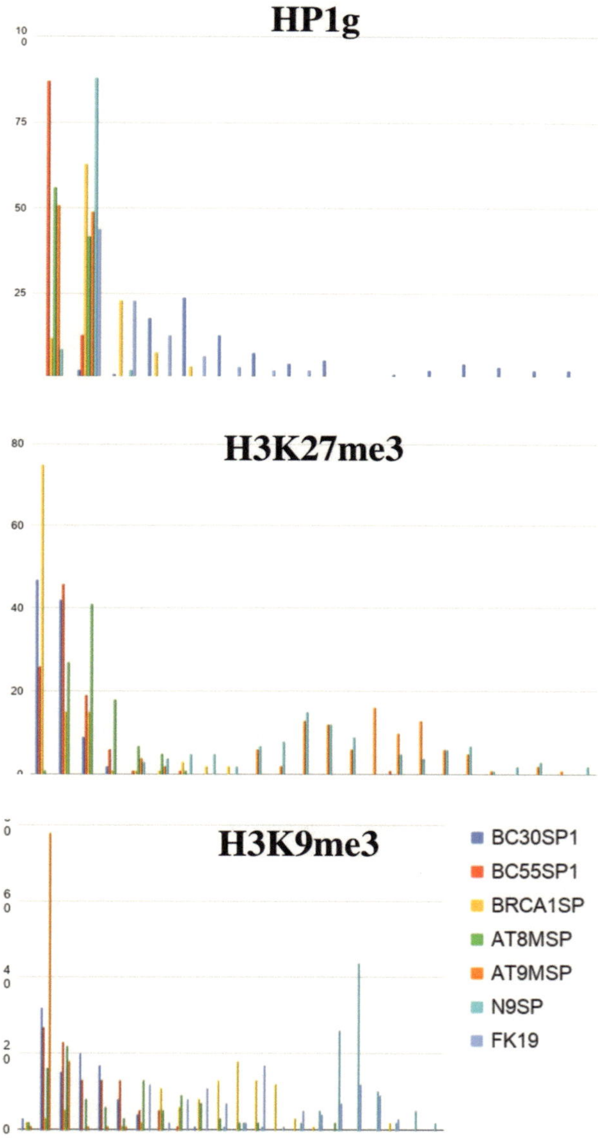

Figure 4.
Overlapping histograms of the distribution of fluorescence intensities of HP1γ, H3K27me3, and H3K9me3 in the studied lines with bin size of 5 RFU.

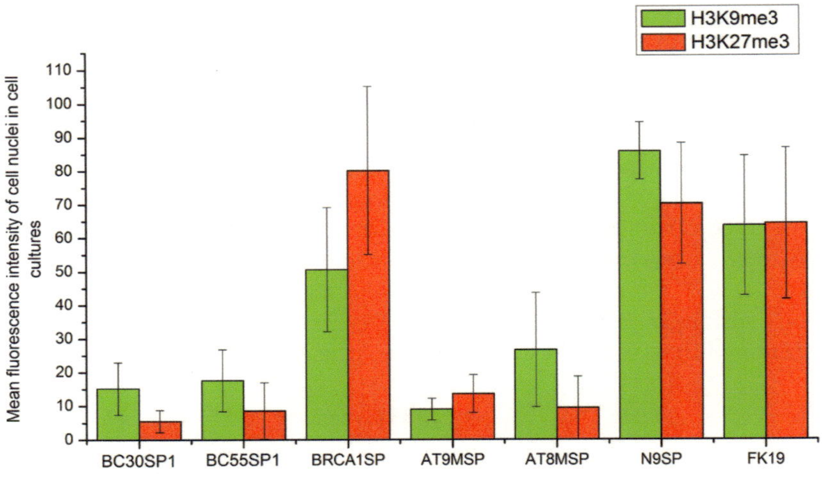

Figure 5.
Average fluorescence intensity of H3K27me3 and H3K9me3 in the nuclei of dermal fibroblasts of the studied lines.

fluorescence intensities were distributed in the BC30SP1 line, in contrast to other lines (**Figure 4**).

3.2 Fluorescence intensity of H3K9me3 and H3K27me3

The average fluorescence intensity levels of H3K9me3 and H3K27me3 in the dermal fibroblast lines of patients with breast cancer significantly differed from the levels in the cells of healthy donors (**Figure 5**). The average levels of H3K9me3 fluorescence intensity in the dermal fibroblast lines BC30SP1 and BC55SP1 virtually did not differ from each other (15.2 and 17.6 RFU) and were significantly lower by 5 and 4 times than the values of healthy donors 30 and 55 years old, respectively, for which these indicators were the highest. Interestingly, in the cells of the BRCA1 gene mutation's carrier, the average value of the H3K9me3 fluorescence intensity was significantly 3 times higher than in the cells of patients with breast cancer, although the data in the cells of mothers of patients with AT did not significantly differ from patients with breast cancer.

Histograms of the distribution of fluorescence intensities of H3K27me3 and H3K9me3 did not reveal two cell subpopulations in any of the studied lines. In all lines, insignificant second peaks of fluorescence intensities were observed (**Figures 6** and **7**). The distribution of fluorescence intensities of H3K27me3 and H3K9me3 revealed something similar to two subpopulations, but with very blurred boundaries. This picture may possibly be associated with age-related epigenetic changes in the cell.

When overlapping the histograms of H3K27me3 and H3K9me3 fluorescence intensity distribution in different cell lines with the same bin size, it can be seen that all lines have a sufficiently wide range of fluorescence intensities of H3K27me3 and H3K9me3, except for the BC30SP1 line, which has a fairly narrow range of fluorescence intensities of H3K27me3. This fact suggests a greater sensitivity of H3K27me3 and H3K9me3 histones to the state of the cell (cell cycle, apoptosis, ptosis, etc.). The widest distribution of fluorescence intensities was shown in the BRSA1SP, AT8MSP, and N9SP lines (**Figure 4**).

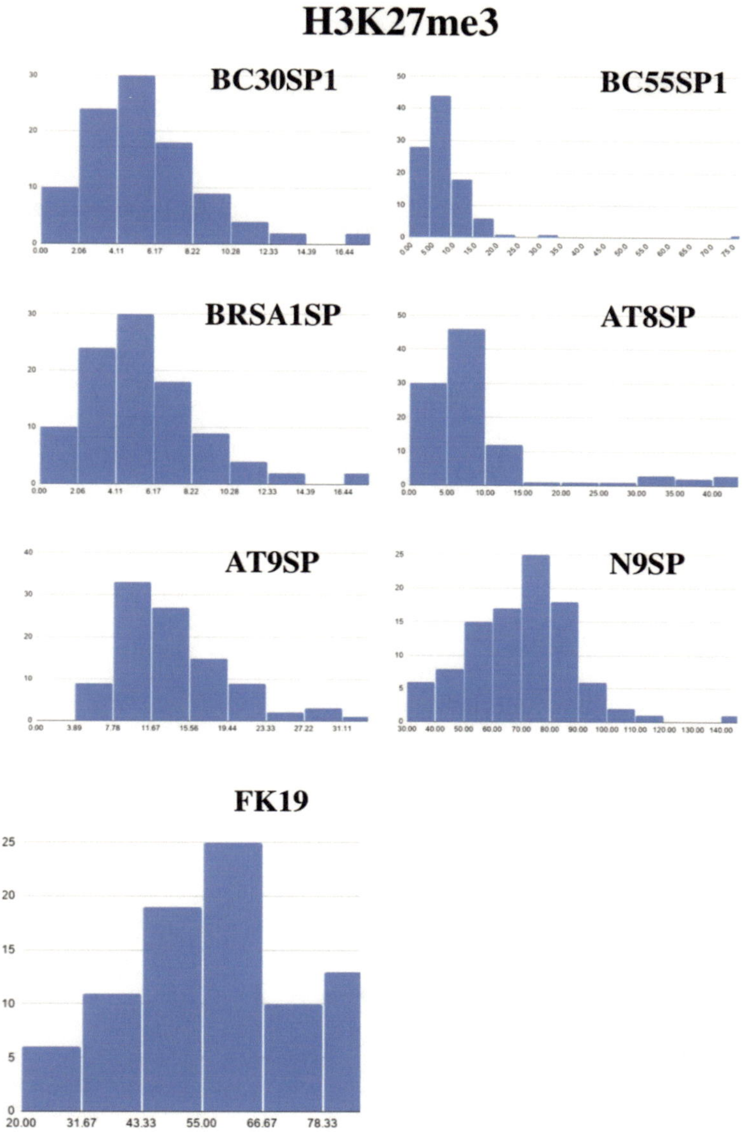

Figure 6.
Histograms of fluorescence intensity distribution of H3K27me3.

3.3 Fluorescence intensity of SIRT1 and SIRT6

The average fluorescence intensity levels of SIRT1 (**Figure 8**) in patients with breast cancer did not significantly differ from the levels in the lines of healthy donors, carriers of the mutations in the *BRCA1* gene, and the mother of one patient with AT. However, the average SIRT1 fluorescence intensity in the line of the second mother of the patient with AT was significantly 13 times higher than in the lines of patients with breast cancer.

The average fluorescence intensity levels of SIRT6 were also the lowest in the lines of patients with breast cancer and patients with a mutation in the *BRCA1*

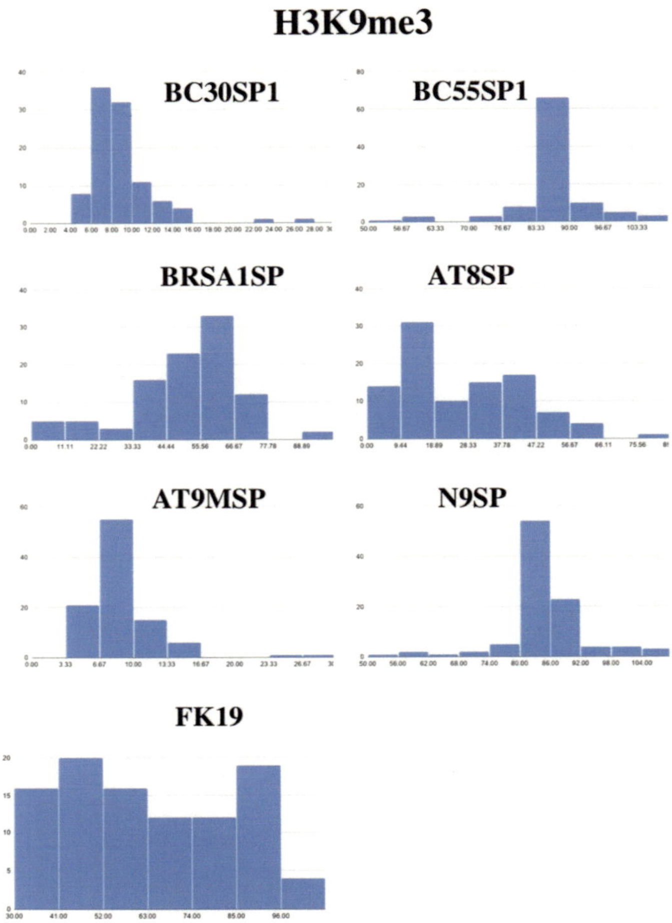

Figure 7.
Histograms of fluorescence intensity distribution of H3K9me3.

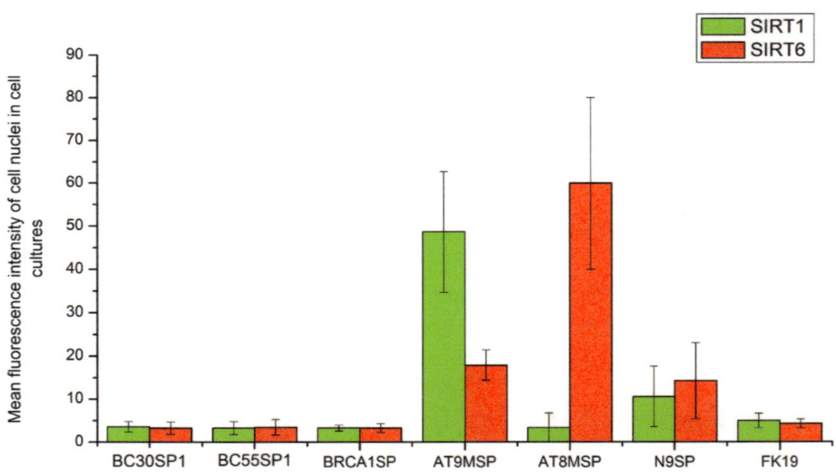

Figure 8.
Average fluorescence intensity of SIRT1 and SIRT6 in the nuclei of dermal fibroblasts of the studied lines.

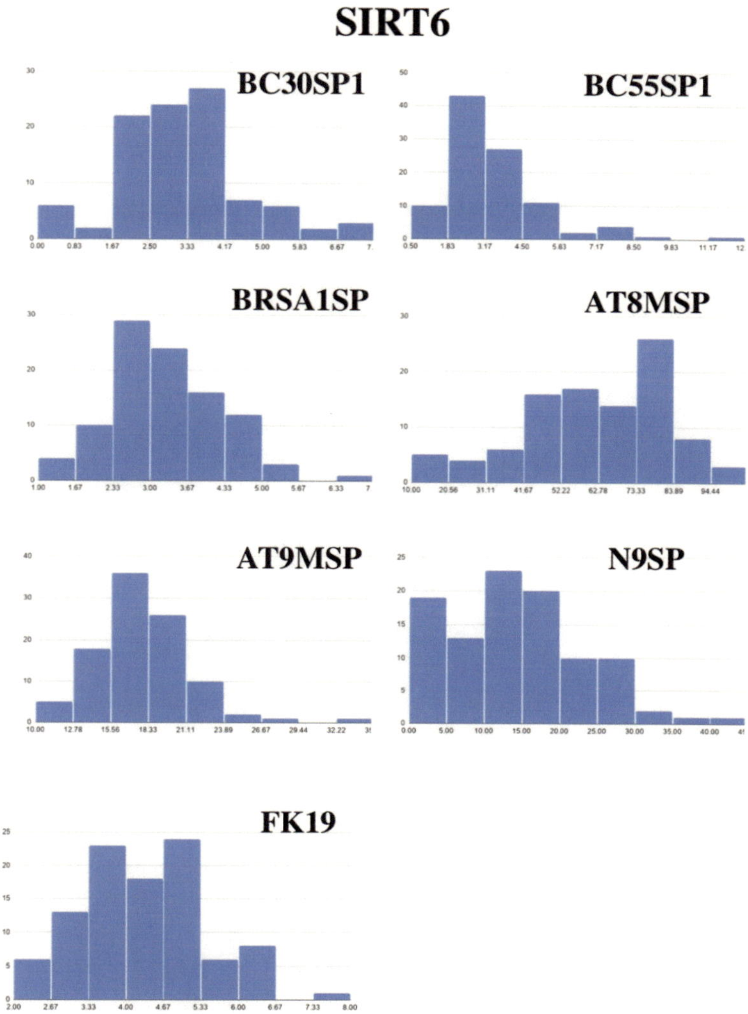

Figure 9.
Histograms of fluorescence intensity distribution of SIRT6.

gene. Significantly, the levels of the average fluorescence intensity in the lines of patients with breast cancer did not differ from the levels of healthy donors, while they were significantly lower by 5 and 18 times the values in the lines of mothers of patients with AT.

Analysis of the histograms of the distribution of the fluorescence intensities of SIRT6 (**Figure 9**) and SIRT1 (**Figure 10**) in cell lines revealed two subpopulations with different distributions in the AT8MSP line according to the SIRT6 fluorescence intensity and in the AT9MSP line according to the SIRT1 fluorescence intensity. In the N9SP and FK19 cell lines, several significant peaks in SIRT6 fluorescence intensity were observed.

When overlapping histograms of SIRT6 and SIRT1 fluorescence intensity distributions in different cell lines with the same bin size, it is possible to observe that the widest range of SIRT6 fluorescence intensity is in the AT8MSP cell line and that of SIRT1 fluorescence intensity is in the AT9MSP cell line (**Figure 11**).

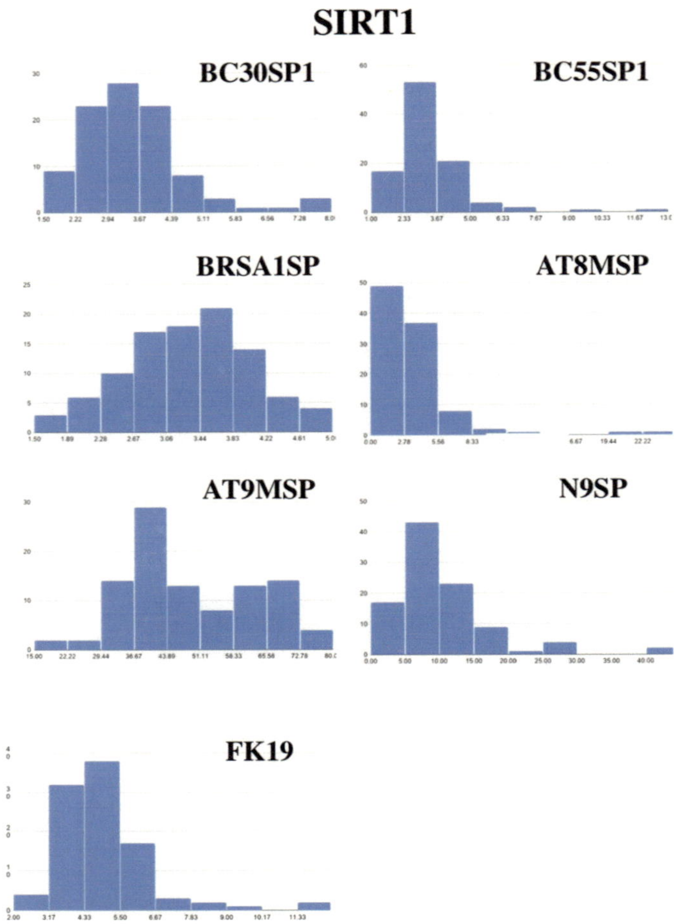

Figure 10.
Histograms of fluorescence intensity distribution of SIRT1.

4. Conclusions

Modulation of chromatin structure through modification of histones is the main epigenetic mechanism and regulator of gene expression. However, the contribution of chromatin properties to the heterogeneity and evolution of tumors remains unknown. Epigenetic labels are critical regulators of chromatin and gene activity. Their role in normal physiology and pathological conditions, including the development of cancer, is still unclear.

High expression of immune control points in the tumor microenvironment plays an important role in suppressing antitumor immunity, which is associated with poor prognosis and cancer progression. Major epigenetic modifications of both DNA and histone may be involved in the activation of immune control points in cancer.

Of course, for a more accurate study of epigenetic differences and features in breast cancer, a wider sample of the studied groups is needed. However, our results allow us to draw some preliminary conclusions in this matter. The results of our work showed significant differences in the lines of patients with breast cancer from the values of healthy donors only in terms of the average fluorescence intensity of H3K9me3 and H3K27me3. Contrary to published data, the average value of

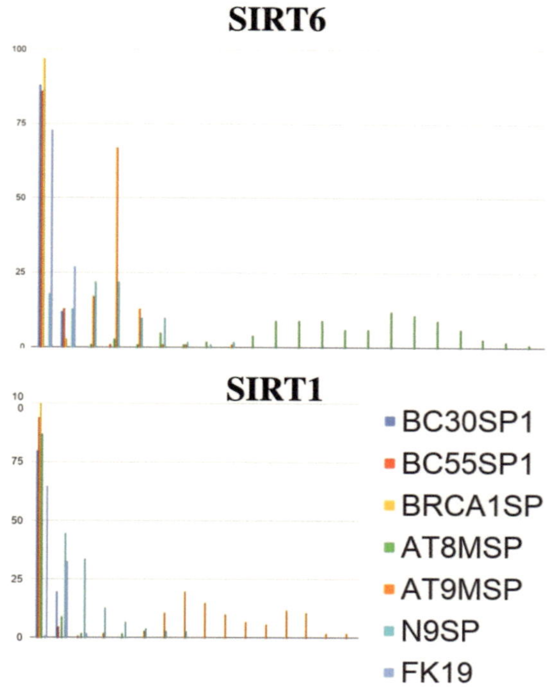

Figure 11.
Overlapping histograms of the distribution of fluorescence intensities of SIRT1 and SIRT6 in the studied lines with bin size of 5 RFU.

H3K27me3 fluorescence intensity in the cells of patients with breast cancer was significantly lower than in cells of healthy donors. Most likely, this is due to the fact that the study was not carried out in tumor cells.

Significant differences between the lines of patients with breast cancer were observed only in the values of the average intensity of HP1γ. The average HP1γ fluorescence intensity in a patient with breast cancer of 30 years was the largest of all the studied lines and in a patient with breast cancer of 55 years the lowest. The HP1γ fluorescence intensity in the line of 30 years old breast cancer patient was 8.25 times higher than in the line of 55 years old patient. Our results suggest that heterochromatin protein is a potential marker for breast cancer.

Thus, our results confirmed data on changes in the amount of H3K9me3 and H3K27me3 proteins in breast cancer.

An analysis of histograms of fluorescence intensity distribution of HP1γ, H3K27me3, H3K9me3, SIRT1, and SIRT6 allowed us to see how differently the fluorescence intensities are distributed in the cell lines and also allowed us to distinguish two cell lines with two subpopulations of the fluorescence intensity of HP1γ (BC55SP1 and BRSASP1). However, for other cell markers, two subpopulations in this line were not detected. A second minor peak in HP1γ fluorescence intensity was detected in the BC30SP1 cell line, but the distribution pattern was different from the distribution in the BC55SP1 line.

Interestingly, a pattern of fluorescence intensity distribution differed in different lines. But in most cases, it was similar within the same cell line in terms of fluorescence intensities of the HP1γ, H3K27me3, H3K9me3, SIRT1, and SIRT6 proteins.

Usage of average fluorescence intensity value as a single characteristic is not sufficient to properly compare different cell lines, because it masks significant variations in fluorescence intensity distribution. Overlapping histograms of fluorescence intensity distribution make it possible to assess variability of a given protein in a population.

The broadest range of fluorescence intensity values was demonstrated by the BRCA1SP (HP1γ fluorescence intensity), AT8SP (SIRT6 fluorescence intensity), and AT9SP (SIRT1 fluorescence intensity) lines.

But on average, across all cell lines, the broadest range of fluorescence intensity values was shown by histones H3K27me3 and H3K9me3. The data obtained suggest that these proteins are more "sensitive" to the state of the cell during normal life cycle and during pathological changes.

Thus, our results confirm changes in the amount of H3K9me3 and H3K27me3 proteins in breast cancer.

Acknowledgements

The work was supported by Grants Councils of the President of the Russian Federation MK-3638.2019.7.

Author details

Mirya Kuranova[1*], Aleksandra Nozdracheva[1], Nadezhda Pleskach[1], Daria Pasekova[2] and Aleksey Manikhas[2]

1 Institute of Cytology, RAS, Saint Petersburg, Russia

2 City Clinical Oncology Center, Saint Petersburg, Russia

*Address all correspondence to: miryakuranova@gmail.com

References

[1] Calcinotto A, Kohli J, Zagato E, Pellegrini L, Demaria M, Alimonti A. Cellular senescence: Aging, cancer, and injury. Physiological Reviews. 2019;**99**(2):1047-1078. DOI: 10.1152/physrev.00020.2018

[2] Anastasiadi Z, Lianos GD, Ignatiadou E, Harissis VH, Mitsis M. Breast cancer in young women: An overview. Updates in Surgery. 2017;**69**:313-317. DOI: 10.1007/s13304-017-0424-1

[3] Carroll JE, Dyk VK, Bower EJ, Scuric Z, Petersen L, Schiestl R, et al. Cognitive performance in survivors of breast cancer and markers of biological aging. Cancer. 2019;**125**(2):298-306. DOI: 10.1002/cncr.31777

[4] Zhang R, Adams DP. Heterochromatin and its relationship to cell senescence and Cancer therapy. Cell Cycle. 2007;**6**(7):784-789

[5] Kim J, Shin Y, Lee S, Kim M, Punj V, Lu FJ, et al. Regulation of breast cancer-induced osteoclastogenesis by MacroH2A1.2 involving EZH2-mediated H3K27me3. Cell Reports. 2018;**24**:224-237

[6] Li QL, Lei PJ, Zhao QY, Li L, Wei G, Wu M. Epigenomic analysis in a cell-based model reveals the roles of H3K9me3 in breast cancer transformation. Epigenomics. 2017;**9**(8):1-16

[7] Healey AM, Hu R, Beck HA, Collins CL, Schnitt JS, Tamimi MR, et al. Association of H3K9me3 and H3K27me3 repressive histone marks with breast cancer subtypes in the nurses' health study. Breast Cancer Research and Treatment. 2014;**147**:639-651

[8] Nair SV, Salhat EH, Taha ZR, John A, Ali RB, Elkord E. DNA methylation and repressive H3K9 and H3K27 trimethylation in the promoter regions of PD-1, CTLA-4, TIM-3, LAG-3, TIGIT, and PD-L1 genes in human primary breast cancer. Clinical Epigenetics. 2018;**10**:78

[9] Liarte S, Alonso-Romero LJ, Nicolás LF. SIRT1 and estrogen signaling cooperation for breast cancer onset and progression. Frontiers in Endocrinology. 2018;**9**:552

[10] Shi L, Tang X, Qian M, Liu Z, Meng F, Zhu WG, et al. A SIRT1-centered circuitry regulates breast cancer stemness and metastasis. Oncogene. 2018;**37**:6299-6315

[11] Jin X, Wei Y, Xu F, Zhao M, Dai K, Shen R, et al. SIRT1 promotes formation of breast cancer through modulating Akt activity. Journal of Cancer. 2018;**9**(11):2012-2023. DOI: 10.7150/jca.24275

[12] Thirumurthi U et al. MDM2-mediated degradation of SIRT6 phosphorylated by AKT1 promotes tumorigenesis and trastuzumab resistance in breast cancer. Science Signaling. 2014;**7**:ra71. DOI: 10.1126/scisignal.2005076

[13] Mahmud Z, Gomes RA, Jin Lee JH, Aimjongjun S, Jiramongkol Y, Yao S, et al. EP300 and SIRT1/6 co-regulate lapatinib sensitivity via modulating FOXO3-acetylation and activity in breast cancer. Cancers. 2019;**11**(8):1067. DOI: 10.3390/cancers11091067

[14] Jerzak K. J, Mancuso T, Eisen a. Ataxia–telangiectasia gene (ATM) mutation heterozygosity in breast cancer: A narrative review. Current Oncology. 2018;**25**(2):e176-e180

Chapter 9

Mammalian Cell Viability Methods in 3D Scaffolds for Tissue Engineering

Benjamin Gantenbein, Andreas S. Croft and Marie Larraillet

Abstract

Modern methods have evolved in tissue engineering to evaluate cell viability (CV) in 3D scaffolds and tissues. These involve either the usage of 3D confocal laser microscopy of live or fixed tissues, or separation of cells from the tissue, either live or fixed, and then their analysis by flow cytometry. Generally, working with live cells has the disadvantage that all the scanning needs to be completed immediately at the end of an experiment. Two different approaches can be distinguished: staining intact cell membranes and staining fixed cells. The entire cytoplasm is stained with amine-reactive dyes (ARDs), these use the principle of dead cell exclusion. Here, we list and compare live-cell versus fixed-cells fluorescence-based methods and also show their limitations, especially when working with autofluorescent or cross-linking materials like silk or genipin-reinforced hydrogels. Microscopic techniques have the advantage over flow cytometry-based methods in that these provide the spatial distribution and morphology of the cells. Calcein AM combined with ethidium-homodimer works for most 3D constructs, where no strong fluorescent background is found on the tissue or scaffold. Frequently, however, concentrations and incubation times need to be adjusted for a specific tissue to ensure diffusion of dyes and optimise emittance for detection.

Keywords: live/dead staining, 3D scaffold, hydrogels, confocal laser scanning, calcein AM, ethidium homodimer-1, cell tracking, mammalian cells, amine-reactive dyes (ARDs), microscope, fixed cell staining, glycosaminoglycan-rich tissues, fibrin, silk, genipin

1. Introduction

1.1 The need for 3D culture models in tissue engineering

There are several areas of applied research where the importance for ex vivo 3D culture models is crucial, especially if 3D organ culture models are to be established using perfusion or non-perfusion characteristics or even under static culture conditions. These concepts apply to all research areas involving organs and tissues in culture. As the authors of this chapter are active in the field of musculoskeletal research, examples will be illustrated using connective tissues or cells isolated from bone and joints. These fields often focus on joint-derived tissues that cause clinical problems for healing such as (i) anterior cruciate ligament

IntechOpen

(ACL) [1–3], (ii) cartilage scaffold engineering [4–7], (iii) intervertebral disc (IVD) regeneration [8–10] and (iv) bone regeneration [11–15]. We demonstrate a subset of fluorescent staining and methods to stain cells from bone and joint-derived tissues: that is, staining mesenchymal stromal cells, chondrocytes and IVD cells in native tissue or in 3D hydrogel-like scaffolds such as fibrin [16], polyethylene glycol (PEG) [17], and also cells on solid materials such as silk [18]. In all of these fields, bioreactor models have been developed to better understand the mechanism of mechanical loading on these tissues. However, the combination of novel smart biomaterials can also be studied together or in direct contact with the organ of interest under more realistic conditions [10, 19–21]. More generally speaking, as the aim is to develop clinically relevant models considering the integrity of tissue and organ explants, judging cell viability (CV) becomes a major issue that needs to be assessed [22]. CV is also the main read-out for new biomaterials and scaffolds to be tested for tissue engineering (TE) and cell therapy applications. Of course, there are other parameters quantified such as DNA content, mitochondrial activity [23, 24] and cell apoptosis [25, 26]. However, in this book chapter, we would like to emphasise CV as a key parameter and summarise the currently most applied methods to stain and visualise living and dead cells. Furthermore, we provide a link of an automated cell counting macro for the integration of ImageJ, a free microscope imaging processing platform to quantify CV [27]. At the end, we would like to present some examples using 3D automated macros for high-throughput screening. The decision whether to analyse cells in a living or dead state has to be made relatively early in the work flow (**Figure 1**). Most fluorescent dyes that are introduced here are commercially available as kits (**Table 1**) [4, 28–31]. However, as many of the patents have expired, most

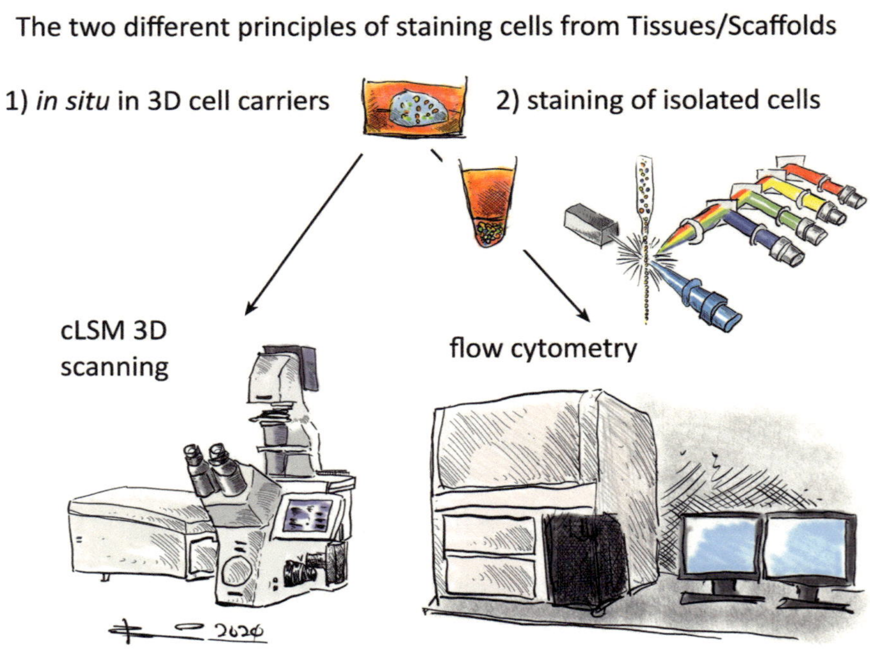

Figure 1.
Overview of tissue engineering principles and judgement for cell viability (point out microscopy or FACS methods). The main decision is whether information on the cell morphology and spatial distribution is essential or whether quantitative numbers of CV are enough. In the latter case, determination of CV by flow cytometry using either fixed or live cells is the choice. However, in these cases the cells need to be isolated from the ECM firstly and then the 3D information and original cell morphology are lost. For certain applications in TE, it is crucial to see into the 3D distribution of living cells in the tissue/scaffold.

Staining	Name	λEx [nm]	λEm [nm]	Providers	Applications and references
Calcein AM	Calcein acetoxymethyl ester (green)	496	516	Sigma-Aldrich, Thermo Fisher Scientific	Live cell staining, cell tracking [4, 16, 28–31, 43]
CellVue™	Claret far red linker kits	655*	675*	Sigma-Aldrich	Live staining [51]
DAPI	4′,6-diamidino-2-phenylindole	340	488	Sigma-Aldrich, molecular probes, Thermo Fisher Scientific	Dead nuclei staining for fixed cells, does not stain live cells
DID	Vybrant™ Cell-Labelling Solutions (far red)	644	665	Thermo Fisher Scientific	Live/dead membrane staining [59]
DIL	Vybrant™ Cell-Labelling Solutions (orange)	549	565	Thermo Fisher Scientific	Live/dead membrane staining [59, 60]
DIO	Vybrant™ Cell-Labelling Solutions (green)	484	501	Thermo Fisher Scientific	Live/dead membrane staining [59]
EthD-1, EtDi	Ethidium homodimer-1	528	617	Sigma-Aldrich	Dead nuclei staining
LuminiCell Tracker™	Nanoparticles	422*	540*	Sigma-Aldrich	In vivo cell tracking
PKH26	Red fluorescent cell linker kits for general cell membrane labelling	551	567	Sigma-Aldrich	Live membrane staining, transplantation studies [53–56]
PKH67	Green fluorescent cell linker kits	490	502	Sigma-Aldrich	Live membrane staining [52, 60]

Kits are available in multiple colours.

Table 1.
An overview of the most commonly applied fluorescent dyes to either determine cell viability (CV) or to track cells in scaffolds or tissues. Also given are whether they are more suitable for live cell imaging or to work with fixed cells.

fluorescent dyes are now also available from a wider range of distributors even as concentrated powders and can be bought for affordable prices. We would like to provide a series of more or less straightforward protocols to trace living cells in tissues, of which, we can provide our own experiences. Finally, we would like to conclude with some, in our view, hard-to-trace examples using biomaterials with strong autofluorescence or difficult optical characteristics. For these biomaterials, scanning may not be easy and no good workarounds exist yet.

1.2 Evaluation of the cell viability in hydrogels, living tissues and organs

In TE and for orthopaedics and other fields involving organ and tissue-oriented research, it is crucial to understand whether cells are alive if seeded into a scaffold and after a specific time of culture. For a cross-disciplinary field in regenerative medicine, it is essential to understand the viability of tissues and cells over time, especially in 3D scaffolds, where diffusion gradients arise through unequal distribution of nutrients, oxygen concentration and pH gradients caused by cellular activity. TE frequently involves scaffold designing that then often needs to be assessed for cytocompatibility and cell viability. Thus, a central question is whether the spatial distribution of cells is essential or whether solely quantitative numbers are sufficient.

Currently, there are a number of assays available to determine CV in tissue or three-dimensional (3D) scaffolds, including lactate dehydrogenase staining (LDH) [32–34], calcein-AM with ethidium homodimer-1 staining, for example Live/Dead®, Ca-AM/EthD-1 and cell counting after scaffold/tissue digestion [35, 36] (**Table 1**). The easy and straightforward combination of Ca-AM/EthD-1 dyes can be used to stain living and dead cells directly in the scaffold or tissue. The Ca-AM is enzymatically hydrolysed into calcein in living cells, turning those into a bright fluorescent green. The EthD-1, on the other hand, is only able to enter cells with a compromised membrane and stains nucleic acid fluorescent red. It should be noted that it is also possible to count cells after scaffold digestion using different dyes or stains, which differentiate living from dead cells, such as Ca-AM or Trypan Blue (TB) [37].

There are many advantages of the LIVE/DEAD viability/cytotoxicity tests: One advantage is that the live cell staining is not dependent on cell proliferation and is a nonradioactive assay unlike thymidine uptake and ^{51}Cr release assays [38–40]. Furthermore, stained cells can be observed using fluorescence illumination and can be counted to determine the percentage of viable (or non-viable) cells. For cells grown in multi-well plates, the overall fluorescence per well can be determined using a fluorescence plate reader. For 3D specimens, confocal laser scanning microscopy (cLSM) has widely been applied. The main advantage of the cLSM is its ability to record the precisely defined optical sections from a 3D sample. This can be achieved by moving the focal plane of the instrument stepwise through the depth of the specimen, whereby a series of optical sections (stacks) can be collected. This provides, of course, more detailed information than from a piece of tissue that contains data from only one focal plane [41, 42]. Because optical sectioning is relatively non-invasive, the 3D distribution and relative spatial relationship of stained living, as well as fixed, cells can be observed with reasonable clarity. Another essential feature is that the slices obtained by the cLSM can be forwarded to automated image analysis and can also be rendered as 3D stacks if the optical plains are sufficiently overlaid (about 20–30% is recommended). These stacks can then also be used to compute surface- or volume-rendered 3D reconstructions of the specimen. Accordingly, data from images can be processed and converted into cell counts and live/dead cell ratios, respectively.

Recently, the question was asked whether there are differences in outcomes among widespread cell viability techniques used in TE. To answer this question, three commonly used methods were recently systematically investigated: (i) classical hand counting by Trypan Blue and hemocytometer (ii) Live/Dead® staining in combination to confocal laser microscopy and (iii) lactate dehydrogenase (LDH) activity staining on histological slices (**Figure 2**) [16].

Generally, CV is defined as

$$CV = \frac{number\ of\ living\ cells}{total\ number\ of\ cells} \tag{1}$$

This formula is the basis to judge CV (Eq. (1)). The following paragraph will introduce the principles of live/dead staining and 3D scanning using cLSM technology, and provide hands-on protocols for staining connective tissues such as joints.

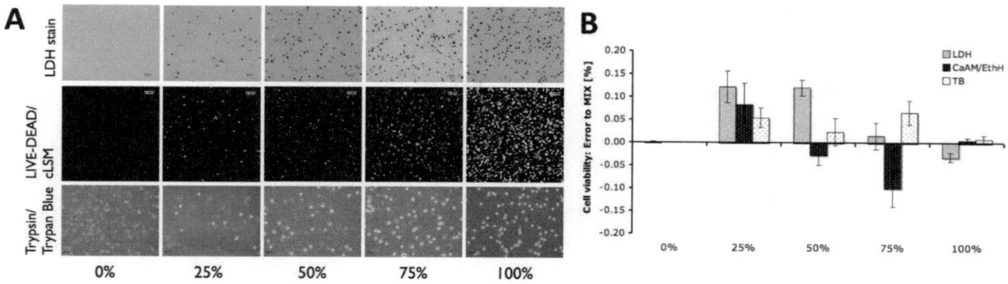

Figure 2.
Representative images from the three different techniques used to estimate cell viability (i.e. LDH, Calcein AM/Ethidium homodimer (CaAM/EthD-1), and Trypan Blue staining after trypsin digestion) for five different cell viability mixtures, called MIX. (A) Exemplary images of cell viability stainings in fibrin 3D carriers using bovine chondrocytes. The pictures represent the outcome of an evaluation staining to control for cell viability (CV) and to investigate into the accuracy of these three prominent techniques to evaluate CV. (B) Error to MIX of the estimated CV in the MIX. It was evident that hand counting and Trypan Blue with hemocytometer is the most accurate method. CV was mostly overestimated with increasing number of living cells, especially with LDH assay. The figure was modified from [16] and reprinted with permission from the publishers. Values are means ± SD, n = 5.

2. Applied protocols for live/dead staining of cells in tissues or scaffolds using cLSM

The following staining protocols using Ca-AM/EthD-1 in 3D carriers and tissues were partially reprinted with permission from the publishers. The protocols were modified and are based on the book chapter by [43].

2.1 Stage for 3D scanning

To ensure optimal optics, it is recommended not to use plastic well-plates from commercial manufacturers for scanning. Instead, a simple stage for 3D scanning with excellent optical characteristics for inverted microscopy can be produced with the following materials. The following protocol has been tested for intervertebral disc tissue of bovine and human origin but has also been successfully used for ligament tissue of human joints and cartilage biopsies.

1. Aluminium plate (dimensions 50 × 80 × 6 mm) (**Figure 3A** and **B**)
2. Metal drill to cut out a circular hole (Ø 22 mm)
3. Coverslip 30 × 50 mm No.1 (e.g. Gerhard Menzel Glasbearbeitungswerk GmbH & Co. KG, Braunschweig, Germany)
4. Nusil© Medical Grade Silicon (MED-1137, Adhesive Silicone Type A, Silicone Technology, Carpinteria, CA, USA)

The Nusil© Medical Grade Silicon is dried for at least 72 hours.

The complete sample holder is washed twice with methanol and dried for 15 minutes before use (**Figure 3**).

2.2 Preparation of staining solutions

Prepare 1 mL of staining solution using serum-free Dulbecco's Modified Eagle Medium (DMEM) or phosphate-buffered saline (PBS) per tissue (~3 × 3 × 3 mm^3).

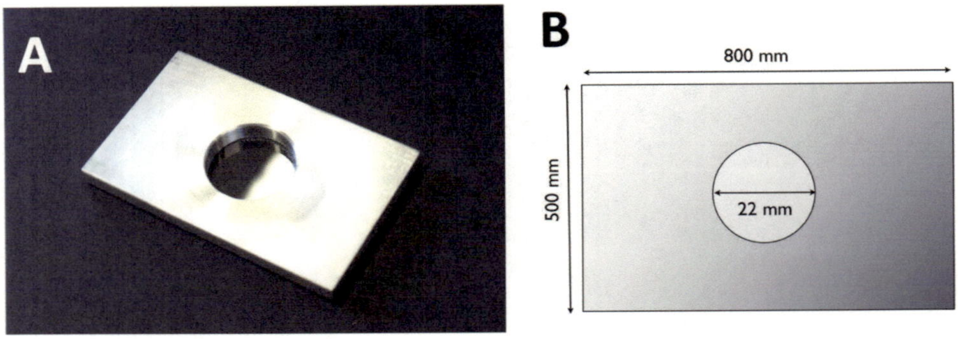

Figure 3.
Customised sample holder for scanning tissues and scaffolds. Dimensions of simple custom-made sample holder (aluminium plate with 22 mm Ø drilled hole) used for improved 3D scanning. (A) Side view. (B) Schematic top view. The picture is modified but based on [43] and reprinted with permission from the publishers.

For the staining of tissue, we used a 10× higher concentration of Ca-AM (10 µM) than usually indicated on live/dead kits. The reason for this increased concentration is the fact that the fluorescent molecule needs a longer time to penetrate into the tissue or carrier, and a higher concentration is needed. However, for the EthD-1, we kept a 1 µM concentration as recommended by several manufacturers. Prepare the dye 'cocktail' directly before staining since fluorescent dyes are unstable in contact with water. Also, keep these freshly prepared dye solutions in the dark if possible.

2.3 Preparation of the tissue samples

IVD tissue from bovine or human origin or tissue of human ACL (in the latter case, both with ethical written consent) can be stained with this protocol, which we provide below. It is likely that incubation times and concentrations need to be adjusted for other tissue types.

1. Discs are then separated into nucleus pulposus and annulus fibrosus tissue by a 6-mm Ø biopsy puncher (Polymed, Glattbrugg, Switzerland).

2. Circular tissues are then further cut into half by a scalpel blade and annulus fibrosus is further dissected into inner and outer annulus fibrosus by eye.

3. Tissues pieces are emerged in 1 mL of a freshly prepared staining solution in a 12-well plate and incubated for 1 h at 37°C in 100% humidity and 5% CO_2.

4. After incubation, tissue bits are transferred into a fresh well-plate and washed with 1 mL of TBSS and checked immediately with the laser scanning microscope. For the imaging, a customised sample holder was used. The preparation of this customised sample holder is shown in **Figure 3**. For all of our live cell imaging, it was recommended to use a custom-made holder, which enables significantly improved optics than culture well-plates plastics on their own.

3. Staining and counting of pre-isolated fixed cells

In the following sections of this chapter, we present novel, unpublished data, and thoughts on the staining of fixed cells and how accurate these kits are to evaluate the CV in scaffolds and tissues. We used the new class of fixable dead stains with the aim that CV could be determined in fixed tissues as an alternative to live/dead cell staining as outlined in the previous section. Fixable dead cell stains referred to as amine-reactive dyes (ARDs) are a class of viability dyes designed for the discrimination between living and dead cells in samples that will be fixed. It is a new class of viability assays using the principle of dead cell exclusion markers. The method is based on the reaction of a fluorescent reactive dye with cellular amines [44, 45]. The reactive dye crosses the compromised membrane of dead cells, thus reacting with free amines present in both the cytoplasm and the cell surface. This reaction leads to an intense fluorescent staining of the dead cells. On the opposite, living cells exclude this dye as their membranes are intact and only the cell-surface amines are available to react with the dye. This results in a significantly lower fluorescence signal, which is due to a fewer amount of amines reacting with the dye, as shown in **Figure 4**. The main advantage of fixable methods lies in the fact that the reaction is irreversible. Therefore, after cell fixation and permeabilisation, the bound dye remains linked with the dead cells and the staining is stable with no loss of fluorescence signal over a certain period of time [44, 45]. Moreover, ARDs are available in a variety of excitation and emission wavelengths, which gives a great advantage in terms of diversity, thus allowing a considerable flexibility when establishing staining protocols [44]. These cells will then be counted most efficiently using flow cytometry.

3.1 Cell viability mixtures

To evaluate different methods to estimate CV in terms of accuracy and precision, the approach is to generate, both, living and dead cell solutions that were then combined with pre-known CV ratios (PREMIX) 100%/0%, 75%/25%, 50%/50%, 25%/75% and 0%/100%. The preparation of the dead cells for the mixtures was performed based on the method by Gantenbein-Ritter et al. [16], where the non-viable cells were obtained by 1 N HCl treatment. It has been shown that cell nuclei stay intact with this method and, thus, can be detected with all methods under investigation. The cells were detached with 1% trypsin, centrifuged (500 g, 5 min) and then

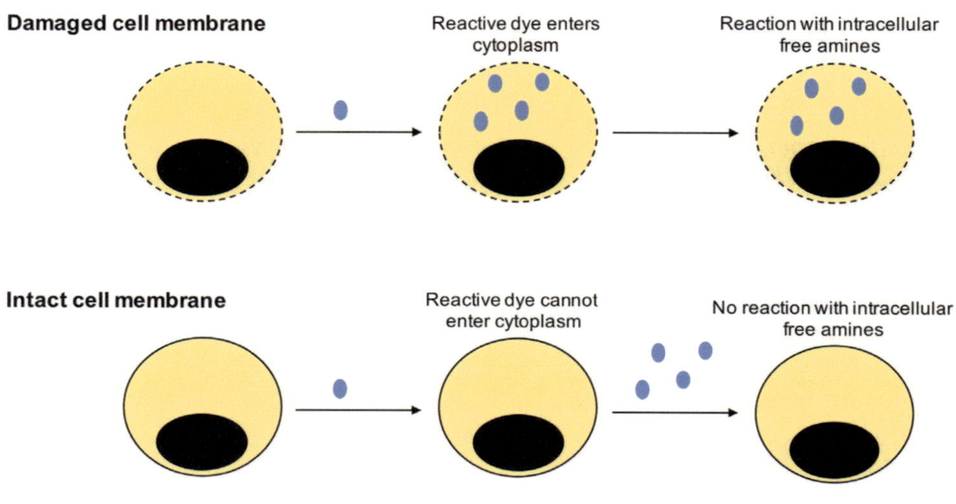

Figure 4.
Principle of action of a fixable cell viability assay using amine-reactive dyes (ARDs). The top path represents cells with damaged membrane where the ARD (in blue) can enter into the cytoplasm. The bottom path shows living cells with intact cell membrane in which case the dye cannot pass. The principle is described in Perfetto et al. [45].

resuspended in PBS. The dying process was achieved by the addition of 1% of 1 N HCl in a PBS solution to the cell suspension followed by 10 min of incubation at room temperature (RT). The cells were then washed and resuspended in PBS. In parallel, living cells were detached, centrifuged (500 g, 5 min) and resuspended in PBS.

3.1.1 Amine-reactive dye (ARD) staining

In the following section, we describe how the ARD staining was evaluated for accuracy, and how it was compared to hand counting using TB and cLSM imaging.

3.1.2 Cell staining with ARD

For the comparison of the fixable versus the live cell imaging, the different cell mixtures were stained using the ViaQuant ™ blue fixable dead cell stain kit from GeneCopoeia (GeneCopeia, inc., Rockville, MD, USA). A cell density of 2×10^6 cells/mL for each mixture was used using ACL-derived ligamentocytes (LCs) and the staining protocol of the manufacturer's instructions was followed. In short, the cell suspension was incubated on ice and protected from light during 30 min in a 1-mL PBS volume where 1 µL of dye was added. The cells were then washed with PBS prior to fixation with 3.7% formaldehyde during 15 min. Finally, the cells were used to prepare the CV ratios as described above for PREMIXes.

3.1.3 Cell labelling and gel casting to evaluate the ARD staining

Each cell suspension was added to 2% agarose (Lonza, Rockland, ME, USA) in a 1:1 ratio of 250 µL volume to obtain a 1% agarose gel. The final cell density was approximately 2×10^6 cells/mL. Subsequently, 27 µL of the agarose-cell mixture was then casted into a custom-made silicone mould (cylinder dimensions ⌀ 4 mm × 2 mm height). After embedding into the gel, the cells were stained with the ViaQuant ™ blue fixable dead cell stain kit and the pellets were incubated for 60 min in 1 mL of PBS containing 1 µL of dye. In order to localise the cells in the 3D carriers, they were counterstained with EthD-1, which labelled all nuclei in red fluorescence. The cells were incubated for 5 min at RT in a 1-mL PBS solution containing 0.5 µM EthD-1-1.

3.1.4 Hand counting method

With the TB method, the number of living and dead cells were manually counted with a Neubauer improved cell counting chamber by taking four repeated measurements of each cell suspension (n = 4/suspension). The CV was computed for each count and then averaged. The hand counting procedure was then repeated after staining with ARD (n = 8/cell mixture). In this case, cells were counted manually using a hemocytometer under a Leica DM IL microscope (Leipzig, Germany) using a blue fluorescence filter to determine the proportion of the two cell populations. Blue fluorescent cells corresponded to dead cells and the nonfluorescent ones to living cells.

3.1.5 cLSM imaging

The quantification of CV after ARD and EthD-1 staining of the 3D agarose carriers was performed using the ImageJ software (v1.51k, NIH) [27] by first obtaining the colocalised cells between the red and blue channels with the plug-in 'colocalisation highlighter'. Colocalised cells between EthD-1 and ARD staining corresponded to the dead cells. For this purpose, the settings had to be adjusted in order to take into account only the bright blue fluorescent cells. The threshold values for both channels were set at 50% and two points were considered as colocalised if their respective intensities were strictly higher than the threshold of their channels. Dead cells were then counted using the plug-in 'nucleus counter'. The number of red cells corresponding to the total number of cells (living and dead) was also computed with the plug-in 'nucleus counter'. Thus, the number of living cells was obtained by subtracting the number of dead cells from the total number of cells.

4. Validation of cell viability methods

It is important to know the accuracy and precision of the most commonly applied cell viability methods in the laboratories. Here, we summarise the results of two studies, one involving the comparison of several live methods, that is, Trypan Blue staining and manual cell counting using hemocytometer (TB), live/dead staining, and cLSM imaging (CA-AM/EthD-1). The second study, we present here, is on the usage of ARD stains and its accuracy for CV in comparison to live/dead staining followed by cLSM and was not published previously.

In the case, where CV is then quantified in 3D using a cLSM microscope, the following procedure was applied: Firstly, the stacks were separated into separated channel images and the CV was determined on each of the images (**Figure 5**). This step was done using the open source macro of ImageJ. In this routine, the red and green cells were quantified separately per single image using a custom-made macro in ImageJ software named 'Cellcounter3D' (deposited at http://imagej.nih.gov/ij/macros/Cellcounter3D.txt) and available free of charge under the GNU licence [16]. The macro consists of a converting step to 8-bit mode and a threshold step that passes a binary image with pixels in the range of 100–255 to the plug-in 'nucleus

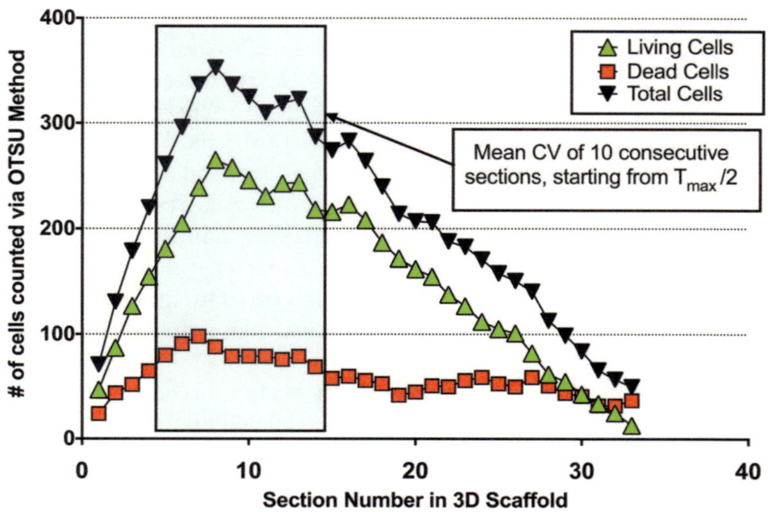

Figure 5.
Progress of cell viability (CV) estimates through a 3D stack using the custom-written 'Cellcounter3D' routine of imageJ and the OTSU method to count the number of green (live cells), red (= dead cells) and the total number of cells in z-stacks of cLSM images. As can be seen, the total number of cells increases after about 5–10 sections inside the scaffold and then reaches a plateau of about 10–15 consecutive optical sections. The first five images from the tissue surface are rejected as these are prone to error due to cell death by tissue/carrier manipulation. In light blue is the window indicated that is then used to estimate the CV of 10 consecutive pictures after the maximum of the total cell number reached $T_{max}/2$. The picture is based on [43] and has been reproduced with permission from the publishers.

counter' (available at https://imagej.nih.gov/ij/plugins/mbf/index.html). This step then uses the 'Otsu' method for particle counting [46]. The minimum and maximum island sizes were set to 7–50 and 15–100 pixels for the red and green channel, respectively. Please, note that these parameters can be modified dependent on the quality of a specific hydrogel/biomaterial and dependent on the cell density and the type of cells seeded. It is important to inspect the segmentation and cell counting progress by eye to judge whether the numbers are meaningful. The intermediate steps like the results of the segmentation progress can be also stored as .tiff or .jpg files for further documentation.

In the following steps, we describe the settings to determine CV using live/dead staining and cLSM technology. The 3D carriers were produced using a mould, made of easy-sterilisable materials, such as silicon membranes or stainless steel. The dimensions of these moulds were in this case of a cylindrical shape of 4 mm in ø × 2 mm in height, corresponding to a volume of 27 μL of carrier. The final chosen cell density was 2×10^6 cells/mL. The 3D carriers, that is fibrin, PEG or 1% agarose, respectively were cut sagittally into halves and incubated in 1-mL high glucose Dulbecco's Modified Eagle Medium (DMEM) without fetal calf serum (FCS) containing 10 μM CaAM and 1 μM EthD-1 (both Fluka, Sigma-Aldrich, Buchs, Switzerland) for 3 h at 4°C followed by 1 h of incubation at 37° C, 5% CO_2 and 100% humidity. The carriers were then scanned from top and bottom surfaces to ~200 μm depth at two random locations per side with a confocal laser scanning microscope (cLSM510, Carl Zeiss). Stacks were taken at 10× magnification at a 512 × 512 pixels resolution (field size of 921.4 × 921.4 μm) with the pinhole at 1 Airy unit and 50% image overlap and 5.8-μm intervals. The proposition was then to quantify CV on a subset of 10 consecutive images, starting at the image with 50% or more of the maximum amount of the total cells per image in a single stack (**Figure 5**). Additionally, the first five sections starting from the surface into the scaffold were also rejected as there cell death occurred through tissue manipulation, which causes a strong bias. To determine this frame, a MatLab routine has been written (MatLab R2020a, MathWorks inc., Natick, MA, US), which picks the analysis-window automatically and summarises the CV results in a compact table. The code is available free of charge at the Mathworks repository under the keyword 'Cell Viability Estimator for 3D Scaffolds' [47]. **Figure 5** illustrates a typical profile through a stack of images for the living, dead and total cells. The cells in this case were bovine chondrocytes from articular cartilage seeded at a cell density of 4 M cells/mL in fibrin hydrogel (Tisseel™, Baxter, Vienna, Austria) according to a customised recipe as described in [16]. These cells were counted then using three different methods on day one of culture under standard conditions. It is important to note that CV can be determined also on z-compressed images, which represent a cumulative summary of the entire stack or on individual scans through the stack as introduced here.

The accuracy of the dead cell fixable viability assay was then determined by computing the relative error between the theoretical cell viability and the hand counting, flow cytometry and cLSM results obtained after the ARD staining. The theoretical viability was based on the hand counting of living and dead cells of the different cell mixtures with the TB method. Then, the relative error (Δx) was computed as followed in Eq. (2):

$$\Delta x = \left| \frac{x_0 - x}{x} \right| \qquad (2)$$

where, x_0 corresponds to the CV measured with any of the to be compared methods, that is, LDH, ARD+FACS, cLSM live/dead and compared to x, the one from the TB assay known as the reference method (=PREMIX).

The accuracy of all the methods was quantified in regards to the cell mixtures by calculating the absolute error (δx) as followed in Eq. (3):

$$\delta x_{i,j} = \overline{x}_{i,j} - p_i \tag{3}$$

where $x_{i,j}$ is the mean CV of each method, j, and p_i are the estimated CV of the prepared cell mixtures, i.

5. Comparisons of CV estimation methods

It was recently found in an earlier study that in a systematic comparison of Trypan Blue, CA-AM/EthD-1 by 3D stack scanning with lactate dehydrogenase activity on cryosections and lastly vs. and averaging [16] and thirdly by comparing lactate dehydrogenase (LDH) activity and ethidium homodimer-1 co-staining and using histological sectioning. What is needed in such studies are known cell premixes (in short MIX) to determine the error. In this study, it became evident that LDH and Ca-AM/EthD-1 methods overestimated the number of living cells with respect to MIX in all cell viability mixtures (except for 0%), whereas TB method always slightly underestimated this number; however, it was clearly the one closest to the MIX values. **Figure 5** illustrates how premixes of known living and dead cells can be produced and then how three particular different live/dead staining (methodologies) were compared. In this comparison, it became evident that optical methods such as cLSM have a relatively high deviation from the MIX as it was overestimating the total number of cells per volume by a factor of about four. The reason for this was attributed to the refraction index of the IVD tissue and/or the hydrogels that might cause an experimental bias.

5.1 Flow cytometry

Flow cytometry analysis of the different CV mixtures was correlated with certain variations of the aimed living/dead cell ratio. The results were different

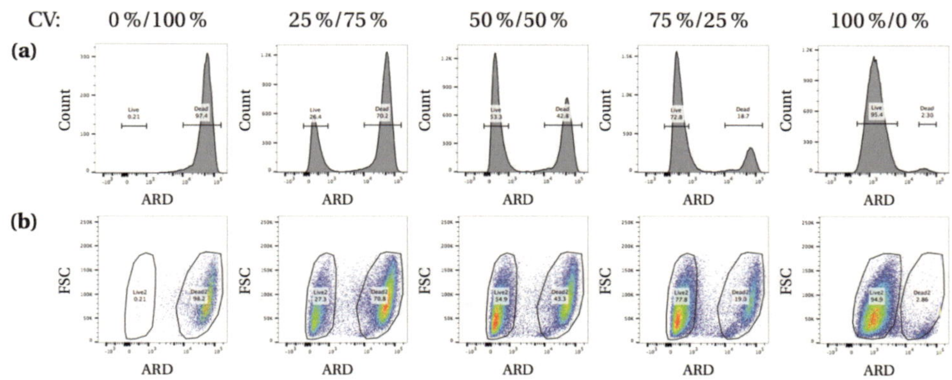

Figure 6.
Flow cytometry results for isolated primary human cells from the anterior cruciate ligament (ACL) cells, which were stained at different PREMIXes of living and dead cells, that is, 100%/0%, 75%/25%, 50%/50%, 25%/75% and 0%/100% and then stained with the ARD. CV can then be inferred from either (a) histogram of the ARD and cell counts with the different peaks depending on the cell populations (FACS-2D plot) or from (b) 2D (with the dimensions plots of the ARD and forward scatter (FSC) with the two distinct populations according to the CV mixture (FACS – histogram plot)). However, these are only minor deviations between the two methods, which arose from the variation in the gating.

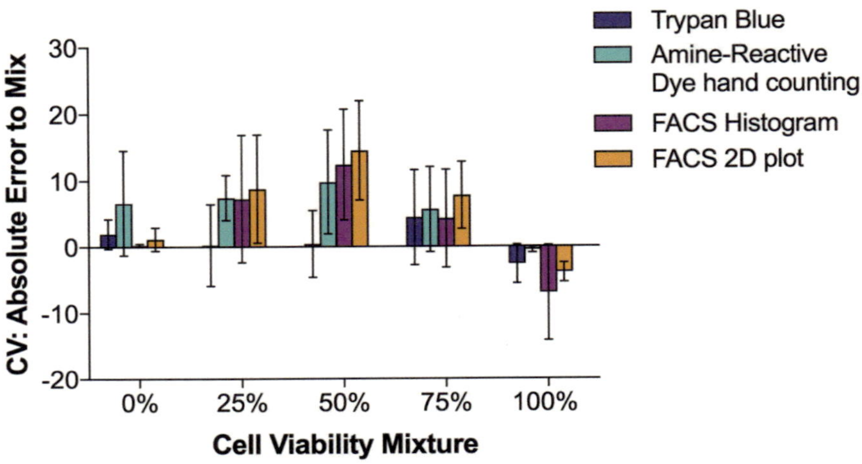

Figure 7.
Results of absolute error relative to the prepared mixtures for CV measurement determined with the four different methods: TB assay, hand counting with ViaQuant™ staining, FACS-histogram and FACS-2D plot illustrating the determination of the gates for live and dead cells. Values are presented as mean ± SD, N = 4.

depending on the representation of the data with either the histogram (**Figure 6a**) or 2D plot (**Figure 6b**), but they stayed in a close range. Both representations revealed an overestimation for the 0%, 25%, 50% and 75% targeted CV, but for 100% CV mixture, the result was underestimated.

5.2 Accuracy of fixable viability assay

The accuracy of the fixable viability methods was tested by comparing the computed viability of the different methods to the theoretical viability, based on the TB assay. All CV mixtures matched the targeted CV with variations among some ratios and methods, as depicted in **Figure 7**.

6. Cell tracking methods: colocalisation of fluorescence

There are study designs that require cell tracing methods. For instance, isolated cells from species in which they are labelled with non-specific membrane dyes or organelle dyes. In this case, cells from species (or an individuum) A are transplanted into species (or an individuum) B and are then monitored over time with respect to cell viability [48, 49]. This could be of particular importance in the case of autologous or allogeneic stem cell research, where for instance, mesenchymal stromal cells (MSCs) are transplanted into live organisms or in vitro using organ culture systems. Here, long-lasting fluorescent dyes are the method of choice. It has been characterised in a number of model systems and has been found to be useful for in vitro cell labelling, in vitro proliferation studies and long-term, in vivo cell tracking [50]. Here, dyes such as PKH26 or PKH67 (PKH26GL, Sigma-Aldrich, Buchs, Switzerland), and CellVue® Claret are commonly used [51–55] (**Table 1**). The half-life for the elution of red-fluorescent lipophilic membrane dye PKH26 from labelled rabbit red blood cells is greater than 100 days. PKH26 was used also to stain extracellular vesicles of mammalian cells [56]. It has been recently found that this particular dye seems to leak into non-stained cells in vitro and in vivo [54]. Furthermore, the staining involves a step, where cells are exposed for about 5 min in a diluent, which is highly hydrophobic and is known to cause cell stress and therefore potentially influences the outcome. Thus, exposure in this staining process needs to be optimised for each cell type (User bulletin of Sigma-Aldrich [57]). Less cytotoxic seems to be, at least from our experience, the carbocyanine dyes like the product from Vybrant™ Cell-Labeling Solutions (Molecular Probes, Thermo Fisher Scientific, inc., a multicolor kit available under cat#V-22889) [58]. There is a notable selection of these dyes, including DiI (λ_{Ex} = 549 nm; λ_{Em} = 565 nm; in blue range emission), CM-DiI, DiO (λ_{Ex} = 481 nm; λ_{Em} = 501 nm; orange-red emission) and DiD (λ_{Ex} = 644 nm; λ_{Em} = 655 nm; dark red range emission). In the case of CM-Dil (λ_{Ex} = 553 nm; λ_{Em} = 570 nm; red range emission), the cells can be traced even after fixing and paraffin embedding [59, 60] (**Table 1**). These dyes do not need to go through a rather stressful staining step in a strong solvent, a feature that is welcome for many cell tracing experiments, where the transplanted tissue cannot be scanned immediately after the experiment on a certain time point. Our previously presented imageJ macro 'Cellcounter3D' (available from: http://imagej.nih.gov/ij/macros/Cellcounter3D.txt) also allows to count multiple single channels. We recommend to use the plug-in 'LSM_Batch_with_colors.txt', which can be also obtained from the ImageJ repository, to split the channels firstly. If multiple dyes and channels are being used, it is recommended to split the channels into single .tiff or .jpg images with increasing indices of the z-stack images. Then, there are also plug-ins for colocalisation, such as the 'colocalisation finder', which was found to be highly useful, as these allow to define the pixels with colocalised stainings. In the case of PKH26 or DID stained cells, thus cells in the red wavelength range for instance in combination of Ca-AM staining, the number of 'yellow' stained cells should be counted. With this open source solution it is possible to distinguish for instance injected viable cells that show Ca-AM staining and cell tracer labelling (green + red stained = yellow), live native cells (green only) and dead injected cells (red only cells), and 4′,6-diamidino-2-phenylindole (DAPI) was used to counterstain for dead cells, thus, blue only cells are to be considered 'native' or autochthonous dead cells. These concepts were applied in studies were stem cells needed to be traced after injection into an IVD organ culture model [48, 61]. Here, it was possible to trace cells inside an organ culture system of an IVD, labelled with the proposed dyes and colours.

7. Real-time microscopy using cell fluorescent dyes

Lately, it has become fashionable to monitor cell proliferation using a selected number of time points but using real-time imaging and using cell tracing/tracking experiments. Several microscopes offer now high-throughput approaches to monitor cell proliferation and cell death over time. Here, two devices are of interest to be mentioned: one is the Nikon Biostation CT (Nikon, Tokyo, Jp) and the Incucyte S3 (Essen BioScience, Ltd., Newark Close, UK) with three fluorescent channels (note: the latest release S5 comes even with five channels). There are also several more affordable devices on the market, which, however, do not allow to track multiple culture vessels and cell conditions at the same time but rather single vessel monitoring over time. Real-time monitoring of cell shapes turned out to be very useful to monitor for instance MSCs undergoing differentiation as it was possible to discriminate differentiating cells according to certain 'features' using segmentation methods and statistical shape modelling [62]. MSCs could also be traced in dependency of their passaging number, that is, in dependence of their senescence [63]. These cell motility pathways were tracked for instance using ImageJ [27] and the open source plugin 'MTrackJ', which allows manual tracking of individual cell trails. More recently, fast microscopes were designed that can handle high-throughput monitoring of multiple wells, and even multiple spots of interest in the well. Here, we can report on the experience with the use of the Incucyte S3 (Essen Bioscience), which in our hands, produced highly satisfactory results for cell tracing in co-culture experiments with direct cell–cell contact. As an example, the cytoplasm green of bone marrow-derived MSCs was stained with IncuCyte® CytoLight Rapid Green Reagent, cat# (cat. No. 4705). Additionally, primary human nucleus pulposus cells (NPCs) of the IVD were stained with IncuCyte® CytoLight Rapid Red reagent (cat. No. 4706). These cells were then seeded in 96-well plates in a 'race' experiment seeding 50:50 with 4000 cells per cell type, and the interactions and the cell proliferation between the two cells was monitored (Video 1). Both cell types were cultured in an osteogenic medium, which was alpha-Modified Eagle Medium (αMEM, Gibco, Thermo Fisher Scientific) supplemented with β-glycerophosphate, dexamethasone and vitamin C to thrive MSCs towards osteogenic differentiation. The presence of red NPCs was added in this experiment to monitor the inhibitory effects of these onto the MSCs undergoing ossification. We have previously shown that IVD cells such as NPCs can inhibit MSCs undergoing ossification by expressing BMP antagonists [64, 65]. As for the fluorescent stainings in this case, it became evident in this experiment that both dyes are washed out in less than 12 h of culture. **Figure 8** illustrates the cytoplasmic stains on MSCs (in green) and NPCs (in red) and how the staining faints after only a few hours by cell divisions in the co-culture. Thus,

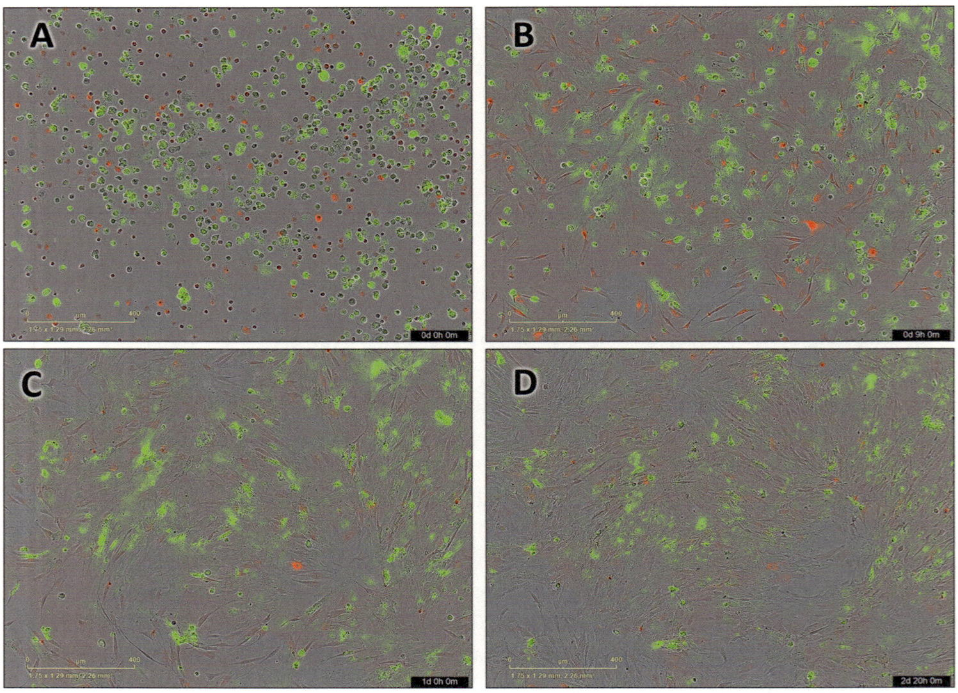

Figure 8.
Phase-contrast microscopic images of time-lapse microscopy of a co-culture of human MSCs (stained in green) with NPCs (stained in red) seeded in a 1:1 ratio. (A) Status immediately when cells were seeded at time point 0; (B) cell populations after 9 hours; C) Cell populations after 24 hours; and (D) cell populations after 2 days and 20 hours in culture. Please, note how the red cytoplasmic staining faints more rapidly over time than the green staining of the MSCs. Fluorescence staining must be optimised for each cell type separately.

these stainings from Essen Bioscience would need to be optimised for long-term experiments for several weeks.

8. Trouble shooting of difficult carriers with strong fluorescent interference

In some cases, unexpected problems were encountered with special biomaterials, which hindered any kind of CV designation as presented above using classical live/dead stain. For instance, staining for cyto-compatibility is often done on 3D hydrogels and other 3D biomaterials that do not allow a transparent view. In the case of *Bombyx* mori silk, for instance, it was found that staining with EthD-1 was not possible due to a very strong autofluorescence of the silk material (**Figure 9A** and **B**). Moreover, integration of genipin, which is a natural cross-linker to increase stiffness of hydrogels, has been proposed [66–69]. The increasing concentration of genipin leads to increasing autofluorescence, noticeable as 'noisy background' on the red fluorescent channel. This can be seen in **Figure 9C**, where MSCs were stained with Ca-AM life staining but the genipin-reinforced fibrin hydrogel was making it impossible to use EthD-1 as the second dye to visualise dead cells. Therefore, DAPI (blue)

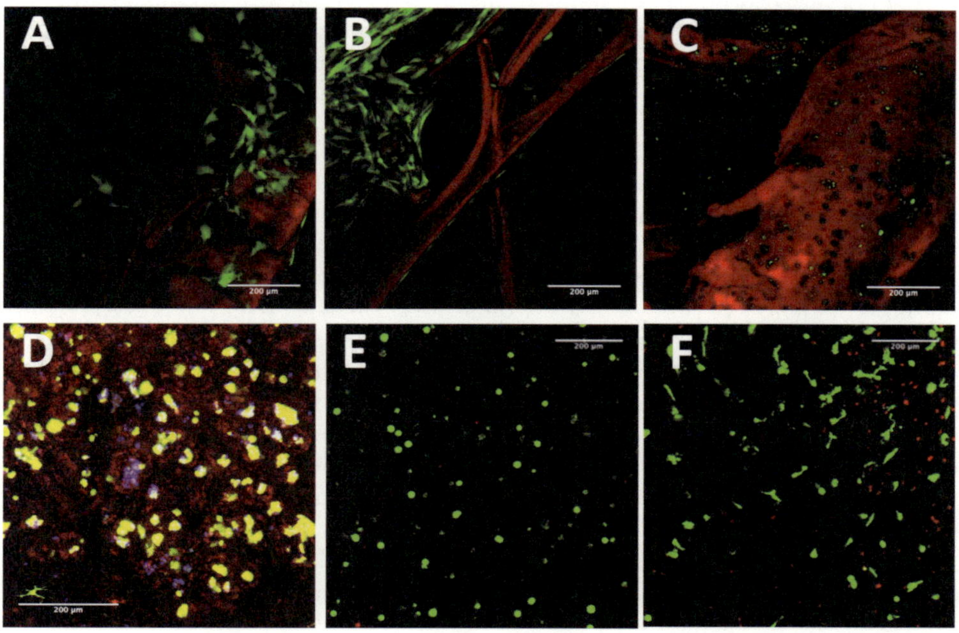

Figure 9.
(A–C). Examples of difficulties to quantify cLSM 3D-scan projections on complex biomaterials. Pictures represent z-stacks of compressed images through a stack of ~200 µm through the respective biomaterials. Primary human-derived bone marrow mesenchymal stromal cells (MSCs) were seeded and cultured for 21 days on (A) silk-fleece from Bombyx mori *(produced from Spintec engineering, GmbH, Aachen, Germany) (in red with strong autofluorescence from the silk material [18]); (B) human MSCs on* B. mori *silk after 21 days of culture containing growth factor GDF6 (naturally expressed by* Baculo-*virus transduced* B. mori *larvae that co-expressed the human growth factor during larval growth); (C) human MSCs cultured in fibrin hydrogel (45 mg fibrinogen/mL, Tisseel™, Baxter) containing 4.5 mg/genipin, a natural cross-linker to stiffen the hydrogel. Notice, the strong red staining resulting from the genipin cross-linker in the hydrogel [67]. (D) Example of an image with co-staining of injected cells into an organ, that is isolated progenitor cells stained with DID (Vybrant™, thermo Fisher scientific) and calcein AM into IVD; (E) human NPCs cultured for 21 days in polethyleneglycol hydrogel (PEG) lacking RGD motive (Q-gel™); (F) human NPCs cultured for 21 days in PEG hydrogel with RGD motive, showing filopodia growth. Cells adhere with filopodia to the matrix and change to a very artificial cell phenotype [17].*

ore an alternate dead staining needs to be taken. The feature of intrinsic fluorescence of genipin was recently proposed as a wanted side effect for *in vivo* studies to track hydrogels [70]. **Figure 9D** is an example of an experiment, where a hydrogel, in this case a methacrylated gellan gum [71, 72], is used as a carrier material and mixed with bovine nucleus pulposus progenitor cells (NPPCs) [73]. This mix was then injected into a bovine IVD organ culture model. The aim was to determine the CV and to identify and distinguish transplanted and native cells. The cells were monitored for live/dead staining after 2 days of organ culture [61]. The cells were labelled with DID and CaAM prior injection and resulted in yellow co-labelled cells. **Figure 9** E and 9F demonstrates the effects of integration of cell adhesion motives by lack (**Figure 9E**) or integration of RGD (**Figure 9F**) into a commercially available hydrogel made of high-molecular weight polyethylene glycol (PEG) (QGel™, QGel inc., Lausanne, Switzerland). Due to the addition of the RGD, NPCs started to develop filopodia, which is a very unnatural cell shape and morphology for these cells. This feature, however, may be very useful if MSCs are cultured in 3D depending on the specific research question under mechanical loading [17].

9. Discussion

CV is a very central parameter in TE methods. However, this parameter is prone to multiple sources of errors. Errors come in not only from the methodology, the experimental design, the experimenter, but also from the technical limitations of the devices. Here, we presented our data for the systematic evaluation of some of the more commonly applied methods.

First of all, there are various methods, which allow inferring cell viability. Here, we discussed direct methods, which use fluorescent dyes such as Ca-AM or similar dyes or combined with membrane dyes. Depending on the research question, a major factor is whether the morphology and the spatial distribution matters (**Figure 1**). Then, fluorescent dyes combined with microscopic techniques are the methods of choice. To get dynamic CV quantification in 3D stacks, it is even recommended to use cLSM technology, which will give a very accurate visualisation of the CV in depth. The disadvantage of this methodology is clearly that the tissue needs to be stained right away after stopping the experiment, and the scanning should take place within 1–2 h after the staining procedure. Samples should be kept if possible at 37°C and the laser exposure inside the cLSM should be minimised.

Alternatively, tissue fixation and subsequent CV quantification has the advantage that any microscopic analyses can be postponed for a future time point.

Either tissue fixation or freezing in O.C.T. compound is an option (Tissue-Tek® O.C.T.™ Compound, AJ Alphen aan den Rijn, The Netherlands). These scaffolds then need to be cut using a cryotom. The samples can then be stained with other live stains using LDH assay (usually a brownish formazan staining) in combination with a dead cell staining, for instance DAPI (blue) or EthD-1 (red).

ARD stains that are applied onto cells did not seem to work for any kind of tissue staining (tested were IVD and ACL tissue). However, if cell morphology does not matter and cells can be successfully isolated by mild digestion (e.g. by pronase/collagenase digestions overnight) [74] without losing the dead cells in the digestion process, this could become a valid alternative to live/dead imaging. However, the protocol for ARD stain works perfectly for flow cytometric applications (see **Figure 6**). We have also evaluated the stability of the ARD for later FACS analysis. Fluorescent ratios and intensities remained unchanged if stored at 4°C after 7 days (our data not shown). Interestingly, CV is overestimated for most methods (**Figures 2** and **7**). As seen for all of our comparisons in terms of accuracy, the TB method was performing best. If the ARD dye was used in hand counting methods, it also overestimated the true CV (**Figure 7**).

It can be complex to determine CV in cell suspensions and in tissue samples, as there are two modes of cell death: necrosis and apoptosis [75, 76]. Necrosis is an uncontrolled cellular death where cells lose their membrane integrity, whereas apoptotic cells die in a controlled manner involving cell shrinkage, nuclear fragmentation and the formation of apoptotic bodies [75]. Most of CV assays such as dye exclusion or intracellular enzyme release rely only upon necrotic loss of membrane integrity, thus omitting the detection of apoptotic cells. To detect apoptotic cells, other fluorescent-based kits are recommended such as the detection of caspase 3 activity [22].

Thus, future methods will consider fluorescent methods in the tissue and directly distinguish apoptosis and necrosis. It could involve the application of assays detecting only apoptotic cells such as it is the case in the TUNEL assay. In this case the detection of cell undergoing apoptosis is achieved through the use of modified nucleotides and enzymes to label DNA fragments. DNA fragmentation is detected by labelling the 3′-hydroxyl termini in the double-stranded DNA breaks that are generated during apoptosis. In the case of flow cytometry study, annexin V would also be a possibility as it is an early marker of apoptosis. It binds to phosphatidylserine, which is a marker of apoptosis when it is present on the outer part of the plasma membrane. It should be noticed that recently a third form of cell death has been defined, which differs from apoptosis and necrosis, which is called 'autophagy'. Autophagy differs from the previous forms in that cells can be killed by starvation from particular growth factors and cellular stresses [75].

Live/dead staining can only discriminate in a limited way between necrosis and apoptotic cell death. Autophagy cannot be distinguished with the fluorescent dyes described in this chapter. It could be feasible to address apoptotic cells using cLSM if certain cell nuclei demonstrate fragmentation of the nuclei. However, most dead cells by image segmentation methods that will be identified will have died through necrosis and show a clearly structured nucleus. As it was mentioned in the introductory chapter, an improvement of the reliability can be obtained by combining various CV methods [22, 77]. Thus, the use of the fixable dead cell stain would be best when used in combination with another CV assay or live and dead staining.

10. Conclusions

- CV can be determined with a wide range of methods allowing to stain mammalian cells. However, which techniques should be used is determined from the question at hand.

- TB method, a relatively cheap and affordable method, was the most accurate method to evaluate PREMIXes in all of the previous investigations.

- Live/dead imaging and cLSM imaging using 3D stacks is a very suitable method to quantify CV in 3D scaffolds. The samples with cells embedded in 3D hydrogel carriers or also native in the tissue can be scanned by cLSM about 300 μm deep into the tissue.

- cLSM scanning requires good optics. The total cell number per volume was overestimated by a factor of four for yet unconfirmed reasons, possibly deviations from the refraction indices of hydrogel-like materials.

- Cells can be traced using membrane dyes, such as PKH26, DID, DIO, DIL or similar dyes. Staining protocols for each specific mammalian cell type might have to be optimised.

- Fixable dead cell staining kit, as for instance the ViaQuant™, is optimised for flow cytometric analyses, and can only be partially used for TE applications. Its application for tissue samples is not possible as the dye cannot specifically stain dead cells.

- Rather, this fixable kit is more suitable for toxicology screening where cells are kept in suspension. It still has some advantages over other viability assays like flexibility of the experimental design in terms of excitation wavelength and stability of the dye after fixation, and also its simple handling.

Acknowledgements

This work was supported by the Swiss National Science Foundation grant number 310030E_192674/1 and by a start-up grant from the Center for Applied Biotechnology and Molecular Medicine (CABMM) to B.G.

The authors thank Daniela D. Frauchiger, Silvan Heeb, and Rahel D. May for the contribution of the images/movies involving silk or genipin-reinforced fibrin hydrogel. They are grateful to László Kupcsik and Marije de Jong (Van der Werf) for the assistance in the development of the ImageJ 'Cellcounter' macro and the MatLab routine 'Cell Viability Estimator for 3D Scaffolds' for CV determination for cLSM analysis. The cytometer equipment was from FACSlab core facility of University of Bern (www.facslab.unibe.ch). The microscopes were provided by the microscope core facility of the University of Bern (www.mic.unibe.ch).

Conflict of interest

The authors declare no conflict of interest.

Abbreviations

ACL	anterior cruciate ligament
αMEM	alpha-modified eagle medium
ARD	amine-reactive dye
Ca-AM	calcein acetoxymethyl ester
Ca-AM/EthD-1	calcein acetoxymethyl/ethidium homodimer-1 = LIVE/DEAD staining
cLSM	Confocal laser scanning microscopy
CV	cell viability
DAPI	4′,6-diamidino-2-phenyl-indole
EthD-1	ethidium homodimer-1
FACS	fluorescence-activated cell sorting
FCS	fetal calf serum
IVD	intervertebral disc
LC	ligamentocyte
LDH	lactate dehydrogenase
LG-DMEM	low-glucose Dulbecco's modified eagle's medium

MSC	mesenchymal stromal cell
PREMIX	the pre-made mixtures of live and dead cells to determine the accuracy of methods
NP	nucleus pulposus
PBS	phosphate-buffered saline
RT	room temperature
SD	standard deviation
TB	Trypan Blue

Author details

Benjamin Gantenbein[1,2]*, Andreas S. Croft[1] and Marie Larraillet[1]

1 Department for BioMedical Research (DBMR), University of Bern, Bern, Switzerland,

2 Department of Orthopaedic Surgery and Traumatology, Inselspital, University of Bern, Bern, Switzerland

*Address all correspondence to: benjamin.gantenbein@dbmr.unibe.ch

References

[1] Hohlrieder M, Teuschl AH, Cicha K, van Griensven M, Redl H, Stampfl J. Bioreactor and scaffold design for the mechanical stimulation of anterior cruciate ligament grafts. Bio-medical Materials and Engineering. 2013;**23**(3):225-237. DOI: 10.3233/BME-130746

[2] Nöth U, Schupp K, Heymer A, Kall S, Jakob F, Schütze N, et al. Anterior cruciate ligament constructs fabricated from human mesenchymal stem cells in a collagen type I hydrogel. Cytotherapy. 2005;**7**(5):447-455. DOI: 10.1080/14653240500319093

[3] Petrigliano FA, McAllister DR, Wu BM. Tissue engineering for anterior cruciate ligament reconstruction: A review of current strategies. Arthroscopy. 2006;**22**(4):441-451. DOI: 10.1016/j.arthro.2006.01.017

[4] Lightfoot A, Martin J, Amendola A. Fluorescent viability stains overestimate chondrocyte viability in osteoarticular allografts. The American Journal of Sports Medicine. 2007;**35**(11):1817-1823. DOI: 10.1177/0363546507305010

[5] Mohanraj B, Hou C, Meloni GR, Cosgrove BD, Dodge GR, Mauck RL. A high throughput mechanical screening device for cartilage tissue engineering. Journal of Biomechanics. 2014;**47**(9):2130-2136. DOI: 10.1016/j.jbiomech.2013.10.043

[6] Stoddart MJ, Grad S, Eglin D, Alini M. Cells and biomaterials in cartilage tissue engineering. Regenerative Medicine. 2009;**4**(1):81-98. DOI: 10.2217/17460751.4.1.81

[7] Behrendt P, Ladner Y, Stoddart MJ, Lippross S, Alini M, Eglin D, et al. Articular joint-simulating mechanical load activates endogenous TGF-β in a highly cellularized bioadhesive hydrogel for cartilage repair. The American Journal of Sports Medicine. 2020;**48**(1):210-221. DOI: 10.1177/0363546519887909

[8] Haglund L, Moir J, Beckman L, Mulligan KR, Jim B, Ouellet JA, et al. Development of a bioreactor for axially loaded intervertebral disc organ culture. Tissue Engineering. Part C, Methods. 2011;**17**(10):1011-1019. DOI: 10.1089/ten.TEC.2011.0025

[9] Rosenzweig DH, Gawri R, Moir J, Beckman L, Eglin D, Steffen T, et al. Dynamic loading, matrix maintenance and cell injection therapy of human intervertebral discs cultured in a bioreactor. European Cells & Materials. 2016;**30**:26-39

[10] Pfannkuche JJ, Guo W, Cui S, Ma J, Lang G, Peroglio M, et al. Intervertebral disc organ culture for the investigation of disc pathology and regeneration - benefits, limitations, and future directions of bioreactors. Connective Tissue Research. 2019;**61**(3-4):1-18. DOI: 10.1080/03008207.2019.1665652

[11] Endres S, Kratz M, Wunsch S, Jones DB. Zetos: A culture loading system for trabecular bone. Investigation of different loading signal intensities on bovine bone cylinders. Journal of Musculoskeletal & Neuronal Interactions. 2009;**9**(3):173-183

[12] Jones DB, Boudriot U, Kratz M, Martens F, Koller K, Smith EL. A trabecular bone and marrow bioreactor. European Cells & Materials. 2001;**1**(suppl 2):53

[13] Luo L, Thorpe SD, Buckley CT, Kelly DJ. The effects of dynamic compression on the development of cartilage grafts engineered using bone marrow and infrapatellar fat pad derived stem cells. Biomedical Materials. 2015;**10**(5):055011. DOI: 10.1088/1748-6041/10/5/055011

[14] Raveling AR, Theodossiou SK, Schiele NR. A 3D printed mechanical bioreactor for investigating mechanobiology and soft tissue mechanics. Methods. 2018;**5**:924-932. DOI: 10.1016/j.mex.2018.08.001

[15] Bourgine PE, Klein T, Paczulla AM, Shimizu T, Kunz L, Kokkaliaris KD, et al. In vitro biomimetic engineering of a human hematopoietic niche with functional properties. Proceedings of the National Academy of Sciences of the United States of America. 2018;**115**(25):E5688–E5695. DOI: 10.1073/pnas.1805440115

[16] Gantenbein-Ritter B, Potier E, Zeiter S, van der Werf M, Sprecher CM, Ito K. Accuracy of three techniques to determine cell viability in 3D tissues or scaffolds. Tissue Engineering. Part C, Methods. 2008;**14**(4):353-358. DOI: 10.1089/ten.tec.2008.0313

[17] Guggisberg S, Benneker LM, Keel MJ, Gantenbein B. Mechanical loading promoted Discogenic differentiation of human mesenchymal stem cells incorporated in 3D-PEG scaffolds with RhGDF5 and RGD. International Journal of Stem cell Research & Therapy. 2015;**2**(006):006. DOI: 10.23937/2469-570X/1410006

[18] Frauchiger DA, Heeb SR, May RD, Wöltje M, Benneker LM, Gantenbein B. Differentiation of MSC and annulus fibrosus cells on genetically engineered silk fleece-membrane-composites enriched for GDF-6 or TGF-β3. Journal of Orthopaedic Research. 2018;**36**(5):1324-1333. DOI: 10.1002/jor.23778

[19] Gantenbein B, Frauchiger DA, May RD, Bakirci E, et al. Developing bioreactors to host joint-derived tissues that require mechanical stimulation. In: Reis RL, Gomes ME, editors. Encyclopedia of Tissue Engineering and Regenerative Medicine. 1st ed. NY: Elsevier; 2019. pp. 261-280

[20] Martin I, Wendt D, Heberer M. The role of bioreactors in tissue engineering. Trends in Biotechnology. 2004;**22**(2):80-86. DOI: 10.1016/j.tibtech.2003.12.001

[21] Wendt D, Jakob M, Martin I. Bioreactor-based engineering of osteochondral grafts: From model systems to tissue manufacturing. Journal of Bioscience and Bioengineering. 2005;**100**(5):489-494. DOI: 10.1263/jbb.100.489

[22] Riss TL, Moravec RA, Niles AL, Duellman S, et al. Cell viability assays. In: Sittampalam GS, Grossman A, Brimacombe K, Arkin M, et al., editors. Assay Guidance Manual. Bethesda (MD): Eli Lilly & Company and the National Center for Advancing Translational Sciences; 2004. DOI: NBK144065 [bookaccession]

[23] Bonnier F, Keating ME, Wróbel TP, Majzner K, Baranska M, Garcia-Munoz A, et al. Cell viability assessment using the alamar blue assay: A comparison of 2D and 3D cell culture models. Toxicology In Vitro. 2015;**29**(1):124-131. DOI: 10.1016/j.tiv.2014.09.014

[24] Xiao J, Zhang Y, Wang J, Yu W, Wang W, Ma X. Monitoring of cell viability and proliferation in hydrogel-encapsulated system by resazurin assay. Applied Biochemistry and Biotechnology. 2010;**162**(7):1996-2007. DOI: 10.1007/s12010-010-8975-3

[25] Plesca D, Mazumder S, Almasan A. Chapter 6 DNA damage response and apoptosis. Methods in Enzymology. 2008;**446**:107-122. DOI: 10.1016/S0076-6879(08)01606-6

[26] Duchi S, Piccinini F, Pierini M, Bevilacqua A, Torre ML, Lucarelli E, et al. A new holistic 3D non-invasive analysis of cellular distribution and motility on fibroin-alginate microcarriers using light sheet

fluorescent microscopy. PLoS One. 2017;**12**(8):e0183336. DOI: 10.1371/journal.pone.0183336

[27] Rasband WS. ImageJ, USA, National Institutes of Health, Bethesda, Maryland, USA. 2020. Available from: http://rsb.info.nih.gov/ij/ [Accessed: 22 May 2020]

[28] Bordbar S, Lotfi Bakhshaiesh N, Khanmohammadi M, Sayahpour FA, Alini M, Baghaban Eslaminejad M. Production and evaluation of decellularized extracellular matrix hydrogel for cartilage regeneration derived from knee cartilage. Journal of Biomedical Materials Research. Part A. 2020;**108**(4):938-946. DOI: 10.1002/jbm.a.36871

[29] Masson-Meyers DS, Bumah VV, Enwemeka CS. A comparison of four methods for determining viability in human dermal fibroblasts irradiated with blue light. Journal of Pharmacological and Toxicological Methods. 2016;**79**:15-22. DOI: 10.1016/j.vascn.2016.01.001

[30] Sanfilippo S, Canis M, Ouchchane L, Botchorishvili R, Artonne C, Janny L, et al. Viability assessment of fresh and frozen/thawed isolated human follicles: Reliability of two methods (trypan blue and calcein AM/ethidium homodimer-1). Journal of Assisted Reproduction and Genetics. 2011;**28**(12):1151-1156. DOI: 10.1007/s10815-011-9649-y

[31] Elson KM, Fox N, Tipper JL, Kirkham J, Hall RM, Fisher J, et al. Non-destructive monitoring of viability in an ex vivo organ culture model of osteochondral tissue. European Cells & Materials. 2015;**29**(0):356-369. DOI: 10.22203/eCM.v029a27

[32] Stoddart MJ, Furlong PI, Simpson A, Davies CM, Richards RG. A comparison of non-radioactive methods for assessing viability in ex vivo cultured cancellous bone: Technical note. European Cells & Materials. 2006;**12**: 16-25. Discussion 16-25. DOI: 10.22203/ecm.v012a02

[33] Rauch B, Edwards RB, Lu Y, Hao Z, Muir P, Markel MD. Comparison of techniques for determination of chondrocyte viability after thermal injury. American Journal of Veterinary Research. 2006;**67**(8):1280-1285. DOI: 10.2460/ajvr.67.8.1280

[34] Chang SW, Chou SF, Wang YH. Ethanol treatment induces significant cell death in porcine corneal fibroblasts. Cornea. 2006;**25**(9):1072-1079. DOI: 10.1097/01.ico.0000254200.69742.b5

[35] Catelas I, Sese N, Wu BM, Dunn JC, Helgerson S, Tawil B. Human mesenchymal stem cell proliferation and osteogenic differentiation in fibrin gels in vitro. Tissue Engineering. 2006;**12**(8):1-12. DOI: 10.1089/ten.2006.12.ft-97

[36] Haschtmann D, Stoyanov JV, Ettinger L, Nolte LP, Ferguson SJ. Establishment of a novel intervertebral disc/endplate culture model: Analysis of an ex vivo in vitro whole-organ rabbit culture system. Spine. 2006;**31**(25):2918-2925. DOI: 10.1097/01.brs.0000247954.69438.ae

[37] Melamed MR, Kamentsky LA, Boyse EA. Cytotoxic test automation: A live-dead cell differential counter. Science. 1969;**163**(864):285-286. DOI: 10.1126/science.163.3864.285

[38] Supino R. MTT assays. Methods in Molecular Biology. 1995;**43**:137-149. DOI: 10.1385/0-89603-282-5:137

[39] Mosmann T. Rapid colorimetric assay for cellular growth and survival: Application to proliferation and cytotoxicity assays. Journal of Immunological

Methods. 1983;**65**(1-2):55-63. DOI: 10.1016/0022-1759(83)90303-4

[40] Oral HB, George AJ, Haskard DO. A sensitive fluorometric assay for determining hydrogen peroxide-mediated sublethal and lethal endothelial cell injury. Endothelium. 1998;**6**(2):143-151. DOI: 10.3109/10623329809072201

[41] Lichtman JW. Confocal microscopy. Scientific American. 1994;**271**(2):40-45

[42] Boyde A. Bibliography on confocal microscopy and its applications. Scanning. 1994;**16**:33-56

[43] Gantenbein-Ritter B, Sprecher CM, Chan S, Illien-Jünger S, Grad S. Confocal imaging protocols for live/dead staining in three-dimensional carriers. Methods in Molecular Biology. 2011;**740**:127-140. DOI: 10.1007/978-1-61779-108-6_14

[44] Perfetto SP, Chattopadhyay PK, Lamoreaux L, Nguyen R, Ambrozak D, Koup RA, et al. Amine reactive dyes: An effective tool to discriminate live and dead cells in polychromatic flow cytometry. Journal of Immunological Methods. 2006;**313**(1-2):199-208. DOI: 10.1016/j.jim.2006.04.007

[45] Perfetto SP, Chattopadhyay PK, Lamoreaux L, Nguyen R, Ambrozak D, Koup RA, et al. Amine-reactive dyes for dead cell discrimination in fixed samples. Current Protocols in Cytometry. 2010;**Chapter 9**(July):1-14. DOI: 10.1002/0471142956.cy0934s53

[46] Otsu N. A threshold selection method from gray-level histograms. IEEE Transactions on Systems, Man, and Cybernetics. 1979;**9**(1):62-66

[47] Gantenbein B. Cell Viability Estimator for 3D Scaffolds for MATLAB. 2020. Available from: https://www.mathworks.com/matlabcentral/fileexchange/75548-cell-viability-estimator-for-3d-scaffolds [Accessed: 22 May 2020]

[48] Chan SCW, Gantenbein-Ritter B, Leung VY, Chan D, Cheung KM, Ito K. Cryopreserved intervertebral disc with injected bone marrow-derived stromal cells: A feasibility study using organ culture. The Spine Journal. 2010;**10**(6):486-496. DOI: 10.1016/j.spinee.2009.12.019

[49] Perez-Cruet M, Beeravolu N, McKee C, Brougham J, Khan I, Bakshi S, et al. Potential of human nucleus pulposus-like cells derived from umbilical cord to treat degenerative disc disease. Neurosurgery. 2019;**84**(1):272-283. DOI: 10.1093/neuros/nyy012

[50] Wallace PK, Tario JD, Fisher JL, Wallace SS, Ernstoff MS, Muirhead KA. Tracking antigen-driven responses by flow cytometry: Monitoring proliferation by dye dilution. Cytometry. Part A. 2008;**73**(11):1019-1034. DOI: 10.1002/cyto.a.20619

[51] Gertner-Dardenne J, Poupot M, Gray B, Fournié JJ. Lipophilic fluorochrome trackers of membrane transfers between immune cells. Immunological Investigations. 2007;**36**:5-6. DOI: 10.1080/08820130701674646

[52] Yin X, Li P, Li Y, Cai Y, Wen J, Luan Q. Growth/differentiation factor-5 promotes in vitro/vivo periodontal specific differentiation of induced pluripotent stem cell-derived mesenchymal stem cells. Experimental and Therapeutic Medicine. 2017;**14**(5):4111-4117. DOI: 10.3892/etm.2017.5030

[53] Yang Y, Mao Y, Wang J, Sun C, Zhang Y, Chen X. In vivo tracing of human amniotic mesenchymal stem cells labeled with PKH26 in rat intrauterine adhesions model. Shengwu Gongcheng Xuebao/Chinese Journal of

Biotechnology. 2018;**34**(10):1660-1667. DOI: 10.13345/j.cjb.180018

[54] Li P, Zhang R, Sun H, Chen L, Liu F, Yao C, et al. PKH26 can transfer to host cells in vitro and vivo. Stem Cells and Development. 2013;**22**(2):340-344. DOI: 10.1089/scd.2012.0357

[55] Ford JW, Welling TH, Stanley JC, Messina LM. PKH26 and 125I-PKH95: Characterization and efficacy as labels for in vitro and in vivo endothelial cell localization and tracking. The Journal of Surgical Research. 1996;**62**(1):23-28. DOI: 10.1006/jsre.1996.0167

[56] Dominkuš PD, Stenovec M, Sitar S, Lasič E, Zorec R, Plemenitaš A, et al. PKH26 labeling of extracellular vesicles: Characterization and cellular internalization of contaminating PKH26 nanoparticles. Biochimica et Biophysica Acta - Biomembranes. 2018;**1860**(6):1350-1361. DOI: 10.1016/j.bbamem.2018.03.013

[57] Sigma-Aldrich: PKH26 Red Fluorescent Cell Linker Kits for General Cell Membrane Labeling. Available from: https://www.sigmaaldrich.com/content/dam/sigma-aldrich/docs/Sigma/Bulletin/pkh26glbul.pdf [Accessed: 22 May 2020]

[58] Molecular Probes: Fisher Thermo Scientific, Inc. Vybrant Cell-Labeling Solutions. Available from: http://www.icms.qmul.ac.uk/flowcytometry/uses/fret/diagrams/mp22885.pdf [Accessed: 22 May 2020]

[59] Lehmann TP, Juzwa W, Filipiak K, Sujka-Kordowska P, Zabel M, Głowacki J, et al. Quantification of the asymmetric migration of the lipophilic dyes, DiO and DiD, in homotypic co-cultures of chondrosarcoma SW-1353 cells. Molecular Medicine Reports. 2016;**14**(5):4529-4536. DOI: 10.3892/mmr.2016.5793

[60] Nagyova M, Slovinska L, Blasko J, Grulova I, Kuricova M, Cigankova V, et al. A comparative study of PKH67, DiI, and BrdU labeling techniques for tracing rat mesenchymal stem cells. In Vitro Cellular & Developmental Biology. Animal. 2014;**50**(7):656-663. DOI: 10.1007/s11626-014-9750-5

[61] Frauchiger D, Tekari A, Benneker LM, Sakai D, Gantenbein B. Proceedings of the ORS. In: The Fate of Allogeneic Injected Angiopoietin-1 Receptor Tie2+ Cells in Intervertebral Disc Organ Culture. San Diego; 19-22 March, 2017

[62] Seiler C, Gazdhar A, Reyes M, Benneker LM, Geiser T, Siebenrock KA, et al. Time-lapse microscopy and classification of 2D human mesenchymal stem cells based on cell shape picks up myogenic from osteogenic and adipogenic differentiation. Journal of Tissue Engineering and Regenerative Medicine. 2014;**8**(9):737-746. DOI: 10.1002/term.1575

[63] Bertolo A, Gemperli A, Gruber M, Gantenbein B, Baur M, Pötzel T, et al. In vitro cell motility as a potential mesenchymal stem cell marker for multipotency. Stem Cells Translational Medicine. 2014;**4**(1):84-90. DOI: 10.5966/sctm.2014-0156

[64] Chan SC, Tekari A, Benneker LM, Heini PF, Gantenbein B. Osteogenic differentiation of bone marrow stromal cells is hindered by the presence of intervertebral disc cells. Arthritis Research and Therapy. 2015;**18**(1):29. DOI: 10.1186/s13075-015-0900-2

[65] Tekari A, May RD, Frauchiger DA, Chan SCW, Benneker LM, Gantenbein B. The BMP2 variant L51P restores the osteogenic differentiation of human mesenchymal stromal cells in the presence of intervertebral disc cells. European Cells & Materials. 2017;**33**:197-210. DOI: 10.22203/eCM.v033a15

[66] Panebianco CJ, DiStefano TJ, Mui B, Hom WW, Iatridis JC. Crosslinker concentration controls TGFβ-3 release and annulus fibrosus cell apoptosis in genipin-crosslinked fibrin hydrogels. European Cells & Materials. 2020;**39**: 211-226. DOI: 10.22203/eCM.v039a14

[67] Frauchiger DA, May RD, Bakirci E, Tekari A, Chan SCW, Wöltje M, et al. Genipin-enhanced fibrin hydrogel and novel silk for intervertebral disc repair in a loaded bovine organ culture model. Journal of Functional Biomaterials. 2018;**9**(3):40. DOI: 10.3390/jfb9030040

[68] Scheibler AG, Götschi T, Widmer J, Holenstein C, Steffen T, Camenzind RS, et al. Feasibility of the annulus fibrosus repair with in situ gelating hydrogels - A biomechanical study. PLoS One. 2018;**13**(12):e0208460. DOI: 10.1371/journal.pone.0208460

[69] Liang HC, Chang WH, Lin KJ, Sung HW. Genipin-crosslinked gelatin microspheres as a drug carrier for intramuscular administration: in vitro and in vivo studies. Journal of Biomedical Materials Research. Part A. 2003;**65**(2):271-282. DOI: 10.1002/jbm.a.10476

[70] Santos-Vizcaino E, Haley H, Gonzalez-Pujana A, Orive G, Hernandez RM, Luker GD, et al. Monitoring implantable immunoisolation devices with intrinsic fluorescence of genipin. Journal of Biophotonics. 2019;**12**(4):e201800170. DOI: 10.1002/jbio.201800170

[71] Silva-Correia J, Gloria A, Oliveira MB, Mano JF, Oliveira JM, Ambrosio L, et al. Rheological and mechanical properties of acellular and cell-laden methacrylated gellan gum hydrogels. Journal of Biomedical Materials Research. Part A. 2013;**101**(12):3438-3446. DOI: 10.1002/jbm.a.34650

[72] Silva-Correia J, Oliveira JM, Caridade SG, Oliveira JT, Sousa RA, Mano JF, et al. Gellan gum-based hydrogels for intervertebral disc tissue-engineering applications. Journal of Tissue Engineering and Regenerative Medicine. 2011;**5**(6):e97–e107. DOI: 10.1002/term.363

[73] Frauchiger DA, Tekari A, May RD, Džafo E, Chan SC, Stoyanov J, et al. FACS is more potent to fish IVD progenitor cells than magnetic and bead-based methods. Tissue Engineering. Part C, Methods. 2019;**25**(10):571-580. DOI: 10.1089/ten.TEC.2018.0375

[74] Lee JT, Cheung KM, Leung VY. Systematic study of cell isolation from bovine nucleus pulposus: Improving cell yield and experiment reliability. Journal of Orthopaedic Research. 2015;**33**(12):1743-1755. DOI: 10.1002/jor.22942

[75] D'Arcy MS. Cell death: A review of the major forms of apoptosis, necrosis and autophagy. Cell Biology International. 2019;**43**(6):582-592. DOI: 10.1002/cbin.11137

[76] Kanduc D, Mittelman A, Serpico R, Sinigaglia E, Sinha AA, Natale C, et al. Cell death: Apoptosis versus necrosis (review). International Journal of Oncology. 2002;**21**(1):165-170

[77] Sellers JR, Cook S, Goldmacher VS. A cytotoxicity assay utilizing a fluorescent dye that determines accurate surviving fractions of cells. Journal of Immunological Methods. 1994;**172**(2):255-256. DOI: 10.1016/0022-1759(94)90112-0

Chapter 10

Detection of Apoptosis in Cancer Cells Using Heat Shock Protein 70 and p53 Antibody Conjugated Quantum Dot Nanoparticles

Lev B. Matyushkin, Olga A. Aleksandrova, Anna O. Drobintseva, Igor M. Kvetnoy, Yuliya S. Krylova, Yaroslav Y. Marchenko, Dmitriy S. Mazing, Vyacheslav A. Moshnikov, Sergey F. Musikhin, Boris P. Nikolaev, Victoriya O. Polyakova, Maxim A. Shevtsov and Ludmila Y. Yakovleva

Abstract

Clinical experience indicates that enhanced level of heat shock protein 70 (Hsp70) and p53 correlates with poor prognosis due to malignant cell overexpression of these proteins in tumor progression. Cadmium selenide quantum dots (QDs) were synthesized in aqueous solution using mercaptopropionic acid and L-cysteine (L-Cys) as ligands. They were conjugated with a monoclonal antibody (Ab) to p53 and cmHp70.1 to Hsp70 for detection of cancer cell apoptosis that was demonstrated in the experiment by fluorescent confocal microscopy both for breast carcinoma cells and for thyroid tissue. It is shown that in comparison with organic dyes, quantum dots have superior photostability of tracking apoptosis in cancer cells for longer time.

Keywords: quantum dot, p53, heat shock protein, conjugation, cancer imaging

1. Introduction

Imaging is the most powerful diagnostic approach in oncology that provides visualization of early stages of cancer and optimizes its treatment. Fluorescent semiconductor nanoparticles (quantum dots (QDs)) play a great role in cell imaging in oncology [1–3]. QDs have fluorescent characteristics that change conventional approaches in medical diagnostics. The general optical features of QDs compared to organic fluorescent probes are broad excitation band, narrow fluorescence band, high quantum yield (including near-infrared region), and stability against photobleaching [4, 5]. Organic fluorescent dyes are usually used to label cells and tissues for both in vitro and in vivo imaging. QDs are more convenient probes for imaging of tumor cell surface receptors by the receptor-mediated recognition.

IntechOpen

Originally proposed back in the early 1990s, synthesis method for colloidal cadmium chalcogenide nanocrystals (CdS, CdSe, CdTe) in high-boiling solvents (the so-called hot injection method) made it possible to carry out controlled synthesis of semiconductor quantum dots with a narrow nanoparticle size distribution and promising optical characteristics [6]. Additional control and modification of optical properties can be achieved using alloying (e.g., $Zn_{1-x}Cd_xS$ QDs) or doping and creation of core-shell heterostructures of type-I (e.g., CdSe/ZnS) or type-II (e.g., CdTe/CdSe) [7, 8]. The technology of colloidal synthesis was extended to other binary compounds, such as lead and silver chalcogenides (for the near-IR range applications), zinc chalcogenides including doped with transition metal atoms (e.g., ZnS:Mn and ZnSe:Mn QDs), III-V semiconductors (InP QDs are actively considered as a less toxic competitor to CdSe-based nanocrystals for the visible range), and group IV semiconductors (e.g., Si, C) [7, 9–12]. Doping of QDs with transition metal ions is a prospective approach for development of bimodal labels for combined magnetic resonance and fluorescence imaging [13]. Another developing system is nanocrystals of ternary I-III-VI compounds ($CuInS_2$, $CuInSe_2$, $AgInS_2$, $AgInSe_2$), as well as quaternary QDs based on them (e.g., Zn-Ag-In-S QDs) [14, 15]. Being another alternative to cadmium-based nanocrystals in bioimaging applications, they tend to have a common drawback of an inherently broad photoluminescence (PL) band. Some nanocrystalline systems, such as I-III-VI QDs and type-II nanoheterostructures (CdTe/CdSe QDs), with long photoluminescence lifetimes can be used to create fluorescent probes suitable for time-gated imaging with improved signal-to-noise ratio [16]. Despite the fact that the methods of synthesis of colloidal nanocrystals in nonpolar solvents are well studied and allow synthesis of highly fluorescent QDs, methods in aqueous medium attract the attention of researchers due to better cost-effectiveness, relative environmental friendliness, the absence of the need to perform additional operations to hydrophilize the surface of the nanoparticles, and the ability to readily tailor biofunctionality according to specific application. The general restriction to this approach is often higher defect densities and, as a consequence, lower fluorescence quantum yields of nanoparticles. Moreover aqueous synthetic routes may be more complicated due to involvement of H_2O, H_3O^+, and OH^- species [17].

Some proteins are the markers of malignant process and can be used as targeting ligands in imaging. The heat shock protein 70 (Hsp70) and tumor suppressor gene p53 have distinguished position among protein markers. Clinical experience indicates that enhanced level of Hsp70 and p53 correlates with poor prognosis due to malignant cell overexpression of these proteins in tumor progression [18]. Previous studies have shown that these proteins are expressed at high levels in oncological diseases such as endometrial cancers, osteosarcomas, renal cell tumors, hepatoma, and glioblastoma [19, 20]; elevated serum levels of Hsp70 are associated with breast cancer [21]. Hsp70 has prognostic significance for diagnostics of endometrial carcinoma as its expression appears to correlate with sex steroid receptor status in cancer [22].

The wild-type p53 is known to be tumor suppressor apoptotic protein that transcriptionally regulates DNA damage and cell transformation [23]. Interaction of p53 with heat shock proteins changes the apoptotic activity and causes the growth of cancer cells in organism [24, 25]. The Hsp70 is chaperone molecule that affects the p53 function by binding client protein at certain sites. The p53 has two binding sites, the most important one is the C-terminal domain. Therefore, the chaperone function of stress proteins is essential for a sustained state of p53 for malignant cell apoptosis. Targeting of Hsp70 and p53 is an important task for developing the cancer treatment strategy. It is well known that the most sensitive and specific moieties for detection of antigen are its antibodies (Ab). The conjugation of QDs with monoclonal antibodies against Hsp70 and p53 opens the way for constructing a suitable platform for assay of malignant organs. Nanoparticles ensure delivery and sufficient accumulation of labels in diseased and apoptotic tissues. Targeted delivery of conjugated antibody to a cell membrane receptor was recently demonstrated in the case of magnetic iron oxide nanoparticles. The magnetic conjugates with Ab against Hsp70 being intravenously injected were uptaken by C6 glioma cells [26]. The targeted delivery of magnetic conjugates against epidermal growth factor was confirmed for mouse melanoma. Although the attachment of monoclonal Ab to p53 at the surface of nanoparticle for molecular imaging is not described, the results of successful targeted detection of p53 gene indicate the possibility of apoptosis imaging in transformed cells [27]. Engineered conjugated QDs have diagnostic potential for molecular imaging of apoptosis usually performed by measurement of annexin V bound to phosphatidylserine on an outer cell membrane of tumor [28]. Phosphatidylserine is expressed on the inner side of normal plasma membrane but transfers onto outer leaflet in apoptosis. Some false-positive results in necrosis are the main obstacle for wide application of this assay. Differentiation of apoptosis from necrosis remains an actual problem for oncological diagnostics today. Radiolabeling, magnetic, and QD conjugation do not solve this problem at full scale [29, 30]. As an alternative approach, p53 and Hsp70 proteins in the complex can serve as pro-apoptotic targets for registration with the help of QDs. The QD associated with antibody to pro-apoptotic p53 protein may be used for diagnostic purposes, in particular for the assessment of tumor progression. The apoptotic response after chemotherapy is the special subject for this approach.

In the present study, CdSe QDs were conjugated with monoclonal antibodies to p53 (p53Ab) and cmHp70.1 to Hsp70. The cmHp70.1-QD and p53Ab-QD conjugates were synthesized using covalent bonding. The biocompatibility and cytotoxicity of the obtained conjugates were evaluated by in vitro studies in various cancer cell lines C6 glioma, U87 human glioma, K562 human leukemia, and B16 melanoma cells using MTT assay. The imaging potential was evaluated on the breast carcinoma cells by fluorescent confocal microscopy. The use of QD conjugates for expression of cancer marker proteins was also investigated in endometrial and thyroid tissue. The unique photophysical properties of CdSe nanoparticles facilitate the observation of cancer cells by laser confocal microscopy method.

2. Methods

2.1 Reagents

Cadmium chloride ($CdCl_2 \cdot 2.5H_2O$), selenium powder (Se, 99.5%), sodium borohydride ($NaBH_4$), mercaptopropionic acid (MPA, 99%), and L-cysteine (L-Cys) were used as purchased without further purification.

2.2 Synthesis of CdSe QDs

Typical synthesis procedure included the following steps. Sodium hydroselenide NaHSe solution (Se precursor) was prepared by reacting of $NaBH_4$ with Se powder in distilled water. It resulted in vigorous reaction accompanied by intensive hydrogen gassing:

$$4NaBH_4 + 2Se + 7H_2O \rightarrow 2NaHSe + Na_2B_4O_7 + 14H_2 \quad (1)$$

To obtain cadmium precursor solution, $CdCl_2{\cdot}2.5H_2O$ was dissolved in distilled water, and the certain amount of MPA was added so that the mixture became turbid. By the dropwise addition of NH_4OH, acidity of the solution was brought up to approximate value pH = 11 and it cleared. It is suggested that moderately alkaline environment facilitates the formation of Cd^{2+}-MPA complexes which act as an immediate reactant. Component molar ratio was adjusted to $[Cd^{2+}]:[Se^{2-}]:[MPA] = 1:0.5:2.4$. Selenium precursor solution was swiftly injected in a solution containing cadmium ions under vigorous stirring at room temperature, followed by rapid heating to 90°C. The QDs were purified by isopropyl alcohol-assisted precipitation and dissolved in phosphate-buffered saline (PBS) buffer. Analogous technique was used to synthesize CdSe QDs capped with L-cysteine.

Luminescence of such particles essentially is associated with radiative recombination via surface states. It is typical for CdSe QDs prepared in an aqueous medium. As compared to QDs, synthesized in an organic medium, such particles have a broader spectrum of luminescence but do not require the additional step of transfer into the aqueous phase. Absorption and photoluminescence spectra are presented in **Figure 1**.

2.3 Characterization of CdSe QDs

The absorption spectra were obtained with ultraviolet-visible-near-infrared spectrophotometer (Ecohim, PE-5400UV) between 190 and 1000 nm, with a spectral resolution of 1 nm. Fluorescence spectra were recorded with a fluorescence spectrophotometer with a solid-state 405 nm laser. Dynamic light scattering (DLS), also known as photon correlation spectroscopy, was used to determine the hydrodynamic diameter of free and conjugated QDs. Diluted samples were used to avoid multiple scattering. The measurements were conducted with Zetasizer (Nano ZS, Malvern Instruments Ltd., UK).

2.4 Evaluation of stability

To assess the stability of the QDs in buffer solutions, CdSe/L-Cys nanoparticles were added in a concentration of 17.5 μM/ml to the following solutions: phosphate-buffered saline, pH = 7.5 (BIOLOT); wash buffer, pH = 8.0 (Dako); and Tris buffer, pH = 5.0 (Sigma). Aliquots of each solution were selected after 60 min, placed on the slides, evaporated, and observed for fluorescent areas using Olympus FV 1000 confocal laser scanning microscope (excitation laser wavelength—405 nm).

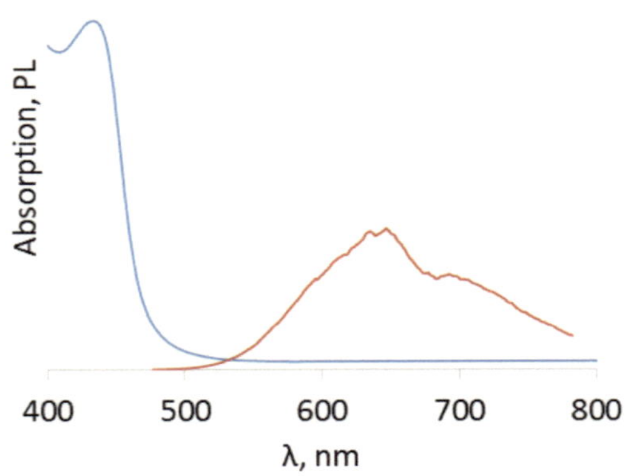

Figure 1.
Characterization of the optical absorption and photoluminescence of CdSe nanoparticles.

2.5 Evaluation of fluorescence intensity

The evaluation of the fluorescence intensity of CdSe/L-Cys QDs samples compared to organic fluorophores FITC (rabbit anti-mouse immunoglobulins, conjugated with FITC, Dako) and Alexa Fluor 568 (donkey anti-mouse immunoglobulins, conjugated with Alexa Fluor 568, Abcam) was done in the following way: 15 μl of the fluorophore was added to the slide and mixed with a fluorescence mounting medium (Dako), which prevents the fluorophores from crystallizing and decolorizing, coverslipped, and subjected to ultraviolet radiation (using a Mercury Vapor Short Arc, 120 W) of 25% intensity for 120-min period. A Niba fluorescent filter (for the UV radiation) was used to observe fluorescence. The samples were scanned with the same laser beam intensity every 15 min. The QD sample was irradiated with 405 nm laser, the FITC sample with 488 nm laser, and the Alexa Fluor sample with 559–568 nm laser. The mean fluorescence intensity was measured in relative units using the FluoView 10 software.

2.6 An assessment of the cytotoxic effect of QDs

The cytotoxicity of the CdSe/L-Cys QDs was assessed in two cell cultures. The first cell culture was a cell line ZR-75-1 (breast carcinoma) and the second—normal human mononuclear lymphocytes (MNL).

The apoptosis induction by a fixed staining procedure with propidium iodide (PI) that stains the nuclei of cells with compromised cell membrane structure was determined to assess the cytotoxic effect of QDs in the ZR-75-1 cell line in 1, 3, 18, and 24 h after administration of QD solution.

Phosphate-buffered saline was added as a control in the same quantities to the cell cultures. The QD samples were added to the growth medium at the concentrations of 175 nM/ml, 1.75 μM/ml, and 17.5 μM/ml. The similar method was used to evaluate MNL culture cytotoxic effect, but the concentration of QDs was 1.15 nM/μl and further evaluated after 48 h of incubation.

2.7 An assessment of the specificity of p53Ab-QD conjugates

The specificity study was conducted with p53Ab-QD conjugate at a 1:50 dilution. The material of the study was normal endometrial cell culture, taken by endometrial pipeline biopsy within the examinations of a patient entering into IVF cycle, as well as fixed and paraffin-embedded thyroid tissue. Endometrial cell culture was grown in Dulbecco's Modified Eagle's Medium (DMEM)/F12 (Gibco) supplemented with 10% fetal bovine serum (Gibco), penicillin, and streptomycin in a humidified atmosphere of 5% CO_2 at 37°C. Apoptosis induction was carried out by adding 200 μM H_2O_2 to the cell culture for 24 h. Normal thyroid tissue possesses a high index of apoptosis. As a control, QD solution of 0.003 M was added in PBS. Immunofluorescence staining was performed using primary antibody (Monoclonal Mouse Anti-Human p53 Protein Clone DO-7, 1:50, Dako) and secondary antibody with Alexa Fluor 647 (Anti-Mouse Alexa Fluor 647, Abcam). The cell nuclei were counterstained with Hoechst stain 33,258 (Sigma); 30 μl volume of each reagent was added per a single slice or well. Fluorescence evaluation was carried out under Olympus FV 1000 confocal laser scanning microscope. The number of stained cells was counted in 10 fields of view at 20× magnification.

2.8 Glioma model for Hsp70 assay

Male nu/nu NMRI mice weighing 28–30 g were purchased in animal nursery "Charles River Laboratories" (Germany). All animal experiments have been approved by the local ethical committee of I.P. Pavlov State Medical University (St. Petersburg, Russia) and were in accordance with institutional guidelines for the welfare of animals.

Animals were anesthetized before mounting in a stereotactic frame (David Kopf Instruments, Tujunda, CA) with 10 mg "Zoletyl-100" (Vibrac sante Animale, France) and 0.2 ml 2% Rometar (Bioveta, Czech Republic) intraperitoneally. U87 cell suspension (1×10^6 cells/ml) in 2 μl was injected into the nucleus caudatus dexter.

For assessment of the accumulation of the nanoparticles in the tumor, animals were randomly divided into four groups (three animals each) on the 14th day after inoculation: (1) i.v. injection of the PBS (control group); (2) i.v. injection of the QD-Isotype IgG1 for 24 h; (3) i.v. injection of QDs-cmHsp70.1 conjugates for 24 h. Following administration of the QDs, the animals were sacrificed, and extracted brains were fixed in 4% formaldehyde. Brain tumor sections were additionally stained with 4,6-diamidino-2-phenylindole solution (DAPI) (Vector Laboratories, Burlingame, CA, USA). Glasses were mounted in DAKO fluorescent mounting medium (Dako North America Inc., USA) and further analyzed on confocal system. Fluorescence images were captured with a Leica TCS SP5 confocal system (Leica Microsystems, Heidelberg, Germany). For evaluation of co-localization, single z-planes were analyzed with Leica confocal software LCSLite (Leica Microsystems, Heidelberg, Germany) and ImageJ 1.37 (Wright Cell Imaging Facility, Toronto, Canada).

3. Experimental results

3.1 Physical characterization of the CdSe QDs

The optical absorption and fluorescence spectra of the colloidal aqueous solution are shown in **Figure 1**. Absorption band edge around 470 nm is located at energies higher than for the bulk CdSe (bandgap 1.74 eV) proving that grown nanostructures are CdSe QDs. The average diameter of the CdSe QDs can be calculated by inputting the wavelength of the lowest energy absorption peak into the following empirical expression [31]:

$$D = 1.6122 \cdot 10^{-9} \lambda^4 - 2.6575 \cdot 10^{-6} \lambda^3 + 1.6242 \cdot 10^{-3} \lambda^2 - 0.4277 \lambda + 41.57, \qquad (2)$$

Here, D (nm) is the mean diameter of the CdSe QDs and λ (nm) is the wavelength of the lowest energy absorption peak, i.e., the first excitonic transition. This expression yielded a diameter of 2 nm, which indicates the formation of ultrasmall QDs. The fluorescence spectrum of the QDs is relatively broad.

3.2 Stability

The carried out observations demonstrated that after the QDs had been incubated in PBS solution and liquid evaporated, the crystalline structure was formed wherein the QDs fluoresce in the orange region of the spectrum. It is found that the QDs remain stable in buffer solutions at room temperature. Herewith the acidity of the solution does not affect the stability of the QDs, since the fluorescence was observed in an hour in all three samples with pH values equal to 8.0, 7.5, and 5.0, respectively. This result is consistent with our previous publication [32].

3.3 Fluorescence intensity of QDs

The comparison analysis of fluorescence intensity of CdSe/L-Cys QDs and organic dyes (**Figure 2**) showed that the maximum brightness of the fluorescence was typical for organic fluorophore FITC in the first 15 min; however, the decrease of brightness (more than threefold) was observed in 15 min. Alexa 568—another organic fluorophore—was more stable than FITC, but its fluorescence intensity also decreased over time. At the same time, the fluorescence intensity of QDs was increasing over time (from 230 to 1000 relative units). Therefore, due to high photostability, QD can be used to monitor fluorescence in objects for long periods of time.

3.4 The assessment of the QD cytotoxicity

The cytotoxicity of the CdSe/L-Cys quantum dots was evaluated using the ZR-75-1 line. The absence of the cytotoxic effect is characteristic for the time intervals of 1, 3, and 18 h. After 24 h, about 4% of cells in the group with injected QDs (concentration of 17.5 mM/ml) accumulated PI, while for the other two groups, these values were 3%. Such a result was not significantly different from the control group, where the number of fluorescent cells scanned was 2 of 100. The number of dead cells in the ZR-75-1 culture did not exceed 3% both in the control and experimental group in all checkpoints. The following results were obtained for a mixed culture of MNL: apoptotic cells were not observed during the first few hours; but long-term (18 h and above) incubation of QDs with MNL proved that the cytotoxic effect increased, and after 48 h, level of apoptosis reached 20%. The increase of cytotoxicity may be associated with the destruction of the organic coating in a cell culture medium. Moreover, MNL were obtained from peripheral blood where lymphocytes do not proliferate without generating special conditions. Therefore, the cytotoxic effect for MNL cell culture may have been more pronounced than for the immortalized ZR-75-1 line. Thus, cytotoxicity was tested both on tumor cells, with greater viability, and on normal human cells, which approximates cytotoxicity assessment for in vivo conditions.

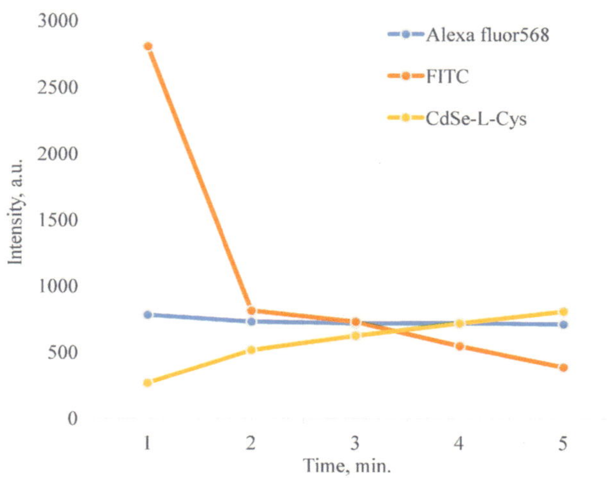

Figure 2.
Dynamic of fluorescence intensity of the CdSe QDs, Alexa Fluor 568, and FITC organic dyes versus time.

3.5 Conjugation of QDs with human antibody against p53 protein (p53Ab)

Conjugation of QDs (CdSe/MPA, 1 mM) and anti-human p53 protein (p53Ab 1:1000 Sigma 0.02 mM) was carried out via water-soluble carbodiimide-mediated linker using CMC (N-Cyclohexyl-N′-(2-morpholinoethyl)-carbodiimide metho-p-toluenesulfonate, Sigma 2 mg/ml (4.7 mM)). 0.2 mM QDs with 0.47 mM CMC was vortexed for 30 min at room temperature. 0.3 μM or 2.2 μM p53Ab and 1× PBS was added to a final volume 210 μl. The conjugation of p53Ab with QDs was performed by incubation in the shaker for 3 h after mixing. Nonconjugated QDs without p53Ab were prepared as the control probe. Nonconjugated QDs were removed from the reaction medium by dialysis with size cutoff limit 12–14 kDa. One probe of p53Ab-QD was ultrafiltrated in Vivaspin 100 kDa filter by centrifugation at 10,000 rpm for 3 min. Protein quantity after ultrafiltration and adjustment to a volume of 100 μl was 0.03 μM by Bradford assay. The prepared conjugated QDs were stable and had fluorescence as for control solutions. Until use, the conjugates were stored protected from light at 4°C.

Successful conjugation is confirmed by DLS (**Figure 3**). The peak in the size distribution of QD conjugates is shifted toward larger diameters than the case of "free" QDs due to the conjugation with antibodies.

The asymmetry of the distribution in the case of conjugates can be explained by the fact that QDs and antibodies can bind to each other in different proportions. The hydrodynamic diameter of the semiconductor particles is typically larger than the diameter calculated from the absorption spectrum due different physical meaning, but the diameter values are of the same order. The PL spectrum was the same for QDs and their conjugates.

3.6 The specificity of p53Ab-QD conjugates

As a result of specificity assessment experiments for the p53Ab-QD conjugates with apoptosis induced in the cell culture, it was found there were a large number of cell nuclei stained with conjugates (**Figure 4A** and **C**).

For the evaluation of p53Ab-QD conjugates, we also used a fixed thyroid tissue (**Figure 4B** and **D**). Such an object of the study was chosen for two reasons: (1) thyroid gland has a high potential of cell renewal (apoptosis index in its parenchyma is about 25%) because of high intensity of synthetic and metabolic processes in normal cells and (2) thyroid pathology often develops during pregnancy and cascade associated with changes in the functioning of the reproductive

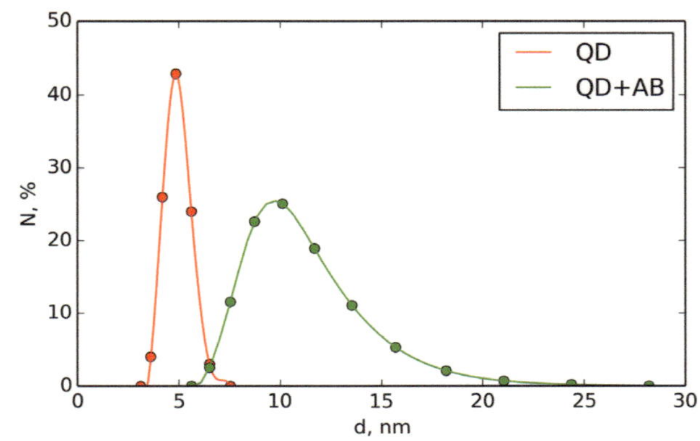

Figure 3.
The size distribution of QDs and QD-p53 antibody in PBS obtained by dynamic light scattering.

system during this period. We observe the expression of the pro-apoptotic protein in the nuclei of thyrocytes.

Thus, the penetration of the p53Ab-QD complexes into the cells was confirmed both in the cell culture and tissue, and specific reaction to the nuclear protein p53 was registered. In comparison with control samples stained with antibodies to p53Ab-Alexa Fluor 647, experimental samples showed a weak fluorescence but sufficient for detection by a confocal microscopy. Stained slides will not discolor over

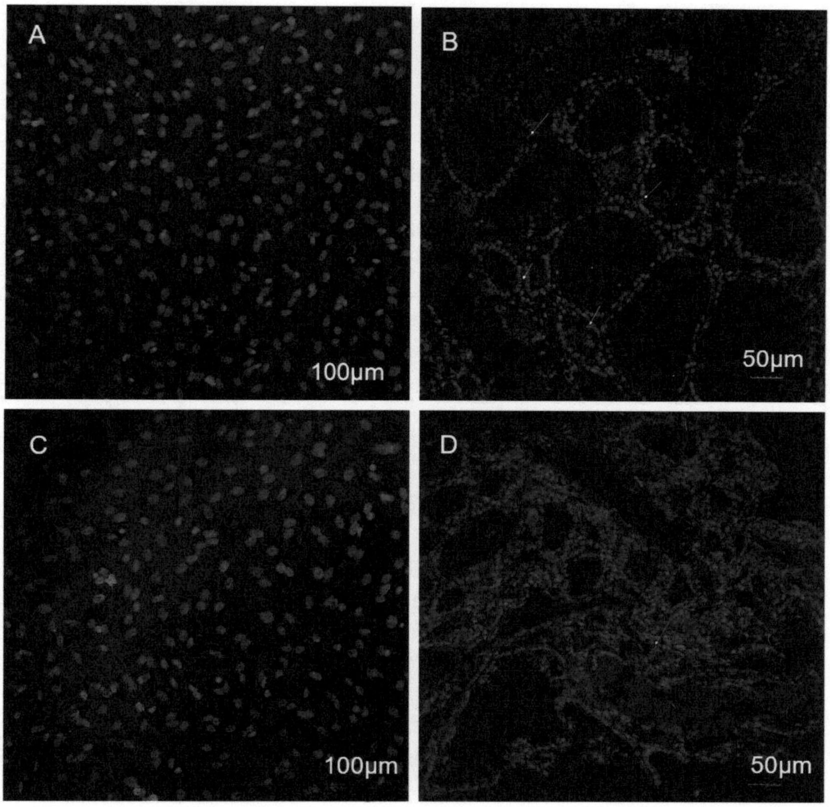

Figure 4.
Confocal laser scanning microscopy images. (A) Control, cell culture stained with p53 antibody-Alexa Fluor 647; (B) control, thyroid tissue stained with p53 antibody-Alexa Fluor 647; (C) cell culture with QD-p53 antibody; (D) thyroid tissue with QD-p53 antibody.

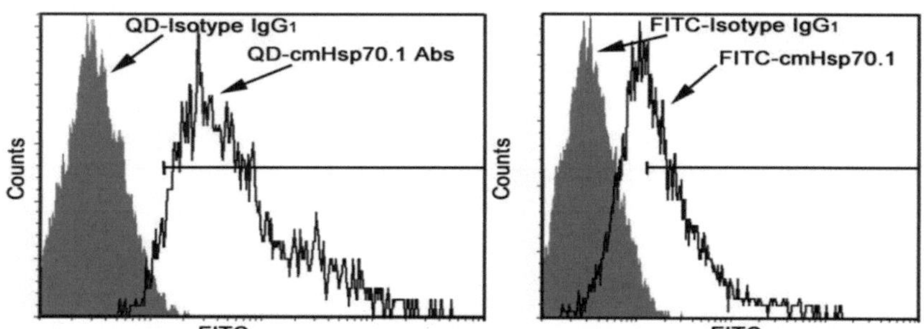

Figure 5.
Flow cytometry analysis of Hsp70-positive cells treated by QD-labeled cmHsp70.1 against Hsp70. The results were presented in comparison to fluorescein isothiocyanate-labeled immunoglobulin IgG1.

time, so preservation of the fluorescence intensity was observed at carrying out repeated scans at 1, 3, and 7 days.

3.7 The specificity of cmHsp70.1-QD conjugates

Flow cytometry analysis of Hsp70-positive cells treated by QD-labeled cmHsp70.1 against Hsp70 is shown on **Figure 5**. The results are presented in comparison to fluorescein isothiocyanate-labeled immunoglobulin IgG1.

4. Discussion

The presented above results suggest that CdSe nanoparticles are suitable for constructing the diagnostic platform for attachment of antibodies against proteins actively participating in apoptosis of malignant cells. Heat shock protein 70 in complex with p53 plays important role in apoptotic cell death. Level estimation of these apoptotic proteins can be adapted for monitoring of early stages of cancer. Whether the interaction of Hsp70 and p53 with cancer cells may be registered with high sensitivity by coupling Ab with QD is a subject of the study. For attachment of antibodies against p53 and Hsp70, QDs were synthesized and studied by optical spectroscopy. The core material is semiconductor CdSe nanocrystal coated by short ligands attached to the nanoparticle surface via sulfhydryl groups, while carboxylic groups of the ligands provide sites for conjugation. The study revealed that QDs remain stable for 1 month regardless of the acidity of the buffer solution.

Synthesized QDs have no cytotoxic effect as confirmed by the study of viability not only on tumor cells but also on normal human cells. CdSe/MPA nanoparticles were successfully conjugated with monoclonal antibodies to p53 and cmHp70.1 to Hsp70. The Ab conjugation results in an increase in the hydrodynamic size of nanoparticle from 4.8 nm to 10 nm. The coupling of Ab to CdSe/MPA nanoparticles is carried out without the changes in fluorescence emission spectra of nonconjugated QDs. The strong photoluminescence at λ = 650 nm allows getting microscopic images unperturbed by background fluorescence of tissues. The amount of Ab per one nanoparticle was similar for Hsp70 and p53 proteins. The covalent coupling of Ab yields random immobilization of Ab at the surface. Some Fab parts of Ab can be in close contact with a surface that restricts immunochemical recognition in ELISA assay. Free access to antigen is determined by chosen synthesis strategy and structure of surface layer. The biological activity depends on the protein density and orientation of Ab on the nanocrystal surface. The large decrease of biological activity follows from a random exposition of epitopes outside of QD [33]. Nonspecific adsorption of conjugated QDs is the additional complication for adequate labeling of apoptotic malignant cells. Experimental data indicate that bioconjugate cmHp70.1-QD shows the specific interaction with cells of C6 glioblastoma line cell. The membrane receptors of C6 cells are appeared to be targeted by QDs coupled with Ab against protein Hsp70. On the basis of experimental data, fluorescent conjugates of Hsp70 are uptaken by surface receptors on abnormal C6 cells. Bioconjugates of QDs for targeted delivery of cancer brain cells were actively studied by numerous researchers. The most known biomolecules for targeted recognition are transferrin, EGFR, some aptamers, RGD peptides, folic acid, etc. Human breast cancer cells and U87MG human glioblastoma cells were robustly labeled by QDs conjugated with RGD peptide [34]. The results of laser scanning confocal microscopy confirm the data earlier obtained for magnetic conjugates SPIONs-Ab-Hsp70 by MRI and NMR methods [26] that QD conjugates are accumulated by C6 glioma cells and brain tumors derived from it. The immunogenic function of Hsp70 results in an induction of immune response to the association of Hsp70 with cytoplasmic peptides/proteins that translocated out of the cell. It should be noted that the pool of p53 protein is not always recognized by antibody in vivo due to its dynamical conformation. p53 protein is an unstable molecule in vitro. For prolongation of its lifetime, the presence of heat shock proteins is essential in vivo. As regards to the endometrial breast carcinoma cell line ZR-75-1 and p53, the conjugated QDs bound only the surface without internalization in the cell. The bioconjugates retain to some extent the affinity to bind an antigen similar to pure monoclonal antibody. The protein complex from a malignant cell is recognized by the immune system as a foreign antigen. The immunogenic regulation agrees well with anti-apoptotic function Hsp70 as protein affiliated with p53 antitumor activity. The molecular chaperone Hsp70 supports p53 tumor suppressor activity by refolding injured molecules. Overexpression of pair p53-Hsp70 by glioma and breast cancer cells makes possible fluorescent visualization with the help of corresponding Ab-QD conjugates. The obtained results are similar to imaging results of the breast cancer MDA-MB-231 cell line treated by CdSe/CdS core-shell magic-sized QD coupled to a specific breast cancer Fab antibody. But incubation of cancer MDA-MB-231 cells in QD solution was followed by intracellular absorption. Our experiment displays only surface interaction of QDs with cell receptors. The intracellular interaction of the CdSe/p53 conjugates was observed in the fixed tissue thyroid gland. Specific recognition of the nuclear protein p53 was registered as evidence of the expression of the pro-apoptotic protein in the nuclei of thyrocytes.

Engineered conjugated QDs have diagnostic potential for molecular imaging of apoptosis usually performed by measurement of annexin—extracellular phosphatidylserine binding. Proteins p53 and Hsp70 in the complex may serve as pro-apoptotic targets for registration with the help of QDs. The QDs associated with antibody to pro-apoptotic protein p53 may be used for diagnostic purposes, in particular for the assessment of tumor progression. The apoptotic response after chemotherapy is the special subject for this approach.

Acknowledgements

This work was supported by the Russian Science Foundation, project 14-15-00324.

Conflict of interest

The authors declare no conflict of interest.

Author details

Lev B. Matyushkin[1], Olga A. Aleksandrova[1], Anna O. Drobintseva[2], Igor M. Kvetnoy[2,3], Yuliya S. Krylova[2], Yaroslav Y. Marchenko[4], Dmitriy S. Mazing[1*], Vyacheslav A. Moshnikov[1], Sergey F. Musikhin[5], Boris P. Nikolaev[4], Victoriya O. Polyakova[2], Maxim A. Shevtsov[6] and Ludmila Y. Yakovleva[4]

1 Saint-Petersburg Electrotechnical University "LETI", Saint-Petersburg, Russia

2 Center of Molecular Biomedicine, Saint-Petersburg Research Institute of Phthisiopulmonology, Saint-Petersburg, Russia

3 Department of Pathology, Saint-Petersburg State University, Saint-Petersburg, Russia

4 State Research Institute of Highly Pure Biopreparations, Saint-Petersburg, Russia

5 Peter the Great St. Petersburg Polytechnic University, Saint-Petersburg, Russia

6 Institute of Cytology of the Russian Academy of Sciences (RAS), Saint-Petersburg, Russia

*Address all correspondence to: dmazing@yandex.ru

References

[1] Gao X, Cui Y, Levenson RM, Chung LW, Nie S. In vivo cancer targeting and imaging with semiconductor quantum dots. Nature Biotechnology. 2004;**22**:969-976. DOI: 10.1038/nbt994

[2] Michalet X, Pinaud FF, Bentolila LA, Tsay JM, Doose S, Li JJ, et al. Quantum dots for live cells, in vivo imaging, and diagnostics. Science. 2005;**307**:538-544. DOI: 10.1126/science.1104274

[3] Radenkovic D, Kobayashi H, Remsey-Semmelweis E, Seifalian AM. Quantum dot nanoparticle for optimization of breast cancer diagnostics and therapy in a clinical setting. Nanomedicine: Nanotechnology, Biology and Medicine. 2016;**12**:1581-1592. DOI: 10.1016/j.nano.2016.02.014

[4] Resch-Genger U, Grabolle M, Cavaliere-Jaricot S, Nitschke R, Nann T. Quantum dots versus organic dyes as fluorescent labels. Nature Methods. 2008;**5**:763-775. DOI: 10.1038/nmeth.1248

[5] Chan WC, Nie S. Quantum dot bioconjugates for ultrasensitive nonisotopic detection. Science. 1998;**281**:2016-2018. DOI: 10.1126/science.281.5385.2016

[6] Murray C, Norris DJ, Bawendi MG. Synthesis and characterization of nearly monodisperse CdE (E = sulfur, selenium, tellurium) semiconductor nanocrystallites. Journal of the American Chemical Society. 1993;**115**(19):8706-8715. DOI: 10.1021/ja00072a025

[7] Rogach AL. Semiconductor Nanocrystal Quantum Dots. Verlag: Wien; 2008. ISBN: 978-3-211-75235-7

[8] Zhang Y, Li Y, Yan XP. Aqueous layer-by-layer epitaxy of type-II CdTe/CdSe quantum dots with near-infrared fluorescence for bioimaging applications. Small. 2009;**5**(2):185-189. DOI: 10.1002/smll.200800473

[9] Yu JS, Kim SH, Man MT, Lee HS. Synthesis and dual-channel optical properties of Mn-doped ZnSe quantum dots. Materials Letters. 2019;**253**:367-371. DOI: 10.1016/j.matlet.2019.07.005

[10] Silva TG, Moura IM, Filho PEC, Pereira MI, Filho CAA, Pereira G, et al. ZnSe: Mn aqueous colloidal quantum dots for optical and biomedical applications. Physica Status Solidi (c). 2016;**13**(7-9):530-533. DOI: 10.1002/pssc.201510300

[11] Hu R, Wang Y, Liu X, Lin G, Tan CH, Law WC, et al. Rational design of multimodal and multifunctional InP quantum dot nanoprobes for cancer: In vitro and in vivo applications. RSC Advances. 2013;**3**(22):8495-8503. DOI: 10.1039/C3RA23169K

[12] Lee SK, McLaurin EJ. Recent advances in colloidal indium phosphide quantum dot production. Current Opinion in Green and Sustainable Chemistry. 2018;**12**:76-82. DOI: 10.1016/j.cogsc.2018.06.004

[13] Jing L, Ding K, Kershaw SV, Kempson IM, Rogach AL, Gao M. Magnetically engineered semiconductor quantum dots as multimodal imaging probes. Advanced Materials. 2014;**26**(37):6367-6386. DOI: 10.1002/adma.201402296

[14] Aldakov D, Lefrançois A, Reiss P. Ternary and quaternary metal chalcogenide nanocrystals: Synthesis, properties and applications. Journal of Materials Chemistry C. 2013;**1**(24):3756-3776. DOI: 10.1039/C3TC30273C

[15] Xu G, Zeng S, Zhang B, Swihart MT, Yong KT, Prasad PN. New generation

cadmium-free quantum dots for biophotonics and nanomedicine. Chemical Reviews. 2016;**116**(19):12234-12327. DOI: 10.1021/acs.chemrev.6b00290

[16] Mandal G, Darragh M, Wang YA, Heyes CD. Cadmium-free quantum dots as time-gated bioimaging probes in highly-autofluorescent human breast cancer cells. Chemical Communications. 2013;**49**(6):624-626. DOI: 10.1039/c2cc37529j

[17] Jing L, Kershaw SV, Li Y, Huang X, Li Y, Rogach AL, et al. Aqueous based semiconductor nanocrystals. Chemical Reviews. 2016;**116**(18):10623-10730. DOI: 10.1021/acs.chemrev.6b00041

[18] Luk JM, Lam CT, Siu AF, Lam BY, Ng IO, Hu MY, et al. Proteomic profiling of hepatocellular carcinoma in Chinese cohort reveals heat shock proteins (Hsp27, Hsp70, GRP78) up-regulation and their associated prognostic values. Proteomics. 2006;**6**:1049-1057. DOI: 10.1002/pmic.200500306

[19] Ciocca DR, Calderwood SK. Heat shock proteins in cancer: Diagnostic, prognostic, predictive, and treatment implications. Cell Stress & Chaperones. 2005;**10**:86-103. DOI: 10.1379/CSC-99r.1

[20] Shevtsov MA, Pozdnyakov AV, Mikhrina AL, Yakovleva LY, Nikolaev BP, Dobrodumov AV, et al. Effective immunotherapy of rat glioblastoma with prolonged intratumoral delivery of exogenous heat shock protein Hsp70. International Journal of Cancer. 2014;**135**:2118-2128. DOI: 10.1002/ijc.28858

[21] Kocsis J, Madaras B, Toth EK, Füst G, Prohászka Z. Serum level of soluble 70-kDa heat shock protein is associated with high mortality in patients with colorectal cancer without distant metastasis. Cell Stress and Chaperones. 2010;**15**:143-151. DOI: 10.1007/s12192-009-0128-7

[22] Nanbu K, Konishi I, Mandai M, Kuroda H, Hamid AA, Komatsu T, et al. Prognostic significance of heat shock proteins HSP70 and HSP90 in endometrial carcinomas. Cancer Detection and Prevention. 1997;**22**:549-555. DOI: 10.1046/j.1525-1500.1998.00069.x

[23] Kern SE, Kinzler KW, Bruskin A, Jarosz D, Friedman P, Prives C, et al. Identification of p53 as a sequence-specific DNA-binding protein. Science. 1991;**252**:1708-1711. DOI: 10.1126/science.2047879

[24] Pinhasi-Kimhi O, Michalovitz D, Ben-Zeev A, Oren M. Specific interaction between the p53 cellular tumour antigen and major heat shock proteins. Nature. 1986;**320**:182-185. DOI: 10.1038/320182a0

[25] Hainaut P, Milner J. Interaction of heat-shock protein 70 with p53 translated in vitro: Evidence for interaction with dimeric p53 and for a role in the regulation of p53 conformation. The EMBO Journal. 1992;**11**:3513. DOI: 10.1002/j.1460-2075.1992.tb05434.x

[26] Shevtsov MA, Yakovleva LY, Nikolaev BP, Marchenko YY, Dobrodumov AV, Onokhin KV, et al. Tumor targeting using magnetic nanoparticle Hsp70 conjugate in a model of C6 glioma. Neuro-Oncology. 2014;**16**:38-49. DOI: 10.1093/neuonc/not141

[27] Lim D, Murali R, Murray MP, Veras E, Park KJ, Soslow RA. Morphological and immunohistochemical reevaluation of tumors initially diagnosed as ovarian endometrioid carcinoma with emphasis on high-grade tumors. The American Journal of Surgical Pathology. 2016;**40**:302-312. DOI: 10.1097/PAS.0000000000000550

[28] van Engeland M, Nieland LJ, Ramaekers FC, Shutte B,

Reutelingsperger CP. Annexin V-affinity assay: A review on an apoptosis detection system based on phosphatidylserine exposure. Cytometry. 1998;**31**:1-9. DOI: 10.1002/(sici)1097-0320(19980101)31:1<1::aid-cyto1>3.0.co;2-r

[29] Shellenberger EA, Sosnovic D, Weissleder R, Josephson L. Magneto/optical annexin V, a multimodal protein. Bioconjugate Chemistry. 2004;**15**:1062-1067. DOI: 10.1021/bc049905i

[30] Lu C, Jiang Q, Hu M, Tan C, Yu H, Hua Z. Preliminary biological evaluation of ^{18}F-FBEM-Cys-annexin V a novel apoptosis imaging agent. Molecules. 2015;**20**:4902-4914. DOI: 10.3390/molecules20034902

[31] Yu WW, Qu L, Guo W, Peng X. Experimental determination of the extinction coefficient of CdTe, CdSe and CdS nanocrystals. Chemistry of Materials. 2003;**15**:2854-2860. DOI: 10.1021/cm034081k

[32] Drobintseva AO, Matyushkin LB, Aleksandrova OA, Drobintsev PD, Kvetnoy IM, Mazing DS, et al. Colloidal CdSe and ZnSe/Mn quantum dots: Their cytotoxicity and effects on cell morphology. St. Petersburg Polytechnical University Journal: Physics and Mathematics. 2015;**1**:272-277. DOI: 10.1016/j.spjpm.2015.11.003

[33] Mahmoud W, Rousserie G, Reveil B, Tabary T, Millot JM, Artemyev M et. al. Advanced procedures for labeling of antibodies with quantum dots. Analytical Biochemistry 2011;416:180-185. DOI: 10.1016/j.ab.2011.05.018

[34] Cai WB, Shin DW, Chen K, Cheysense O, Cao QZ, Wang SX. Peptide labeled near infrared quantum dots for imaging tumour vasculature in living subjects. Nano Letters. 2006;**6**:669-676. DOI: 10.1021/nl052405t

Chapter 11

Fluorescence Methods for the Analysis of Microtubule/Microfilament Involvement in the Regulation of Endothelial Barrier Function

Anton S. Shakhov and Irina B. Alieva

Abstract

The endothelial monolayer is located on the inner surface of blood vessels and provides an important barrier function controlling the transport of metabolites and nutrients through the vessel wall, both toward the circulating blood and in the direction of the underlying tissues. This function is provided by cytoskeletal structures that generate contractile and tensile forces, existing in equilibrium in the intact endothelium. In the case of cytoskeleton rearrangements, there are changes in the shape of cells and the formation of intercellular spaces, which lead to endothelial dysfunction. Deep understanding of endothelial barrier function maintaining is a crucial problem because this phenomenon is common for a number of pathological states and diseases (inflammation, asthma, sepsis, acute lung injury, diabetes, etc.) and can lead to severe organ dysfunction, as well as a complication upon the treatment by a number of anticancer pharmacological drugs. Microtubules and the actin cytoskeleton function in cooperation in normal endothelium and under conditions of the barrier loss. In this review, we describe the application of modern fluorescence methods for investigation and analysis of the individual characteristics of cytoskeletal elements whose reorganization affects endothelial permeability, to emphasize the role of microtubules/microfilament crosstalk in EC barrier regulation.

Keywords: fluorescence analysis, fluorescent microscopy, confocal laser scanning microscopy, structural illumination microscopy, endothelium, endothelial cells, endothelial barrier function, endothelial cytoskeleton, microtubules, microfilaments

1. Introduction

Endothelium—cell monolayer locolazing on the inner vascular surface is formed by specialized cells: endothelial cells (EC) or endotheliocytes. Endothelium functions as a selective permeable barrier between vascular blood flow and tissue fluid in organs comprising these vessels. It is involved in the regulation of macromolecular transport and blood cell movement through the vessel wall. The integrity of vascular endothelium in any organ is critical for its functioning. Thus, corneal endothelium ensures its normal permeability and exchange with the connective stroma [1]. Vascular endothelium in lungs regulates the movement of fluids, macromolecules, and leukocytes into alveolar interstitial and air spaces [2, 3].

Lung endothelium lining the vessels (aorta, arteria, and others) are in close contact with each other, rendering the vascular wall into a tight barrier, which control such diverse processes as vascular tone, homeostasis, adhesion of platelets, and leukocytes to the vascular wall and permeability of vascular wall for cells and fluids [4, 5]. The integrity of the pulmonary EC monolayer, therefore, may be a critical requirement for the preservation of pulmonary function. This barrier is dynamic and highly prone to the regulation, by various stimuli, of physiological and pathological origin. Any breach within the EC barrier ends up in leakage of fluid from the lumen of the vessels into the plant tissue and/or alveolar lumen, severely impairing gas exchange. Disruption of endothelial barrier occurs during disease states like acute lung injury (ALI) and its more severe form, acute respiratory distress syndrome (ARDS), which remains a significant explanation for morbidity and mortality with an overall fatality rate of 30–40% [6], results in the uncontrolled movement of fluid and macromolecules into the interstitium and pulmonary air spaces causing pulmonary edema [7]. Literature and our own data have proved that normal functioning of the endothelial barrier is provided by the balance between contracting and stretching forces generated by EC cytoskeleton [4, 8, 9] and cell adhesive structures are involved in their relationship [9–11].

In order to understand how such a complicated and multi-regulated system works, one should characterize its structure as a whole system (as well as each component separately) in normal conditions and study the complex changes that occur during barrier dysfunction and also investigate their reorganization in dynamics. The main focus of this chapter is fluorescence methods using for investigation of EC, including living endotheliocytes, and we will demonstrate what results have been achieved due to their applying.

2. Endoteliocyte cytoskeleton: interaction of components performing barrier function

Cellular cytoskeleton is a complex of fibrillar components that structurally and functionally ensures the normal course of numerous vital functions of the cell. The cytoskeleton provides the maintenance of the cell shape and all its changes during the life cycle, as well as the native architecture and intracellular organization of its organelles. The components of the cytoskeleton are involved in transport processes during the interphase and they guarantee successive mitotic phases passing in the process of cell division. Endothelial cell cytoskeleton, similarly to the cytoskeletons of the other cell types, consists of three types of filaments—microtubules, actin filaments (microfilaments), and intermediate filaments.

The components of the cytoskeleton are composed of various proteins. Two types of them—microtubules and microfilaments—are monoprotein structures: microtubule walls are polymerized from alpha- and beta-tubulin heterodimers, and microfilaments are built from actin in the cells of a wide variety of different tissues. Unlike tubulin-polymerized microtubules and actin filaments intermediate filaments do not have a highly conservative protein structure. Intermediate filaments consist of different proteins, their composition in the cell depends on the tissue type in which they are located. Intermediate filaments have a different protein composition not only in different tissues but also at different stages of differentiation as well as at different stages of embryonic development. Intermediate filaments consisting of different proteins are expressed in different cell types. Moreover, in some types of cells, intermediate filaments consisting of different proteins can be present simultaneously.

Microtubules are an obligatory component of the cytoskeleton of all cells, they are highly dynamic polar biopolymer structures. As all other cytoskeleton components microtubules are multifunctional structures involved in various cellular processes, including directed cytoplasmic transport of vesicles and signaling molecules and changes in the cell shape during mitotic division as well as in interphase (during spreading, polarization, or movement).

Like the microtubules, actin filaments are liable for changes in cell shape and mobility. In contrast the intermediate filaments are mainly responsible for cell shape. Reorganization of the endotheliocyte cytoskeleton consisting of actin filaments, microtubules, and intermediate filaments and also the subsequent change within the cell shape creates a structural basis for the increase in vascular permeability. Historically, the contribution of various components of the cytoskeleton to the change in the shape of endotheliocytes during the development of barrier function has been studied in an unequal measure. In addition, the role of intermediate filaments is still poorly understood. The apparent role of actin filaments has been studied rather actively [2, 3, 12], whereas the involvement of the microtubule system during this process was revealed significantly later [13, 14]. The main achievement of recent years is that the understanding that each one component of the cell cytoskeleton is functioning with interrelations and their interactions are coordinated on different levels of the regulation. Moreover, cytoskeletal components are more or less interacting with adhesive structures of the cell; this complicated and finely regulated interaction allows the cells to implement the great variety of cellular, tissue, and organ functions. Within the present review, we shall analyze interactions of the cytoskeletal elements during changes in the endothelial permeability and emphasize the role of dynamic microtubules in the regulation of the barrier function of the endothelium.

As mentioned previously, intermediate filaments consisting of different proteins are expressed in different cell types. The homology of intermediate filament proteins is sometimes no more than 20%; nevertheless, based on biochemical, immunological, and structural similarities, five different types of intermediate filaments are distinguished: keratins (there are two types of proteins—type I (acidic keratins) and type II (basic keratins) —for the assembly of keratin intermediate filaments, proteins of both types are required—they form heteropolymers). Type III intermediate filament proteins include four proteins: desmin, vimentin, peripherin, and glial acidic protein. Each of these four proteins can form homopolymers, but besides that they are also able to form heteropolymers with other type III. Type IV intermediate filament proteins are expressed mainly in nerve cells: they include α-internexin and a triplet of neurofilament proteins: NF-L, NF-M, NF-H (neurofilament light, medium, and heavy proteins). The protein nestin, first discovered in the precursors of nerve cells, is sometimes referred to as a special type of intermediate filaments proteins; however, based on its structural features, nestin can be classified as type IV. Type V intermediate filament proteins include nuclear lamins. Vimentin filaments are the most abundant intermediate filaments of endothelial cells, and they are the least investigated, not only in endotheliocytes but also in other types of cells [15, 16]. Vimentin filaments are relatively stable structures, and their network remains almost unchanged in vitro in the process of monolayer formation [16]. Presumably, vimentin filaments might be involved in cytoskeletal interactions mediately: they might be crucial for centrosome positioning [17] and thus influence the microtubule system architecture. It seems that actin and vimentin filaments can interact structurally because they are connected to the biochemical level [18, 19]. Thus, the tail domain of vimentin is shown to interact directly with actin filaments [19], but up to now, no vimentin binding domain of fibrillary actin has been found. Data on microtubules and intermediate filaments coordination are still desultory; likely, kinesin [20] and a few other proteins [21, 22] can act as adaptor proteins between vimentin and microtubules. This insufficient attention given to intermediate filaments will be easily explained because actin filaments are generally assumed the key player crucial for normal endothelial permeability.

Actin polymerization is extremely dynamic in the endothelial cells. This dynamicity allows actin structures to rapidly rearrange and alter the static phenotype characterized by a large cortical actin ring and a minimal number of stress fibers to the so-called activated phenotype characterized by a deficient layer of cortical actin (sometimes up to its complete absence) with numerous stress fibers compressing the cell. Moreover, just this explains why it was originally thought that the actin system should be the central cytoskeletal component responsible for the functioning of the endothelium and development of its dysfunction [4]. A vital role of actin filaments in barrier dysfunction development was shown in many works. Notably, treatment of cultured cells with cytochalasin D that disrupt the actin cytoskeleton induced a drastic increase in the endothelial monolayer permeability [23], whereas treatment with phalloidin inhibiting actin depolymerization prevented the looks of the barrier dysfunction [24]. The interaction of actin with many binding proteins (filamin, vinculin, etc.) and adhesive structures chargeable for the barrier function also seems to evidence its key role within the development of endothelial dysfunction [4].

As polymer actin microfilaments are highly dynamic structures, their exchange time at the cell edge is just a few seconds [25]; nevertheless, they effectively regulate the cell shape and under in vitro conditions retain a stretched cell state. Focal [26, 27] and intercellular contacts [13, 14, 28, 29] also are able to rapidly, during several minutes, assemble and disassemble. Actin filaments are actively involved in the interactions with adhesive structures of the cell [5, 30]. It is generally assumed that the intercellular gap formation is regulated just by the balance of competing compressing forces, which curtail the cell toward its center, and forces generated by adhesive intercellular junctions and focal cell-substrate adhesions jointly regulating the cell shape [31]. The two competing forces in this model are related through actin fibers, which interact with the various adhesive molecules of the focal adhesions and intercellular adherent junctions of membranes.

However, recent works have shown that actin filaments do not seem to be instantly involved in the manifestation of endothelial barrier dysfunction, while microtubules are just the structure liable for the primary stages of this process [9, 32]. Thus, thrombin, which is produced on the surface of damaged cells from prothrombin circulating in the blood and inducing blood coagulation, can disturb the barrier function of the endothelium both in vitro and in vivo. It was shown that thrombin theatment leads to increase the permeability of the endothelium that is associated with a rapid decrease in the number of microtubules on the cell periphery near the margin and the microtubule system reorganization in the inner cytoplasm of endothelial cell during the first minutes of exposure. Actin stress-fibers are formed step-by-step, and therefore the maximal effect is observed only 30 min after treatment with thrombin [9, 29]. Thus, the development of microtubule response was faster than the reorganization of the actin filament system liable for the subsequent changes in the shape of the cells during the development of barrier dysfunction. Permanent depolymerization of microtubules after nocodazole exposure also induced an increase in the permeability of the endothelial cell monolayer and triggered a molecular cascade resulting in barrier dysfunction [9, 33].

The increase in the endothelial barrier permeability under nocodazole treatment is directly associated with the degree of depolymerization of peripheral microtubules. Endothelial barrier permeability increase after nocodazole treatment is directly associated with the degree of peripheral microtubules depolymerization. Normal endothelial barrier permeability can be disturbed even on minimal disruption of peripheral microtubules when the system of actin filaments remains intact and no morphological changes are observed [9]. The contraction of the cell body within the absence of stress fibers is surprising, but not unique fact, as already described earlier [31]. In the great majority of cases, despite this fact, barrier dysfunction is accompanied by stress fibers formation not depending on factors responsible for disturbance of the endothelial permeability, either drugs or pharmacological preparations [13, 14, 34] or neurohumoral factors [13, 35]. Thus, the disruption of peripheral microtubules was shown to be necessary and sufficient for arising of endothelial barrier dysfunction [9, 32].

Microtubules, the primary target of recent investigation, are involved in the most active dynamic interaction with both actin filaments and adhesive structures of various types of cells. During recent decades, these interactions were actively studied. The interaction of microtubules with actin filaments was found to be important for functional activities of various cells [31, 36], including endotheliocytes [37, 38], and microtubules dynamic properties are especially important for such interactions [9, 39].

Microtubules are highly (but less then microfilaments) dynamic cytoskeleton structures. The ends of individual microtubule are continuously growing or shortening even in cases when no visual changes are observed in the cell cytoplasm. Microtubule ends are growing or disassembling at distances of several microns [40, 41], and the microtubule system is continuously exchanging with the cytoplasmic pool of dissolved tubulin with exchange time of 5–20 min [42, 43]. Behavior described was named "dynamic instability" [40, 41], because each individual microtubule, as discriminated from the whole system, is not in a stationary state. However, notwithstanding the dynamic behavior of individual microtubules, changes in the microtubule system organization occur rather slowly even in the cells adapted to the movement. At the same time in nonmotile or poorly motile cells the microtubule system is maintained virtually unchanged in cell space and time. The assembly of microtubules is characterized by attachment of newly polymerized fragments of microtubules with GTP-β-tubulin. Because β-tubulin gradually hydrolyzes its GTP upon incorporation into the microtubule, GDP-β-tubulin dominates in the remaining a part of microtubules. Plus-end proteins specifically bind to the GTP-enriched fragments of microtubules [44, 45]. Microtubule plus-end proteins represent an extensive and varied group, and the most specific role in microtubules polymerization belongs to end-binding (EB) proteins family. Two proteins from the family—EB1 and EB3—interact with tubulin directly. Structural proteins CLIP115 and CLIP 170 belong to the cytoplasmic linker proteins family, bind EB1 and EB3. These proteins alternately bind with structural proteins of the CLASP family (CLIP-associated proteins). Microtubule tips complex of plus-end proteins play a crucial role in regulation of intracellular traffics and are directly involved in the interaction of microtubules with the cell boundary and the structures localized in this region [45–47].

It is thought that just the dynamic instability of microtubule ends and their ability to change frequently the assembly/disassembly stages allows the microtubules to locally modulate the dynamics of cell contacts [26, 27, 48, 49] and also to manage the dynamics of intercellular junctions [27, 38, 50, 51]. Moreover, microtubules can control internal cell organization.

It is thought that just the dynamic instability of microtubule ends and their ability to change frequently the assembly/disassembly stages allows the microtubules to locally modulate the dynamics of cell contacts—in moving fibroblasts through a direct targeted interaction of plus-ends of microtubules with the focal cell-substrate adhesions ("targeting" of cell-substrate adhesions) [26, 27, 48, 49] and also to manage the dynamics of intercellular junctions [27, 38, 50, 51]. Moreover, microtubules can control the organization of actin skeleton inducing local changes in the acto-myosin contractibility on the ends of stress fibers [52]. Later, it had been found that changes within the dynamics of microtubules in the zone of contacts depend on the presence of paxillin [53, 54], and original model that connected the asymmetric distribution of focal adhesions and the asymmetric distribution of microtubule phases of catastrophes (i.e. transitions to the assembly) in the region of adhesion sites was proposed [55]. The leading role of microtubules-actin filaments interaction in development of the barrier dysfunction of the pulmonary endothelium is played by Rho-dependent mechanisms [13, 14, 29, 33].

3. Fluorescense methods using for investigation of endotheliocytes

3.1 Immunofluorescence microscopy

Traditionally, immunochemical labeling of proteins of interest and the intracellular structures formed by them is used for structural studies. After fixation and immunostaining, it is possible to investigate cells by widefield immunofluorescence microscopy. The method was well known and widely used from the middle of the last century; with its help, numerous studies, including the investigation of cell cytoskeleton components, were performed. Using immunofluorescence analysis, the authors showed that cytoplasmic actin in endotheliocytes is organized in structures of three types: membrane cytoskeleton, cortical ringlike structures, and stress fibers [31]. Using double immunofluorescent staining of endothelial cells, mutual arrangement of microtubules and actin stress fibers has been described (**Figure 1a**) as well as the positional relationship of microtubules and specific endothelial VE-cadherin cell-cell contacts (**Figure 1b**).

Active creation of new specific antibodies and synthesis of recent highly effective and resistant to fading fluorescent labels for their detection make the method widely used nowadays.

We present our protocol for double immunofluorescence labeling of cytoskeletal components (microtubules, actin filaments, and vimentin intermediate filaments) adapted for endothelial cells. To quantify the obtained data, we have developed methods for analyzing the fluorescence intensity of immunostained samples [11, 16], which are also given below.

3.1.1 Cell cultures

Two cell lines of human endothelium, hybrid EA.hy926 cells and HPAEC (endothelium of human lung artery), were used in the experiments. EA.hy926 cells were obtained by fusion of primary endotheliocytes isolated from human umbilical vein and thioguanine-resistant clone of A549 cells (human lung carcinoma). According to ATTC and the manufacturer's database, EA.hy926 cells that have undergone more than 100 division cycles manifest all the basic properties common for differentiated endothelium. The cells were grown in DMEM medium at 37°C and 5% CO2. EA.hy926 cells were used in experiments at the sixth to tenth passages. HPAEC cells (Clonetics BioWhittaker Inc., United States) were the primary culture

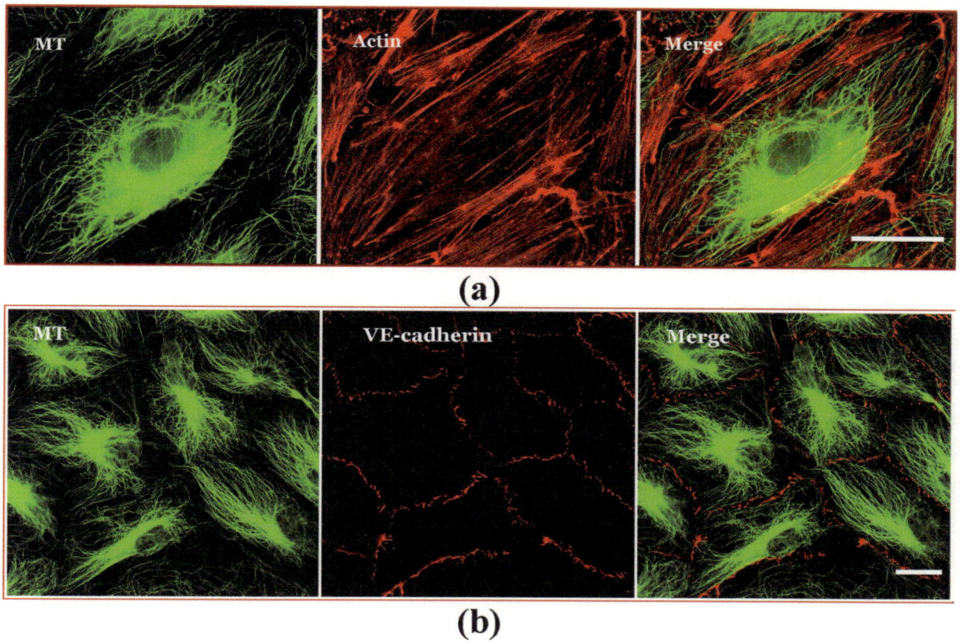

Figure 1.
(a) Double immunofluorescent staining of endothelial cells of human pulmonary artery with antibodies against β-tubulin (green) for detecting microtubules (MT) and with antibodies against actin (red) for detecting actin stress-fibers. (b) Double immunofluorescent staining of endothelial cells of the human pulmonary artery with antibodies against β-tubulin (green) and VE-cadherin (red). Stained preparations were analyzed under a Nikon Eclipse TE2000 microscope (Nikon Intech Co., Japan). Images were recorded using a 60x-objective with Hamamatsu ORCA-2 digital cooled P3C camera (Hamamatsu Photonics, Japan) with MetaView software (Universal Imaging, USA). Scale bar: 20 μm. (from [56]; with modification).

established from human pulmonary artery. The cells were maintained in EGM-2 medium (Clonetics BioWhittaker Inc., United States) at 37°C and 5% CO2.

3.1.2 Immunofluorescent labeling

Cells were fixed with three different methods: (1) for subsequent labeling by antibodies against β-tubulin and for dobble-labeling by antibodies against β-tubulin and β-actin, the cells were fixed with methanol (Sigma, United States) at -20°C and for 8 min. (2) For subsequent dobble-labeling by antibodies against vimentin and β-actin, the cells were fixed for 10 min at room temperature with formalin (Sigma) (4% solution in phosphate buffer solution [PBS], pH 6.8), washed thrice in PBS for 10 min, permeabilized for 15 min with Triton X-100 (Sigma) (0.1% solution in PBS). For fluorescence back ground degradation, the cells were treated with sodium borohydride NaBH4 (Sigma) (0.2% solution in PBS, 30 min) and washed thrice in PBS for 10 min prior to labeling with antibodies. For immunofluorescent staining, the samples were incubated with primary (30 min, at 37°C) and secondary antibodies (30 min, at 37°C). (3) Actin isoforms were visualized in cells fixed with 1% paraformaldehyde (Sigma, United States) in DMEM with HEPES buffer for 15 min, washed with PBS, and postfixed with methanol for 5 min. The same fixation procedure was used for dobble-labeling by antibodies against β-tubulin and γ-actin.

Vimentin intermediate filaments were labeled with antibodies to vimentin (BD Biosciences, United States, dilution 1:200); actin filaments, with antibodies against β-actin (Invitrogen, United States, dilution 1: 200); and microtubules, with monoclonal murine antibodies to β-tubulin (Covance, United States, dilution 1: 200).

Actin isoforms were visualized with mouse monoclonal antibodies to cytoplasmic β- and γ-actin isoforms (Dugina et al., 2009). Cells were washed with DMEM containing 20 mM HEPES at 37°C, fixed in 1% PFA in prewarmed (37°C) DMEM (30 min) and postfixed with methanol (-20°C) for 5 min. Cells were incubated with the following primary monoclonal antibodies: to β-actin (mAb 4C2, IgG1) and to γ-actin (mAb 2A3, IgG2b). Depending on the purpose of the study, as secondary antibodies, FITC- and TRITC-conjugated goat antimouse IgG1, TRITC-conjugated goat anti-mouse IgG2b (Southern Biotechnology, Associates, Birmingham, AL), or Alexa-Fluor-488-conjugated goat anti-mouse IgG (Molecular Probes, Invitrogen) were used in dilution 1:1000.

Coverslips were mounted on object glasses by using Mowiol as the embedding medium. To preserve samples, edges of coverslips were covered with nail polish.

Fluorescent mages were acquired on the confocal microscope LSM510 (Zeiss, Oberkochen, Germany) equipped with oil-immersion objectives (Plan-Neofluar 63 1.4 and PlanFluar 100 1.45, Zeiss). Single optical sections were scanned with ~1 μm thickness near the basal level of the cell. Stacks with z-step of 0.3–0.5 μm were collected as serial optical sections. Images were processed using Adobe Photoshop 7.0 (Adobe Inc., USA) software.

3.1.3 Imaging and image processing

Immunofluorescently stained endothelial monolayer semples were examined under a Nikon Eclipse TE2000 microscope (Nikon Intech Co., Japan) supplied with a 60/1.4 objective. We preliminary scanned of the samples, and the most spread cells were selected for the following analysis.

Images were collected with Hamamatsu ORCA-2 (Hamamatsu Photonics, Japan) digital cooled CCD camera supported with MetaView software (Universal Imaging, USA). The resolution of 12-bit digital images was 9 pixel/μm. Image processing was performed using MetaMorph (Universal Imaging, USA), ImageJ software (NIH), and Adobe Photoshop 7.0 (Adobe Inc., USA) softwares.

3.1.4 Quantitative assay of the actin filament system

Reorganization of actin cytoskeleton structure was analyzed by fluorescence intensity per unit area. At first, we chose a 10 μm^2 area on the captured images in the region of leading edge and contacts with neighboring cells. In this region, the average fluorescence intensity was measured using the ImageJ software. The chosen regions were not 5 μm from the cellular edge. For the quantitative analysis of actin filaments changes, the following parameters were used: relative area occupied by actin filaments, thickness of actin fibers, and length of actin fibers. To estimate the relative area on digital images, we used the following algorithm: the cell contour was outlined, then the cell area occupied by ImageJ (function "Polygon selection"), and the outlined areas were measured (function "Measure"). The ratio of the area occupied by actin filaments was calculated to the total cell area. The obtained data were exported into Excel for statistical analysis and for the construction of histograms. To measure the thickness of actin fibers on cell digital images in the ImageJ, the line was drawn perpendicular to the long fibril axis, and the line of the obtained line corresponded to the thickness of the actin fibers in the place of measurement.

3.1.5 Quantitative assay of the microtubule system

Changes in the microtubule system organization were assessed with the ImageJ software by counting the number of microtubule ends in the area of free cellular lamella and in the zone of cell-cell (intercellular) contacts. The measurements were done in areas of 10 μm length and 2 μm width. Two approaches were used: (1) the microtubule number was counted directly if microtubules were well distinguished visually (at the lamella edge with a rather thin cytoplasm). (2) If the microtubule density prevented their precise identification, the microtubule number was counted with the ImageJ software by the peaks of fluorescence on digital images. Fluorescence peaks higher the background and corresponding to the microtubule position were registered on a section with 10 μm length and 2 μm width. Fluorescence peaks 25% or higher than the background value were taken for the assay of the microtubule number. Statistical treatment was done with the Sigma Plot 12.5 (SPSS Science, United States) and Excel (Microsoft Corp., United States) software. The significant difference between analyzed samples was assayed with the t-test (Student criterion) or nonparametric U-test (Wilcoxon-Mann-Whitney criterion).

3.2 Confocal laser scanning microscopy

The appearance of confocal laser scanning microscopy allowed researchers to get better quality images with high resolution in the plane of the analyzed sample, along the XY-axis.

Confocal laser scanning microscopy is a kind of light optical microscopy, which has significant contrast and spatial resolution compared to classical light microscopy, which is achieved by using a point aperture (pinhole) located in the image plane and restricting the flow of background scattered light emitted not from the focal plane of the lens [57]. This allows one to obtain a series of images at various depths of the focal plane inside the sample (the so-called optical sectioning of the sample by depth) and then reconstruct a three-dimensional image of the sample from these series.

In endothelial cells, actin cytoskeleton is represented by actin stress fibers [58] (**Figure 2**), cortical actin filament network, and membrane skeleton [31]. Confocal microscopy was the method that made it possible to characterize these structures and describe their mutual arrangement in the cell volume. The latter is located beneath the plasma membrane and reaches the width of up to several nanometers. It is composed of relatively short actin filaments, spectrin, ankyrin, and other proteins [31]. The membrane skeleton determines membrane architecture and provides membrane mechanical stability. It includes spectrin and spectrin-stabilizing proteins, such as F-actin, α-catenin, adducin. The cortical actin network consists of long actin filaments stabilized by actin-binding proteins (cortactin, filamin, spectrin, WASP, VASP, etc.) [31]. The membrane skeleton is mainly involved in maintaining cell shape, whereas the cortical network interacts with adjacent cells and extracellular matrix. Stress fibers containing short microfilaments with alternating polarity are found across the entire cell cytosol. Stress fibers are bundles of actin filament bound to α-actinin and other actin-binding proteins. In endothelial cells, actin stress fibers are mainly involved in enabling cell motility and contraction [31].

Similar to actins in other cell types, β- and γ-actin isoforms in endothelial cell can assemble in various intracellular structures. In human pulmonary artery endothelial cells (HPAECs) and human umbilical vein EA.hy926 cells, β-actin was found mostly in stress fibers [11, 16].

Staining with specific anti-β-actin antibodies detected cytosolic bundles of stress fibers and spheroid microparticles, whereas γ-actin formed branched cortical actin network. Moreover, the density of the γ-actin network was higher in EA.hy926 cells compared to HPAECs [16].

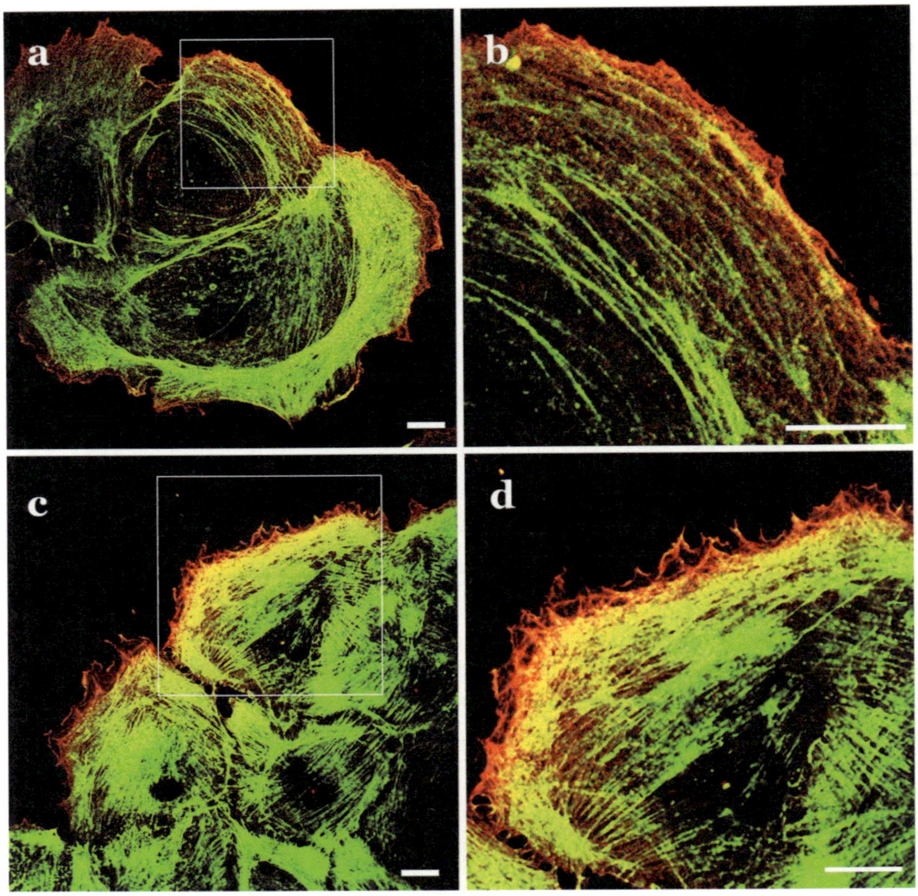

Figure 2.
Double immunofluorescence staining of β- and γ-actins in aorta endothelial cells in vitro. (a, b) In spread endothelial cells, β-actin forms circular bundles, whereas γ-actin forms lamellar microfilament meshwork. (b) Large magnification of boxed area on panel (a); (a) single optical x/y section across the basal cell level (scale bar, 10 μm). (c, d) β- and γ-actins in the leading edge of polarized endothelial cells moving into the experimental wound. Extensive γ-actin network is discovered in the leading edge protrusions of moving cells, whereas β-actin localizes in the microfilament bundles and in the area of cell-cell contacts. (d) Large magnification of boxed area on panel (c); (c) single optical x/y section across basal cell level. Scale bar, 10 μm. (from [58]; with modification).

For the analysis of β- and γ-actins in spread endothelial cells (**Figure 2**) immunofluorescent images were acquired using a confocal microscope (LSM510, Zeiss, Oberkochen, Germany) equipped with oil-immersion objectives (Plan-Neofluar 63 1.4 and PlanFluar 100 1.45, Zeiss). Single optical sections were scanned with ~1 μm thickness in the basal level of the cell. Stacks of serial optical sections (with z-step of 0.3–0.5 μm) were collected. Images were processed using Adobe Photoshop software.

Confocal microscopy as the method allowed reconstructing the position of the analyzed structures in the cell volume; however, the Z-axis resolution was insufficient, which stimulated the further development of microscopic technology.

3.3 Structural illumination microscopy

Structured illumination microscopy (SIM) is a relatively new high resolution method of microscopy analysis objects, which overcomes the shortcomings of confocal laser scanning microscopy. The main goal that SIM solves as the modern light microscopy method is to enhance spatial resolution. The main innovation of the method is "patterned light"—for the observation of semples not direct, but so-called structural illumination is used. SIM method provides illuminating the sample with patterned light and using software to analyze the information in Moiré fringes outside the normal range of observation [59]. Reconstruction software deciphers the images at about 2-fold higher resolution than the diffraction limit, approximately equal to 100 nm. Significant advantages of SIM has over other microscopy methods are (1) SIM can be used for imaging thicker sections; (2) for 3D imaging; and (3) for live-cell imaging. Additionally, it is possible to increase the image quality with photostable and more bright and dyes as well as precise targeting. Fluorescent proteins are commonly used for SIM investigations of live cells in addition to multiplexing with organic dyes and Qdot® probes. SIM itself, as well as in combination with other methods of analysis (proximity ligation assay, etc.), allows not only to describe the relative position of the structures of interest, but also—due to their high resolution—to make conclusions about their colocalization.

Here, we represent the application of modern fluorescence methods for investigation and analysis of the individual characteristics and interrelation of dynamic cytoskeletal components—microtubules and actin microfilament in the regulation of endothelial permeability (**Figure 3**).

Microtubules were labeled with monoclonal murine antibodies to β-tubulin (Covance, United States), in dilution 1:200, actin filaments were labeled with antibodies to β-actin (Invitrogen, United States), in dilution 1:200. Alexa-Fluor-568-conjugated (ThermoFisher Sci) or Alexa-Fluor-488-conjugated (Molecular Probes, Invitrogen) secondary antibodies were used (both in dilution 1:1000).

For the samples, preparation cells were mounted in Mowiol 4-88 (Calbiochem). To prevent photobleaching, 1% DABCO solution was added to the mounting dye. Cells and examined on a Nikon N-SIM (Nikon) equipped with NA oil immersion objective 100x/1.49, 488 nm and 561 nm diode laser excitation. For image aqusition, stacks with z-steps of 0.12 μm were used. Image stacks were acquired with EMCCD camera (iXon 897, Andor, effective pixel size 60 nm), herewith, exposure conditions were adjusted to get typical yield about 5000 max counts (16-bit raw image) while keeping bleaching minimal. Serial optical sections of the same cell taken in widefield mode were deconvolved using the AutoQuant blind deconvolution algorithm. Image acquisition, SIM image reconstruction, and data alignment were preformed using NIS-Elements 4.2 software (Nikon).

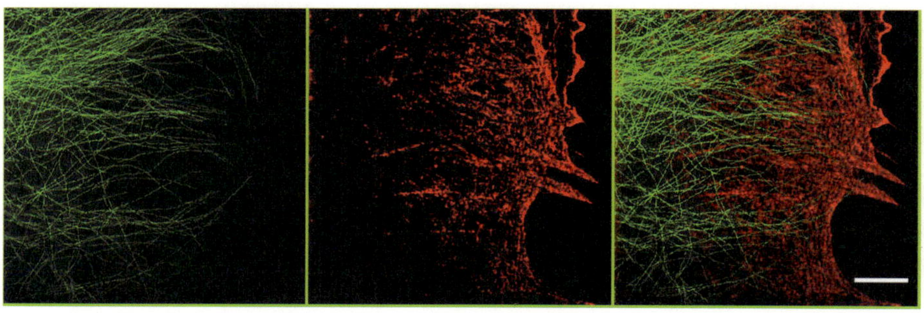

Figure 3.
Microtubules and γ-actin in human venous endothelial cells. Super resolution SIM (structured illumination microscopy) image of cells stained with antibodies against β-tubulin (green) and γ-actin (red). Scale bar, 5 μm. (from: [58] With modification).

3.4 Real-time study of endothelial cells microtubule organization and dynamics

Here, we have already mentioned several times that microtubules are highly dynamic cell component, and the results of our previous experiments demonstrated the critical involvement of the microtubule disassembly in induced endothelial barrier dysfunction [13, 29, 60]. Our previous data allowed us to assume that microtubule dynamics is an early event in the circuit of the reactions leading to the changes in pulmonary endothelial cells barrier permeability [29]. Taking into account the active dynamic properties of microtubules, as well as the ability of microtubule plus-ends tips to act as targeting factors, we proposed to use the microtubule system as a target for the action of substances that cause endothelial barrier dysfunction and on this basis to create a model for studying factors potentially capable to disrupt endothelial layer permeability. For the examination of microtubule behavior in the cytoplasm of endothelial cells, we created and described living cellular model established in physiologically relevant human pulmonary artery endothelial cells (protocol is given below). This model is suitable for a real-time study of microtubule system organization and reorganization in details as well as individual microtubules dynamics (including growth rate of microtubule plus-ends) in the quiescent human endothelial cell monolayer [10] in the normal condition of cultivation as well as under the experimental treatment (in the cells treated with the agents compromising/enhancing endothelial barrier) (**Figure 4**) [9]. Using this model, we were able to obtain several key parameters of microtubule organization in the endothelium such as a ratio between stable and dynamic microtubule subpopulations, direct measurement of microtubule growth rates, and their difference in single endothelial cells and the cells grown as a monolayer [10]. In our opinion, this cellular model would also allow us to study the involvement of microtubules in the barrier-protective/compromising mechanisms activated in pulmonary endothelium by various pharmacological agents of interest.

In this study, we aimed to understand how microtubules contribute to the dynamic reorganization of the endothelial cytoskeleton, we established an endothelial cells model expressing chimeric protein EB3-GFP—microtubule plus ends protein EB3 (a protein that is located on microtubule plus-ends) fused to the fluorescent protein GFP. Using this EB3-GFP microtubule plus-ends marker, we were able to measure microtubule growth rate in different parts of endothelial cell—at the centrosome region and near the cell periphery of a single human endothelial cells or in the endothelial monolayer where cells are adjacent to each other and form cell-cell contacts. This model allows us to show microtubule dynamics heterogeneity of within endothelial cells and to demonstrate that (a) the majority of the endothelial cells microtubules are dynamic and (b) the plus-ends growth rate is highest in the area of the centrosome location, in the internal region cytoplasm. Interestingly, but growth rate of microtubule plus-ends decreases from the cell center toward the periphery. The quantitative data we obtained in the experiments with the single cells and the cells growing in monolayer allow us to suggest the existing mechanism(s) of local regulation of microtubule plus-ends growth in endotheliocytes. It was shown that in the internal cytoplasm of edotheliocytes in the monolayer, microtubule growth rate is lower than that of single endothelial cells suggesting the regulatory effect of cell-cell contacts.

Unexpectedly, the dynamics of microtubules growing from the centrosome in singly located on the substrat surface with no neighbors turned out to be especially original—centrosomal microtubule growth rate distribution in single endothelial cells indicated the presence of two subpopulations of microtubules. One of the populations, the most numerous, included microtubules with "normal" (similar to those in monolayer endothelial cells) and the second population included "fast"

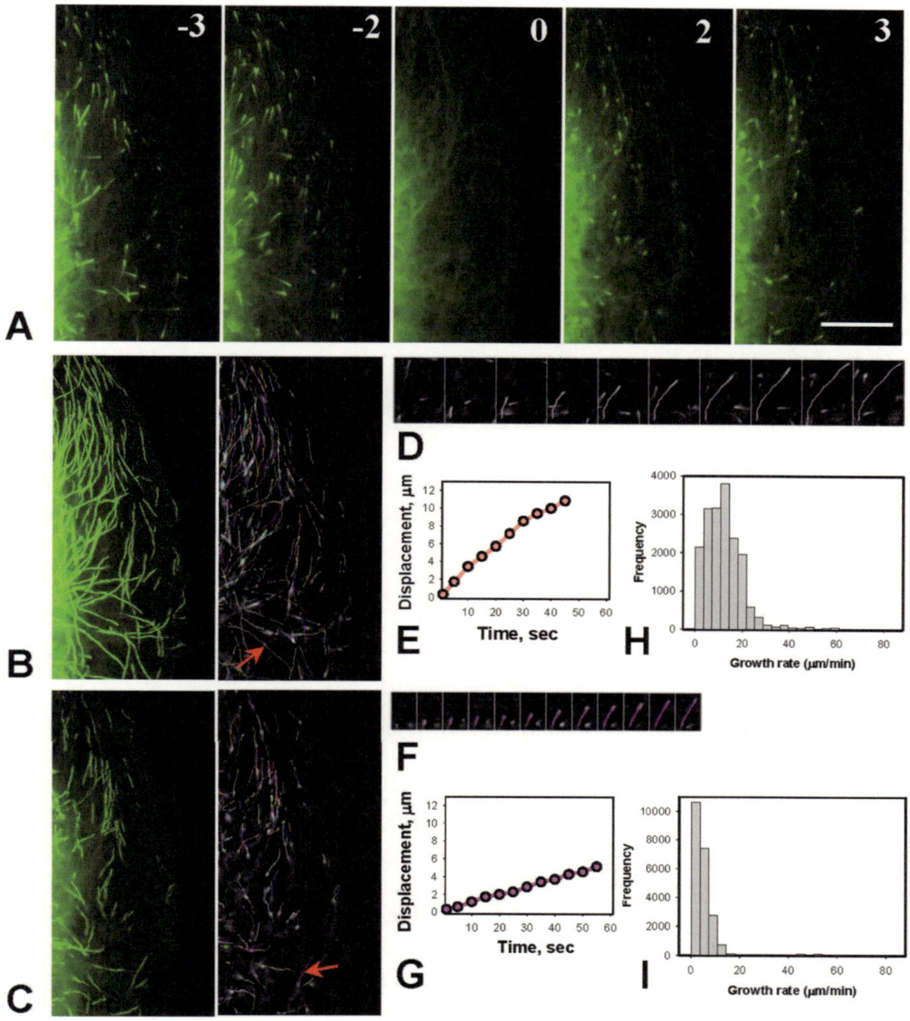

Figure 4.
Real-time study of microtubule dynamics in EB3-GFP-transfected endothelial cells. (A-C). EB3-GFP was used as a marker of growing microtubule plus ends. EB3-GFP movement was examined to analyze the changes of microtubule dynamics during barrier dysfunction EC expressing EB3-GFP were selected for the analysis by time-lapse microscopy (images were acquired every 1 s). (a) EB3-GFP tracks 3 and 2 min before and after nocodazole treatment. EB3-GFP is presented at microtubule plus-ends during growth phases but disappears after transition from growth to pause or shortening phase at the moment of nocodazole application (0 min). Time (min) marked in right top corner. Scale bar, 10 mm. (B, C) EB3 tracks projection obtained by EB3-GFP patches displacement on time-lapse series during 60 seconds (left, tracks projection; right, EB3 tracks are colored individually); between 2 and 3 min before (B) and after (C) nocodazole application. (D, F) Individual microtubules plus end displacement before (D) and after (F) nocodazole treatment (5 s interval between frames). (E) Microtubule plus-ends displacement (red arrow) quantification data are shown in D. (G) Microtubule plus-ends displacement (red arrow) quantification data are shown in F. (H, I) Histograms of microtubule growth rate distribution were obtained by tracking EB3-GFP comets at microtubule plus-ends in HPAEC near the cell margin (H) before nocodazole treatment (mean growth rate, 12.27 ± 0.62 µm/min) and (I) after nocodazole treatment (mean growth rate, 5.76 ± 0.05 µm/min). Thus, the obtained quantitative data allow us to conclude that plus-ends growth rate decreased. (from: [9] With modification).

microtubules, with significantly higher growth rates (threefold as much). Real-time study of endothelial cell microtubule organization and dynamics allowed indicating functional interactions between cell-cell contacts and microtubules. Later in the experiments with nocodazole (anti-microtubule agent, shifting the equilibrium of tubulin polymerization reaction and leading to depolymerization of cytoplasmic microtubules), it was possible to show that direct microtubules depolymerization by nocodazole initiates the cascade of barrier dysfunction reactions. It turned out that the short-term loss of endothelial barrier function occurs even at the minimal destruction of peripheral microtubules at the time when actin filament system remains intact. Specifically, here, we demonstrate that under these experimental conditions, microtubule dynamics of endothelial cells is very sensitive to external treatment: microtubule plus-ends growth rate has changed and significantly decreased near the cell periphery (**Figure 4**) [9].

3.4.1 Protocol for expression construct and transfection of plasmid

To analyze microtubule dynamics and to calculate the microtubule growth rates in living HPAEC cells, we used expression vectors encoding EB3-GFP (kind gift of Dr. I. Kaverina with permission from Dr. A. Akhmanova), which serves as a marker of growing microtubule plus-ends [61]. For transfection of plasmids into human artery endothelial cells, Effectene transfection reagent (Qiagen Inc., USA) was used according to the manufacturer's protocol. Transfected cells were selected for imaging by GFP fluorescence.

3.4.2 Video microscopy of EB3-GFP-transfected cells

To acquire imaging, human artery endothelial cells were cultured on glass-bottomed dishes with No.1S coverslips (Iwaki, Japan). Images were collected with a PC-based DeltaVision optical sectioning system using one of the following objectives (Olympus): PlanApo 100x/1.40 NA oil or PlanApo 60x/1.40 NA oil ph 3. Images were acquired with a cooled CCD camera (Hamamatsu Photonics) with an appropriate ND filter, binning of pixels, exposure time, and time intervals. Fluorescence signals were visualized using the Endow GFP bandpass emission filter set (41,017, Chroma) for GFP imaging. Quantitative analysis of the microtubule dynamics was carried out on time-lapse movies of cells expressing EB3-GFP. Microtubule growth rates were obtained by tracking EB3-GFP comets at microtubule plus-ends (the shooting frequency was 1 s/frame) using ImageJ software linked to an Excel spreadsheet. Statistical analysis was performed using Sigma Plot 12.5 (SPSS Science, USA).

Modern methods of super-resolution microscopy (stochastic optical reconstruction microscopy [STORM], photo-activated localization microscopy [PALM], and fluorescence photo-activation localization microscopy [FPALM]) especially in combination with another techniques (in particular, Proximity Ligation Assay, etc.) will allow the investigation of living cells structure, so we hope see new interesting results on the cytophysiology of endotheliocytes.

Acknowledgements

This work was supported by Russian Foundation for Basic Research (Grants no. 9-04-00363; 12-04-00488; 15-04-08550; 18-29-09082) to I. B. A. and MSU Development Program (PNR 5.13).

Author details

Anton S. Shakhov* and Irina B. Alieva
A.N. Belozersky Institute of Physico-Chemical Biology, Lomonosov Moscow State University, Moscow, Russia

*Address all correspondence to: antshakhov@gmail.com

References

[1] Jalimarada SS, Shivanna M, Kini V, Mehta D, Srinivas SP. Microtubule disassembly breaks down the barrier integrity of corneal endothelium. Experimental Eye Research. 2009;**89**:333-343. DOI: 10.1016/j.exer.2009.03.019

[2] Garcia JG, Davis HW, Patterson CE. Regulation of endothelial cell gap formation and barrier dysfunction: Role of myosin light chain phosphorylation. Journal of Cellular Physiology. 1995;**163**:510-522. DOI: 10.1002/jcp.1041630311

[3] Lum H, Malik AB. Mechanisms of increased endothelial permeability. Canadian Journal of Physiology and Pharmacology. 1996;**74**:787-800. DOI: 10.1139/y96-081

[4] Dudek S, Garcia J. Cytoskeletal regulation of pulmonary vascular permeability. Journal of Applied Physiology. 2001;**91**:1487-1500. DOI: 10.1152/jappl.2001.91.4.1487

[5] Bazzoni G, Dejana E. Endothelial cell-to-cell junctions: Molecular organization and role in vascular homeostasis. Physiological Reviews. 2004;**84**:869-901. DOI: 10.1152/physrev.00035.2003

[6] Ware LB, Matthay MA. The acute respiratory distress syndrome. The New England Journal of Medicine. 2000;**342**:1344-1349. DOI: 10.1056/NEJM200005043421806

[7] Ermert L, Bruckner H, Walmrath D, Grimminger F, Aktories K, Suttorp N, et al. Role of endothelial cytoskeleton in high-permeability edema due to botulinum C2 toxin in perfused rabbit lungs. American Journal of Physiology-Lung Cellular and Molecular Physiology [Internet]. 1995;**268**:L753-L761. DOI: 10.1152/ajplung.1995.268.5.L753

[8] Bogatcheva NV. Verin. AD. The role of cytoskeleton in the regulation of vascular endothelial barrier function. Microvasc. Research. 2009;**76**:202-207. DOI: 10.1016/S0026-2862(09)00021-1

[9] Alieva IB, Zemskov EA, Smurova KM, Kaverina IN, Verin AD. The leading role of microtubules in endothelial barrier dysfunction: Disassembly of peripheral microtubules leaves behind the cytoskeletal reorganization. Journal of Cellular Biochemistry. 2013;**114**:2258-2272. DOI: 10.1002/jcb.24575

[10] Alieva IB, Zemskov EA, Kireev II, Gorshkov B, Wiseman D, Black SM, et al. Microtubules growth rate alteration in human endothelial cells. Journal of Biomedicine & Biotechnology. 2010;**2010**:1-10. DOI: 10.1155/2010/671536

[11] Shakhov AS, Dugina VB, Alieva IB. Reorganization of actin and microtubule systems in human vein endothelial cells during intercellular contact formation. Cell Tissue Biology. 2015;**9**:299-309. DOI: 10.1134/S1990519X15040112

[12] Garcia JGN, Verin AD, Schaphorst KL. Regulation of thrombin-mediated endothelial cell contraction and permeability. Seminars in Thrombosis and Hemostasis. 1996;**22**:309-315. DOI: 10.1055/s-2007-999025

[13] Birukova AA, Birukov KG, Smurova K, Adyshev D, Kaibuchi K, Alieva I, et al. Novel role of microtubules in thrombin-induced endothelial barrier dysfunction. The FASEB Journal. 2004;**18**:1879-1890. DOI: 10.1096/fj.04-2328com

[14] Birukova AA, Smurova K, Birukov KG, Kaibuchi K, Garcia JGN, Verin AD. Role of rho GTPases in thrombin-induced lung vascular endothelial cells barrier dysfunction.

Microvascular Research. 2004;**67**:64-77. DOI: 10.1016/j.mvr.2003.09.007

[15] Bruneel A, Labas V, Mailloux A, Sharma S, Vinh J, Vaubourdolle M, et al. Proteomic study of human umbilical vein endothelial cells in culture. Proteomics. 2003;**3**:714-723. DOI: 10.1002/pmic.200300409

[16] Shakhov AS, Verin AD, Alieva IB. Reorganization of endothelial cells cytoskeleton during formation of functional monolayer in vitro. Cell Tissue Biology. 2014;**8**:138-151. DOI: 10.1134/S0006297919040035

[17] Morgan JT, Pfeiffer ER, Thirkill TL, Kumar P, Peng G, Fridolfsson HN, et al. Nesprin-3 regulates endothelial cell morphology, perinuclear cytoskeletal architecture, and flow-induced polarization. Molecular Biology of the Cell. 2011;**22**:4324-4334. DOI: 10.1091/mbc.E11-04-0287

[18] Cary RB, Klymkowsky MW, Evans RM, Domingo A, Dent JA, Backhus LE. Vimentin's tail interacts with actin-containing structures in vivo. Journal of the Cell Science. 1994;**107** (Pt 6):1609-1622

[19] Esue O, Carson A, Tseng Y, Wirtz D. A direct interaction between actin and vimentin filaments mediated by the tail domain of vimentin. Journla of Biological Chemistry [Internet]. 2006;**281**:30393-30399. DOI: 10.1074/jbc.M605452200

[20] Kreitzer G, Liao G, Gundersen GG. Detyrosination of tubulin regulates the interaction of intermediate filaments with microtubules in vivo via a kinesin-dependent mechanism. Molecular Biology of the Cell. 1999;**10**:1105-1118. DOI: 10.1091/mbc.10.4.1105

[21] Dráberová E, Dráber P. A microtubule-interacting protein involved in coalignment of vimentin intermediate filaments with microtubules. Journal of Cell Science [Internet]. 1993;**106**(Pt 4):1263-1273

[22] Liao G, Gundersen GG. Kinesin is a candidate for cross-bridging microtubules and intermediate filaments. The Journal of Biological Chemistry. 1998;**273**:9797-9803. DOI: 10.1074/jbc.273.16.9797

[23] Shasby DM, Shasby SS, Sullivan JM, Peach MJ. Role of endothelial cell cytoskeleton in control of endothelial permeability. Circulation Research. 1982;**51**:657-661. DOI: 10.1161/01.res.51.5.657

[24] Phillips PG, Lum H, Malik AB, Tsan MF. Phallacidin prevents thrombin-induced increases in endothelial permeability to albumin. American Journal of Physiology. Cell Physiology. 1989;**257**(3 Pt 1):C562-C567. DOI: 10.1152/ajpcell.1989.257.3.C562

[25] Amann KJ, Pollard TD. Cellular regulation of actin network assembly. Current Biology. 2000;**10**:R728-R730. DOI: 10.1016/s0960-9822(00)00751-x

[26] Kaverina I, Krylyshkina O, Small JV. Microtubule targeting of substrate contacts promotes their relaxation. The Journal of Cell Biology. 1999;**146**:1033-1043. DOI: 10.1083/jcb.146.5.1033

[27] Stehbens S, Wittmann T. Targeting and transport: How microtubules control focal adhesion dynamics. The Journal of Cell Biology. 2012;**198**:481-489. DOI: 10.1083/jcb.201206050

[28] Adams CL, Chen Y-T, Smith SJ, James NW. Mechanisms of epithelial cell–cell adhesion and cell compaction revealed by high-resolution tracking of E-cadherin–green fluorescent protein. The Journal of Cell Biology. 1998;**142**:1105-1119. DOI: 10.1083/jcb.142.4.1105

[29] Smurova KM, Birukova A, Verin AD, Alieva IB. Microtubule

system in endothelial barrier dysfunction: Disassembly of peripheral microtubules and microtubule reorganization in internal cytoplasm. Cell Tissue Biology. 2008;**2**:45-52. DOI: 10.1134/S1990519X08010070

[30] Dejana E, Bazzoni G, Lampugnani MG. Vascular endothelial (VE)-cadherin: Only an intercellular glue? Experimental Cell Research. 1999;**252**:13-19. DOI: 10.1006/excr.1999.4601

[31] Prasain N, Stevens T. The actin cytoskeleton in endothelial cell phenotypes. Microvascular Research. 2009;**77**:53-63. DOI: 10.1016/j.mvr.2008.09.012

[32] Smurova KM, Verin AD, Alieva IB. Inhibition of RHO-kinase depends on factors that modify endothelial permeability. Cell Tissue Biology. 2011;**5**:221-227

[33] Smurova KM, Birukova AA, Verin AD, Alieva IB. Dose-dependent effect of nocodazole on endothelial cell cytoskeleton. Biochemistry (Moscow) Supplement Series A: Membrane and Cell Biology. 2008;**2**:119-127. DOI: 10.1134/S1990519X11030138

[34] Verin AD, Birukova A, Wang P, Feng L, Becker P, Birukov K, et al. Microtubule disassembly increases endothelial cell barrier dysfunction: Role of MLC phosphorylation. American Journal of Physiology-Lung Cell and Molecular Physiology. 2001;**281**:565-574. DOI: 10.1152/ajplung.2001.281.3.L565

[35] Amerongen GP, van Nieuw A, Vermeer MA, van Hinsbergh VWM. Role of RhoA and Rho kinase in lysophosphatidic acid–induced endothelial barrier dysfunction. Arteriosclerosis, Thrombosis, and Vascular Biology. 2000;**20**:e127-e133. DOI: 10.1161/01.atv.20.12.e127

[36] Meng W, Mushika Y, Ichii T, Takeichi M. Anchorage of microtubule minus ends to adherens junctions regulates epithelial cell-cell contacts. Cell. 2008;**135**:948-959. DOI: 10.1016/j.cell.2008.09.040

[37] Lee T-YJ, Gotlieb AI. Microfilaments and microtubules maintain endothelial integrity. Microscopy Research and Technique. 2003;**60**:115-127. DOI: 10.1002/jemt.10250

[38] Brieher WM, Yap AS. Cadherin junctions and their cytoskeleton(s). Current Opinion in Cell Biology. 2013;**25**:1-8. DOI: 10.1016/j.ceb.2012.10.010

[39] Komarova YA, Huang F, Geyer M, Daneshjou N, Garcia A, Idalino L, et al. VE-cadherin signaling induces EB3 phosphorylation to suppress microtubule growth and assemble adherens junctions. Molecular Cell. 2012;**48**:914-925. DOI: 10.1016/j.molcel.2012.10.011

[40] Mitchison T, Kirschner M. Dynamic instability of microtubule growth. Nature. 1984;**312**:237-242. DOI: 10.1038/312237a0

[41] Mitchison T, Kirschner M. Microtubule assembly nucleated by isolated centrosomes. Nature. 1984;**312**:232-237. DOI: 10.1038/312232a0

[42] Vorobjev IA, Rodionov VI, Maly IV, Borisy GG. Contribution of plus and minus end pathways to microtubule turnover. Journal of Cell Science. 1999;**112**:2277-2289

[43] Vorobjev I, Malikov V, Rodionov V. Self-organization of a radial microtubule array by dynein-dependent nucleation of microtubules. Proceedings of the National Academy of Sciences of the United States of America. 2001;**98**:10160-10165. DOI: 10.1073/pnas.181354198

[44] Galjart N. Plus-end-tracking proteins and their interactions at microtubule ends. Current Biology. 2010;**20**:R528-R537. DOI: 10.1016/j.cub.2010.05.022

[45] Akhmanova A, Steinmetz MO. Microtubule +TIPs at a glance. Journal of Cell Science. 2010;**123**:3415-3419. DOI: 10.1242/jcs.062414

[46] Gundersen GG, Gomes ER, Wen Y. Cortical control of microtubule stability and polarization. Current Opinion in Cell Biology. 2004;**16**:106-112. DOI: 10.1016/j.ceb.2003.11.010

[47] Carramusa L, Ballestrem C, Zilberman Y, Bershadsky AD. Mammalian diaphanous-related formin Dia1 controls the organization of E-cadherin-mediated cell-cell junctions. Journal of Cell Science. 2007;**120**:3870-3882. DOI: 10.1242/jcs.014365

[48] Efimov A, Kharitonov A, Efimova N, Loncarek J, Miller PM, Andreyeva N, et al. Asymmetric CLASP-dependent nucleation of noncentrosomal microtubules at the trans-golgi network. Developmental Cell. 2007;**12**:917-930. DOI: 10.1016/j.devcel.2007.04.002

[49] Kaverina I, Straube A. Regulation of cell migration by dynamic microtubules. Seminars in Cell & Developmental Biology. 2011;**22**:968-974. DOI: 10.1016/j.semcdb.2011.09.017

[50] Shewan AM, Maddugoda M, Kraemer A, Stehbens SJ, Verma S, Kovacs EM, et al. Myosin 2 is a key Rho kinase target necessary for the local concentration of E-cadherin at cell – cell contacts. 2005;**16**:4531-4542. DOI: 10.1091/mbc.e05-04-0330

[51] Stehbens SJ, Paterson AD, Crampton MS, Shewan AM, Ferguson C, Akhmanova A, et al. Dynamic microtubules regulate the local concentration of E-cadherin at cell-cell contacts. Journal of Cell Science. 2006;**119**:1801-1811. DOI: 10.1242/jcs.02903

[52] Small JV, Kaverina I. Microtubules meet substrate adhesions to arrange cell polarity. Current Opinion in Cell Biology. 2003;**15**:40-47. DOI: 10.1016/s0955-0674(02)00008-x

[53] Efimov A, Schiefermeier N, Grigoriev I, Ohi R, Brown MC, Turner CE, et al. Erratum: Paxillin-dependent stimulation of microtubule catastrophes at focal adhesion sites. Journal of Cell Science. 2008;**121**:405; 196-204. DOI: 10.1242/jcs.012666

[54] Broussard JA, Webb DJ, Kaverina I. Asymmetric focal adhesion disassembly in motile cells. Current Opinion in Cell Biology. 2008;**20**:85-90. DOI: 10.1016/j.ceb.2007.10.009

[55] Efimov A, Kaverina I. Significance of microtubule catastrophes at focal adhesion sites. Cell Adhesion & Migration. 2009;**3**:285-287. DOI: 10.4161/cam.3.3.8858

[56] Alieva IB. Role of microtubule cytoskeleton in regulation of endothelial barrier function. Biokhimiya/Biochemistry. 2014;**79**:964-975. DOI: 10.1134/S0006297914090119

[57] Pawley JB. Handbook of Biological Confocal Microscopy. 3rd ed. Berlin: Springer; 2006. p. 985. DOI: 10.1007/978-0-387-45524-2

[58] Shakhov AS, Dugina VB, Alieva IB. Structural features of actin cytoskeleton required for endotheliocyte barrier function. The Biochemist. 2019;**84**:358-369. DOI: 10.1134/S0006297919040035

[59] Gustafsson MGL, Shao L, Carlton PM, Wang CJR, Golubovskaya IN, Cande WZ, et al. Three-dimensional resolution doubling in wide-field fluorescence microscopy

by structured illumination. Biophysical Journal. 2008;**94**:4957-4970. DOI: 10.1529/biophysj.107.120345

[60] Birukova AA, Smurova K, Birukov KG, Usatyuk P, Liu F, Kaibuchi K, et al. Microtubule disassembly induces cytoskeletal remodeling and lung vascular barrier dysfunction: Role of rho-dependent mechanisms. Journal of Cellular Physiology. 2004;**201**:55-70. DOI: 10.1002/jcp.20055

[61] Stepanova T, Slemmer J, Hoogenraad CC, Lansbergen G, Dortland B, De Zeeuw CI, et al. Visualization of microtubule growth in cultured neurons via the use of EB3-GFP (end-binding protein 3-green fluorescent protein). The Journal of Neuroscience. 2003;**23**:2655-2664. DOI: 10.1523/JNEUROSCI.23-07-02655.2003

Chapter 12

Fluorescent Nanoagents for Biomedical Applications

Dmitry Korolev, Maria Istomina, Anton Belorus, Artem Brovko, Dmitry Sonin, Galina Shulmeyster, Natalya Evreinova and Vyacheslav Moshnikov

Abstract

Fluorescence imaging is a promising method widely used for the evaluation of the biodistribution and accumulation of various fluorescent agents cross-linked to the drug for effective therapy control. This work presents the methods for the functionalization of nanomaterials to provide them with fluorescent properties. The first of these methods is a unique technology for producing porous silicon with fluorescent properties. The second approach demonstrates linking of the fluorophores to inorganic nanoparticles (NP) using a spacer molecule ending with a functional group. For all these examples of fluorophores, biodistribution studies were performed with the fluorescent imaging system IVIS Lumina LT III (PerkinElmer, USA). It was noted that the size of particles and the method of their injection affect the distribution and accumulation in organs. The resulting materials can be used to develop platforms for theranostics.

Keywords: fluorescence, imaging, biodistribution, nanoagents, fluorophores, nanomaterials, theranostics

1. Introduction

Fluorescent imaging is nowadays being widely developed in biology, experimental and practical medicine, and both on cells in vitro and on living organisms in vivo [1]. The main advantages of this method are an increased contrast and the usage of several fluorescent agents working at different wavelengths. Among the other features of the method are low concentrations of agents, simplicity, and relative cheapness of the hardware design.

It happens, often, that the area of localization and distribution of the studied substances after injection is unknown. The application of the fluorescence imaging in preclinical studies allows us to noninvasively track the total biodistribution, localization, and accumulation of the injectable drugs with a fluorescent label complementary to the studied structures [2–7].

Fluorescence tomography has been used in practical medicine for about 50 years. Currently, there are only two fluorescent agents in the world, indocyanine green (ICG) and fluorescein (FLN), as well as their various chemical modifications, that are allowed for clinical use in vivo [8]. Fluorophores such as rhodamines B, 6G, 110, and 123 and thioflavins T and S are widely used in in vitro experiments.

Fluorescein and indocyanine green are primarily used for retinal angiogram investigations, which require an incomparably high dose of the agent. Herewith, the toxic effects of these fluorophores are not observed in vivo [9]. For molecular imaging, a reduced concentration of the fluorescent agent, which has minimal toxicity, is usually used [8].

Recently, active and activated complex fluorescent agents have been employed [10]. Active agents consist of fluorescent dyes conjugated with nanoobjects, enzymes, monoclonal antibodies, etc. Also, the liposomal forms of agents are used [11]. Activated agents contain a fluorophore in their composition, the fluorescence of which is suppressed during the storage and manipulation. The fluorescence radiation is obtained by chemical suppression of the quenching factor [12]. An example of this effect is the dimerization of ICG. In this case, it is possible to destroy dimers that are not capable of fluorescence by interacting with the blood protein—albumin [13].

ICG-labeled monoclonal antibodies to tumor cells allow the visualization of tumors, both in experimental (in vivo and ex vivo tomography) and in clinical practices during surgical interventions [14]. Optical tomography with implemented green indocyanine enables to identify the sentinel lymph nodes and preserving regional lymph nodes without metastases when removing a malignant tumor [15]. Liposomal ICG can be used for the evaluation of the drainage function of the lymphatic system and detect the spread of metastases by lymph drainage [16]. Physical adsorption of IGG and monoclonal antibodies on the free surface of liposomes is possible [17]. In this case, there is an increased quantum yield compared to the free, i.e., unbound ICG. Liposomes with incorporated polyethylene glycol (PEG) conjugate with ICG can be used for photoacoustic tomography [18]. In [19], the authors found that green indocyanine could be embedded in the membrane wall of lipids of liposomes. Herewith, it is possible to store such a drug at a temperature of +4°C for a fairly long period (more than 1 year).

The simplest and well-established nanoscale carriers of fluorophores are silica nanoparticles (NPS) and magnetic nanoparticles (MNP), which can be used in platforms for theranostics [20]. In [21], it was shown that ICG could be encapsulated in a polymer shell, which, in turn, could be electrostatically conjugated with monoclonal antibodies later on. These nanoscale capsules are suitable for both photodynamic therapy and tumor cell differentiation. In [22], nanoparticles of composites containing an amorphous matrix of calcium phosphate and silicon oxide with adsorbed ICG or FLN are shown. These composites were conjugated with monoclonal antibodies and were targeted delivered to various types of tumor cells. It was noted that the resulting composites can overcome the blood–brain barrier (BBB). Silicon oxide nanoparticles with adsorbed ICG could be represented as construction for theranostics since they have fluorescent and photoacoustic contrast and can be used for hyperthermia of tumors. In [23], chitosan-genipin nanoparticles in a PEG shell with inclusions of green indocyanine were developed for hyperthermia of tumors. Along with oncology, photodynamic therapy using ICG can be employed to treat actinic keratosis, infections, and acne [24].

However, the clinical use of ICG in high concentrations can cause side effects such as itching, hives, sweating, headache, and anaphylactic shock [25].

The review [26] shows that in 2011, the number of publications stored in the PubMed system concerning the use of ICG for imaging of tissue ischemia is four times less than the number of works that demonstrated the use of ICG in oncology and six times less than the number of works in the field of ophthalmology. The number of works using ICG in the field of phlebology and plastic surgery is growing rapidly [27].

Another fluorophore, and, at the same time, cardioprotector, that can be used in vivo is quinacrine (QIC), which is also called mepacrine and acrichine. QIC was synthesized in the 1930s of the twentieth century as a fluorescent dye. Its antimalarial and antiparasitic activities were soon discovered. In the last few decades, there have been articles showing a positive effect of mepacrine on postischemic myocardial recovery [28, 29]. When injected intravenously in certain doses (for rats >0.5 mg/kg per min), this fluorophore shows a strong hypotensive effect, so it is preferable to use quinacrine encapsulated in liposomes.

Since the discovery of the photoluminescence phenomena of porous silicon in 1990 [30], luminescent (porous) silicon nanoparticles have been obtained by several methods [31–35]. Some of them can be used biologically [34, 35]. Electrochemically etched porous silicon has a significant potential for biological applications due to its biocompatibility [36], biodegradability [37], encoding for multiplex detection [38], and tunable porous nanostructure and usage for drug delivery [39]. For in vivo applications, silicon nanoparticles have attractive physical, chemical, and biological properties compared to heavy metal-containing quantum dots, which have been shown to be toxic in biological environments. In addition, silicon is an important element in human organism, and the biodegradation product of porous silicon, orthosilicic acid ($Si(OH)_4$), is a form absorbed by the human body and is naturally found in numerous tissues. In addition, silicic acid injected to humans is effectively excreted from the body with the urine [40].

This work presents in vitro and in vivo studies of the fluorescent dyes and luminescent porous silicon nanoparticles (PLPor-Si) using the IVIS Lumina III imager (PerkinElmer, USA). These objects are promising for theranostics applications, due to their ability for targeted drug delivery, and their own luminescence enables to control both the accumulation and the degradation of nanoparticles.

2. Synthesis and functionalization of nanomaterials

2.1 Immobilization of fluorescent dyes on the surface of nanoparticles

Fluorescent dyes are attached to nanoobjects for three different reasons. The first reason is an increased circulation time in the body. The second reason is caused by the fact that nanoobjects, due to their specific size, can stay in the ischemic area [41]. This enables to visualize the damage zone. The third and the most interesting reason is related to the possibility of simultaneous attachment of a fluorophore, a medicinal substance, and a guide ligand to nanoobjects. Such a nanoparticle turns into a theranostic structure [42].

Immobilization of fluorescent dyes on the surface of nanoparticles of any nature is carried out using an insertion—spacer molecule. For the chemical attachment of a fluorophore, the spacer must end with a functional group (**Figure 1**). Sometimes the material of the nanoparticle itself has a functional group, for instance, chitosan nanoparticles [43–45].

In practice, different types of spacers, ending in a functional group, are used. This occurs since the different fluorescent dyes have different functional groups and, hence, can react, that is, superstructure, on fragments of molecules ending in certain functional groups. For example, fluorophores that have acid groups in their composition (carboxyl, sulfo-, sulfonyl, and so on) are better immobilized to a radical that has an amino group, and vice versa, fluorescent dyes containing an amino group easily interact with acid groups. Besides, there are spacers (e.g., glycerine) that can transform and form a covalent bond with a functional group (for instance, amino-). Moreover, they can be both primary and secondary amines.

For the spacer synthesis, its length is of great importance. For instance, the fluorophore, in order to avoid quenching the fluorescence, should be as far as possible remoted from the magnetic nanoparticle. This is regulated by the length of the insertion molecule, i.e., the spacer. To immobilize the fluorophores on magnetic nanoparticles, albumin can be used as a large protein with different functional groups and chitosan as a biopolymer with a large molecular weight (**Figure 2**).

Another popular method to prolong the action period of fluorophores is their encapsulation in liposomes (**Figure 3**). In addition, the liposome itself may contain a drug substance and a targeting ligand and serve as an object of theranostics [46].

Chitosan is one of the most promising biodegradable polymers. Chitosan-based nanoparticles are used for directed drug delivery [43–45]. Systems for the controlled release of medicines are being developed on the chitosan basis as well [47]. There is a fundamental opportunity of using chitosan nanoparticles (NPC) as markers for in vitro and in vivo imaging in those areas of the body, where the vascular permeability changes dramatically. For this purpose, they are coated with fluorescent dyes, such as indocyanine green, fluorescein, or colloidal quantum dots [48, 49].

Chitosan is a natural biopolymer, with every monomer containing a primary amino group. This makes it an extremely profitable material for the coordination-ion immobilization of fluorophores, which are organic acids (fluorescein and indocyanine green). The scheme of their immobilization is shown in **Figure 4**.

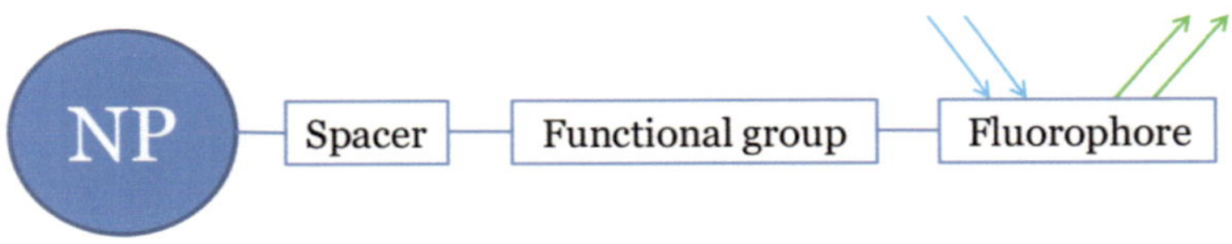

Figure 1.
Scheme of fluorophore immobilization on the nanoparticle surface.

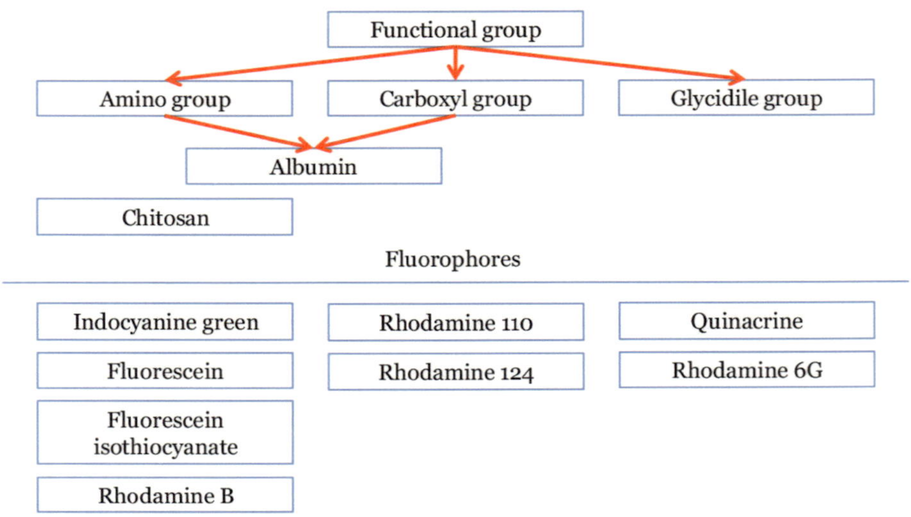

Figure 2.
Classification of functional groups and fluorophores.

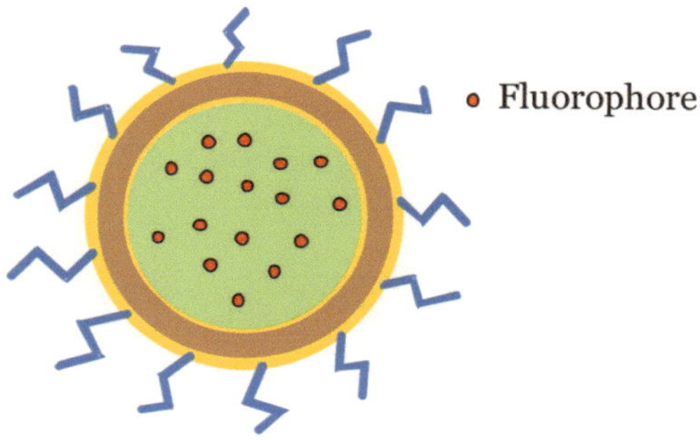

Figure 3.
Fluorophore in a liposomal form.

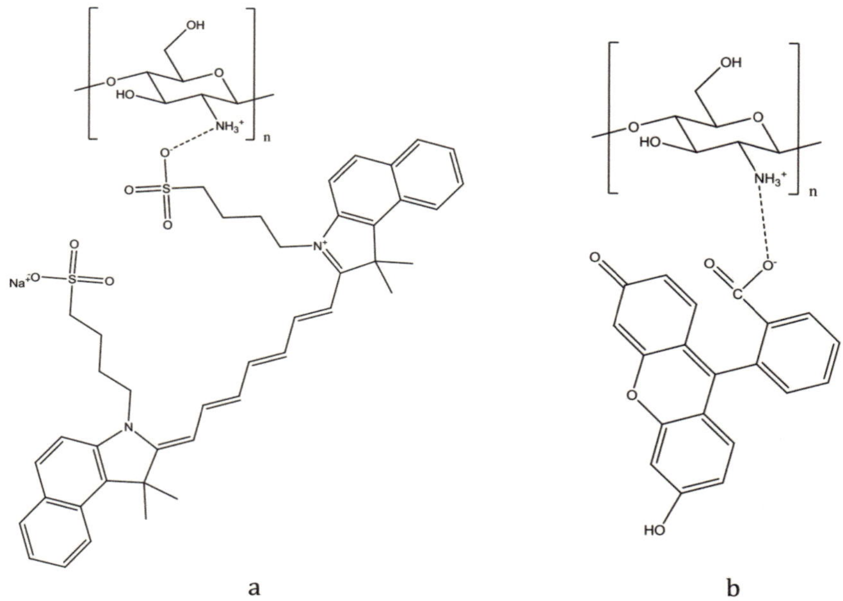

Figure 4.
Chitosan conjugation.

Thus, if one prepares nanoparticles from chitosan, it is possible to immobilize these fluorophores on them and get a fluorescent nanoagent.

If the nanoparticle does not contain a natural functional group, then it is synthesized on the surface.

The traditional method of such synthesis on the surface of functional amino group nanoparticles is the treatment of (3-aminopropyl)triethoxysilane (APTES). The reaction is usually conducted in an organic solvent at its boiling point with an intensive mixing [50]. Common solvents applied for the synthesis are benzene, toluene, cyclohexane, or monoatomic alcohols. APTES reacts with the hydroxylated surface of the nanoparticle (**Figure 5**). Since the hydroxylated surface is inherent for magnetite and silica nanoparticles, these nanoparticles are used for the amination of APTES.

Coordination-ion immobilization of FLN on the APTES-synthesized spacer gives an extremely weak chemical bond, which is hydrolyzed in aqueous solutions. Therefore, fluorescein is immobilized covalently using the carbodiimide method. **Figure 6** shows a method for immobilizing FLN using diisopropylcarbodiimide on a spacer obtained from APTES.

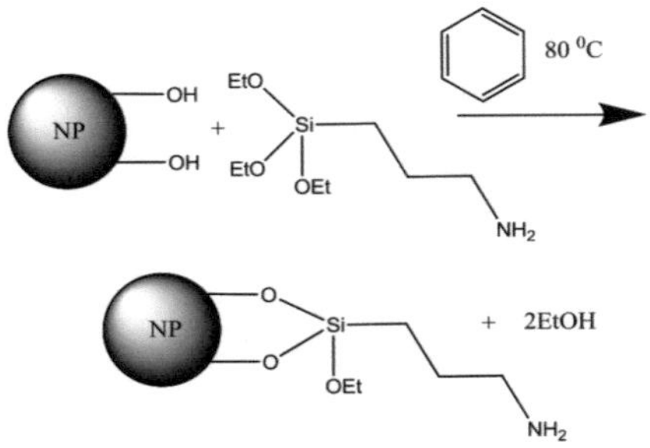

Figure 5.
The scheme of amino-spacer synthesis.

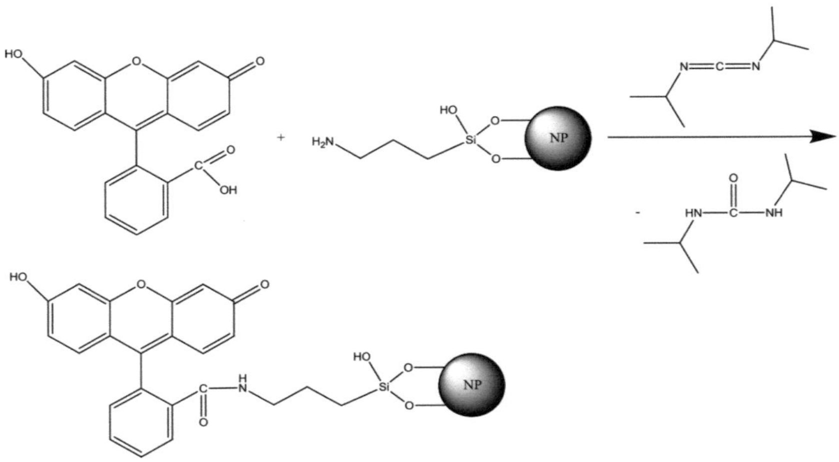

Figure 6.
Scheme of covalent immobilization of fluorescein on an amino spacer using diisopropylcarbodiimide (DIPC).

This is the simplest and most effective way to immobilize fluorescein on the surface of laminated nanoparticles. It is based on the application of the FLN—fluoresceinisothiocyanate (FITC) derivative. By processing nanoparticles with a spacer containing an amino group, under standard conditions in a dimethyl sulfoxide (DMSO) solvent, the amino group forms a covalent bond with the FITC isothiol group (**Figure 7**). The effectiveness of this immobilization is much higher than the reaction using FLN and DIPC. This was confirmed by the fluorescence measurements of the obtained samples. For FITC immobilization, it was 3.44×10^{11} [p/s]/[μW/cm^2] and for FLN immobilization using DIPC 1.01×10^{11} [p/s]/[μW/cm^2], respectively (**Figure 8**).

In contrast, while ICG is immobilized on an amine spacer obtained from APTES, the sulfogroup forms a strong ionic bond with the amine group, which can only be destroyed by an alcoholic solution of alkali. The scheme of such immobilization is demonstrated in **Figure 9**.

2.2 Synthesis of luminescent nanoparticles of porous silicon

Porous silicon layers were obtained by electrochemical etching of a single-crystal, (100)-oriented n-type silicon with a resistivity of 1–4.5 Ω/cm, at a constant current density of 40 mA/cm^2 for 20 min using an aqueous-alcohol electrolyte based on 48% hydrofluoric acid. The obtained porous film was removed by dipping

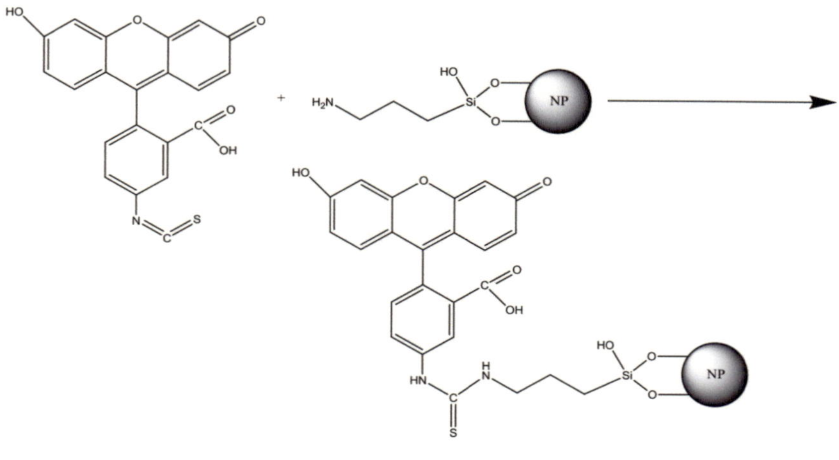

Figure 7.
Scheme of covalent immobilization of fluorescein isothiocyanate on an amino spacer.

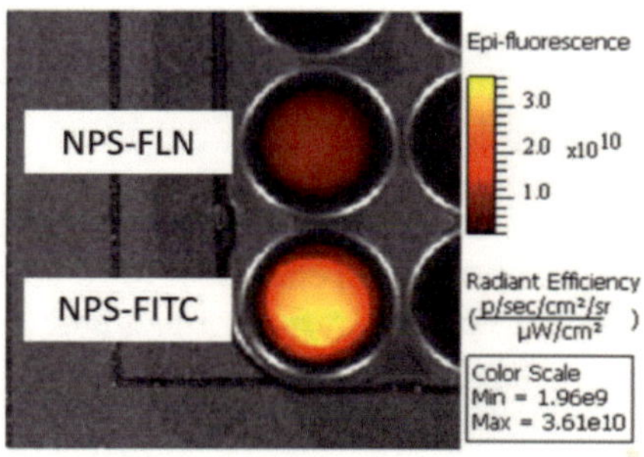

Figure 8.
Comparison of in vitro fluorescence FLN and FITC covalently immobilized on NPS.

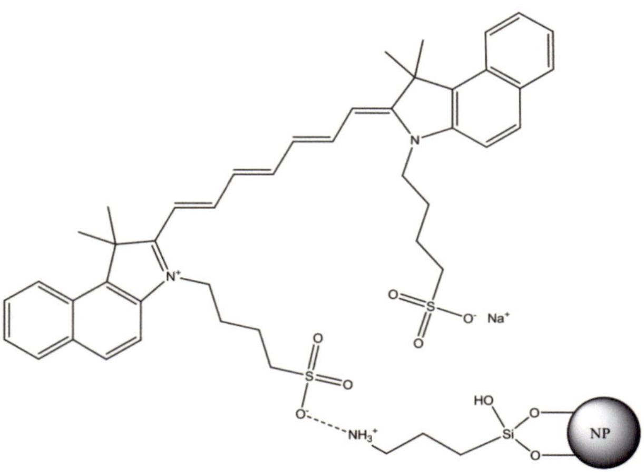

Figure 9.
Scheme of coordination-ion immobilization of green indocyanine on an amino spacer.

the plate in a solution with deionized water, with sequential exposure to ultrasound ("Saphir" ultrasound bath) for 30 min. The resulting suspension was filtered through the syringe filtration membranes (Millipore) with a diameter of less than 0.3 microns. The filtered suspension was further processed using a 600-watt local

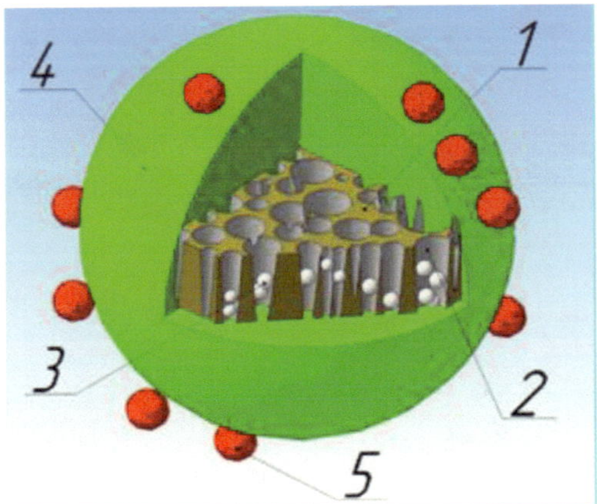

Figure 10.
Model of a system for targeted drug delivery based on photoluminescent porous silicon nanoparticles: (1) PLPor-Si nanoparticles; (2) oxidized layer (SiO_2 layer); (3) dosage form (payload); (4) biopolymer shell (dextran); (5) homing molecules (vitamin B9).

ultrasonic homogenizer (SONOPULS) in an interval pulse mode for 30 min. At the next step, the particles were filtered through a 0.22-μm filtration membrane (Millipore). Afterwards, the nanoparticles were incubated in deionized water for ~2 weeks to activate their photoluminescent glow. To remove the dissolved silicic acids, the resulting dispersion was centrifuged at 13,000 rpm for 30 min, and a supernatant containing silicic acids and nanoparticles smaller than 20 nm in size was removed [51].

2.3 Functionalization of PLPor-Si nanoparticles with gentamicin

The functionalization of photoluminescent nanoparticles of porous silicon with gentamicin (PLPor-Si-Gent) was performed by impregnation: 0.5 ml of gentamicin solution (40 mg/ml) for an intravenous injection was added to 1.5 ml of the dispersion of porous silicon nanoparticles in deionized water. The sample was then subjected to ultrasound treatment for 15 min.

2.4 Functionalization of PLPor-Si nanoparticles with dextran

To cover the obtained photoluminescent nanoparticles derived in previous sections, a dextran shell with 1 ml of dispersion with nanoparticles (PLPor-Si or PLPor-Si-Gent) was used and mixed with 1 ml of aliquot of water containing 100 mg of dextran (MW ~20,000, Sigma). The mixture was stirred for 24 h and washed three times every 8 h using a centrifugal filter (100,000 molecular weight cutoff; Millipore) for the Eppendorf tube at 13,000 rpm for 30 min. The resulting nanoparticles (DPLPor-Si or DPLPor-Si-Gent) were poured into 2 ml Eppendorf tubes for the further injection to male rats of the Wistar stock. An example of a system for targeted drug delivery based on photoluminescent porous silicon nanoparticles is shown in **Figure 10**.

3. Characterization and study of fluorescence of PLPor-Si nanoparticles

Scanning electron microscopy (SEM) was used to control the morphology and the particle size of photoluminescent porous silicon using the Mira Tescan II scanning electron microscope. The research was carried out with the detection of secondary and reflected electrons; the typical values of the accelerating voltage were 5–20 kV. Typical magnification values were in the range of 10,000–200,000. The control of the biodistribution and the study of the morphology of por-Si particles after internalization of the rats were performed on the liver and myocardial tissue fragments. Tissue samples were dried, and a thin layer of carbon was applied to the surface of the samples immediately before the SEM investigation [52].

To control and to verify the presence of photoluminescent PLPor-Si nanoparticles in the tissues after internalization, several experiments were previously performed on the injection of unloaded shell-free carriers (PLPor-Si) using a method similar to that for carrier particles loaded with medication (PLPor-Si-Gent) and coated with dextran (DPLPor-Si and DPLPor-Si-Gent) [53]. The study of the fragments of the rat's liver and heart tissues using the SEM method after an intravenous injection of por-Si particles (from the first set) revealed the presence of electron-dense particles with a shape close to spherical. It is worth noticing that the distribution of the particles in the tissues is uneven and the clusters of particles are observed near to the openings of the large ducts (**Figure 11**). The diameter of the detected particles corresponds well to the range of diameters of the first set of the particles: 60–80 nm the main part of the particles, with some larger (up to 185 nm) particles and a fraction of the smaller particles (≈30–40 nm).

The photon cross-correlation method (PCCS) was employed to determine the size of nanoparticles obtained using the Nanophox analyzer from Sympatec (Germany). Before the dispersion measurement procedure, the samples were further broken up with a 600-W ultrasonic homogenizer for 5 min. Later on, the solutions with a mechanical dispenser were taken and poured into the cuvette of the analyzer. The concentration of solutions was selected in a way that the average intensity of the scattered light for each sample was the same. The absence of multiple light scattering was visually controlled. The processing was performed with the DynaLS software: correlator channels from 30 to 200 and search boundaries from 1 to 10,000 nm. **Table 1** shows the size distribution of the nanoparticles from sets 1 to 4 determined by photon cross-correlation spectroscopy.

The study of the photoluminescence of the samples of porous silicon nanoparticles (PLPor-Si, PLPor-Si-Gent, DPLPor-Si, DPLPor-Si-Gent) obtained in the work was conducted at room temperature using an automated spectral complex assembled based on the MDR_23 and FEU_79 monochromator operating in the

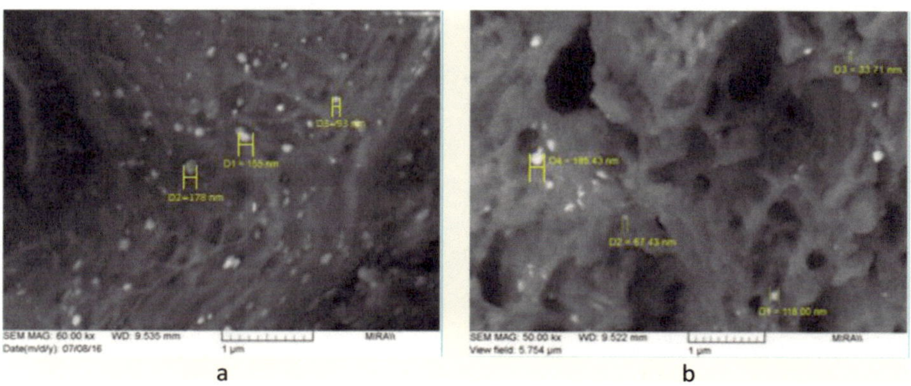

Figure 11.
Rat's heart tissue (myocardium) (a) and liver (b) at a single magnification. Diffuse distribution of electron-dense centers with the size corresponding to the size of the particles from injected dispersion PLPor-Si of the first set (80–190 nm).

photon-counting mode. Photoluminescence was excited by the radiation with a wavelength of 405 nm. **Figure 12** shows the photoluminescence spectra obtained during the experiment 5 days after receiving.

One can see that the process of functionalization of nanoparticles (PLPor-Si) leads to a redistribution of intensity in the photoluminescence band of the initial samples from the first set with almost complete preservation of its position and

Number of the sets		Synthesis conditions				Diameter of the particles (nm)	
		Silicon type	J_A, mA/cm^2	t_A, min	t_{US}, min	Value	Measuring method
1	PLPor-Si	n-Si	40	20	30	≈20–190	PCCS SEM
2	DPLPor-Si					≈250–300	PCCS
3	PLPor-Si-Gent					≈180–250	
4	DPLPor-Si-Gent					≈250–330	

Table 1.
Conditions of the synthesis and the parameters of the silicon porous nanoparticles.

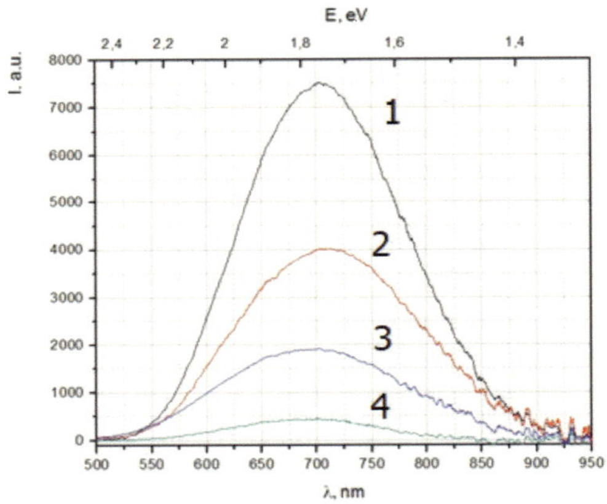

Figure 12.
The photoluminescence spectra: (1) PLPor-Si; (2) DPLPor-Si; (3) PLPor-Si-gent; and (4) DPLPor-Si-gent.

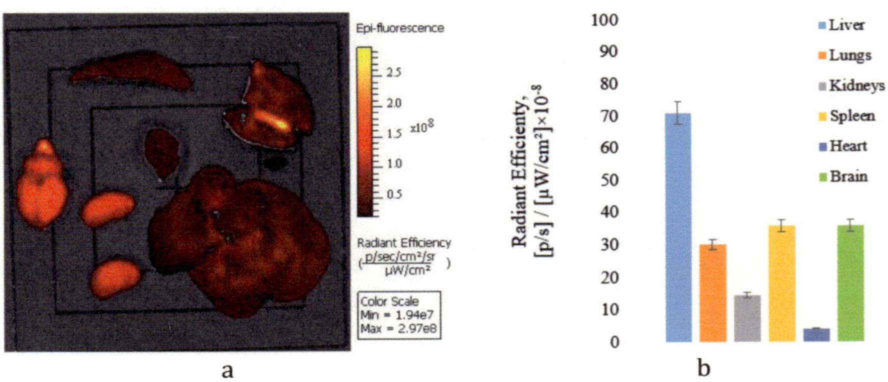

Figure 13.
Detection of PLPor-Si (second series) ex vivo (a); fluorescence intensity of the rat's organs (b).

width. Coating of the second set of samples with the dextran-2 (DPLPor-Si) drives to the formation of a thicker surface layer, which, in turn, causes a decrease in the intensity of FL almost twice as compared to the original samples from the first set. The functionalization of the surface of nanoparticles by the drug (gentamicin) third set (PLPor-Si-Gent) leads to a more intensive passivation of the surface by the drug molecules, which reduces the intensity of the photoluminescence line by more than three times. The combination of drug and dextran functionalization methods for the fourth set of samples (DPLPor-Si-Gent) reduces the luminescence intensity by an order of magnitude, due to a significant increase in the surface layer and passivation of the surface by drug molecules.

The fluorescent emission was measured using a fluorescent imager IVIS Lumina LT III (PerkinElmer, USA). To investigate the overall biodistribution and potential application of DPLPor-Si, a 1-ml solution with 0.5 mg/ml concentration of DPLPor-Si was injected into the tail vein of a Wistar stock rat. After 24 h from the moment of injection, the organs were removed for evaluation. Fluorescent emission was recorded on the standard embedded filters (for excitation, 500 nm, and for emission, 650 nm). **Figure 13** demonstrates the fluorescent analysis of ex vivo organ distribution in a male Wistar rat 24 h after an intravenous injection of the nanoparticles from second set.

Ex vivo organ biodistribution presents a steady accumulation of DPLPor-Si nanoparticles in the liver (**Figure 13b**). It is important to mention that after the intravenous injection, DPLPor-Si penetrates the blood–brain barrier and there is retention of the particles in the brain. Such particles have great potential for research in the fields of neurology and neurosurgery.

4. Use of fluorophore-labeled nanoparticles for fluorescent imaging in preclinical studies

Fluorescence of nanoparticles was measured using a fluorescence imager. Herein, **Figure 14a** shows a photograph of fluorescein (left) and indocyanine green (right) immobilized on chitosan microparticles. As can be seen, the fluorophores are completely chemosorbed on the particles, and the chemical bond with chitosan amino groups is not hydrolyzed, which is confirmed by the absence of a free dye in the solution.

Figure 14b reveals a samples' image taken with the IVIS Lumina LT fluorescent tomography in "spectral unmixed" mode. This mode allows to take multiple frames with different excitation wavelengths and various registration cutoff filters with a consequent combination as a result. Samples are shown from left to right: chitosan particles without fluorophore, particles with FLN, and particles with ICG. To induce fluorescence the embedded light filters were used (for FLN, 465 nm, and for ICG, 745 nm). Registration was performed with built-in FITS and ICG filters, respectively, suitable for the release of induced radiation of the corresponding fluorescent dyes and following the documentation for the device.

The use of nanoparticles labeled with different fluorophores enables to simultaneously or sequentially introduce them into the body of laboratory animals and study their biodistribution. Moreover, particles with different size and origin can be labeled with different fluorophores and examined simultaneously. As a demonstration of this, a Petri dish with the organs of laboratory animals obtained on a fluorescent tomography in the "spectral unmixed" mode is shown in **Figure 15**. Before euthanasia and removal of the organs, the animal underwent an intravenous injection of the suspension with the nanoparticles of chitosan labeled with fluorescein and indocyanine green consistently.

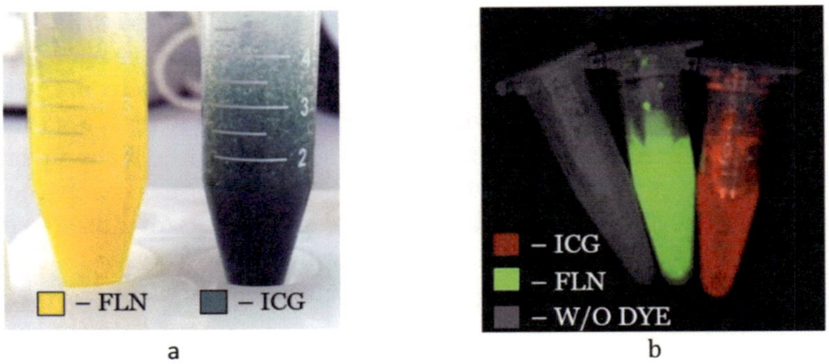

Figure 14.
IImage of chitosan microparticles with immobilized fluorescein (FLN) and indocyanine green (ICG): (a) in daylight; (b) on a fluorescent tomography of samples in comparison with a control sample without dye in the "spectral unmixed" mode.

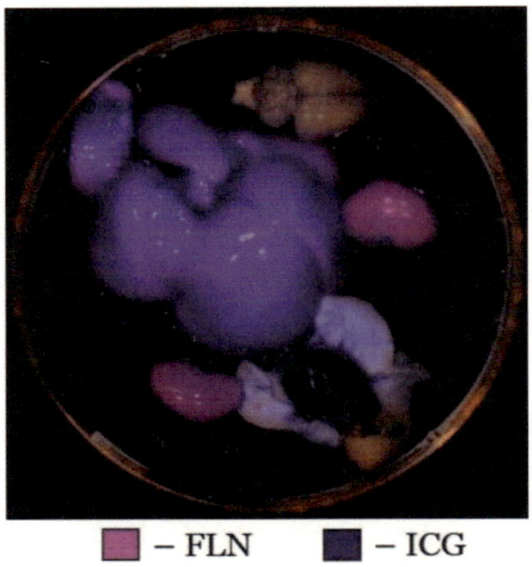

Figure 15.
Rat's organs, after the introduction of chitosan nanoparticles labeled with fluorescein and indocyanine green; image from a fluorescent tomography in the "spectral unmixed" mode.

It is worth noticing that nanoparticles labeled with fluorescein are mainly visible in the kidneys while particles labeled with green indocyanine are localized in the liver and lungs. Chitosan nanoparticles labeled with fluorophores do not overcome the blood–brain barrier. At the same time, one can see that only the autofluorescence inherent to NADH is visible in the brain.

The biodistribution of chitosan nanoparticles in vivo was evaluated over a period of 30 min. The following time intervals were selected: initial state, 5, 10, 15, 20, 25, and 30 min. After catheterization of the femoral vein of the laboratory animal, NPC-ICG was injected. In other words, the image registration in each animal was performed initially, before the introduction of suspensions, 5 min after the introduction, and then every 5 min for half an hour (**Figure 16a**). Afterwards, the animals were removed from the experiment by increasing the dose of anesthetic and removing the organs for examination: heart, lungs, liver, spleen, and kidneys.

The evaluation of the biodistribution based on the obtained fluorescence in vivo from the surface of the entire animal body revealed a maximum accumulation of the nanoparticles had occurred only in one area of the body (**Figure 16a**) with the increase of the intensity at the same area with time (**Figure 16b**). An autopsy of the animal and the removal of the listed organs, which were placed in a Petri dish with

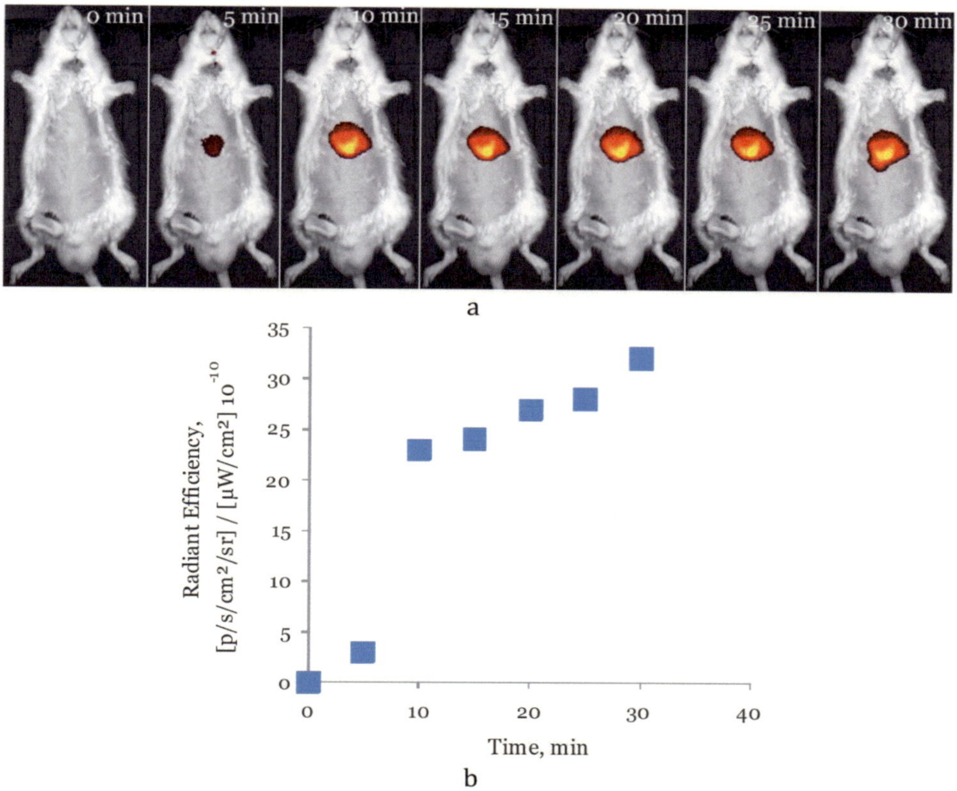

Figure 16.
Fluorescent imaging in vivo distribution NPC-ICG administered intravenously (a); the dependence of the fluorescence intensity distribution on time (b).

consequent analysis on a fluorescent imager, were performed. It was found that the greatest accumulation of NPC-ICG particles occurred in the liver (**Figure 17**). As expected, the data indicates the elimination of nanoobjects by the liver and confirms the previously obtained data using histochemical staining by Grocott on the predominant accumulation of NPC in the rats' liver [54].

Male rats from Wistar were used to estimate the overall biodistribution of NPS-FLN and liposomal quinacrine drug (LQD). NPS-FLN and LQD solutions with a volume of 2 ml were injected intravenously through the tail vein; then the animals were anesthetized with isoflurane anesthesia and placed in the fluorescent analyzer. As for negative control, a rat given with an intravenous saline solution of 2 ml volume was used. The in vivo visualization of the biodistribution was recorded every 5 min for 30 min (**Figure 18**). After removing the animals from the experiment by humane euthanasia with an overdose of inhaled anesthesia of isoflurane, organ sampling was performed (**Figure 19**) for ex vivo quantification of particle accumulation. The fluorescence intensity from the area of interest (ROI) was calculated using the Living Image software (Living Image 4.5.5, PerkinElmer Inc.).

After the evaluation of the overall in vivo distribution of NPS-FLN nanoparticles over time, no local areas of fluorescence accumulation were found. The same situation was observed in the case of the LQD sample. As expected, that happened because fluorescein (ex, 490 nm; em, 525 nm) and quinacrine (ex, 490 nm; em, 525 nm) had the same wavelength of radiation, which was damped by autofluorescence of the biological tissues. Therefore, such fluorescent dyes are more often used for ex vivo and in vitro evaluation.

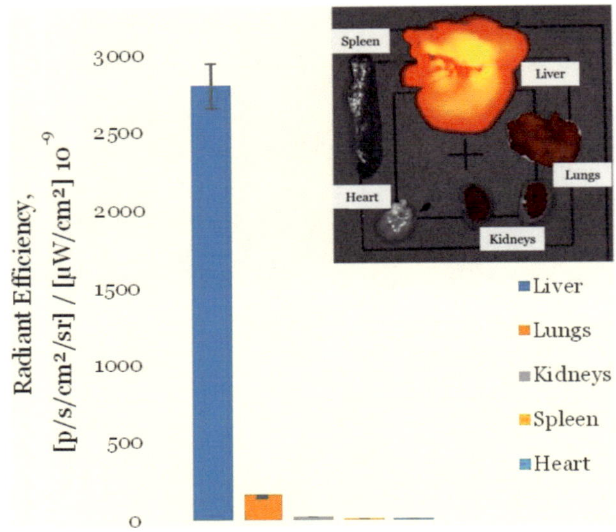

Figure 17.
Images ex vivo and fluorescence intensity of the five major organs, liver, lungs, kidneys, spleen, and heart seized from the rat with introduced NPS-ICG.

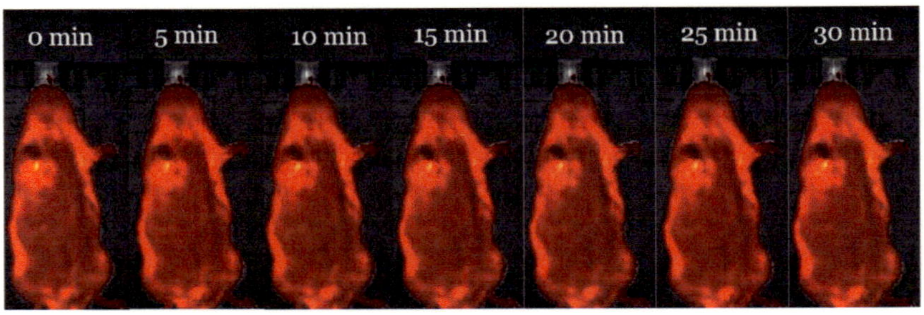

Figure 18.
Fluorescent imaging in vivo distribution NPS-FLN administered intravenously.

The quantitative biodistribution of the intensity of the fluorescent glow NPS-FLN and LQD in the bodies is illustrated in **Tables 2** and **3**. The accumulation of NPS-FLN particles mainly occurs in the liver, lungs, spleen, and kidneys. In the case of LQD, accumulation was observed in the lungs, which is justified by the size of the liposomes themselves, the average size of which is 1 μm. The accumulation or deposition of liposomal points occurs by mechanical filtration through the capillary channel of the lung after an intravenous injection [55, 56].

The NPC-ICG biodistribution study (**Figure 20**) was conducted similarly to the one presented above. Another fluorescent dye (ICG) with the emission outside the autofluorescence of the tissue range was used, so the emission falls into the transparency window of biological objects [57–59].

Depending on the method of intravenous injection of nanoparticles conjugated with the fluorescent dye ICG, a significant difference in the estimation of the fluorescence intensity of the total biodistribution over time is observed. For example, when the NPC-ICG conjugate is directly inserted into the catheterized femoral artery, there is an increase in the intensity of the fluorescent glow in the liver region over time (**Figure 16a**). On the other hand, intravenous injection through the tail vein (**Figure 20**) causes a general glow of the entire surface of the animal. After 5 min, and then during the entire experiment, an accumulation of nanoparticles

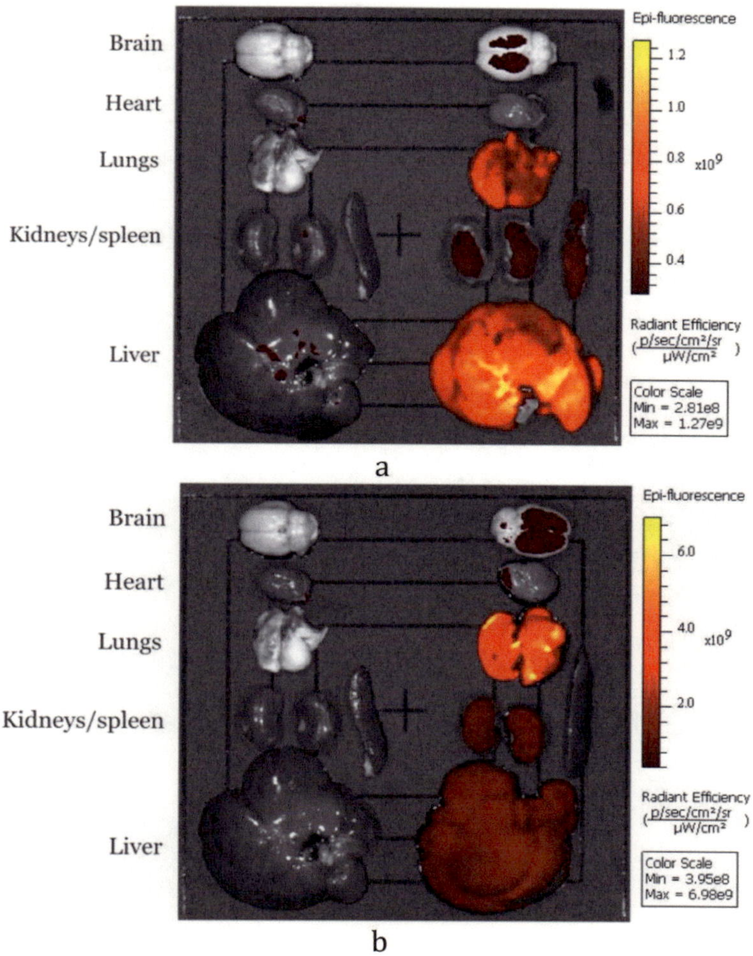

Figure 19.
Ex vivo fluorescence intensity of the five major organs, liver, lungs, kidneys, spleen, and heart seized from the rat with introduced NPS-FLN (a) and LQD (b).

Organ	Total radiant efficiency [p/s]/[μW/cm^2] × 10^{-9}		
	Control	NPS-FLN	The difference with the control
Brain	1.56	6.55	4.99
Heart	0.94	1.25	0.31
Lungs	1.56	25.50	23.94
Kidneys	3.01	5.23	2.22
Spleen	0.82	6.72	5.90
Liver	25.00	108.00	83.00

Table 2.
Total radiant efficiency of organs with control and with introduced NPS-FLN.

in the liver occurs, and a decrease in the intensity of fluorescence (**Figure 20b**) takes place. As shown in the previous experiment on NPS-ICG biodistribution, the accumulation of such conjugates occurs in the liver, lungs, kidneys, and spleen.

Organ	Total radiant efficiency [p/s]/[μW/cm^2] × 10^{-9}		
	Control	LQD	The difference with the control
Brain	4.39	14.10	9.71
Heart	1.16	3.91	2.75
Lungs	1.56	169.00	167.44
Kidneys	2.95	16.40	13.45
Spleen	1.02	1.97	0.95
Liver	27.80	207.00	179.2

Table 3.
Total radiant efficiency of organs with control and with introduced LQD.

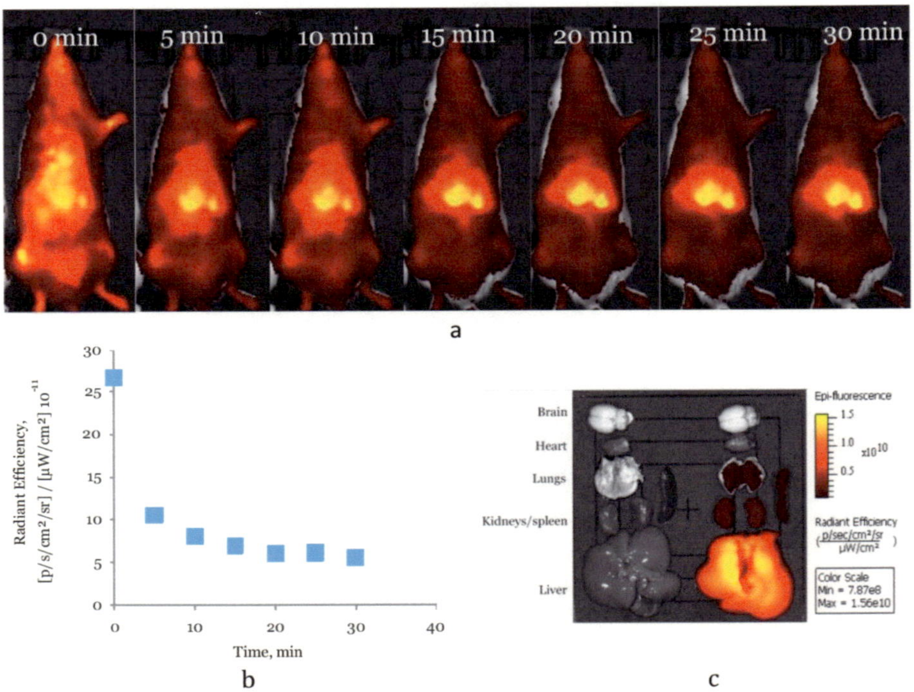

Figure 20.
Fluorescent imaging in vivo distribution of NPS-ICG administered intravenously (a); the dependence of the fluorescence intensity distribution on time (b); ex vivo biodistribution of NPS-ICG (c).

Organ	Total radiant efficiency [p/s]/[μW/cm^2] × 10^{-9}		
	Control	NPS-ICG	The difference with the control
Brain	3.01	4.47	1.46
Heart	1.55	2.62	1.07
Lungs	5.85	46.40	40.55
Kidneys	1.02	32.40	31.38
Spleen	2.67	21.30	18.63
Liver	12.80	2240.00	2227.20

Table 4.
Total radiant efficiency of organs with control and with introduced NPS-ICG.

Ex vivo biodistribution in organs shows the accumulation of the particles mostly in the liver. The numerical distribution of NPS-ICG accumulation is shown in **Table 4**.

It is important to mention that nanoparticles conjugated with the fluorescent dye ICG are suitable for in vivo visualization of biodistribution and accumulation in the area of interest since its emission lies in the transparency window of biological tissues [60, 61]. Further modification and creation of agents for theranostic based on such dyes is promising for fluorescence tomography in preclinical studies.

5. Conclusion

Nanomaterials conjugated with fluorophores have a great perspective for application of in vitro and in vivo fluorescence tomography in preclinical studies on laboratory animals. In comparison with pure fluorophores, they have several advantages. First, these conjugates have the effect of prolongation allowing performing a visualization for at least 30 min. Non-bound fluorophores can metabolize within 1–2 min. Second, the usage of nanoparticles of different nature and physical size enables the visualization of different organs, due to their accumulation in different parts of the body. Finally, such conjugates can be used to create platforms for theranostics, which will suit well for the early diagnosis and therapy for use in personalized medicine.

Local drug delivery systems based on the photoluminescent nanoparticles of porous silicon (PLPor-Si) obtained by electrochemical anodic etching were developed. The developed functionalization methods allowed the encapsulation of the photoluminescent nanoparticles on a porous silicon basis with the drug gentamicin (PLPor-Si-Gent) and obtaining the shell systems for local drug delivery using a biocompatible dextran polymer (DPLPor-Si or DPLPor-Si-Gent).

Photon cross-correlation spectroscopy applied to a series of porous silicon samples before the internalization determined the average size of the particles. For instance, for the samples of first series (PLPor-Si), the average size of the hydrodynamic diameter of the studied nanoparticle dispersions was determined by a range from 20 to 190 nm, which is confirmed by the measurements of the scanning electron microscopy of this series. The analysis of the fragments of the rat's liver and heart tissue performed by the SEM method after an intravenous injection of the first set nanoparticles yielded the presence of electron-dense particles with a shape close to spherical. The distribution in the tissues was nonuniform with the clusters observed near to the openings of the large ducts. The diameter of the detected particles corresponds well to the range of diameters of the first set particles: the major part of the particles (60–80 nm), with some larger (up to 185 nm) particles and a fraction of smaller particles (≈30–40 nm).

The functionalization processes of the porous silicon nanoparticles with the molecules of the drug gentamicin and the dextran-based shell structure led to an increase in the average hydrodynamic diameter of the studied samples of second, third, and fourth sets. In this case, smaller particles aggregated to the following systems (herein the range of average hydrodynamic diameters): 250–300, 180–250, and 250–330 nm for second (DPLPor-Si), third (PLPor-Si-Gent), and fourth (DPLPor-Si-Gent) sample sets, respectively.

The photoluminescence spectra of the samples obtained in this work showed a process of functionalization of the nanoparticles (PLPor-Si). It drove to a redistribution of intensity in the photoluminescence band of the initial samples from set 1 with almost complete preservation of its position and width, at which the peak of the spectra fell in the wavelength range from 700 to 710 nm.

The samples coated with the dextran-2 (DPLPor-Si) form a thicker surface layer, which results in a decrease in the intensity of FL almost twice as compared to the original first set. The functionalization of the surface of nanoparticles with the drug (gentamicin) third sets (PLPor-Si-Gent) entails the more intensive passivation of the surface by the drug molecules, which reduces the intensity of the photoluminescence line by more than (a) factor of 3. The combination of drug and dextran functionalization methods for samples from the fourth set (DPLPor-Si-Gent) dramatically reduces the luminescence intensity (by an order of magnitude), due to a significant increase in the surface layer and passivation of the surface by drug molecules.

The analysis of the biodistribution in the organs of male rats of the Wistar stock obtained by fluorescence spectroscopy of the samples from the second set after the internalization process showed an intense accumulation of DPLPor-Si nanoparticles in the liver, which in turn correlates with the data of scanning electron microscopy for these samples. After intravenous injection, DPLPor-Si penetrates the blood–brain barrier, and the particles concentrate in the brain.

Thus, the resulting local drug delivery systems which are able to circulate in the body have a huge potential for the diagnosis and treatment of various diseases. Such systems have a vast list of advantages like the ability to safely remove from the body for a given period of time after they perform their diagnostic or therapeutic function, the capability to be detected in vivo, a low level of toxicity, and a potential to overcome various complex barriers within the body, such as histohematic.

Conflict of interest

The authors declare no conflict of interest.

Author details

Dmitry Korolev[1,2], Maria Istomina[1,3]*, Anton Belorus[3], Artem Brovko[4], Dmitry Sonin[1,2], Galina Shulmeyster[1], Natalya Evreinova[1,5] and Vyacheslav Moshnikov[3]

1 Almazov National Medical Research Centre, Saint Petersburg, Russia

2 Pavlov First Saint Petersburg State Medical University, Saint Petersburg, Russia

3 Saint Petersburg Electrotechnical University "LETI", Saint Petersburg, Russia

4 Tel Aviv University, Tel Aviv, Israel

5 Saint Petersburg State Institute of Technology, Saint Petersburg, Russia

*Address all correspondence to: ist_maria@mail.ru

References

[1] Wu Q, Merchant FA, Castleman KR. Microscope Image Processing. New York: Academic Press; 2008

[2] Pic E, Pons T, Bezdetnaya L, Leroux A, Guillemin F, Dubertret B, et al. Fluorescence imaging and whole-body biodistribution of near-infrared-emitting quantum dots after subcutaneous injection for regional lymph node mapping in mice. Molecular Imaging and Biology. 2010;**12**:394-405. DOI: 10.1007/s11307-009-0288-y

[3] Walsh EM, Cole D, Tipirneni KE, Bland KI, Udayakumar N, Kasten BB, et al. Fluorescence imaging of nerves during surgery. Annals of Surgery. 2019;**270**:69-76. DOI: 10.1097/SLA.0000000000003130

[4] Jung JS, Jo D, Jo G, Hyun H. Near-infrared contrast agents for bone-targeted imaging. Tissue Engineering and Regenerative Medicine. 2019;**16**:443-450. DOI: 10.1007/s13770-019-00208-9

[5] Lichtman JW, Conchello JA. Fluorescence microscopy. Nature Methods. 2005;**2**:910-919. DOI: 10.1038/nmeth817

[6] Rao J, Dragulescu-Andrasi A, Yao H. Fluorescence imaging in vivo: Recent advances. Current Opinion in Biotechnology. 2007;**18**:17-25. DOI: 10.1016/j.copbio.2007.01.003

[7] Etrych T, Lucas H, Janoušková O, Chytil P, Mueller T, Mäder K. Fluorescence optical imaging in anticancer drug delivery. Journal of Controlled Release. 2016;**226**:168-181. DOI: 10.1016/j.jconrel.2016.02.022

[8] Kobayashi H, Ogawa M, Alford R, Choyke PL, Urano Y. New strategies for fluorescent probe design in medical diagnostic imaging. Chemical Reviews. 2010;**110**:2620-2640. DOI: 10.1021/cr900263j

[9] Butner RW, McPherson AR. Adverse reactions in intravenous fluorescein angiography. Annals of Ophthalmology. 1983;**15**:1084-1086

[10] Ntziachristos V. Fluorescence molecular imaging. Annual Review of Biomedical Engineering. 2006;**8**:1-33. DOI: 10.1146/annurev.bioeng.8.061505.095831

[11] Neri D, Carnemolla B, Nissim A, Leprini A, Querzè G, Balza E, et al. Targeting by affinity-matured recombinant antibody fragments of an angiogenesis associated fibronectin isoform. Nature Biotechnology. 1997;**15**:1271-1275. DOI: 10.1038/nbt1197-1271

[12] Weissleder R, Tung CH, Mahmood U, Bogdanov A. In vivo imaging of tumors with protease-activated near-infrared fluorescent probes. Nature Biotechnology. 1999;**17**:375-378. DOI: 10.1038/7933

[13] Sonin D, Papayan G, Pochkaeva E, Chefu S, Minasian S, Kurapeev D, et al. In vivo visualization and ex vivo quantification of experimental myocardial infarction by indocyanine green fluorescence imaging. Biomedical Optics Express. 2017;**8**:151. DOI: 10.1364/boe.8.000151

[14] Ogawa M, Kosaka N, Choyke PL, Kobayashi H. In vivo molecular imaging of cancer with a quenching near-infrared fluorescent probe using conjugates of monoclonal antibodies and indocyanine green. Cancer Research. 2009;**69**:1268-1272. DOI: 10.1158/0008-5472.CAN-08-3116

[15] Papayan GV, Akopov AL. Fluorescence diagnostics in the near-IR: Apparatus, application. Journal of Optical Technology. 2016;**83**:536. DOI: 10.1364/jot.83.000536

[16] Proulx ST, Luciani P, Derzsi S, Rinderknecht M, Mumprecht V, Leroux JC, et al. Quantitative imaging of lymphatic function with liposomal indocyanine green. Cancer Research. 2010;**70**:7053-7062. DOI: 10.1158/0008-5472.CAN-10-0271

[17] Portnoy E, Lecht S, Lazarovici P, Danino D, Magdassi S. Cetuximab-labeled liposomes containing near-infrared probe for in vivo imaging. Nanomedicine: Nanotechnology, Biology, and Medicine. 2011;**7**:480-488. DOI: 10.1016/j.nano.2011.01.001

[18] Beziere N, Lozano N, Nunes A, Salichs J, Queiros D, Kostarelos K, et al. Dynamic imaging of pegylated indocyanine green (ICG) liposomes within the tumor microenvironment using multi-spectral optoacoustic tomography (MSOT). Biomaterials. 2015;**37**:415-424. DOI: 10.1016/j.biomaterials.2014.10.014

[19] Kraft JC, Ho RJY. Interactions of indocyanine green and lipid in enhancing near-infrared fluorescence properties: The basis for near-infrared imaging in vivo. Biochemistry. 2014;**53**:1275-1283. DOI: 10.1021/bi500021j

[20] Muthu MS, Leong DT, Mei L, Feng SS. Nanotheranostics—Application and further development of nanomedicine strategies for advanced theranostics. Theranostics. 2014;**4**:660-677. DOI: 10.7150/thno.8698

[21] Yu J, Javier D, Yaseen MA, Nitin N, Richards-Kortum R, Anvari B, et al. Self-assembly synthesis, tumor cell targeting, and photothermal capabilities of antibody-coated indocyanine green nanocapsules. Journal of the American Chemical Society. 2010;**132**:1929-1938. DOI: 10.1021/ja908139y

[22] Barth BM, Sharma R, Altinoğlu EI, Morgan TT, Shanmugavelandy SS, Kaiser JM, et al. Bioconjugation of calcium phosphosilicate composite nanoparticles for selective targeting of human breast and pancreatic cancers in vivo. ACS Nano. 2010;**4**:1279-1287. DOI: 10.1021/nn901297q

[23] Song X, Wu H, Li S, Wang Y, Ma X, Tan M. Ultrasmall chitosan-genipin nanocarriers fabricated from reverse microemulsion process for tumor photothermal therapy in mice. Biomacromolecules. 2015;**16**:2080-2090. DOI: 10.1021/acs.biomac.5b00511

[24] Giraudeau C, Moussaron A, Stallivieri A, Mordon S, Frochot C. Indocyanine green: Photosensitizer or chromophore? Still a debate. Current Medicinal Chemistry. 2014;**21**:1871-1897. DOI: 10.2174/0929867321666131218095802

[25] Kaiser M, Yafi A, Cinat M, Choi B, Durkin AJ. Noninvasive assessment of burn wound severity using optical technology: A review of current and future modalities. Burns. 2011;**37**:377-386. DOI: 10.1016/j.burns.2010.11.012

[26] Alander JT, Kaartinen I, Laakso A, Pätilä T, Spillmann T, Tuchin VV, et al. A review of indocyanine green fluorescent imaging in surgery. International Journal of Biomedical Imaging. 2012;**2012**:1-26. DOI: 10.1155/2012/940585

[27] Marshall MV, Rasmussen JC, Tan I-C, Aldrich MB, Adams KE, Wang X, et al. Near-infrared fluorescence imaging in humans with indocyanine green: A review and update. The Open Surgical Oncology Journal. 2020;**2**:12-25. DOI: 10.2174/1876504101002010012

[28] Otani H, Engelman RM, Breyer RH, Rousou JA, Lemeshow S, Das DK. Mepacrine, a phospholipase inhibitor. A potential tool for modifying myocardial reperfusion injury. Journal

of Thoracic and Cardiovascular Surgery. 1986;**92**:247-254. DOI: 10.1016/S0022-5223(19)35905-7

[29] van Bilsen M, van der Vusse GJ, Willemsen PHM, Coumans WA, Roemen THM, Reneman RS. Effects of nicotinic acid and mepacrine on fatty acid accumulation and myocardial damage during ischemia and reperfusion. Journal of Molecular and Cellular Cardiology. 1990;**22**:155-163. DOI: 10.1016/0022-2828(90)91112-K

[30] Canham LT. Silicon quantum wire array fabrication by electrochemical and chemical dissolution of wafers. Applied Physics Letters. 1990;**57**:1046-1048. DOI: 10.1063/1.103561

[31] Heinrich JL, Curtis CL, CREDO GM, Sailor MJ, Kavanagh KL. Luminescent colloidal silicon suspensions from porous silicon. Science. 1992;**255**:66-68. DOI: 10.1126/science.255.5040.66

[32] Wilson WL, Szajowski PF, Brus LE. Quantum confinement in size-selected, surface-oxidized silicon nanocrystals. Science. 1993;**262**:1242-1244. DOI: 10.1126/science.262.5137.1242

[33] Mangolini L, Kortshagen U. Plasma-assisted synthesis of silicon nanocrystal inks. Advanced Materials. 2007;**19**:2513-2519. DOI: 10.1002/adma.200700595

[34] Wang L, Reipa V, Blasic J. Silicon nanoparticles as a luminescent label to DNA. Bioconjugate Chemistry. 2004;**15**:409-412. DOI: 10.1021/bc030047k

[35] Li ZF, Ruckenstein E. Water-soluble poly(acrylic acid) grafted luminescent silicon nanoparticles and their use as fluorescent biological staining labels. Nano Letters. 2004;**4**:1463-1467. DOI: 10.1021/nl0492436

[36] Bayliss SC, Heald R, Fletcher DI, Buckberry LD. The culture of mammalian cells on nanostructured silicon. Advanced Materials. 1999;**11**:318-321. DOI: 10.1002/(SICI)1521-4095(199903)11:4<318::AID-ADMA318>3.0.CO;2-Z

[37] Canham LT. Bioactive silicon structure fabrication through nanoetching techniques. Advanced Materials. 1995;**7**:1033-1037. DOI: 10.1002/adma.19950071215

[38] Cunin F, Schmedake TA, Link JR, Li YY, Kom J, Bhatia SN, et al. Biomolecular screening with encoded porous-silicon photonic crystals. Nature Materials. 2002;**1**:39-41. DOI: 10.1038/nmat702

[39] Salonen J, Kaukonen AM, Hirvonen J, Lehto VP. Mesoporous silicon in drug delivery applications. Journal of Pharmaceutical Sciences. 2008;**97**:632-653. DOI: 10.1002/jps.20999

[40] Popplewell JF, King SJ, Day JP, Ackrill P, Fifield LK, Cresswell RG, et al. Kinetics of uptake and elimination of silicic acid by a human subject: A novel application of 32Si and accelerator mass spectrometry. Journal of Inorganic Biochemistry. 1998;**69**:177-180. DOI: 10.1016/S0162-0134(97)10016-2

[41] Galagudza M, Korolev D, Postnov V, Naumisheva E, Grigorova Y, Uskov I, et al. Passive targeting of ischemic-reperfused myocardium with adenosine-loaded silica nanoparticles. International Journal of Nanomedicine. 2012;**7**:1671-1678. DOI: 10.2147/IJN.S29511

[42] Kalash RS, Lakshmanan VK, Cho CS, Park IK. Theranostics. Biomaterials Nanoarchitectonics. William Andrew Publishing; 2016:197-215. DOI: 10.1016/B978-0-323-37127-8.00012-1

[43] De Salamanca AE, Diebold Y, Calonge M, García-Vazquez C, Callejo S, Vila A, et al. Chitosan nanoparticles as a potential drug delivery system for the ocular surface: Toxicity, uptake mechanism and in vivo tolerance. Investigative Ophthalmology and Visual Science. 2006;**47**:1416-1425. DOI: 10.1167/iovs.05-0495

[44] Li L, Chen D, Zhang Y, Deng Z, Ren X, Meng X, et al. Magnetic and fluorescent multifunctional chitosan nanoparticles as a smart drug delivery system. Nanotechnology. 2007;**8**:405102. DOI: 10.1088/0957-4484/18/40/405102

[45] Prabaharan M. Chitosan-based nanoparticles for tumor-targeted drug delivery. International Journal of Biological Macromolecules. 2015;**72**:1313-1322. DOI: 10.1016/j.ijbiomac.2014.10.052

[46] Lila ASA, Ishida T, Allen TM. Liposomal nanomedicines. In: Torchilin V, editor. Handbook of Nanobiomedical Research: Fundamentals, Applications, and Recent Developments. Vol. 3. New Jersey: World Scientific; 2014. pp. 1-53. DOI: 10.1142/9789814520652_0001

[47] Li P, Dai YN, Zhang JP, Wang AQ, Wei Q. Chitosan-alginate nanoparticles as a novel drug delivery system for nifedipine. International Journal of Biomedical Sciences. 2008;**4**:221-228

[48] Istomina MS, Korolev DV, Pochkaeva EI, Mazing DS, Moshnikov VA, Gareev KG, et al. Study of the possibility of using of colloidal quantum dots based on $AgInS_2/ZnS$ for fluorescence imaging in comparison with fluorophores fixed on the surface of the nanoparticles. Translational Medicine. 2017;**4**:56-65. DOI: 10.18705/2311-4495-2017-4-4-56-65

[49] Istomina MS, Pochkaeva EI, Sonin DL, Pechnikova NA, Postnov VN, Mazing DS, et al. Research of the peculiarities of colloidal quantum dots of $AgInS_2/ZnS$ and chitosan nanoparticles labeled with indocyanine green as the fluorescent labels for biomedical applications. Regional Blood Circulation and Microcirculation. 2018;**17**:74-82. DOI: 10.24884/1682-6655-2018-17-1-74-82

[50] Korolev DV, Shulmeister GA, Romanova TN, Postnov VN. Investigation of the effectiveness of various organic solvents in the synthesis of aminospacer on magnetic nanoparticles. Biotech. 2018;**2**:40-44

[51] Belorus AO, Bespalova K, Spivak YM. Morphology and internal structure of porous silicon powders in dependence on the conditions of post-processing. In: Proceedings of the 2016 IEEE North West Russia Section Young Researchers in Electrical and Electronic Engineering Conference (EIConRusNW 2016). Institute of Electrical and Electronics Engineers, Inc.; 2016. pp. 23-28. DOI: 10.1109/EIConRusNW.2016.7448108

[52] Spivak YM, Belorus AO, Somov PA, Tulenin SS, Bespalova KA, Moshnikov VA. Porous silicon nanoparticles for target drag delivery: Structure and morphology. Journal of Physics: Conference Series. 2015;**643**:12022. DOI: 10.1088/1742-6596/643/1/012022

[53] Spivak YM, Belorus AO, Panevin AA, Zhuravskii SG, Moshnikov VA, Bespalova K, et al. Porous silicon as a nanomaterial for disperse transport systems of targeted drug delivery to the inner ear. Technical Physics. 2018;**63**:1352-1360. DOI: 10.1134/S1063784218090207

[54] Sonin D, Pochkaeva E, Zhuravskii S, Postnov V, Korolev D, Vasina L, et al. Biological safety and biodistribution of chitosan nanoparticles. Nanomaterials. 2020;**10**:810. DOI: 10.3390/nano10040810

[55] Wei Y, Zhao L. Passive lung-targeted drug delivery systems via intravenous administration. Pharmaceutical Development and Technology. 2014;**19**:129-136. DOI: 10.3109/10837450.2012.757782

[56] Abra RM, Hunt CA, Lau DT. Liposome disposition in vivo vi: Delivery to the lung. Journal of Pharmaceutical Sciences. 1984;**73**: 203-206. DOI: 10.1002/jps.2600730214

[57] Croce AC, Bottiroli G. Autofluorescence spectroscopy and imaging: A tool for biomedical research and diagnosis. European Journal of Histochemistry. 2014;**58**:320-337. DOI: 10.4081/ejh.2014.2461

[58] WhittingtonNC, WrayS. Suppression of red blood cell autofluorescence for immunocytochemistry on fixed embryonic mouse tissue. Current Protocols in Neuroscience. 2017;**81**:2.28.1-2.28.12. DOI: 10.1002/cpns.35

[59] Lerchenberger M, Al Arabi N, Gallwas JKS, Stepp H, Hallfeldt KKJ, Ladurner R. Intraoperative near-infrared autofluorescence and indocyanine green imaging to identify parathyroid glands: A comparison. International Journal of Endocrinology. 2019;**2019**:1-7. DOI: 10.1155/2019/4687951

[60] Smith AM, Mancini MC, Nie S. Bioimaging: Second window for in vivo imaging. Nature Nanotechnology. 2009;**4**:710-711. DOI: 10.1038/nnano.2009.326

[61] Inyushin M, Meshalkina D, Zueva L, Zayas-Santiago A. Tissue transparency in vivo. Molecules. 2019;**24**:1460. DOI: 10.3390/molecules24132388

Chapter 13

Getting Charged for the Mitochondrial Uptake

Ivo Crnolatac

Abstract

Mitochondria are hot! First of all, literally. Recent results showed that locally in the functional human mitochondria, the temperature rises close to 50°C. The research was based on the thermosensitive fluorescent dye that targets mitochondria. Figuratively, mitochondria are even hotter, in terms of undergoing research exploring their role in energy production, cell signalling, programmed cell death and biosynthesis. Fluorescent probes and dyes are not restricted only to imaging of these fascinating organelles; they are used to monitor the mitochondrial membrane potential, pH and the redox status. The membrane potential is essential in the process of ATP production as it constitutes more than 80% of electrochemical proton motive force used in this process. Thus, observing mitochondrial membrane potential is crucial in most of the mitochondrial research. This imaging should usually be performed with minimal invasiveness and damage to the mitochondria and mitochondrial function. There are only a few fluorescent probes for mitochondrial membrane potential monitoring currently in use as valuable indicators of cells' functional status. Those probes show varying degrees of interference with cell or mitochondrial metabolism and photo-induced damage. In this chapter, the characteristics, advantages and disadvantages of such probes will be discussed. The mechanisms of uptake of these fluorescent probes will be explained.

Keywords: mitochondria, fluorescent probes, morphology, membrane potential, fluorescent sensor, superoxide anion, hydrogen peroxide

1. Introduction

Previously, mitochondria were studied exclusively in terms of bioenergetics. However, the so-called powerhouse of the cell, apart from producing adenosine triphosphate (ATP) by oxidative phosphorylation, takes part in numerous metabolic pathways. Apoptosis signalling [1, 2], ageing [3, 4], calcium homeostasis [5, 6] and cell development [7, 8] are also responsibilities of this remarkable cell organelle. Thus, to no surprise, the pathogenesis of many diseases stems from mitochondrial dysfunction, for instance, Alzheimer type dementia [9] and other neurodegenerative diseases [10] as well as cancer [11]. The relation between mitochondria and cancer goes well beyond excessive reactive oxygen species (ROS) production by Complex I and Complex III of the electron transfer chain (ETC). The differences in mitochondrial activity between healthy and cancer cells include metabolic variances [12], mostly due to the shift to glycolytic metabolism. Various cancers reportedly have mutations and alternations in the mtDNA, with the D-loop region of mtDNA bearing most of these mutations [12]. The mutations in nuclear DNA that encodes for mitochondrial proteins are also observed in cancer cells. However, there is strong indication that those mutations are the result of intergenomic cross-talk between mitochondria and nucleus and also originate from mutations in mtDNA [13]. Another fascinating feature of mitochondria is the interchanging processes of fusion and fission. This mitochondrial dynamics enables the control of the metabolic pathways mentioned at the beginning of this paragraph [14]. Shifting of the fission-fusion equilibrium is also the way of adaptation to physiological changes and environmental stress [14].

Although studying the biology and biochemistry of mitochondria is captivating and rewarding, the main topic in this chapter will be the fluorescent probes and dyes targeting mitochondria, the comprehensive overview of available fluorescent compounds that allow characterisation of mitochondria, and study of its functionality. Sections will cover main groups of compounds according to their applications in mitochondrial cell biology and biochemistry. Mitochondrial markers are used to fluorescently label the mitochondria and enable visualisation by different fluorescence-based imaging techniques. Monitoring and measuring mitochondrial membrane potential is likewise enabled by specific fluorescent probes. Superoxide indicators will also be covered in a separate section, as will the calcium flux and autophagy probes. Some aspects of mitochondrial uptake will be explained, as that bears an exceptional value in the design of the new potential probes. In the final section, some new perspective probes will be presented, showing exciting features and the possibility of simultaneous dual applicability.

2. Mitochondrial morphology probes

As mentioned in the introductory part, mitochondria are obviously very dynamic organelles. They are undergoing cycles of fusion and fission, transport, processes of biogenesis and mitophagy. These processes are crucial for regulating signalling pathways, regulating mitochondrial turnover, maintaining ETC and mitochondrial homeostasis. Fusion-fission and biogenesis-mitophagy dynamics regulate the morphology, number and transient subcellular localisation of mitochondria. Fluorescent mitochondrial trackers and sensors, enabling real-time monitoring and imaging, were always in demand. Improvements in microscopy have not only instigated more detailed observation and in-depth exploration of mitochondrial dynamics, but also sparked the interest in new sensors.

2.1 Mitochondria-specific fluorescence proteins

The scientists studying mitochondrial dynamics, abundance and morphology in live cells often use specifically targeted fluorescent proteins (FPs). Although the mitochondria evolved from *Alphaproteibacteria* in an endosymbiotic manner within a primordial eukaryotic cell [15], and they possess the whole mechanism for transcription and translation [16], only 13 mitochondrial proteins are produced in the mitochondria [17]. Most of the mitochondrial proteins (>2000) are imported from the cytosol, and the genes encoding them are in the nucleus [18]. So are the mitochondrial targeting signals (MTSs), short (<70) amino acid sequences that direct the synthesised proteins to the specific mitochondrial part. The FP gene is fused with a gene encoding MTS. There are few MTSs that are used frequently. For instance, E1 alpha pyruvate dehydrogenase [19] and cytochrome C oxidase VIII [20] are used by companies for ready-made fluorescent mitochondria targeting reagents [21, 22] and fused plasmids [23]. Beside fluorescent protein and MTS encoding genes, the commercially available constructs contain promoters (usually

Cytomegalovirus, CMV), driving the expression of FP, and antibiotic resistance genes, enabling selective pressure and thus allowing the creation of stable cell lines. Transfection or transduction of these vectors into the targeted cells is achieved using the usual methodology, depending on the cells. The FP methodology is safe as these constructs do not replicate in mammalian cells. It can be used for live cell imaging; however fluorescent signal can also be observed after cell fixation procedure, thus rendering the FPs approach applicable for fixed cells, if the FP expression is achieved before fixation. Flow cytometry using mitochondria targeted GFP has also been reported [24]. The advantages of this technique are: the possibility of targeting specific compartment within mitochondria by choice of targeting signals [25], fewer adverse effects on the observed cells and mitochondria, compared to synthetic fluorescent probes, and relatively stable fluorescence. Disadvantages are related to transfection/transduction efficacy, low levels of FP expression and possible mistargeting, caused probably by the sheer size of the FP. The example of photomicrography depicting HeLa cells stably expressing FP in the mitochondria is presented in **Figure 1**.

2.2 Fluorescent probes targeting mitochondria

As mentioned, fused FPs exploit the fact that their fusion partners, MTSs, are actively transported to mitochondria. Small synthetic fluorescent dyes cannot use that trick. As the mechanism of mitochondrial uptake relates to all small molecules targeting mitochondria, it will be deliberated in more detail in a separate section. **Figure 2** shows commonly used, commercially available compounds [27, 28] and some of the mitochondrial probes designed, synthesised and investigated in collaboration with partners in our laboratory [29–31].

The advantages of this methodology are the ease of use and versatile applicability. Small molecule probes can be used for live cell imaging, fixed cells imaging as well as for flow cytometry. The procedure is as easy as adding the probe in desired concentration to the medium, incubating the cells for designated time and imaging (**Figures 3** and **4**), cell fixation followed by imaging (**Figure 5**) or performing flow cytometry experiment. **Figure 4** portrays cells that are treated with multiple fluorescent probes before

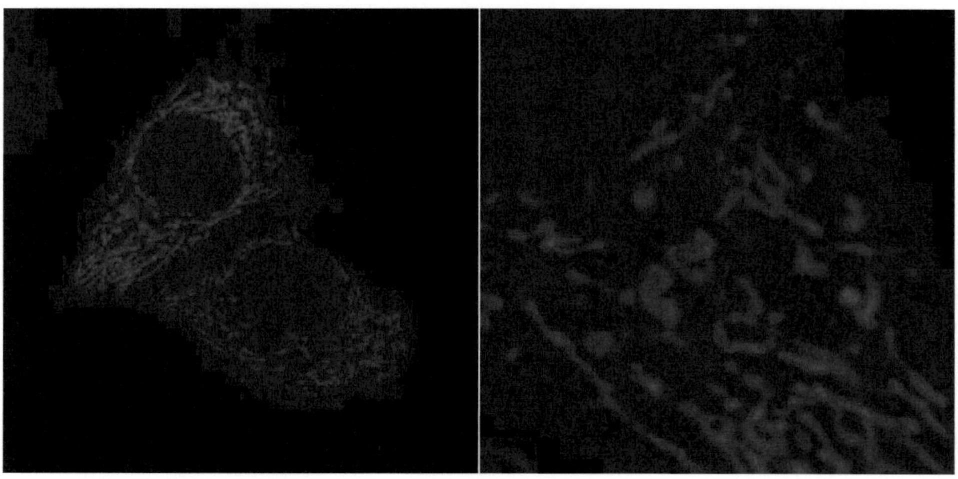

Figure 1.
mitoCFP transfected HeLa cells, expressing mitochondria targeted cyan fluorescence protein (CFP) were seeded on glass-bottom petri dishes (MatTek, USA). After 24 h of cultivation at 37°C in a humidified atmosphere with 5% CO_2*, cells were analysed using a confocal microscope (Leica TCS SP2 AOBS), with* λ_{EX} *= 458 nm, emission was detected with a bandpass filter at 470–510 nm [26].*

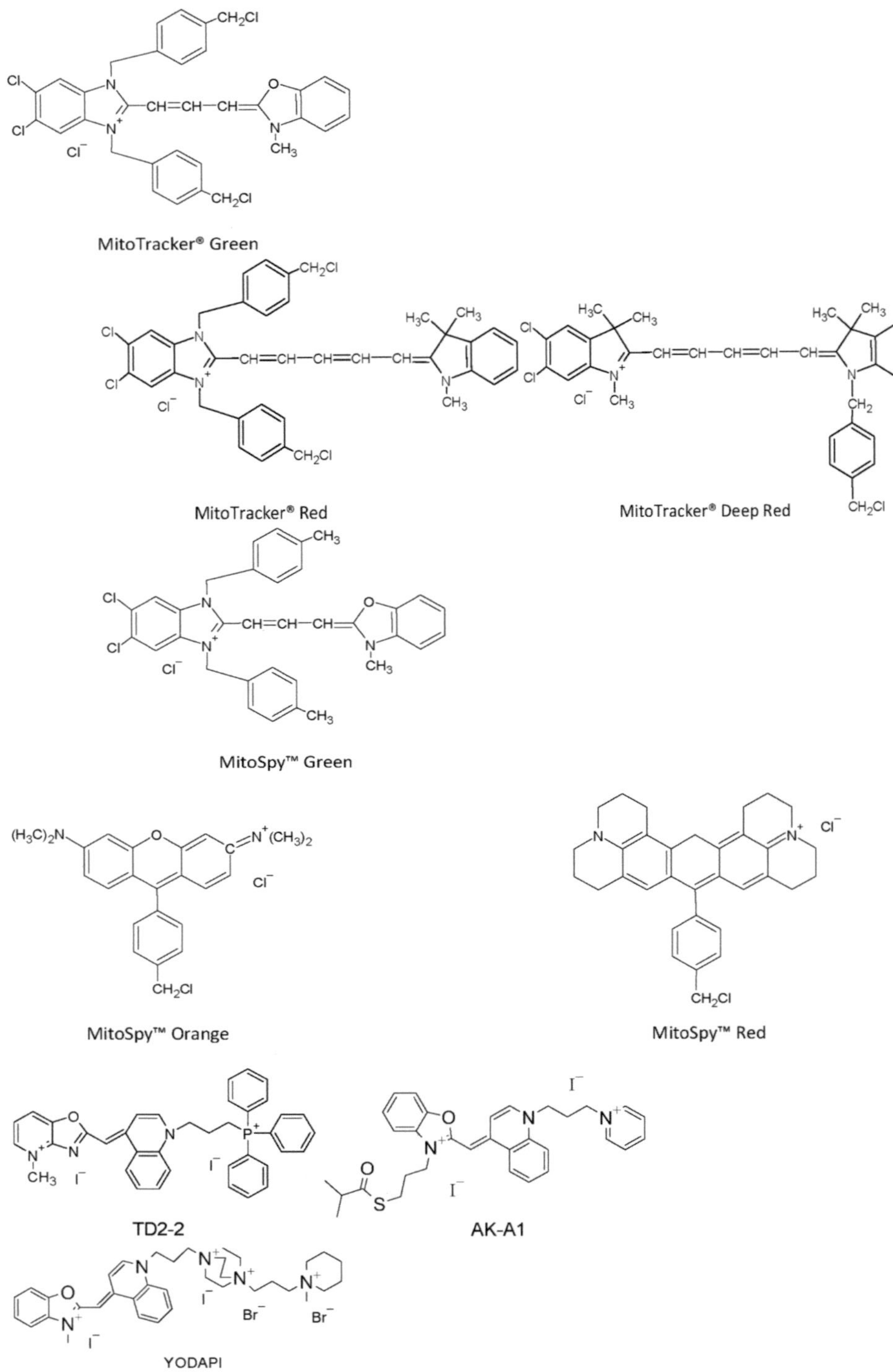

Figure 2.
Structures of the commonly used, commercially available mitochondrial fluorescent probes and some of our cyanine probes that successfully target mitochondria.

imaging. Such methodology is used for colocalisation purposes, to verify the localisation of one of the colocalising partners (**Figures 4(A)** and **6**), or to visualise mitochondria in relation to other cellular organelles and cellular processes (other

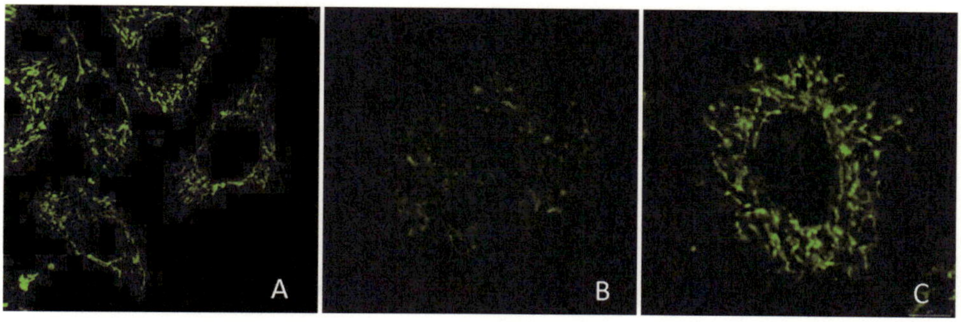

Figure 3.
Live cell images of the cells after exposure to the mitochondria targeting fluorescent probes. (A) HeLa cells incubated for 120 min with 1×10^{-6} Mol/dm^3 TD2-2. λ_{EX} = 514 nm and λ_{EM} = 540–620 nm. (B) H460 cells incubated for 120 min with 1×10^{-6} Mol/dm^3 AK-A1, λ_{EX} = 483 nm and λ_{EM} = 499–522 nm. (C) H460 cells incubated for 120 min with 1×10^{-6} Mol/dm^3 of YODAPI, λ_{EX} = 482 nm and λ_{EM} = 498–535 nm.

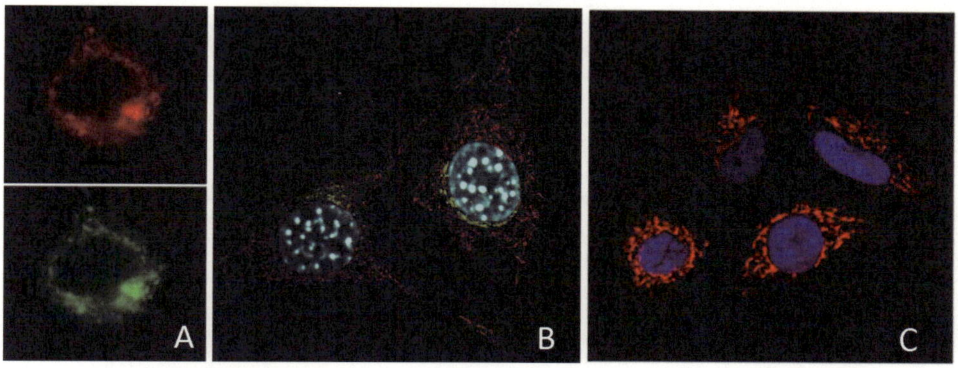

Figure 4.
(A) Images of an NIH 3 T3 cell showing colocalisation of the CoroNa™ Red sodium indicator with the MitoTracker® Green FM mitochondrial marker. A cell loaded with both dyes was imaged consecutively using omega optical bandpass filter set XF41 for CoroNa™ Red sodium indicator and set XF23 for MitoTracker® Green FM. (B) Live NIH 3 T3 cells labelled with probes for mitochondria, Golgi and the nucleus. Mitochondria were labelled with MitoTracker® Red FM, Golgi with BODIPY® FL ceramide, and the nucleus with Hoechst 33342. The image was deconvolved using Huygens software (scientific volume imaging). (C) HeLa cells were co-transduced with CellLight® Nucleus-CFP and incubated overnight. Following staining with MitoTracker® deep red, cells were imaged on a Zeiss 710 confocal microscope. Images and captions are displayed with the permission of Thermo Scientific.

images in **Figure 4**). As the visual comparison of the colocalisation images is subjective due to the different intensity of the fluorescence signal emitted by the two dyes, simply overlapping the images is usually not sufficient for evaluation. The colocalisation is often quantified using a pixel-by-pixel summation of the products of the intensities of two fluorophores, the Pearson correlation coefficient (PCC, r) [32]. In the case of green pixels Gi and red pixels Ri:

$$r = \frac{\sum(Ri - Rav) * (Gi - Gav)}{\sqrt{\sum(Ri - Rav)^2 * \sum(Gi - Gav)^2}} \tag{1}$$

The software packages available for processing microscopy images usually have implemented options for PCC calculation, and so does the most frequently used open-source software, ImageJ [33]. However, PCC, due to the photon noise is not flawless. This noise can be diminished by extending the image acquisition time, by binning adjacent pixels or by summing replicate images. Still, the replicate-based

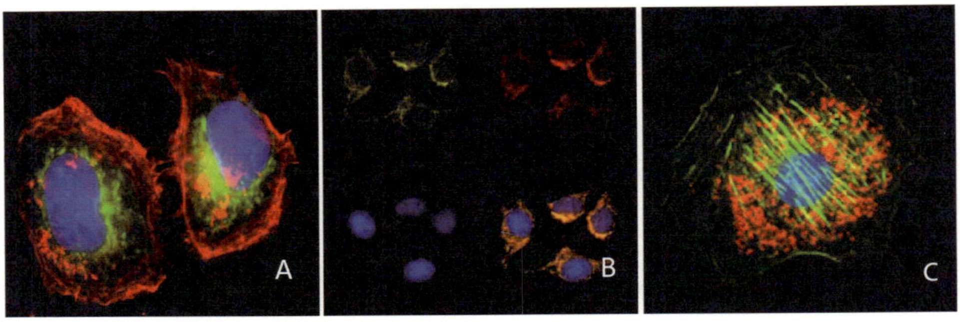

Figure 5.
*(A) HeLa cells were stained with 250 nM of **MitoSpy™ Green** FM (green) for 20 min and fixed with 4% paraformaldehyde (PFA) for 10 min. Then the cells were stained with Alexa Fluor® 594 phalloidin for 20 min (red) and counterstained with DAPI (blue). The image was captured with a 60× objective. (B) HeLa cells were stained with 250 nM of **MitoSpy™ Orange** CMTMR (yellow) for 20 min, fixed with 4% paraformaldehyde (PFA) for 10 min, and permeabilised with 0.1% triton X-100 for 10 min. Then the cells were stained with anti-cytochrome C Alexa Fluor® 647 (red) and counterstained with DAPI (blue). All three images were merged in the bottom right panel. The image was captured with a 60× objective. (C) NIH3T3 cells were stained with 100 nM of **MitoSpy™ red** (red) for 20 min at 37°C, fixed with 1% paraformaldehyde (PFA) for 10 min at room temperature, and permeabilised with 1X True Nuclear™ Perm Buffer for 10 min at room temperature. Then the cells were stained with flash Phalloidin™ NIR 647 (green) for 20 min at room temperature and counterstained with DAPI (blue). The image was captured with a 60× objective. Images and captions are displayed with the permission of BioLegend.*

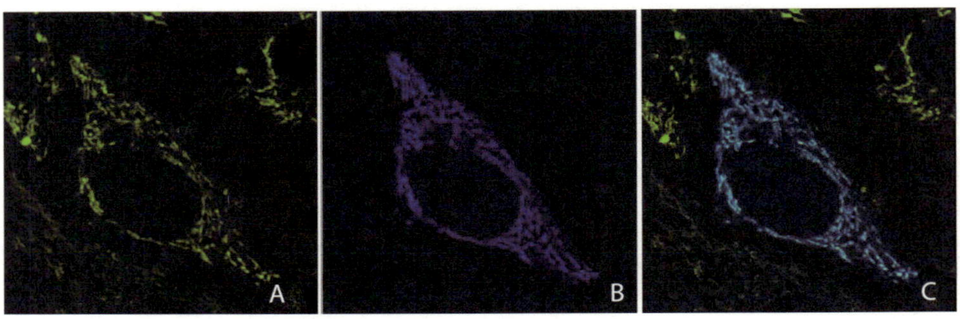

Figure 6.
Confocal photofluoromicrographs of HeLa cells expressing mitochondria targeted CFP and incubated for 120 min with compound TD2-2. (A) Fluorescence emitted by the TD2-2, λ_{EX} = 514 nm and λ_{EM} = 540–620 nm. (B) Fluorescence emitted by CFP, λ_{EX} = 458 nm and λ_{EM} = 470–510 nm. (C) Represents overlayed images (A) and (B).

noise correction correlation (RBNCC) offers excellent results without implementing complicated or time-consuming actions [34]. Shortly, in the case of green and red channels, two consecutive images are taken, and correction factor (r_{GR}) is calculated from those two consecutive green/red channel images. Noise-corrected PCC (r_{corr}) is the product of the mean of four green/red combinations with the above-explained correction, as described in **Figure** 7.

$$r_{corr} = r_{GR} * \frac{(r_{G1R1} + r_{G1R2} + r_{G2R1} + r_{G2R2})}{4} \quad (2)$$

Multiple staining can also be achieved for live cell imaging (**Figures 4** and **6**). In general, these compounds are all hydrophobic, and therefore usually not soluble in aqueous media, but they are readily soluble in dimethylsulphoxide (DMSO). The concentration of the DMSO stock solutions should be high enough not to exceed the cell culture tolerance on DMSO. The DMSO content in the medium that does not

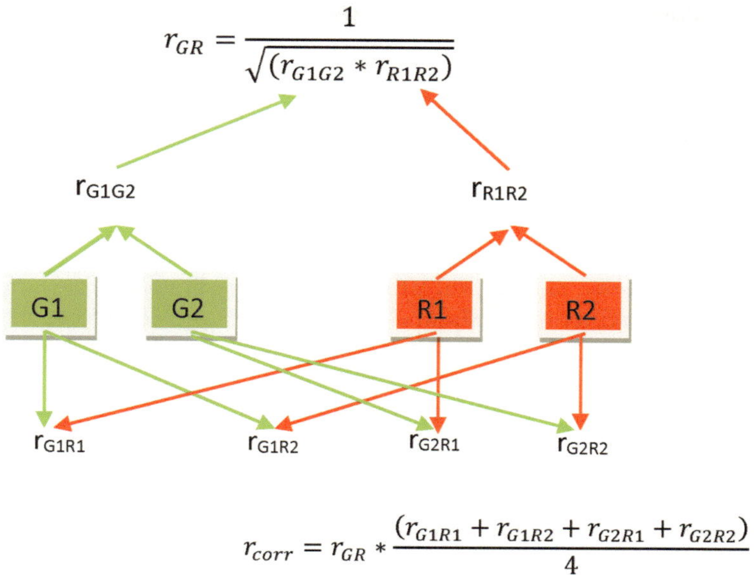

Figure 7.
Schematic description of RBNCC calculation.

affect the cells is cell type dependent, but as a rule of thumb, it should not exceed 0.1–0.5%. Various detrimental effects on the mitochondria and, consequently, on cells are the main disadvantages of these probes, along with non-specific accumulation (**Figure 3(C)**, nucleus) and photobleaching [35].

3. Charge, lipophilicity: keys for the mitochondrial uptake

Synthetic fluorescent probes have to enter the cell and navigate towards mitochondria, pass outer and inner mitochondrial membranes to enter the matrix. So what is the driving force, which attracts and internalises these compounds? The quest for the answer starts with the inspection of the structural similarities. Most of the presented compounds in **Figure 2** are, in fact, cyanine dyes, but that has more to do with their tunable spectroscopic and fluorescence properties [36]. The main features determining them for mitochondrial uptake are lipophilicity and charge. They all are highly lipophilic cations (LC), some of them multiply charged.

Despite the net charge, LCs are lipid-soluble and can move through phospholipid bilayers. As opposed to hydrophilic cations which, unless actively transported by ionophores or carrier proteins, cannot permeate biomembranes. Moving the hydrophilic cation from the aqueous environment into the lipophilic core of the membrane requires too much energy. The activation energy for the movement of LC through the hydrophobic barrier of a biological membrane is lower than for hydrophilic cations. The activation energy for the displacement of a solvated cation from the aqueous milieu to the membrane core consists of electrostatic interactions and hydrophobic forces. The principal electrostatic interaction constituent is Born energy (W_B), the enthalpic penalty of removing the solvating water molecules from the cation upon entering the membrane core [37].

$$W_B = \frac{q^2}{8\pi\varepsilon_0 r}\left(\frac{1}{\varepsilon_1} - \frac{1}{\varepsilon_2}\right) \tag{3}$$

In this equation, ε_0 is the vacuum permittivity, ε_1 is the dielectric constant of membrane core (~2), ε_2 is the dielectric constant of water (~80), q is the electrical charge per mole of the cation and r is the ionic radius. Thus, the Born energy (W_B) in kJ/mol for a cation with Z charges is given by:

$$W_B = 339 * \frac{Z^2}{r}$$

So the enthalpy needed to transport the cation to the membrane is inversely proportional to the ionic radius. When the charge is distributed on the large surface area, as in all of the structures in **Figure 2**, the cation is less likely to be intensely solvated, as the electric field at the surface is not strong and less water polarisation arises. Apart from Born energy, there are two other electrostatic forces related to the transport of the cations through the biological membranes that are significantly less prominent: dipole energy, caused by the electrical potential within the bilayer, a result of the orientation of the dipoles in phospholipids and image energy, caused by the electrostatic forces at the interface [38]. Those are the forces that work against the transport of the LC to the membrane. The force that lowers the activation energy of the transport and attracts LCs is caused by the increase of entropy due to the loss of the water structure, upon moving a large hydrophobic surface from water to lipid environment. Here, again, the large surface area plays a beneficial role for the transport, contributing with 92 J/mol per Å^2 of the solvent accessible area [18].

The energy profile for the passage of the LC through the phospholipid bilayer, depicted in **Figure 8**, is assembled, taking into consideration the electrostatic forces and hydrophobic effect. The potential energy wells at both interface surfaces of the membrane are due to the fact that hydrophobic interactions arise with the proximity to the membrane, while the repulsive electrostatic forces increase upon moving through the membrane [40]. Thus, LCs are adsorbed to the membrane surface (usually in monolayers), before fast passage through the core of the membrane to the potential well on the other side followed by slower desorption from the

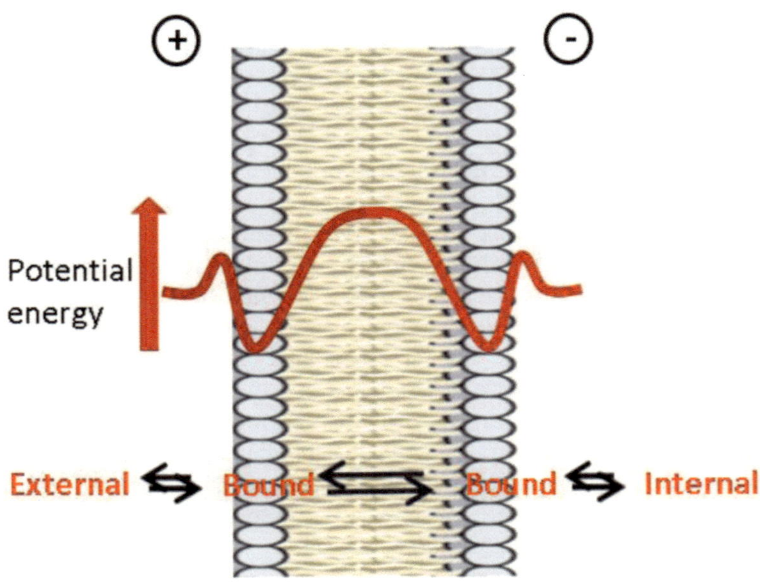

Figure 8.
The energy profile for the passage of the LC through the phospholipid bilayer [39].

membrane. The adsorption can be described by the Langmuir isotherm model assuming the independent binding of the LC to the binding sites on the membrane; in high LC concentrations, repulsive forces between LC molecules have to be considered. The mechanism is spontaneous ($\Delta G_r < 0$), entropy driven ($-T\,\Delta S_r > \Delta Hr$) and occurs even in the absence of mitochondrial electrochemical proton motive force (Δp) or its major constituent membrane potential ($\Delta\psi_m$, constitutes for >80% of Δp). The $\Delta\psi_m$ is present on the cell membrane (30–60 mV) and inner mitochondrial membrane (150–180 mV) [41], causing several hundredfold accumulation (~10 fold per every 60 mV of $\Delta\psi_m$) of the LC in the mitochondrial matrix following the Nernst equation:

$$\Delta\psi_m = \frac{2.303RT}{F}\log_{10}\left(\frac{[cation_{in}]}{[cation_{out}]}\right)$$

where $\Delta\psi_m$ is the membrane potential, R is the universal gas constant: R = 8.314 J/Kmol, F is the Faraday constant, the number of coulombs per mole of electrons: F = 96485.33 C/mol.

4. Membrane potential probes

The flux of protons in the inner membrane of the mitochondria (IMM) is proportional to the mitochondrial respiration rate and can be compared to the simple electrical circuit as depicted in **Figure 9**. Using that analogy, we can describe the Δp as the voltage in the Ohm's law.

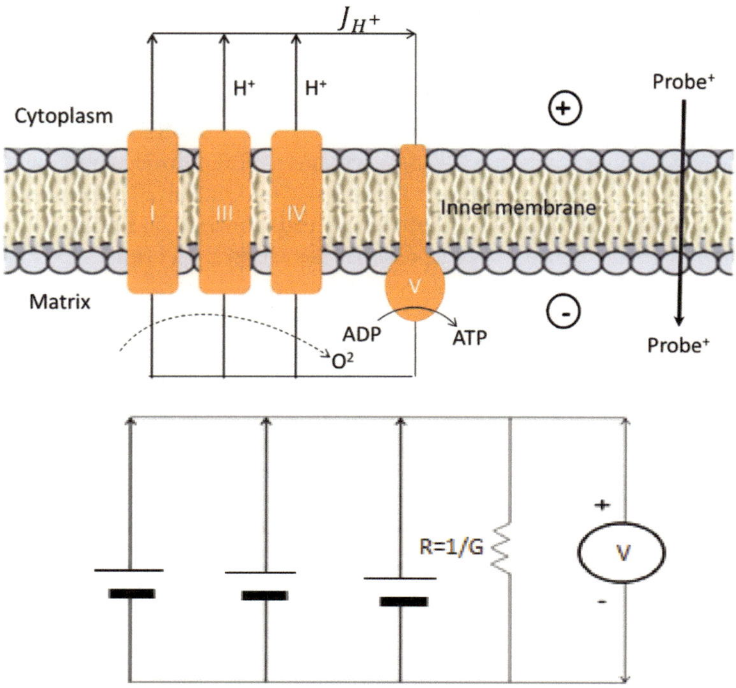

Figure 9.
The similarity of the mitochondrial proton flux and the electrical circuit. (UP) Proton flux across the inner mitochondrial membrane, protons are concentrated in the intermembrane space by the protein complexes I, III and IV, to drive the ATP production upon return via complex V. (DOWN) Simple electrical current circuit.

Ohm's law:

$$I = V * G \tag{4}$$

where I is the electrical current, V is the potential difference and G is the conductance, that is reciprocal to resistance.

Proton flux equation:

$$J_{H^+} = \Delta p * C_m H^+ \tag{5}$$

where J_{H^+} is proton current, Δp is the electrochemical proton motive force and $C_m H^+$ is the membrane conductance of protons [42, 43].

The series of reductive reactions of the protein complexes I–IV that comprise the ETC results in the accumulation of H^+ outside of the membrane. Through the ATP generating F1/F0 ATP synthase (Complex V), these protons return to the mitochondria, producing ATP in the process and concluding the ETC. The Δp is the overall force driving protons back into the mitochondrial matrix. It consists of the electrical or charge gradient, $\Delta\psi_m$, and chemical or concentration gradient, ΔpH, the difference between pH in the matrix and intermembrane space, which usually amounts to -0.5 up to -1 units (the mitochondrial matrix is alkaline) [44]. Thus, Δp can be calculated via the equation expressing proton electrochemical gradient, as follows (at 37°C) [43]:

$$\Delta p = \Delta\psi_m - 61 * \Delta pH \tag{6}$$

In physiological homeostatic conditions, $\Delta\psi_m$ amounts to 150–180 mV, and with ΔpH somewhere between -1 and -0.5 units, Δp falls in range 180–220 mV.

The above noted equation emphasises essential distinction, the probes described in this section are all related to $\Delta\psi_m$, evaluating the charge gradient across the IMM, but do not pertain to ΔpH. The list of fluorescent probes used to monitor the $\Delta\psi_m$ comprises of rhodamine (rhodamine 123—R123, tetramethylrhodamine methyl ester—TMRM and tetramethylrhodamine ethyl ester—TMRE), and cyanine derivatives (5,5′,6,6′-tetrachloro 1,1′,3,3′-tetramethylbenzimidazolyl carbocyanine iodide—JC-1; 3,3′-dihexyloxa carbocyanine iodide—DiOC6(3)). **Figure 10** depicts representatives of these two groups of compounds, both LCs, taking advantage of the compliance with the membrane passage described in Section 3 of this chapter and the charge gradient described above.

There are two experimental scenarios when the measurement of the $\Delta\psi_m$ is considered: first, when the real-time changes in $\Delta\psi_m$, as a response to some treatment, needs to be monitored. The great advantage of this experimental setup is the fact that the sample before treatment can serve as a baseline. Another scenario can be described as the evaluation of change in $\Delta\psi_m$ caused by a chronic experimental

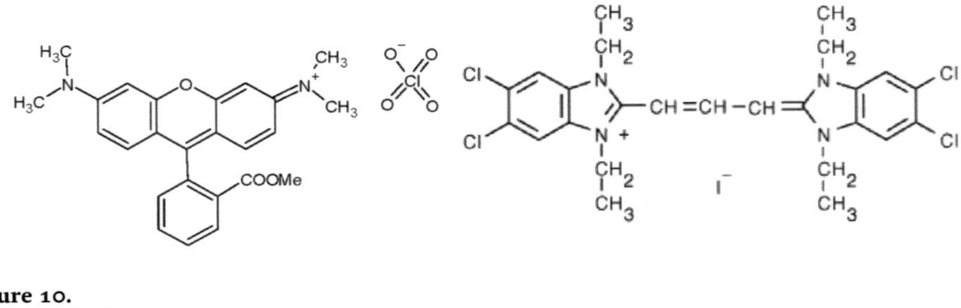

Figure 10.
TMRM and JC-1 structures.

treatment such as chronic exposure to a chemical, or gene manipulation. This methodology is usually implemented on the larger population of cells/mitochondria, showing a change on average. In this case, the baseline or control group has to be prepared from the non-treated sample. The choice of the probe depends on the experimental set-up. Further considerations include the mode of the usage and the choice of the readout methodology, which altogether impacts the selection of the probe. The probe can be used in turn-off (quenching) mode or turn-on fluorescence mode. For the quenching mode, high dye concentrations have to be used (up to several μM), so upon the uptake by mitochondria the autoquenching takes place, as the intramitochondrial concentration is such that the aggregation occurs, decreasing the fluorescence signal. After loading the mitochondria, the externally caused depolarisation, for example by treatment with carbonyl cyanide-p-trifluoro methoxy phenyl hydrazine (FCCP), will result in disaggregation and unquenching of the dye, increase of the fluorescent signal. Contrariwise, hyperpolarisation of mitochondria (oligomycin treatment) will result in higher accumulation of the probe and further quenching of the fluorescence emission. Quenching mode can be used only for the first type of scenario, where acute real-time changes in $\Delta\psi_m$ are observed [45]. Instead of the standard, proposed depolarisation or hyperpolarisation treatments, the investigated agent can be used to evaluate its effect on $\Delta\psi_m$.

In turn-on fluorescence mode, probe concentrations used are lower (up to 50 nM) to prevent the aggregation and the quenching of the fluorescence. Thus, more polarised mitochondria accumulate more probe and emit more fluorescence and the mitochondria with lower $\Delta\psi_m$ store lower concentrations of the probe and emit less fluorescence. In this mode, however, both experimental scenarios can be employed; acute changes can be implemented after the probe loading while chronic treatments should be performed before probe loading. There are several protocols published by research groups [46, 47] and the manufacturers/distributors of the probes disclose the protocols pertained to their specific probe. However, when using these probes, researchers must be aware of the detrimental effects that these probes can evoke. These effects can be caused by photodynamic generation of singlet oxygen or other reactive oxygen species or simply by ETC inhibition. **Table 1** shows the spectral characteristics and practical considerations needed to determine the optimal probe for the specific experiment [45].

When performing a $\Delta\psi_m$ evaluation experiment, regardless of the mode and experimental scenarios, all the influences that can affect the loading and fluorescence signal of the probe should be considered. As mentioned above, when acute real-time changes are measured (Scenario 1), the baseline can be the output before the change infliction. Therefore, special precautions should be considered in performing chronic scenario measurements. The probe concentration in mitochondria is affected by (a) probe concentration in the bath; (b) cell membrane potential; (c) mitochondrial size/mass (surface to volume ratio); (d) loading time and, of course, (e) $\Delta\psi_m$. Except for the measured $\Delta\psi_m$, all other factors should be controlled and should not change during measurement. Nevertheless, in a turn-on fluorescence mode, using lowest possible dye concentrations and leaving the probe in the bath during measurement (no wash-out step) should in general result in reproducible assessments of post-treatment $\Delta\psi_m$ across different samples, if all other above-mentioned factors are fixed or controlled.

5. pH probes

Considering that 80% of the Δp is due to $\Delta\psi_m$, researchers usually focus on measurement, control and investigation of$\Delta\psi_m$ and disregard the relatively small

Probe	Spectral characteristics	Experimental considerations
TMRM, TMRE Probes of choice for slow resolving acute study scenarios or measuring pre-existing $\Delta\psi_m$	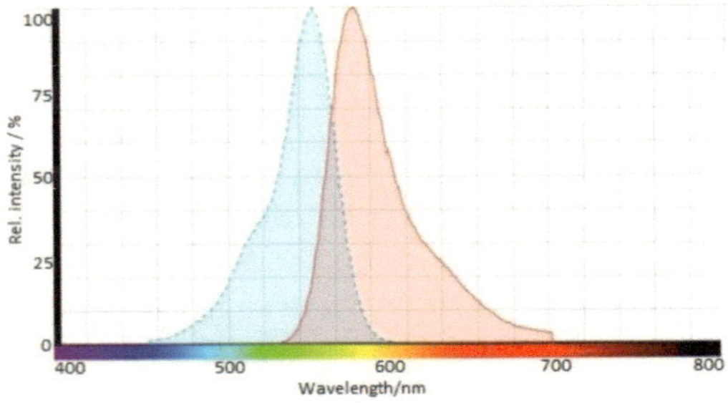	• TMRM shows lowest ETC inhibition [47] • Lowest possible conc. if used in turn-on mode < 30 nm; quenching mode use 50–100 nm • Fast mitochondria loading and equilibration • Both acute and chronic study scenarios
R-123 Best for fast resolving acute studies, turn-off mode	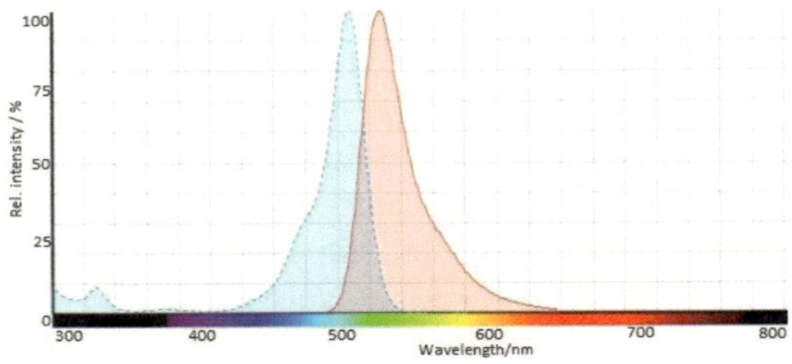	• Preferably used in quenching mode (1–10 μM), monitoring acute changes in $\Delta\psi_m$ after loading and washing out the excess • Note: ETC inhibition slightly less than TMRE but more than TMRM
JC-1 Best for a qualitative distinction of the polarisation state (apoptosis studies, flow cytometry or microscopy)	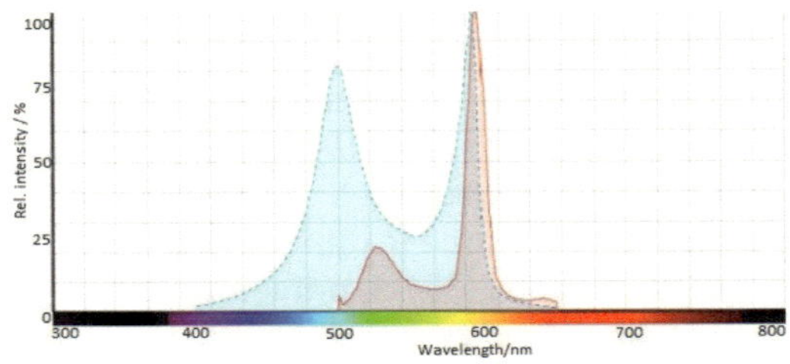	• Monomer and J-aggregated form of the probe exhibit different emission maxima (515 and 590 nm, relatively, see spectra), allowing dual colour ratiometric assessment of $\Delta\psi_m$ • Best if loaded after exp. treatment and left in the bath during imaging to prevent redistribution of the probe • Note: aggregate formation can be affected by factors other than $\Delta\psi_m$, such as surface-to-volume ratios and H_2O_2. JC-1 likely requires longer equilibration times than reported
$DiOC_6(3)$ Best for flow cytometry	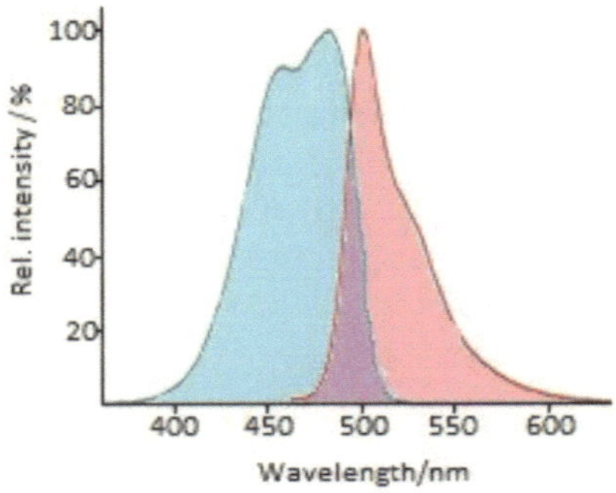	• Most widely used for flow cytometry studies • Should be used in very low concentrations (<1 nM) due to respiration toxicity [48]

Table 1.
Mitochondrial membrane potential probes, spectral characteristics and experimental considerations.

contribution of ΔpH. Thus, in standard physiological conditions, ΔpH accounts for 20% of the Δp, but when $\Delta\psi_m$ fluctuates for whatever reason, Δp remains constant due to the compensation by ΔpH. This compensation can extend to a full amount, and Δp can be maintained solely by ΔpH contribution [49] and *vice versa*—when the chemical gradient of protons is diminished by extrinsic addition of ionophore (nigericin), Δp is maintained by $\Delta\psi_m$ only [50]. Hence, to fully understand and evaluate the mitochondrial state, both contributions must be considered. In consort with the proton motive force contribution, ΔpH regulates mitochondrial calcium, sodium and potassium fluxes as Ca^{2+}—H^+ [51]; Na^+—H^+ and K^+—H^+ [52] exchange transporters depend on ΔpH [53]. The Pi—H^+ phosphate cotransporter, responsible for the importation of phosphate ion for the ATP synthesis, also relies on ΔpH [54]. The uncoupling proteins, H+ channels that uncouple the oxidative phosphorylation from the ATP synthesis and the permeability transition pores that initiate the cell death by opening their large ion channels, allowing nonselective influx across the IMM, depend on both, ΔpH and $\Delta\psi_m$.

So, the chemical proton gradient, ΔpH, is a sort of an underappreciated smaller associate of the electrical, charge gradient $\Delta\psi_m$. However, the reason why this contribution to the proton motive force has been 'neglected' is not only the scale of impact but also more challenging measurement. The search for the optimal pH sensor for mitochondria is an ongoing task. The ideal sensor should efficiently target mitochondria, exhibit no toxicity or phototoxicity, display wide dynamic range with the prompt and reversible response to changes in pH and show slightly alkaline pK_a, as the mitochondrial matrix pH is in 7.5–8.5 range. Furthermore, they should be ratiometric to allow compensation for the different cell morphologies and sensor uptake. Historically, mitochondrial pH was measured using radioactively tagged weak acids and bases [55], but the introduction of fluorescent pH-sensitive probes allowed more detailed measurements, with spatial, single-cell resolution [56, 57]. These pH sensors are based on fluorescein structure and are not mitochondria targeted. Thus, if used for a whole cell measurement, cells should also be loaded with mitochondrial morphology probes to distinguish the fluorescence pH-related signal from mitochondrial and cytosolic areas. The photodynamic properties of such sensors are also a concern, as they usually produce ROS upon light excitation. The main representative is carboxy SNARF 1 (**Figure 11**), the cell permeant, pH-sensitive dye, with pK_a of 7.5 and consequently suitable for measuring the pH in the range 7–8. The fluorescence emission can be measured ratiometrically using emission bands at 580 and 640 nm. Visually, upon the increase of the pH, the emission shifts from yellow-orange towards deep red fluorescence. The fluorescence

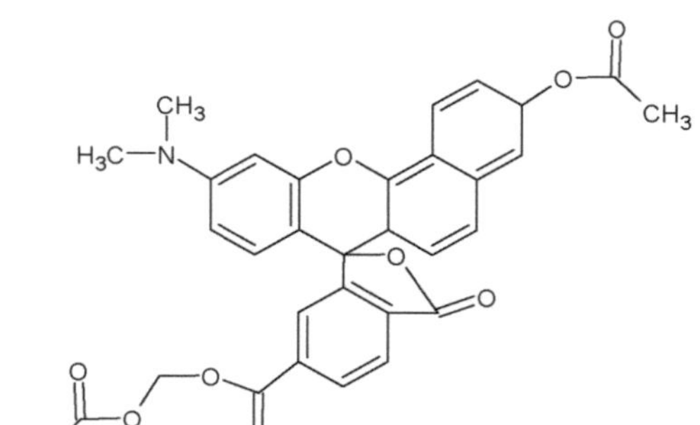

Figure 11.
Spiro[7H-benzo[c]xanthene-7,1′(3H)-isobenzofuran]-ar′-carboxylic acid, 3-(acetyloxy)-10-(dimethylamino)-3′-oxo-, (acetyloxy)methyl ester or shorter 6-carboxy SNARF-1, Acetoxymethylester, Acetat.

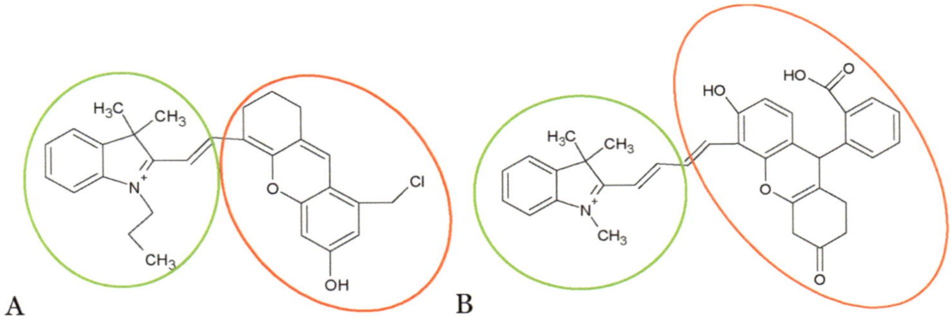

Figure 12.
(A) HXPI-Cl mitochondrial pH sensor [60], (B) FDI mitochondrial pH sensor [61].

measurements can be performed using flow cytometer, fluorescence plate reader or fluorescence microscopy.

Targeting mitochondria is not a problem for engineered protein constructs bearing MTS. Considering that the fluorescence properties of FPs naturally depend on the protonation state of its chromophore, located inside the ß-barrel of the protein, they can be transformed into pH sensors [58, 59].

As the interest for mitochondrial pH measurement gained momentum [44], so did the pursuit for new and improved pH sensors that would efficiently target mitochondria. There are more new sensors synthesised and published [60, 61] but the general idea of their design is similar. They usually consist of a pH-sensitive moiety (red circled in **Figure 12**) unsaturated (di)alkene bridge and an indole quarternary ammonium moiety, LC, responsible for mitochondria targeting (green circled in **Figure 12**). The (di)alkene bridge affords ratiometric measurements by broadening the absorption and emission spectra.

6. Redox sensors

ROS were mentioned in this chapter more than once, in relation to the phototoxic probes that invoke production of singlet oxygen and ROS, and as a by-product of mitochondrial metabolism. Mitochondria are the major ROS producers in the cell, being a part of physiological or pathological processes [62, 63]. The ROS that is first produced in mitochondria is superoxide anion (O_2^-), a product of mitochondrial respiratory chain. However, the main signalling molecule is H_2O_2 generated by manganese superoxide dismutase, which is abundantly concentrated in the mitochondrial matrix and quickly converts O_2^- to H_2O_2. Due to its short life and poor diffusion, O_2^- is not a prominent signalling molecule. When respiratory complex III generates O_2^- and releases it into the intermembrane space, it can diffuse to the cytosol or gets converted to H_2O_2 by the Cu,Zn-superoxide dismutase, which is present there [64]. Mitochondrial redox signalling mechanism also involves the regulation of the concentration of formed H_2O_2 in the matrix, via peroxiredoxins (Prx3 and Prx5) and glutathione peroxidase 1 (Gpx1) [65]. Peroxiredoxins employ mitochondrial thioredoxin 2 reduction system, while Gpx1 engage glutathione for H_2O_2 reduction. Nitric oxide (NO) is another ROS that modulates mitochondrial function. It is generated by NO synthase, once it accumulates in the matrix [66]. The ability to affect modification of the protein, causing reversible alteration of its functionality, renders H_2O_2 and NO efficient biological messengers. Modification of a thiol on a cysteine residue by H_2O_2 oxidation into a disulphide, changing the activity of the protein, is an example of a signalling cascade. Once the levels of the redox signal return to the baseline concentrations, the modification is reversed, and the protein regains its function [67]. Numerous reviews describe the redox signalling and redox pathologies in mitochondria, and further explanation would fall outside of the scope of this chapter.

The mitochondrial fluorescent redox sensors are listed in **Table 2** [86]. They again use the same two strategies for mitochondrial targeting, LCs or MTSs.

The different principles of fluorescence response provoked by ROS are shown in the series of figures below. ROS can instigate the cleavage of the masking groups

Name	Targeting group	Ex/Em (nm)	Selectivity	Measuring mode
Lipophilic cations				
MitoSOX [68, 69]	Triphenylphosphonium ion (TPP)	510/580	O_2^-, OH·, $ONOO^-$	Turn-on, irreversible
MitoPY1 [70]		515/543	H_2O_2, $ONOO^-$, OCl^-	Turn-on, irreversible
SSH-Mito [71]		338/462 and 545	Thiols	Emission ratiometric (F545/F462 increase), irreversible
NpFR2 [72]		488/545	Total ROS	Turn-on, reversible
MitoAR/ MitoHR [73]	rhodamine	553/574	OH·, $ONOO^-$, OCl^-	Turn-on, irreversible
RhoSS [74]		500/530	Thiols	Turn-on, irreversible
FRR2 [75]		460 and 550/590	Total ROS	Excitation ratiometric (F550/F460 increase), reversible
Cy-O-EB [76]	cyanine	768/794	H_2O_2	Turn-on, reversible
pep3-NP1 [77]	styryl	455(442)/555 (646)	H_2O_2	Emission ratiometric (F646/F555 increase), irreversible
Fluorescent protein constructs carrying MTS				
roGFP1 and roGFP2 [78, 79]	Pyruvate dehydrogenase (COX4)	395 and 475 (GFP1) 490 (GFP2)/508	General redox status	Excitation ratiometric (F395/F475 (F490) increase), reversible
HyPer [80]	Two copies human COX8	420 and 500/516	H_2O_2	Excitation ratiometric (F500/F420 increase), reversible
rxYFP [81]	COX4	512/523	General redox status	Turn-off upon oxidation, reversible
Grx1-roGFP2 [82]	Non-human ATP synthase prot 9	395 and 488/508	GSH/GSSG redox couple	Excitation ratiometric (F395/F488 increase), reversible
Orp1-roGFP2 [83]	Human COX8	395 and 488/508	H_2O_2	Excitation ratiometric (F395/F488 increase), reversible
HyPerRed [84]	Two copies human COX8	560/605	H_2O_2	Turn-on, reversible
Sulphonyl fluorescein [85]	Human COX8	495/519	O_2^-	Turn-on, irreversible

OH·, hydroxyl radical.

Table 2.
Mitochondrial redox sensors.

(**Figures 13** and **14**), quenching moieties (**Figure 15**), but also transform the sensor structure into a rigid form that exhibits high fluorescence (**Figure 16**) and allow fluorescence resonance energy transfer (FRET) pairing (**Figure 17**).

Sulphonyl fluorescein, the small molecule sensor, targets mitochondria with 25 amino acid-long MTS. Upon reaction with superoxide anion, fluorescein is released (**Figure 18**). This confirms that MTSs can also facilitate the transport of small molecules to the mitochondria.

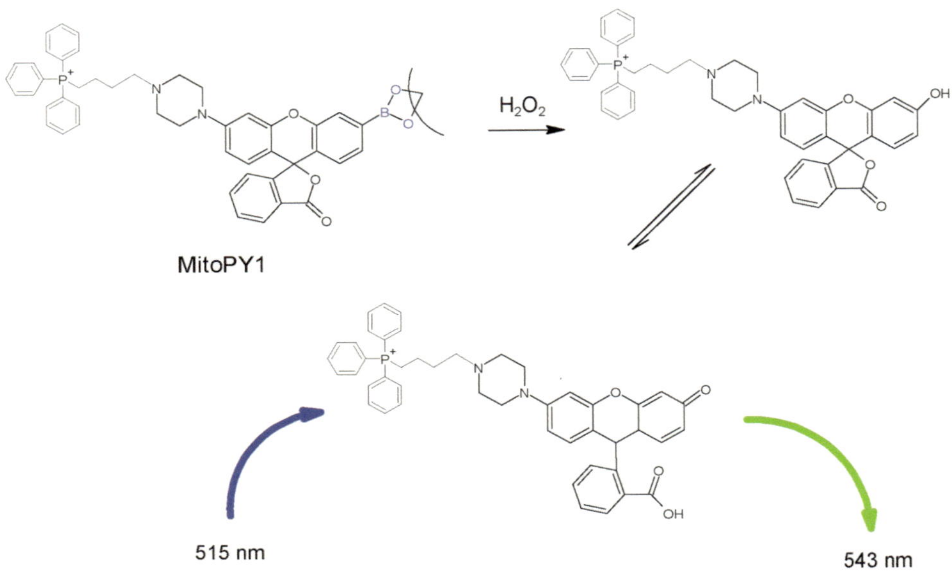

Figure 13.
MitoPy1 redox sensor, upon reaction with H_2O_2 or peroxynitrite ($ONOO^-$), the boronate masking group is cleaved, initiating tautomerisation and subsequent fluorescence response. Since the boronate cleavage is not reversible, the fluorescence signal is also irreversibly turned on.

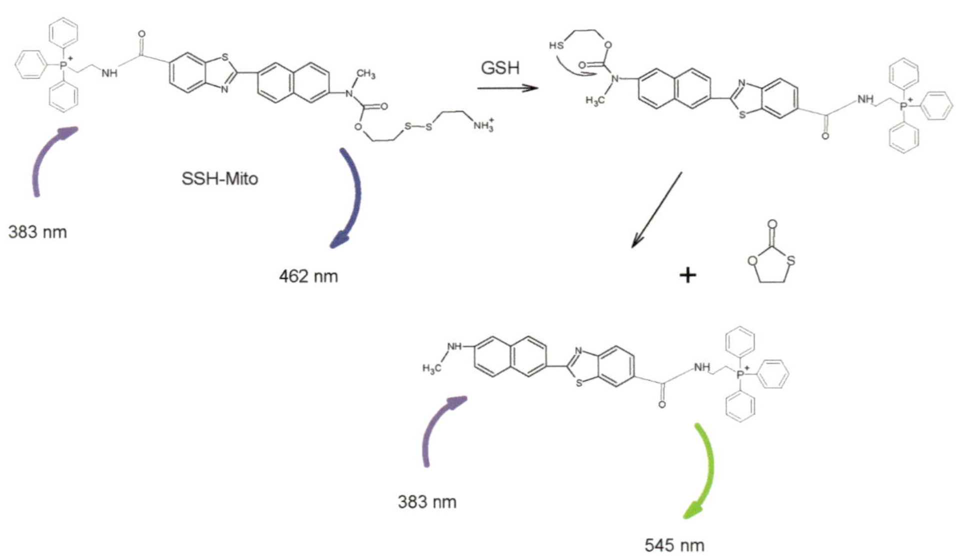

Figure 14.
SSH-Mito with a redox-sensitive disulphide bond, reduction of this bond instigates reaction sequence that results in a redshift of the fluorescence emission.

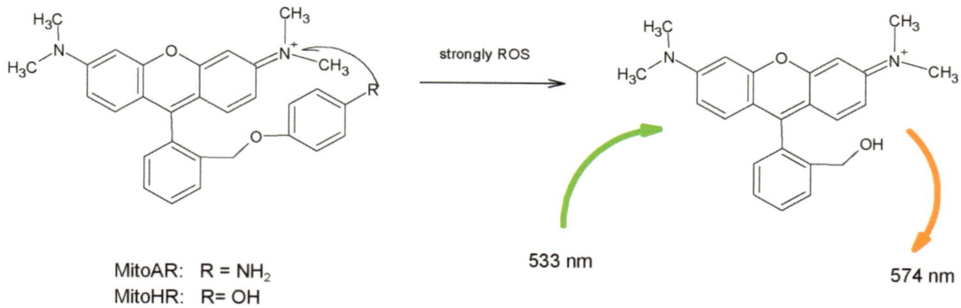

Figure 15.
The fluorescence quencher is cleaved by highly reactive oxygen species from the rhodamine derivatives MitoAR and MitoHR.

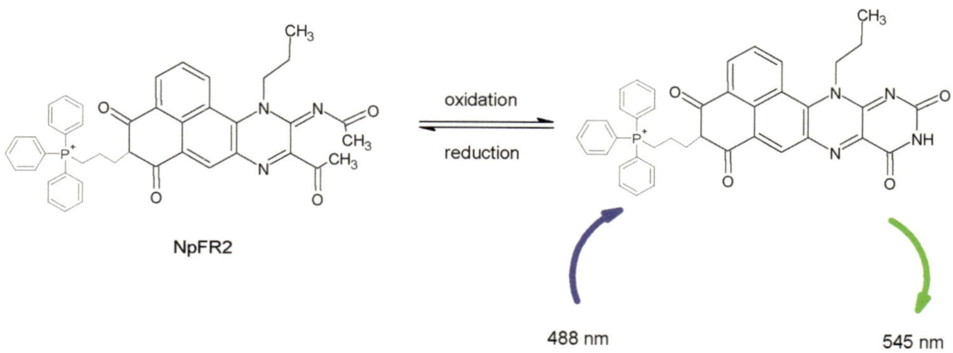

Figure 16.
NpFR2 upon oxidation is reversibly transformed from a flexible bent structure to a rigid highly fluorescent one.

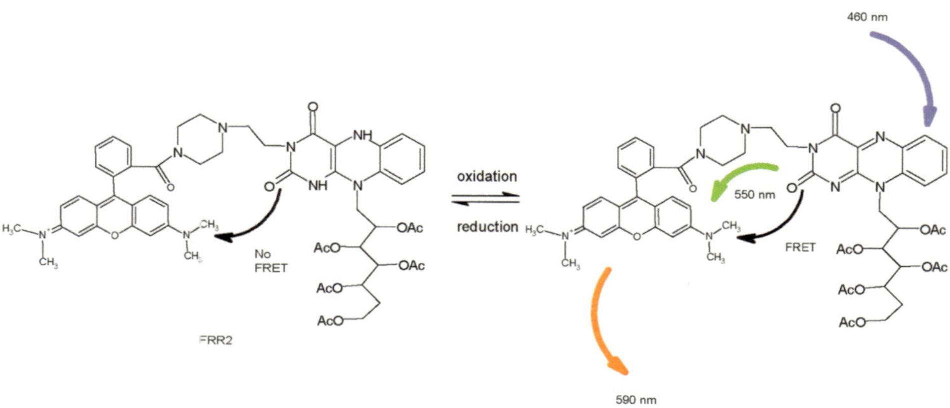

Figure 17.
The FRR2 sensor reports on ROS presence by FRET pairing, with riboflavin as a FRET donor and rhodamine as an acceptor.

7. (Instead of) Conclusion

The main types of fluorescent probes and sensors are described in the above sections. The new ones should report on characteristics that are not yet covered or have dual signalling capabilities. The fluorescent temperature sensor was mentioned in the abstract of this chapter. The sensor, MitoThermo Yellow, is, in fact, a

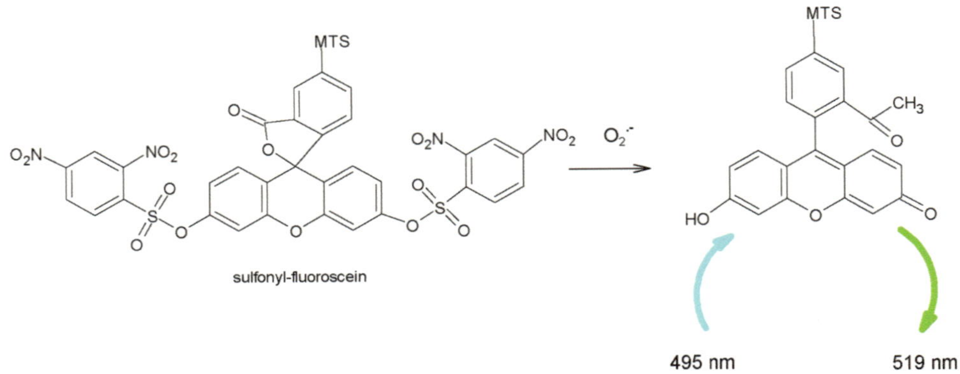

Figure 18.
Sulphonyl fluorescein, small molecule fluorescent sensor, targeting mitochondria through MTS.

rhodamine analogue [87, 88]. The pentamethine analogue Mito-V reports on intramitochondrial viscosity [89]. There are others, that were not yet mentioned. For example, fluorescent ion probes like Ca^{2+}, Na^{+} and K^{+} sensors are important and frequently used to evaluate mitochondrial physiology. However, they are not actually mitochondria targeted, so the experimental set-up includes colocalisation staining with one of the mitochondrial morphology probes (**Figure 4(A)**). There is also a whole field of mitochondria targeted fluorescent methodology that was not covered, fluorescence-based immunohistochemistry (immunofluorescence) because it exceeds the scope of this chapter. Mitochondrial physiology and dysfunction are always going to be a topic of great interest. The search for the new dyes, sensors and probes will probably be driven by the technological developments in the detection of the signalling molecules, membrane potentials and morphological characteristics of mitochondria. Regarding fluorescence detection, 'the future' is already here, the technology allowing the resolution to go beyond the diffraction limit is already in use. Stimulated emission depletion microscopy (STED), proposed in 1994 [90], brought its inventors Nobel Prize in chemistry 20 years later and the first ready-made STED microscopes are coming to the market 25 years later. Mitochondrial STED imaging is still a novelty, but the STED microscopy studies on mitochondria are growing in numbers [91, 92]. The desired characteristics for STED dyes are not that different from characteristics of standard fluorescence probes. Brightness, low toxicity/phototoxicity and efficient targeting, those are features needed for all types of fluorescent imaging. Photostability is also a requirement for all types of imaging. Still, for STED microscopy, due to high laser beam energies involved, the dye has to be almost entirely resistant to photodegradation. Another difference is that for STED, excitation/emission should be in the far-red, near-infrared range. Having tunable absorption/emission properties and relatively high photostability, previously mentioned cyanine dyes are excellent candidates for this type of imaging. In whatever direction the technology takes the research, targeting mitochondria will probably always depend on MTSs or LCs. However, with this previously unthinkable resolutions, the spatial positioning of a mitochondrial targeting moiety and a fluorophore attached to it does not necessarily have to be the same. The quest for new mitochondrial fluorescent probes is ongoing.

Acknowledgements

I would like to extend my gratitude to Dr. Marijana Radić Stojković and Dr. Lidija-Marija Tumir from Ruđer Bošković Institute for their help with the text.

Thanks

Greatest thanks to my family, Nina, Hana and Marko, they are my inspiration always. However, this time their patience was tested more thoroughly as this chapter was written during the quarantine due to the Covid-19 pandemic.

Author details

Ivo Crnolatac
Ruđer Bošković Institute, Zagreb, Croatia

*Address all correspondence to: icrnolat@irb.hr

References

[1] Susin SA, Zamzami N, Kroemer G. Mitochondria as regulators of apoptosis: Doubt no more. Biochimica et Biophysica Acta. 1998;**1366**:151-165

[2] Bhola PD, Letai A. Mitochondria—Judges and executioners of cell death sentences. Molecular Cell. 2016;**61**: 695-704. DOI: 10.1016/j.molcel.2016.02.019

[3] Lenaz G. Role of mitochondria in oxidative stress and ageing. Biochimica et Biophysica Acta (BBA)—Bioenergetics. 1998;**1366**:53-67. DOI: 10.1016/S0005-2728(98)00120-0

[4] Sun N, Youle RJ, Finkel T. The mitochondrial basis of aging. Molecular Cell. 2016;**61**:654-666. DOI: 10.1016/j.molcel.2016.01.028

[5] Ichas F, Mazat JP. From calcium signaling to cell death: Two conformations for the mitochondrial permeability transition pore. Switching from low-to high-conductance state. Biochimica et Biophysica Acta. 1998;**1366**:33-50

[6] Bravo-Sagua R, Parra V, López-Crisosto C, Díaz P, Quest AFG, Lavandero S. Calcium transport and signaling in mitochondria. In: Terjung R, editor. Comprehensive Physiology. Hoboken, NJ, USA: John Wiley & Sons, Inc.; 2017. pp. 623-634. DOI: 10.1002/cphy.c160013

[7] Mitra K, Wunder C, Roysam B, Lin G, Lippincott-Schwartz J. A hyperfused mitochondrial state achieved at G1-S regulates cyclin E buildup and entry into S phase. Proceedings of the National Academy of Sciences. 2009;**106**:11960-11965. DOI: 10.1073/pnas.0904875106

[8] Salazar-Roa M, Malumbres M. Fueling the cell division cycle. Trends in Cell Biology. 2017;**27**:69-81. DOI: 10.1016/j.tcb.2016.08.009

[9] Bonilla E, Tanji K, Hirano M, Vu TH, DiMauro S, Schon EA. Mitochondrial involvement in Alzheimer's disease. Biochimica et Biophysica Acta (BBA)—Bioenergetics. 1999;**1410**:171-182. DOI: 10.1016/S0005-2728(98)00165-0

[10] Schapira AHV. Mitochondrial involvement in Parkinson's disease, Huntington's disease, hereditary spastic paraplegia and Friedreich's ataxia. Biochimica et Biophysica Acta (BBA)—Bioenergetics. 1999;**1410**:159-170. DOI: 10.1016/S0005-2728(98)00164-9

[11] Wallace DC. Mitochondria and cancer. Nature Reviews. Cancer. 2012; **12**:685-698. DOI: 10.1038/nrc3365

[12] Modica-Napolitano JS, Kulawiec M, Singh KK. Mitochondria and human cancer. Current Molecular Medicine. 2007;7:121-131. DOI: 10.2174/156652407779940495

[13] Singh KK, Kulawiec M, Still I, Desouki MM, Geradts J, Matsui S-I. Inter-genomic cross talk between mitochondria and the nucleus plays an important role in tumorigenesis. Gene. 2005;**354**:140-146. DOI: 10.1016/j.gene.2005.03.027

[14] Lee S, Kim S, Sun X, Lee J-H, Cho H. Cell cycle-dependent mitochondrial biogenesis and dynamics in mammalian cells. Biochemical and Biophysical Research Communications. 2007;**357**: 111-117. DOI: 10.1016/j.bbrc.2007.03.091

[15] Gray MW, Burger G, Lang BF. Mitochondrial evolution. Science. 1999; **283**:1476-1481. DOI: 10.1126/science.283.5407.1476

[16] D'Souza AR, Minczuk M. Mitochondrial transcription and translation: Overview. Essays in Biochemistry. 2018;**62**:309-320. DOI: 10.1042/EBC20170102

[17] Ott M, Herrmann JM. Co-translational membrane insertion of mitochondrially encoded proteins. Biochimica et Biophysica Acta. 2010; **1803**:767-775. DOI: 10.1016/j.bbamcr.2009.11.010

[18] Schmidt O, Pfanner N, Meisinger C. Mitochondrial protein import: From proteomics to functional mechanisms. Nature Reviews. Molecular Cell Biology. 2010;**11**:655-667. DOI: 10.1038/nrm2959

[19] PDHA1. Pyruvate Dehydrogenase E1 Component Subunit Alpha, Somatic Form, Mitochondrial Precursor—Homo Sapiens (Human). PDHA1 Gene & Protein. Available from: https://www.uniprot.org/uniprot/P08559

[20] COX8A. Cytochrome C Oxidase Subunit 8A, Mitochondrial Precursor—Homo Sapiens (Human)—COX8A Gene & Protein. Available from: https://www.uniprot.org/uniprot/P10176

[21] CellLight™ Mitochondria-GFP, BacMam 2.0. Available from: http://www.thermofisher.com/order/catalog/product/C10600

[22] Mitochondria Cyto-Tracer™, pCT-Mito-GFP (MSCV) | System Biosciences. Available from: https://systembio.com/shop/pct-mito-gfp-pmscv-mitochondria-cox8-tag/

[23] Addgene: Fluorescent Protein Guide: Subcellular Localization. Available from: https://www.addgene.org/fluorescent-proteins/localization/

[24] Sirk DP, Zhu Z, Wadia JS, Mills LR. Flow cytometry and GFP: A novel assay for measuring the import and turnover of nuclear-encoded mitochondrial proteins in live PC12 cells. Cytometry. Part A. 2003;**56**:15-22. DOI: 10.1002/cyto.a.10084

[25] Friedman JR, Lackner LL, West M, DiBenedetto JR, Nunnari J, Voeltz GK. ER tubules mark sites of mitochondrial division. Science. 2011;**334**:358-362. DOI: 10.1126/science.1207385

[26] Crnolatac I, Tumir L-M, Lesev NY, Vasilev AA, Deligeorgiev TG, Mišković K, et al. Probing the structural properties of DNA/RNA grooves by sterically restricted phosphonium dyes; screening of dye cell toxicity and uptake. ChemMedChem. 2013;**8**:1093-1103. DOI: 10.1002/cmdc.201300085

[27] Organelle Stains Products, Cell Stains. Thermo Fisher Scientific. Available from: https://www.thermofisher.com/search/results?query=organelle+stains&persona=Catalog&navId=16601&refinement Action=true&focusarea=Suchen+in+Alle

[28] MitoSpy™. Mitochrondrial Probes. Available from: https://www.biolegend.com/en-us/mitospy?gclid=CjwKCAjw7e_0BRB7EiwAlH-goFmlW06US19mt9eRXmaSL4BYT_z527E2E-icIoP1YmYiWjUq_AvOvhoCemEQAvD_BwE

[29] Šarić A, Crnolatac I, Bouillaud F, Sobocanec S, Mikecin A-M, Mačak Šafranko Ž, et al. Non-toxic fluorescent phosphonium probes to detect mitochondrial potential. Methods and Applications in Fluorescence. 2017;**5**:015007. DOI: 10.1088/2050-6120/aa5e64

[30] Kurutos A, Orehovec I, Tomašić Paić A, Crnolatac I, Horvat L, Gadjev N, et al. New series of non-toxic DNA intercalators, mitochondria targeting fluorescent dyes. Dyes and Pigments. 2017;**148**:452-459. DOI: 10.1016/j.dyepig.2017.09.049

[31] Kurutos A, Orehovec I, Saftić D, Horvat L, Crnolatac I, Piantanida I, et al. Cell penetrating, mitochondria targeting multiply charged DABCO-cyanine dyes. Dyes and Pigments. 2018;**158**:517-525. DOI: 10.1016/j.dyepig.2018.05.035

[32] Manders EMM, Verbeek FJ, Aten JA. Measurement of co-localization

of objects in dual-colour confocal images. Journal of Microscopy. 1993; **169**:375-382. DOI: 10.1111/j.1365-2818.1993.tb03313.x

[33] Schneider CA, Rasband WS, Eliceiri KW. NIH image to ImageJ: 25 years of image analysis. Nature Methods. 2012;**9**:671-675. DOI: 10.1038/nmeth.2089

[34] Adler J, Pagakis SN, Parmryd I. Replicate-based noise corrected correlation for accurate measurements of colocalization. Journal of Microscopy. 2008;**230**:121-133. DOI: 10.1111/j.1365-2818.2008.01967.x

[35] Demchenko AP. Photobleaching of organic fluorophores: Quantitative characterization, mechanisms, protection. Methods and Applications in Fluorescence. 2020;**8**:022001. DOI: 10.1088/2050-6120/ab7365

[36] Armitage BA. Cyanine dye–DNA interactions: Intercalation, groove binding, and aggregation. In: Waring MJ, Chaires JB, editors. DNA Binders and Related Subjects. Topics in Current Chemistry. Berlin, Heidelberg: Springer; 2005. pp. 55-76. DOI: 10.1007/b100442

[37] Honig BH, Hubbell WL, Flewelling RF. Electrostatic interactions in membranes and proteins. Annual Review of Biophysics and Biophysical Chemistry. 1986;**15**:163-193. DOI: 10.1146/annurev.bb.15.060186.001115

[38] Flewelling RF, Hubbell WL. The membrane dipole potential in a total membrane potential model. Applications to hydrophobic ion interactions with membranes. Biophysical Journal. 1986;**49**:541-552. DOI: 10.1016/S0006-3495(86)83664-5

[39] Ross MF, Kelso GF, Blaikie FH, James AM, Cochemé HM, Filipovska A, et al. Lipophilic triphenylphosphonium cations as tools in mitochondrial bioenergetics and free radical biology. Biochemistry (Moscow). 2005;**70**: 222-230. DOI: 10.1007/s10541-005-0104-5

[40] Ketterer B, Neumcke B, Läuger P. Transport mechanism of hydrophobic ions through lipid bilayer membranes. Journal of Membrane Biology. 1971;**5**: 225-245. DOI: 10.1007/BF01870551

[41] Alberts B, Johnson A, Lewis J, Raff M, Roberts K, Walter P. Molecular Biology of the Cell. 4th ed. New York, USA: Garland Science; 2002

[42] Nicholls DG, Ward MW. Mitochondrial membrane potential and neuronal glutamate excitotoxicity: Mortality and millivolts. Trends in Neurosciences. 2000;**23**:166-174. DOI: 10.1016/s0166-2236(99)01534-9

[43] David NG, Ferguson SJ. Bioenergetics. 4th ed. Amsterdam: Academic Press; 2013

[44] Santo-Domingo J, Demaurex N. The renaissance of mitochondrial pH. The Journal of General Physiology. 2012;**139**: 415-423. DOI: 10.1085/jgp.201110767

[45] Perry SW, Norman JP, Barbieri J, Brown EB, Gelbard HA. Mitochondrial membrane potential probes and the proton gradient: A practical usage guide. BioTechniques. 2011;**50**:98-115. DOI: 10.2144/000113610

[46] Chazotte B. Labeling mitochondria with JC-1. Cold Spring Harbor Protocols. 2011;**2011**. DOI: 10.1101/pdb.prot065490

[47] Scaduto RC, Grotyohann LW. Measurement of mitochondrial membrane potential using fluorescent rhodamine derivatives. Biophysical Journal. 1999;**76**:469-477

[48] Rottenberg H, Wu S. Quantitative assay by flow cytometry of the mitochondrial membrane potential in

intact cells. Biochimica et Biophysica Acta. 1998;**1404**:393-404. DOI: 10.1016/s0167-4889(98)00088-3

[49] Nicholls DG. Commentary on: 'Old and new data, new issues: The mitochondrial Δψ' by H. Tedeschi. Biochimica et Biophysica Acta (BBA)—Bioenergetics. 2005;**1710**:63-65. DOI: 10.1016/j.bbabio.2005.09.002

[50] Lambert AJ, Brand MD. Superoxide production by NADH:ubiquinone oxidoreductase (complex I) depends on the pH gradient across the mitochondrial inner membrane. The Biochemical Journal. 2004;**382**:511-517. DOI: 10.1042/BJ20040485

[51] Jiang D, Zhao L, Clapham DE. Genome-wide RNAi screen identifies Letm1 as a mitochondrial Ca^{2+}/H^+ antiporter. Science. 2009;**326**:144-147. DOI: 10.1126/science.1175145

[52] Nowikovsky K, Froschauer EM, Zsurka G, Samaj J, Reipert S, Kolisek M, et al. The LETM1/YOL027 gene family encodes a factor of the mitochondrial K+ homeostasis with a potential role in the wolf-Hirschhorn syndrome. The Journal of Biological Chemistry. 2004;**279**: 30307-30315. DOI: 10.1074/jbc.M403607200

[53] Bernardi P. Mitochondrial transport of cations: Channels, exchangers, and permeability transition. Physiological Reviews. 1999;**79**:1127-1155. DOI: 10.1152/physrev.1999.79.4.1127

[54] Palmieri F. The mitochondrial transporter family (SLC25): Physiological and pathological implications. Pflügers Archiv. 2004; **447**:689-709. DOI: 10.1007/s00424-003-1099-7

[55] Hoek JB, Nicholls DG, Williamson JR. Determination of the mitochondrial protonmotive force in isolated hepatocytes. The Journal of Biological Chemistry. 1980;**255**:1458-1464

[56] Lemasters JJ, Chacon E, Ohata H, Harper IS, Nieminen AL, Tesfai SA, et al. Measurement of electrical potential, pH, and free calcium ion concentration in mitochondria of living cells by laser scanning confocal microscopy. Methods in Enzymology. 1995;**260**:428-444. DOI: 10.1016/0076-6879(95)60156-2

[57] SNARF pH Indicators. Available from: http://tools.thermofisher.com/content/sfs/manuals/mp01270.pdf

[58] Kneen M, Farinas J, Li Y, Verkman AS. Green fluorescent protein as a noninvasive intracellular pH indicator. Biophysical Journal. 1998;**74**:1591-1599

[59] Abad MFC, Di Benedetto G, Magalhães PJ, Filippin L, Pozzan T. Mitochondrial pH monitored by a new engineered green fluorescent protein mutant. The Journal of Biological Chemistry. 2004;**279**:11521-11529. DOI: 10.1074/jbc.M306766200

[60] Li X, Hu Y, Li X, Ma H. Mitochondria-immobilized near-infrared ratiometric fluorescent pH probe to evaluate cellular mitophagy. Analytical Chemistry. 2019;**91**:11409-11416. DOI: 10.1021/acs.analchem.9b02782

[61] Li G, Zhang B, Song X, Xia Y, Yu H, Zhang X, et al. Ratiometric imaging of mitochondrial pH in living cells with a colorimetric fluorescent probe based on fluorescein derivative. Sensors and Actuators B: Chemical. 2017;**253**:58-68. DOI: 10.1016/j.snb.2017.06.065

[62] Murphy MP. How mitochondria produce reactive oxygen species. The Biochemical Journal. 2009;**417**:1-13. DOI: 10.1042/BJ20081386

[63] Balaban RS, Nemoto S, Finkel T. Mitochondria, oxidants, and aging. Cell. 2005;**120**:483-495. DOI: 10.1016/j.cell.2005.02.001

[64] Okado-Matsumoto A, Fridovich I. Subcellular distribution of superoxide

dismutases (SOD) in rat liver: Cu,Zn-SOD in mitochondria. The Journal of Biological Chemistry. 2001;**276**: 38388-38393. DOI: 10.1074/jbc.M105395200

[65] Cox AG, Winterbourn CC, Hampton MB. Mitochondrial peroxiredoxin involvement in antioxidant defence and redox signalling. The Biochemical Journal. 2009;**425**: 313-325. DOI: 10.1042/BJ20091541

[66] Moncada S, Erusalimsky JD. Does nitric oxide modulate mitochondrial energy generation and apoptosis? Nature Reviews. Molecular Cell Biology. 2002;**3**:214-220. DOI: 10.1038/nrm762

[67] Eaton P. Protein thiol oxidation in health and disease: Techniques for measuring disulfides and related modifications in complex protein mixtures. Free Radical Biology and Medicine. 2006;**40**:1889-1899. DOI: 10.1016/j.freeradbiomed.2005.12.037

[68] Zielonka J, Kalyanaraman B. Hydroethidine- and MitoSOX-derived red fluorescence is not a reliable indicator of intracellular superoxide formation: Another inconvenient truth. Free Radical Biology and Medicine. 2010;**48**:983-1001. DOI: 10.1016/j.freeradbiomed.2010.01.028

[69] Robinson KM, Janes MS, Beckman JS. The selective detection of mitochondrial superoxide by live cell imaging. Nature Protocols. 2008;**3**: 941-947. DOI: 10.1038/nprot.2008.56

[70] Dickinson BC, Chang CJ. A targetable fluorescent probe for imaging hydrogen peroxide in the mitochondria of living cells. Journal of the American Chemical Society. 2008;**130**:9638-9639. DOI: 10.1021/ja802355u

[71] Lim CS, Masanta G, Kim HJ, Han JH, Kim HM, Cho BR. Ratiometric detection of mitochondrial thiols with a two-photon fluorescent probe. Journal of the American Chemical Society. 2011; **133**:11132-11135. DOI: 10.1021/ja205081s

[72] Kaur A, Brigden KWL, Cashman TF, Fraser ST, New EJ. Mitochondrially targeted redox probe reveals the variations in oxidative capacity of the haematopoietic cells. Organic & Biomolecular Chemistry. 2015;**13**: 6686-6689. DOI: 10.1039/C5OB00928F

[73] Koide Y, Urano Y, Kenmoku S, Kojima H, Nagano T. Design and synthesis of fluorescent probes for selective detection of highly reactive oxygen species in mitochondria of living cells. Journal of the American Chemical Society. 2007;**129**:10324-10325. DOI: 10.1021/ja073220m

[74] Pires MM, Chmielewski J. Fluorescence imaging of cellular glutathione using a latent rhodamine. Organic Letters. 2008;**10**:837-840. DOI: 10.1021/ol702769n

[75] Kaur A, Jankowska K, Pilgrim C, Fraser ST, New EJ. Studies of hematopoietic cell differentiation with a ratiometric and reversible sensor of mitochondrial reactive oxygen species. Antioxidants & Redox Signaling. 2016; **24**:667-679. DOI: 10.1089/ars.2015.6495

[76] Xu K, Qiang M, Gao W, Su R, Li N, Gao Y, et al. A near-infrared reversible fluorescent probe for real-time imaging of redox status changes in vivo. Chemical Science. 2013;**4**:1079-1086. DOI: 10.1039/C2SC22076H

[77] Wen Y, Liu K, Yang H, Liu Y, Chen L, Liu Z, et al. Mitochondria-directed fluorescent probe for the detection of hydrogen peroxide near mitochondrial DNA. Analytical Chemistry. 2015;**87**:10579-10584. DOI: 10.1021/acs.analchem.5b03326

[78] Hanson GT, Aggeler R, Oglesbee D, Cannon M, Capaldi RA, Tsien RY, et al. Investigating mitochondrial redox potential with redox-sensitive green

fluorescent protein indicators. The Journal of Biological Chemistry. 2004; **279**:13044-13053. DOI: 10.1074/jbc. M312846200

[79] Waypa GB, Marks JD, Guzy R, Mungai PT, Schriewer J, Dokic D, et al. Hypoxia triggers subcellular compartmental redox signaling in vascular smooth muscle cells. Circulation Research. 2010;**106**:526-535. DOI: 10.1161/CIRCRESAHA.109.206334

[80] Belousov VV, Fradkov AF, Lukyanov KA, Staroverov DB, Shakhbazov KS, Terskikh AV, et al. Genetically encoded fluorescent indicator for intracellular hydrogen peroxide. Nature Methods. 2006;**3**: 281-286. DOI: 10.1038/nmeth866

[81] Østergaard H, Henriksen A, Hansen FG, Winther JR. Shedding light on disulfide bond formation: Engineering a redox switch in green fluorescent protein. The EMBO Journal. 2001;**20**:5853-5862. DOI: 10.1093/emboj/20.21.5853

[82] Gutscher M, Pauleau A-L, Marty L, Brach T, Wabnitz GH, Samstag Y, et al. Real-time imaging of the intracellular glutathione redox potential. Nature Methods. 2008;**5**:553-559. DOI: 10.1038/ nmeth.1212

[83] Albrecht SC, Barata AG, Großhans J, Teleman AA, Dick TP. In vivo mapping of hydrogen peroxide and oxidized glutathione reveals chemical and regional specificity of redox homeostasis. Cell Metabolism. 2011;**14**:819-829. DOI: 10.1016/j.cmet.2011.10.010

[84] Ermakova YG, Bilan DS, Matlashov ME, Mishina NM, Markvicheva KN, Subach OM, et al. Red fluorescent genetically encoded indicator for intracellular hydrogen peroxide. Nature Communications. 2014;**5**:1-9. DOI: 10.1038/ncomms6222

[85] Si F, Liu Y, Yan K, Zhong W. A mitochondrion targeting fluorescent probe for imaging of intracellular superoxide radicals. Chemical Communications. 2015;**51**:7931-7934. DOI: 10.1039/C5CC01075F

[86] Yang K, Kolanowski JL, New EJ. Mitochondrially targeted fluorescent redox sensors. Interface Focus. 2017;**7**: 20160105. DOI: 10.1098/rsfs.2016.0105

[87] Arai S, Suzuki M, Park S-J, Yoo JS, Wang L, Kang N-Y, et al. Mitochondria-targeted fluorescent thermometer monitors intracellular temperature gradient. Chemical Communications (Cambridge, England). 2015;**51**: 8044-8047. DOI: 10.1039/c5cc01088h

[88] Chrétien D, Bénit P, Ha H-H, Keipert S, El-Khoury R, Chang Y-T, et al. Mitochondria are physiologically maintained at close to 50°C. PLOS Biology. 2018;**16**:e2003992. DOI: 10.1371/journal.pbio.2003992

[89] Jiang N, Fan J, Zhang S, Wu T, Wang J, Gao P, et al. Dual mode monitoring probe for mitochondrial viscosity in single cell. Sensors and Actuators B: Chemical. 2014;**190**:685-693. DOI: 10.1016/j.snb.2013.09.062

[90] Hell SW, Wichmann J. Breaking the diffraction resolution limit by stimulated emission: Stimulated-emission-depletion fluorescence microscopy. Optics Letters. 1994;**19**: 780-782. DOI: 10.1364/OL.19.000780

[91] Ishigaki M, Iketani M, Sugaya M, Takahashi M, Tanaka M, Hattori S, et al. STED super-resolution imaging of mitochondria labeled with TMRM in living cells. Mitochondrion. 2016;**28**: 79-87. DOI: 10.1016/j.mito.2016.03.009

[92] Wang C, Taki M, Sato Y, Tamura Y, Yaginuma H, Okada Y, et al. A photostable fluorescent marker for the superresolution live imaging of the dynamic structure of the mitochondrial cristae. PNAS. 2019;**116**:15817-15822. DOI: 10.1073/pnas.1905924116

Chapter 14

Confocal Laser Scanning Microscopy of Living Cells

Alexey Moshkov

Abstract

Living cells are distinguished in their ability to keep the structural stability in the changing environment by means of energy supply. Understanding the mechanisms of energy conversion is impossible without knowledge of the physiological processes involved. This question is addressed to confocal laser scanning microscopy (CLSM) that has become a leader among the group of light microscopy method. The high resolution in CLSM images is reached without deconvolution procedures. Linear characteristics of its variable parameters are important for quantitative image analysis. These properties of CLSM are indispensable in the study of mitochondrial membrane potential (MMP) using fluorescent dyes. Lipophilic rhodamine dyes freely pass the plasma membrane and accumulate in the mitochondria according to electrical potential difference. CLSM images of TMRE (tetramethylrhodamine, ethyl ester)-stained U937 cells were analyzed by standard image processing procedures in ImageJ software. These procedures allow the creation of mask, applying it to original image. Average fluorescence intensity in selected regions was used as a measure of MMP changes during phytohemagglutinin treatment. This value was decreased by 34% after 2 h of lectin treatment. Some cells after mitogenic stimulation completely lost MMP. Deregulation of mitochondrial calcium handling and changes of cytoplasmic monovalent ion concentration are considered as mechanisms of MMP decrease during lymphocyte stimulation.

Keywords: confocal microscopy, mitochondrial membrane potential, image processing, image analysis, phytohemagglutinin, lymphocyte stimulation

1. Introduction

Before cultured cells were available, the main objects of cell physiology were nerve and blood cells. The study of giant axons showed that excitability and irritability are the most important properties of living cells. Russian physiologist E. Bauer used these properties to affirm the principle of "stable nonequilibrium" [1]. This principle explains that in the changing environment, living systems maintain their structural stability through energy supply. Instead of "energy," however, he uses the term "work" performed by system against equilibrium. Physiological implication of this principle was found upon the study of nonequal redistribution of Na^+ and K^+ ions in cuttlefish neuron during impulse propagation [2].

Physiological studies of living cells are always done in conjunction with light microscopy methods. Later these methods got resolution sufficient to study cell structures in molecular scale. The maximum resolution that allows observation of hexagonal pattern on diatom algae is reached in basic microscopy by shutting the condenser [3]. Near-field scanning optical microscopy (NSOM) overcomes the diffraction limit by using optical fiber for both illumination and scanning [4]. NSOM was able to visualize small antibody-labeled domains on the plasma membrane of living human skin fibroblasts [5]. The small thickness of these cells uses conventional epi-fluorescence microscopy to study the mitochondrial membrane potential (MMP) [6, 7] using the protocol of automated image analysis of mitochondrial shape originally tested by confocal microscopy [8, 9]. The same functionality but greater performance resides in the method of imaging flow cytometry. The method allows fluorescence quantification using spot masks that are created independently on pixel intensity. This method was applied for the characterization of MMP changes involved in the malignant transformation of cancer stem cells. MMP together with such parameters as glucose uptake, superoxide-anion production, and mitochondria mass served as indicator of tumorigenesis [10].

Besides using CCD camera in conventional fluorescence microscopy, video microscopy suggests built-in image processing algorithms that cause inadvertent effect on pixel intensity, but with additional calibration procedure, absolute intensities are also measurable [11]. Instead of these algorithms, conventional microscopy uses deconvolution to improve the quality of images as in confocal microscopy [12]. For example, deconvolution was applied to wide-field microscopy images to resolve DNA replication units that are studied only by electron and structured illumination microscopy [13]. Unlike these later methods, confocal and classical light microscopy possesses the ability to study these structures in the living cells.

Discrimination of depth is a property of confocal microscopy that relies to confocal diaphragm [14] and can be checked by measuring the point spread function (PSF), which is defined as the image of a single point [15]. To see the changes on the image of real object, which contains convoluted PSFs, is impossible unless deconvolution is applied [12]. Special image processing algorithm is also applied in confocal microscopy for quantitative fluorescence analysis. This algorithm was applied in quantitative studies of mitochondrial $[Na^+]$ ($[Na^+]_m$) and Ca^{2+} ($[Ca^{2+}]_m$) and mitochondrial pH (pH_m) in MDCK cells stained with one of the ion-sensitive dyes [16–18]. Obtaining information about cellular organelles such as mitochondria requires creation of mask.

The same analytical approach was applied for the study of MMP changes caused by phytohemagglutinin (PHA) treatment in histiocytic lymphoma U937 cells. The image of TMRE-stained cells was used for both creating the mask and obtaining quantitative data. The value of average intensity obtained from masked image is independent of the wide variation of MMP. Therefore, this image can serve as a measure of MMP. To consider the effect of large mitochondrial depolarization that occurs in cells after PHA treatment, data were presented as integrated density divided by the number of cells on the image.

PHA is a lectin that stimulates growth and division of lymphocytes in the process known as transformation [19]. Previous studies using indirect isotopic method have not found considerable MMP change during mitogenic stimulation [20]. At the same time, lectins are affecting all known metabolic pathways. Early response in lymphocytes to PHA treatment includes stimulation of glycolysis [21], activation of pentose-phosphate cycle [22], and activation of pyruvate dehydrogenase (PDH) [23] that is a key enzyme to connecting glycolysis and citric acid cycle.

2. Principles of confocal laser scanning microscopy (CLSM)

The history of CLSM began when Marvin Minsky include in his patent letter the scheme with epi-illumination with two pinholes that play the same role as in "double focusing" scheme which have already been tested with a stage-scanning device [24]. Historical information and many technical details are not covered in the present paper but can be found in an encyclopedia article written by Amos and colleagues [14]. From a biological point of view, the development of CLSM began in 1987 when microscopes with laser beam deflection system were able to scan living kidney epithelium cells [25] and spinal cord neurons [26].

The development of CLSM ranging from 1987 to 1995 is distinctive by exponential rise of citations for the query "microscopy AND confocal" (**Figure 1**). At this period, the number of studies retrieved by the query "microscopy NOT confocal" shows a rather decelerating trend. In 1995, the number of confocal microscopy studies started to grow linearly, while many microscopic studies remain unpublished. The second period ended with the new millennium when linear trend of CLSM publications was accompanied by extensive growth of other microscopic studies. The third period is therefore noticeable by concomitant development of all microscopic methods. During this period, confocal microscopy constitutes one sixth of microscopic papers. Indeed, this share would be greater if query is restricted only to cytological area. It can be concluded that the first period caused a 10-year delay in the development of other microscopical methods, while the second period caused its rapid growth.

Ideal confocal microscope has detector pinhole that is small enough to suppress the diffraction in emission light path. This made its resolution equally dependent from excitation and emission light paths [14]. Lateral resolution of real confocal microscope is determined by the illuminating pinhole that participates in the formation of PSF [15]. Therefore, lateral resolution of confocal microscope is given by formula

$$\mathrm{FWHM}_{lateral} = \frac{0.51\lambda_{ex}}{NA} \quad (1)$$

where FWHM is the distance between the points on distribution where its intensity is half of that of the peak intensity, λ_{ex} is the wavelength of excitation light, and NA is the numerical aperture of objective. At NA = 1.42 and λ_{ex} = 488 nm, lateral resolution of confocal microscope is determined as 0.175 μm, while at λ_{ex} = 561 nm, it reduced to 0.2 μm.

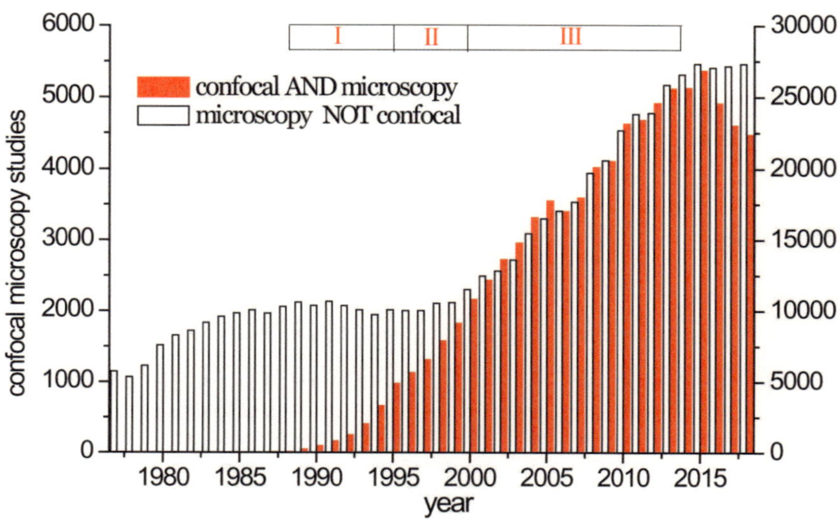

Figure 1.
Relation between the studies done by confocal or other microscopical methods from 1972 to 2018 following the records from PubMed. Trends of publications by year were downloaded from PubMed web page [27] after placing queries: "confocal and microscopy" and "confocal not microscopy" and pressing the button "timeline" if necessary. Three periods in CLSM development are marked on the graph (see text for discussion).

In real confocal microscope, the detector pinhole determines its axial resolution. The difference in axial resolution between confocal and conventional microscopy can be demonstrated using thin fluorescent sheet. By switching microscope to epi-fluorescent mode, no changes in fluorescence intensity are observed upon moving the test object out of focus. Back to the confocal microscopy mode, fluorescence intensity will decrease in sigmoid manner, the slope of which will depend on pinhole diameter [14]. Thickness of optical section is then given by the formula

$$\mathrm{FWHM} = \frac{0.67\lambda_{ex}}{n - \sqrt{n^2 - NA^2}}\sqrt{1 + \mathrm{AU}^2} \quad (2)$$

where n is the refractive index of immersion medium (n = 1.52), NA is the numerical aperture of objective lens, and AU is the pinhole diameter in Airy units (or simply "Airy"). Eq. (2) predicts that FWHM of optical section will increase from 0.47 to 1.06 μm by changing AU from 1 to 3 units. Similarly changing AU from 1 to 0.6 units will decrease FWHM from 0.47 to 0.39 μm, i.e., by 0.08 μm. These theoretical calculations almost coincide with the data obtained in the real test.

In the current study, we use a step-function fluorescent object for testing the limits of axial resolution of CLSM. The test object was prepared as described in [14] but with the following modifications. Fluoroshield™ mounting media containing 20 μg/ml rhodamine 6G was placed between a microscopic slide and 24 mm squire coverslip and left under 1 kg weight until drying. Fluorescence intensity profiles were obtained by scanning fluorescent object in X,Z direction at different pinhole sizes. As shown in **Figure 2A**, an increase of pinhole diameter from 1.05 to 3.0 Airy widens fluorescence intensity profile and fluorescence background. Taking the advantage of normal distribution of fluorescence values, we use Gauss function for obtaining FWHM values of intensity curves. The effect of pinhole diameter ranging from 0.45 to 4 Airy units on profile width is presented in **Figure 2B**.

Changes of profile width are correlated with the changes in the optical thickness upon variation of pinhole diameter up to the 3 Airy units. Namely, changing of the pinhole diameter from 1.05 to 3 Airy units increases the width of intensity profile by 0.6 μm. Changing pinhole in reverse order, from 1.05 to 0.65 Airy units, decreases the width of intensity profile by 0.1 μm (**Figure 2B**).

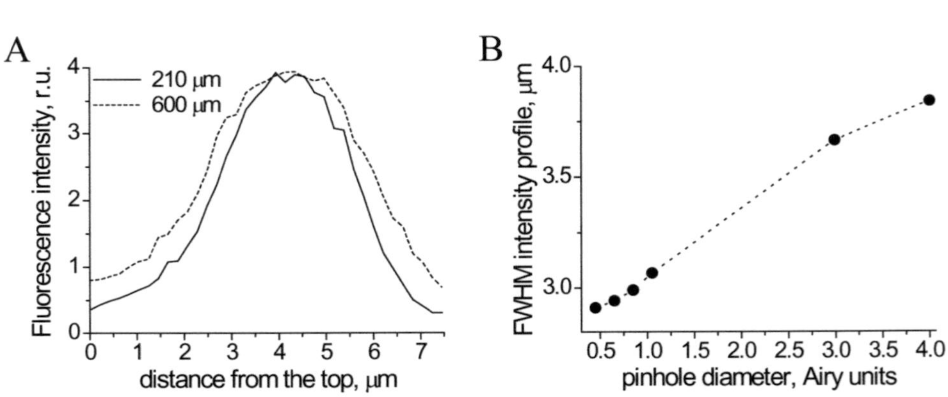

Figure 2.
Effect of pinhole diameter on the width of fluorescence profile recorded across a layer of Fluoroshield™ media containing 20 μg/ml rhodamine G. (A) Fluorescent intensity profiles of X,Z optical sections are registered by Olympus FV3000 confocal microscope equipped with 60x/1.42 plan apochromat objective. Fluorescence was excited at 488 nm and collected from 500 to 600 nm. Two fluorescent profiles obtained at pinhole diameters of 210 and 600 μm (corresponding to 1.05 and 3 Airy units) are shown. One curve is the mean of seven plots generated using ImageJ program. Fluorescence intensity is kept constant by adjustment of laser power. (B) FWHM is derived from the intensity profiles obtained at pinhole diameters of 0.45, 0.65, 0.8, 1.05, 3, and 4 Airy units.

The absolute resolution values are usually different from those predicted by Eqs. (1) and (2). Practically it is determined by the measurement of PSF, which is generated by microspheres with a diameter of 170 nm in lateral and axial planes. In one practical study, FWHM was determined as 0.32 and 1.9 μm in X,Y or X,Z directions, respectively [18]. The high value of axial FWHM depth is explained by the necessary use of high pinhole diameter in a study of mitochondrial pH in MDCK cells. Pinhole diameter of about three Airy units was applied in this study to attenuate laser power and minimize its photo-damaging effect on living cells [18, 28].

The resolution of CLSM in large extent depends on the quality of objective lens. Using the correct objective in CLSM is especially important because lens parameters are assumed in the design of particular confocal system. Choosing lens with the same nominal parameters but designed for other microscopes results in more than twice a decrease of resolution [29]. In addition, high-magnification oil-immersion objectives work correctly in media with the refractive index being very close to the refractive index of living cells. Any mismatch of the refractive index decreases the quantity of excited fluorescence and, therefore, resolution of confocal system. Water immersion objectives allow working with deeply lying cells [30].

3. Using CLSM in studies of mitochondria

Earlier confocal systems have difficulties in using multiple dyes. Therefore, studies were done in parallel samples assuming that experimental conditions equally apply for both dyes. SNARF-1 (seminaphtorhodafluor-1, free form) was used for the study of pH of luminal solution along colon crypt. Using CLSM localization SNARF-1 was compared with the localization of dye Lucifer yellow [30]. Pseudo-ratiometric approach utilizes cells stained simultaneously with two dyes localized in the same cellular compartment. MitoTracker Green (MTG) resides in the mitochondria and therefore can be colocalized with the dye of interest [16–18]. As the latter were used one of the following dyes, Rhod-2 (AM), CoroNa Red, or SNARF-1 (AM), accumulated in the mitochondria and giving information on cytoplasmic concentrations of $[Na^+]_m$, $[Ca^{2+}]_m$, and pH_m, respectively. By using Fura-2 (AM), it was shown that during metabolic inhibition, the main source for transient increase in cytoplasmic Ca^{2+} concentration ($[Ca^{2+}]_c$) is the mitochondria [16], and only functionally active mitochondria can have buffering capacity to Ca^{2+} [31].

A large number of studies are directed to understand mitochondrial heterogeneity, which is manifested as the differences in shape and size of the mitochondria in a single cell [8, 9] or morphological differences between cell types [32]. Some cell mitochondria appear as network but in others as discrete individuals [32]. Fission and fusion that maintain the dynamic structure of mitochondrial network are mechanisms involved in the regulation of cell proliferation and apoptosis [33]. These processes are sensitive to cell metabolic state and MMP. Elongated mitochondria are dominant in mouse embryonic fibroblasts grown in conditions promoting oxidative phosphorylation. The addition of glucose suppresses elongation and causes fragmentation [34]. In the study, reviewed in the next section, the attempt was made to uncover the dynamic nature of MMP changes and its role in maintaining mitochondrial network structure.

4. Using CLSM in quantitative MMP studies

In studies of MMP, it is important to consider the thickness of optical section because it determines the affectivity of detection of mitochondrial fluorescence.

FWHM of optical section that is calculated by Eq. (2) is almost fit to the dimensions of the mitochondria in U937 cells. Electron microscopic studies follow that the mitochondria in U937 cells are spheroidal in structures with a diameter of 0.6–0.8 μm [35]. Therefore, CLSM is effectively collecting fluorescence from the mitochondria of TMRE-stained U937 cells. The mitochondria in lymphocytes have a diameter of only 0.3–0.4 μm [19, 36, 37], and the affectivity of collection of mitochondrial fluorescent signal is less than U937 cells. The mitochondrial shape is also considered in choosing the right procedure for image processing. Thus, different procedures are required for the study of mitochondria in U937 cells or in skin fibroblasts having threadlike appearance [38].

Using CLSM in the study of MMP, it is possible to select for analysis the whole cell area. This selection includes nonmitochondrial compartment that constitutes about 95–98% of the total cell volume [10, 20]. Its signal can be set as a background and subtracted from total fluorescence. The example of this approach is found in a study of MMP changes in lamprey hepatocytes during prespawning migration. As can be expected, CLSM gave results similar to that of flow cytometry [39]. The reason why this approach is chosen by authors is the absence of parallel control samples that are necessary for analytical procedures involving image processing.

Another simplified approach is based on staining of cells with TMRE and MTG and using MTG as reference. For example, TMRE/MTG fluorescence ratio was decreased in the muscle cells of zebrafish embryo after chronical treatment by rotenone [40]. The utilization of MTG as reference dye however is limited because its uptakes in some types of cells depend on MMP [41]. Therefore, MTG is generally used for the creation of mask [16–18]. By combination of these approaches, it was shown that both TMRE/MTG ratio and MTG signal increased during malignant transformation of mesenchymal stem cells [10].

Good substitute for MTG is mitochondria-specific green fluorescence proteins (GFP). MitoAcGFP1 located in mitochondrial matrix was used for the study of mitochondrial $[Ca^{2+}]_m$ [6]. Photoactivatable GFP was used for tracking individual mitochondria in order to relate their MMP to fission events [42, 43]. Data obtained in later studies lead authors to the conclusion that depolarization triggers fission of the mitochondria and fission serves as the quality control for its functional properties [43]. However, the applicability of GFP is limited by insufficient level of expression in some cell types [32, 42]. Rhodamine ester dyes were also used as reference dye in the study of $[Ca^{2+}]_m$ [31]. Evidence was provided that TMRM (tetramethylrhodamine, methyl ester) is not liable to auto-quenching and therefore suitable both for the creation of mask and determination of MMP [6]. This is especially important because data obtained using dye rhodamine 123 do not allow their interpretation in terms of MMP.

CLSM experiments with living cells are preceded by cell staining. During staining, TMRE dye almost completely uptakes the cells; hence, it is important to keep constant the dye-to-cell ratio. This ratio should also be equal in control and treated cells. During image acquisition it is possible to use previously saved settings. The function of "auto-exposure" can be used only once with living cells at the beginning of the study [44]. The reason for this is to keep cells from overexposure from laser irradiation. The optimum settings however allow keeping the dynamic range of detector maximum and preventing most of the pixel on image from saturation. Fluorescence intensity is almost linearly dependent on laser power and nearly exponentially—on voltage of photomultiplier tube (PMT). Therefore, it is generally recommended to set PMT voltage first and then adjust laser excitation intensity [44]. With calibration, PMT voltage also can be varied and used for adjustment [18, 28].

5. Digital image processing in quantitative MMP studies

Examples of image processing that can be found in the earliest studies are threshold, gradient filtering, and segmentation. These procedures are applied to images resulted from summing of intensity along the Z-axis [26]. Graphical filters are specific instruments used to eliminate noise or other unwanted information from image. "Rolling ball" algorithm uses rank operators to remove pixels exceeding the local background level and replaces them with pixels of neighborhood intensity. The processed image is then subtracted from the original image [45]. "Top hat" filter is used for processing such complex structures as mitochondrial network in human skin fibroblast [8, 9]. "Rolling ball" filter is suitable for images of U937 cells containing the mitochondria of elliptical shape.

5.1 Image processing in quantitative MMP studies

For demonstration of image processing in our study, U937 cells were treated with PHA (30 μg/ml, 2 h) and stained with 25 nM of TMRE dye. Cell suspension was placed in Plexiglas holder and scanned in the middle plane of the most cells. Histogram equalization was used to check the amount of TMRE in nonmitochondrial compartment. This was done by running command "enhance contrast" in ImageJ program. Apart from the mitochondria and cytoplasm, small fluorescence is present in the nucleus (**Figure 3A** and **B**). To remove background fluorescence, images were processed by "subtract background" algorithm in ImageJ program that utilizes "rolling ball" operator [46]. On the next step, the processed image is subjected to thresholding, which is used to suppress variations at the background level [45]. ImageJ program allows obtaining quantitative data after setting a threshold and execution of commands "create selection" and "analyze stack." However,

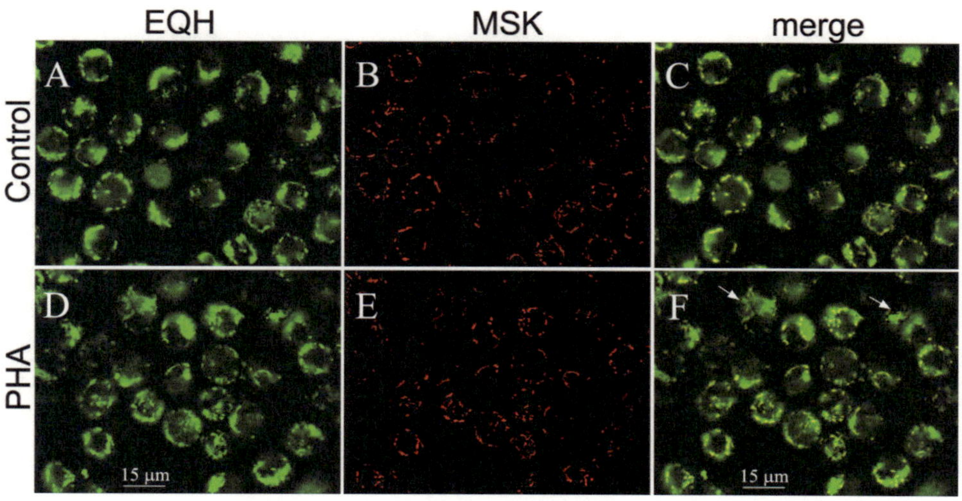

Figure 3.
Demonstration of using image processing for quantitative study of MMP changes in U-937 cell during PHA treatment. Cells were incubated without (A–C) or with p-PHA (30 μg/ml) for 2 h (D–F) in atmosphere of 5% CO2 and stained with 25 nM TMRE for 30 min at 37°C. Excess of media was removed by centrifugation. 20 μl of cell suspension is placed on coverslip and a 1% gelatin solution is attached to Plexiglas holder sealed from another side with coverslip. Fluorescence was excited by 561 nm diode laser and collected at 575–675 nm through pinhole of 178 mm (0.73 Airy units). (A, D), raw images after histogram equalization (EQH); (B, E), masked version of this images (MSK); (C, F), merged images of MSK (red pseudo-color), and EQH (green pseudo-color). Red pixels colocalized with green pixels in an area of high intensity except the blurred regions that are present in some cells. Arrows indicate small aggregates of agglutinated cells.

for demonstration of validity of this procedure, binary mask (mask) was applied to histogram-equalized image (HEQ) rather than to original image (RAW). This operation gives possibility to see variations within regions, which appeared uniform in the original image [45]. The procedure of histogram equalization is similar to the linear contrast stretch used in other works [6]. The resulting masked images (MSK) contain nonzero pixels that correspond to mitochondrial compartment (**Figure 3B** and **E**). Histogram equalization emphasizes mitochondrial heterogeneity and proves the absence of fluorescence in cytoplasmic compartment. However, HEQ images do not give possibility to see the difference of intensities between control- and PHA-treated cells because changes of intensity caused by this procedure are specific to each image [45].

Using merged images it is possible to see similarity between MSK and nonzero regions of HEQ image. As it can be seen in **Figure 3C** and **F**, colocalization is present in all regions except blurred parts that are present in both control- and PHA-treated cells. Good correlation between MSK and HEQ is also seen by "Color inspector 3D" plug-in in ImageJ program. The presence of cell groups that mark the beginning of agglutination process was noticed on the image of PHA-treated cells (**Figure 3F**, arrows). At this period, however, agglutinated cells are lying in plane of imaging available for study.

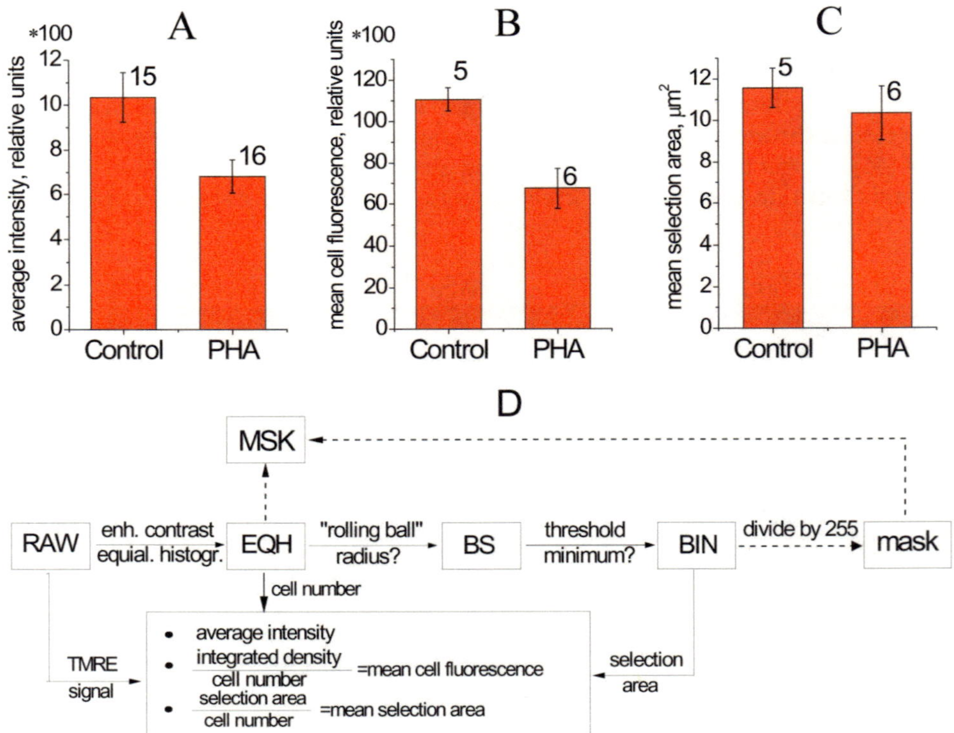

Figure 4.
Quantification of MMP changes in U937 cells during PHA treatment (30 μg/ml, 2 h). (A–C) quantitative data retrieved from images acquired in one (B, C) and three (A) experiments. Data are expressed as means ±SD. The number of images that is used in the analysis is shown near each bar. (A) Average TMRE fluorescence intensity found in selection area. (B) Mean cell fluorescence calculated as integrated density divided by the number of cells. (C) Mean area of selection calculated as total selection area divided by the number of cells. (D) Protocol of imaging processing and analysis. Most important operations are shown by arrows, while its results are indicated in boxes: RAW, original image; EQH, original image after histogram equalization; BS and BIN, images after running "subtracted background" and "threshold" commands, respectively; mask, mask image; MSK, image obtained by multiplying EQH image by mask image. Dot lines show the operations used for the creation of masked image (MSK).

5.2 Study of MMP changes during PHA stimulation

For quantitative analysis of MMP changes, two analytical values were used. Average intensity is a measure of absolute MMP, but it underestimates changes of MMP if they occur below the threshold level (**Figure 4A**). The relative value is obtained by the division of integrated density to cell number considering the presence of cells with completely depolarized MMP. This value is applied more specifically and therefore was named "mean cell fluorescence" (**Figure 4B**). The decrease of average intensity caused by PHA treatment is lower than the corresponding decrease of mean cell fluorescence by 5%. This dissimilarity shows the presence of cells whose fluorescence changed below the detection limit. The changes in the selection area are calculated as the total selection area on images divided by the number of cells attributing to these fluorescent images (**Figure 4C**).

The relation between these analytical options can be understood from the schematic representation of full analytical process shown in **Figure 4D**. It started from original image (RAW), the duplicate of which passes three steps until it resulted in the generation of mask or the selection of congruent regions on RAW image with output of numerical results. Histogram equalization, background subtraction, and thresholding were done similarly to those published in earlier works [6, 28]. The method presented in **Figure 4D** also gives possibility to control mitochondrial selection area and compare it with already known morphological data. From the data presented in **Figure 4C** and from known value of cell area (120 μm^2), it can be concluded that selected area constitutes about 10% of cell area. This is close, for example, to a value of 8% in tumorigenic cell lines obtained by electron-microscopic examination [10].

6. Role of mitochondria in response of lymphoid cells to PHA treatment

The decrease of average fluorescence intensity by 34% after 2 h of PHA treatment is attributed to the changes of absolute MMP value. While our data were obtained on the early stage of stimulation, they can be correlated with data on metabolic shift that occurs in lymphoid cells after prolonged treatment. These data suggest that the role of the mitochondria in total energy balance decreases during stimulation. It was shown that glycolytic activity in lymphocytes increases by 36 times but the respiratory activity is only by 43% [21]. In thymocytes that obtain energy mainly by glucose oxidation, lectin treatment leads to the deceleration of this pathway [47]. On the contrary, in mesenchymal stem cell, mitochondrial activity is increased during malignant transformation [10].

The difference in selection area between control- and PHA-treated cells (**Figure 4C**) suggests that MMP in some cells is completely lost. These changes could be related to the changes of $[Ca^{2+}]_c$ because it is known that $[Ca^{2+}]_c$ is increased after mitogenic stimulation [48]. The mitochondria play essential role in the regulation of $[Ca^{2+}]_c$ [31, 49]. During metabolic inhibition in MDCK cells, $[Ca^{2+}]_c$ is transiently increased. Restoration of normal $[Ca^{2+}]_c$ occurs by coupled action mitochondrial Na^+/Ca^{2+} and Na^+/H^+ exchangers. Metabolic inhibition decreases proton gradient on mitochondrial membrane [18] and reduces outward Na^+ and Ca^{2+} movement mediated by these transporters. As a consequence, Na^+ and Ca^{2+} concentration in mitochondrial matrix is steadily increased [16, 17].

Similar situation may exist in stimulated lymphoid cells where the magnitude of proton electrochemical gradient is compromised by the changes of pH_c. The Na^+ content in lymphocytes is increased rapidly after PHA treatment with maximum of 2 h [50]. Both K^+ content and influx are also increased during mitogenic stimulation [50].

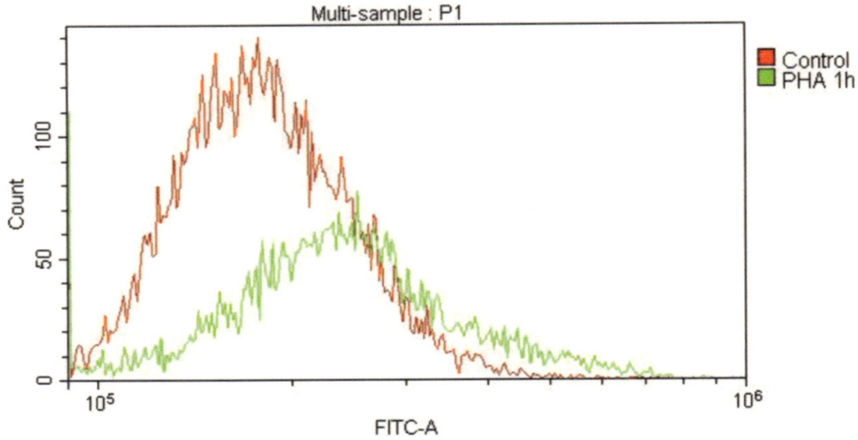

Figure 5.
*Effect of PHA treatment (30 µg/ml, 1 h) on fluorescence of ANG dye. Cells stained with ANG-AM (1 µM) during the last 40 min of PHA treatment (37°C). ANG fluorescence was excited using a 488 nm laser, and emission was detected in the FITC channel with a 525/40 nm bandpass. Data were analyzed using CytExpert 2.0 Beckman Coulter software. The major cell populations were selected for analysis using forward/side scatter plot. Data were obtained using the same samples as data presented in **Figures 3** and **4B** and **C**.*

Therefore, the increase of cytoplasmic $[Na^+]$ ($[Na^+]_c$) in response to lectin treatment is mediated by Na^+/H^+ exchanger. This view agrees with the increase of pH_c in both thymocytes and lymphocytes after lectin treatment [51, 52]. To test whether these changes take place in U937 cells during PHA stimulation, we performed investigation of $[Na^+]_c$ by Na-sensitive dye Asante Natrium Green-2 (ANG). As followed from the flow cytometric data, the fluorescence of ANG was increased after 1 h of PHA treatment (**Figure 5**). The mean ANG fluorescence increased by PHA treatment from 1.89E+05 to 2.59E+05 relative units. Assuming that changes of fluorescence are proportional to the changes of concentration, $[Na^+]_c$ will increase from 30 mM in resting U937 cells to 38 mM upon PHA treatment.

The demonstrated increase of $[Na^+]_c$ in PHA-stimulated U937 cells may be considered as part of the regulatory mechanism involved in the increase of $[Ca^{2+}]_m$. The increase of $[Ca^{2+}]_m$ in many cell types causes activation of mitochondrial dehydrogenases [49] and accounts for the rapid activation of mitochondrial PDH in response to lectin action [23, 53]. The increase of $[Ca^{2+}]_m$ can also lead to mitochondrial injury that was confirmed by serious ultra-structural changes observed in human lymphocytes after 3 days of lectin treatment [36, 54]. The deleterious effect of lectins on mitochondrial morphology, however, was absent in other experiments [19, 37, 54]. Controversy of results obtained in these studies can be explained by the complexity of regulatory mechanisms controlling mitochondrial Ca^{2+} and Na^+ homeostasis. This reason can also be applied in our study in which a large variation in the number of cells with complete mitochondrial depolarization was noticed.

7. Conclusions

The role of the mitochondria in energy supply is determined by the functional state of the cells. This problem is usually addressed by biochemical, polarographic and optical methods which measure ATP production and oxygen consumption in the living cell. In many studies, it is also corroborated by parallel measurement of mitochondrial membrane potential (MMP). This article describes fluorescent method of measuring MMP in cancer cell using potential-sensitive dye TMRE. This method can distinguish fluorescence of TMRE in mitochondrial and nonmitochondrial compartment by using standard software for image analysis.

Acknowledgements

I acknowledge TS Goryachaya for maintaining a cell culture and ND Aksenov for assistance with flow cytometry investigations. I am very grateful to JP Battersby for discussions leading to the current study and to GI Shtein for technical consultations. I am also thankful to Prof. VS Saakov for his help in the writing of this manuscript.

Author details

Alexey Moshkov
Institute of Cytology of Russian Academy of Sciences, St. Petersburg, Russia

*Address all correspondence to: moshov_alexey@mail.ru

References

[1] Frank GM, Tigyi J, Shnol SE, Bauer ES, editors. Theoretical Biology (Reprint of 1935 Edition with Preface, a Biographical and Critical Essay). Budapest: Akadémiai Kiadó; 1982. p. 295

[2] Keynes RD, Lewis PR. The sodium and potassium content of cephalopod nerve fibers. The Journal of Physiology. 1951;**114**:151-182. DOI: 10.1113/jphysiol.1951.sp004609

[3] Lacey AJ. Basic optical microscopy. In: Lacey AJ, editor. Light Microscopy in Biology. A Practical Approach. 2nd ed. New York: Oxford University Press; 1999. pp. 1-43

[4] Van Hulst NF. Near-field optical microscopy. In: Lacey AJ, editor. Light Microscopy in Biology: A Practical Approach. 2nd ed. New York: Oxford University Press; 1999. pp. 341-371

[5] Hwang J, Gheber LA, Margolis L, Edidin M. Domains in cell plasma membranes investigated by near-field scanning optical microscopy. Biophysical Journal. 1998;**74**:2184-2190. DOI: 10.1016/S0006-3495(98)77927-5

[6] Distelmaier F, Koopman WJ, Testa ER, de Jong AS, Swarts HG, Mayatepek E, et al. Life cell quantification of mitochondrial membrane potential at the single organelle level. Cytometry. Part A. 2008;**73**:129-138. DOI: 10.1002/cyto.a.20503

[7] Koopman WJ, Distelmaier F, Esseling JJ, Smeitink JA, Willems PH. Computer-assisted live cell analysis of mitochondrial membrane potential, morphology and calcium handling. Methods. 2008;**46**:304-311. DOI: 10.1016/j.ymeth.2008.09.018

[8] Koopman WJ, Verkaart S, Visch HJ, et al. Inhibition of complex I of the electron transport chain causes O_2^{-}• -mediated mitochondrial outgrowth. American Journal of Physiology. Cell Physiology. 2005;**288**:C1440-C1450. DOI: 10.1152/ajpcell.00607.2004

[9] Koopman WJ, Visch HJ, Smeitink JA, Willems PH. Simultaneous quantitative measurement and automated analysis of mitochondrial morphology, mass, potential, and motility in living human skin fibroblasts. Cytometry. Part A. 2006;**69**:1-12. DOI: 10.1002/cyto.a.20198

[10] Lonetto G, Koifman G, Silberman A, Attery A, Solomon H, Levin-Zaidman S, et al. Mutant p53-dependent mitochondrial metabolic alterations in a mesenchymal stem cell-based model of progressive malignancy. Cell Death and Differentiation. 2019;**26**:1566-1581. DOI: 10.1038/s41418-018-0227-z

[11] Weiss DG. Video microscopy. In: Lacey AJ, editor. Light Microscopy in Biology: A Practical Approach. 2nd ed. New York: Oxford University Press; 1999. pp. 73-149

[12] Shaw PJ. Comparison of widefield/deconvolution and confocal microscopy for three-dimensional imaging. In: Pawley JB, editor. Handbook of Biological Confocal Microscopy. 3d ed. Boston: Springer; 2006. pp. 453-467. DOI: 10.1007/978-0-387-45524-2_23

[13] Deng X, Zhironkina OA, Cherepanynets VD, Strelkova OS, Kireev II, Belmont AS. Cytology of DNA replication reveals dynamic plasticity of large-scale chromatin fibers. Current Biology. 2016;**26**:2527-2534. DOI: 10.1016/j.cub.2016.07.020

[14] Amos B, McConnell G, Wilson T. Chapter 2.2: Confocal Microscopy. In: Egelman E, editor. Comprehensive Biophysics. 1st ed. Amsterdam: Elsevier;

2012. pp. 3-23. DOI: 10.1016/B978-0-12-374920-8.00203-4

[15] Shaw PJ. Introduction to confocal microscopy. In: Lacey AJ, editor. Light Microscopy in Biology: A Practical Approach. 2nd ed. New York: Oxford University Press; 1999. pp. 45-71

[16] Smets I, Caplanusi A, Despa S, Molnar Z, Radu M, VandeVen M, et al. Ca^{2+} uptake in mitochondria occurs via the reverse action of the Na^{+}/Ca^{2+} exchanger in metabolically inhibited MDCK cells. American Journal of Physiology. Renal Physiology. 2004;**286**:F784-F794. DOI: 10.1152/ajprenal.00284.2003

[17] Baron S, Caplanusi A, van de Ven M, Radu M, Despa S, Lambrichts I, et al. Role of mitochondrial Na^{+} concentration, measured by CoroNa red, in the protection of metabolically inhibited MDCK cells. Journal of the American Society of Nephrology. 2005;**16**:3490-3497. DOI: 10.1681/ASN.2005010075

[18] Balut C, van de Ven M, Despa S, Lambrichts I, Ameloot M, Steels P, et al. Measurement of cytosolic and mitochondrial pH in living cells during reversible metabolic inhibition. Kidney International. 2008;**73**:226-232. DOI: 10.1038/sj.ki.5002632

[19] Marshall WH, Roberts KB. The growth and mitosis of human small lymphocytes after incubation with a phytohaemagglutinin. Quarterly Journal of Experimental Physiology and Cognate Medical Sciences. 1963;**2**:146-155. DOI: 10.1113/expphysiol.1963.sp001645

[20] Brand MD, Felber SM. Membrane potential of mitochondria in intact lymphocytes during early mitogenic stimulation. Biochemical Journal. 1984;**217**:453-459. DOI: 10.1042/bj2170453

[21] Pachman LM. The carbohydrate metabolism and respiration of isolated small lymphocytes. In vitro studies of normal and phytohemagglutinin stimulated cells. Blood. 1967;**30**:691-706. DOI: 10.1182/blood.V30.6.691.691

[22] Hedeskov CJ. Early effects of phytohemagglutinin on glucose metabolism of normal human lymphocytes. Biochemical Journal. 1968;**110**:373-380. DOI: 10.1042/bj1100373

[23] Beachy JC, Goldman D, Czech MP. Lectins activate lymphocyte pyruvate dehydrogenase by a mechanism sensitive to protease inhibitors. Proceedings of the National Academy of Sciences of the United States of America. 1981;**78**:6256-6260. DOI: 10.1073/pnas.78.10.6256

[24] Minsky M. Memoir on inventing the confocal scanning microscope. Scanning. 1988;**10**:128-138. DOI: 10.1002/sca.4950100403

[25] White JG, Amos WB. Confocal microscopy comes of age. Nature. 1987;**328**:183-184. DOI: 10.1038/328183a0

[26] Aslund N, Liljeborg A, Forsgren PO, Wahlsten S. Three-dimensional microscopy using the PHOIBOS scanner. Scanning. 1987;**9**:227. DOI: 10.1002/sca.4950090603

[27] PubMed [Internet]. Bethesda, MD: National Library of Medicine (US). 1946. Available from: https://www.ncbi.nlm.nih.gov/pubmed/ [Accessed: 10 March 2020]

[28] van de Ven M, Balut C, Baron S, Smets I, Steels P, Ameloot M. Analysis of mitochondrial pH and ion concentrations. Methods in Molecular Biology. 2010;**591**:275-309. DOI: 10.1007/978-1-60761-404-3_17

[29] Zucker R. Confocal microscopy system performance: Axial resolution. Microscopy Today. 2004;**12**:38. DOI: 10.1017/S1551929500051816

[30] Chu S, Brownell WE, Montrose MH. Quantitative confocal imaging along the crypt-to surface axis of colonic crypts. American Journal of Physiology. Cell Physiology. 1995;**269**:C1557-C1564. DOI: 10.1152/ajpcell.1995.269.6.C1557

[31] McKenzie M, Duchen MR. Impaired cellular bioenergetics causes mitochondrial calcium handling defects in MT-ND5 mutant cybrids. PLoS One. 2016;**11**:e0154371. DOI: 10.1371/journal.pone.0154371

[32] Collins TJ, Bootman MD. Mitochondria are morphologically heterogeneous within cells. Journal of Experimental Biology. 2003;**206**:1993-2000. DOI: 10.1242/jeb.00244

[33] Rehman J, Zhang HJ, Toth PT, Zhang Y, Marsboom G, Hong Z, et al. Inhibition of mitochondrial fission prevents cell cycle progression in lung cancer. The FASEB Journal. 2012;**26**:2175-2186. DOI: 10.1096/fj.11-196543

[34] Mishra P, Carelli V, Manfredi G, Chan DC. Proteolytic cleavage of Opa1 stimulates mitochondrial inner membrane fusion and couples fusion to oxidative phosphorylation. Cell Metabolism. 2014;**19**:630-641. DOI: 10.1016/j.cmet.2014.03.011

[35] Snigirevskaya ES, Moshkov AV, Yurinskaya VE, Vereninov AA, Komissarchik YY. Ultrastructural and X-ray microanalysis of U-937 cells in hypertonia-induced apoptosis. Cell and Tissue Biology. 2015;**9**:96-109. DOI: 10.1134/S1990519X15020091

[36] Inman DR, Cooper EH. Electron microscopy of human lymphocytes stimulated by phytohaemagglutinin. Journal of Cell Biology. 1963;**19**:441-445. DOI: 10.1083/jcb.19.2.441

[37] Johnson FR, Roberts KB. The growth and division of human small lymphocytes in tissue culture: An electron microscopic study. Journal of Anatomy. 1964;**98**:303-311

[38] Koopman WJ, Distelmaier F, Hink MA, Verkaart S, Wijers N, Fransen J, et al. Inherited complex I deficiency is associated with faster protein diffusion in the matrix of moving mitochondria. American Journal of Physiology. Cell Physiology. 2008;**294**:C1124-C1132. DOI: 10.1152/ajpcell.00079.2008

[39] Savina MV, Konovalova SA, Zubatkina IS, Nikiforov AA. Reversible metabolic depression in lamprey hepatocytes during prespawning migration: Dynamics of mitochondrial membrane potential. Comparative Biochemistry and Physiology Part B: Biochemistry and Molecular Biology. 2011;**160**:194-200. DOI: 10.1016/j.cbpb.2011.08.007

[40] Byrnes J, Ganetzky R, Lightfoot R, Tzeng M, Nakamaru-Ogiso E, Seiler C, et al. Pharmacologic modeling of primary mitochondrial respiratory chain dysfunction in zebrafish. Neurochemistry International. 2018;**117**:23-34. DOI: 10.1016/j.neuint.2017.07.008

[41] Buckman JF, Hernandez H, Kress GJ, Votyakova TV, Pal S, Reynolds IJ. MitoTracker labeling in primary neuronal and astrocytic cultures: Influence of mitochondrial membrane potential and oxidants. Journal of Neuroscience Methods. 2001;**104**:165-176. DOI: 10.1016/s0165-0270(00)00340-x

[42] Twig G, Graf SA, Wikstrom JD, Mohamed H, Haigh SE, Elorza A, et al. Tagging and tracking individual networks within a complex mitochondrial web with photoactivatable GFP. American Journal of Physiology. Cell Physiology. 2006;**291**:C176-C184. DOI: 10.1152/ajpcell.00348.2005

[43] Twig G, Elorza A, Molina AJ, et al. Fission and selective fusion govern mitochondrial segregation and elimination by autophagy. EMBO Journal. 2008;**27**:433-446. DOI: 10.1038/sj.emboj.7601963

[44] Brown CM. Fluorescence microscopy—Avoiding the pitfalls. Journal of Cell Science. 2007;**120**:1703-1705. DOI: 10.1242/jcs.03433

[45] Russ JC. The Image Processing Handbook. 3d ed. Boca Raton: CRC Press; 1999. p. 771

[46] Ferreira T, Rasband WS. ImageJ User Guide IJ1.46r [Internet]. Available from: https://imagej.nih.gov/ij/docs/guide/146-29.html#sub:Subtract-Background... [Accessed: 18 April 2020]

[47] Brand KA, Hermfisse U. Aerobic glycolysis by proliferating cells: A protective strategy against reactive oxygen species. FASEB Journal. 1997;**11**:388-395. DOI: 10.1096/fasebj.11.5.9141507

[48] Whitney RB, Sutherland RM. Characteristics of calcium accumulation by lymphocytes and alterations in the process induced by phytohemagglutinin. Journal of Cellular Physiology. 1973;**82**:9-20. DOI: 10.1002/jcp.1040820103

[49] Denton RM, McCormack JG. On the role of the calcium transport cycle in heart and other mammalian mitochondria. FEBS Letters. 1980;**119**:1-8. DOI: 10.1016/0014-5793(80)80986-0

[50] Marakhova II, Vereninov AA, Toropova FV, Vinogradova TA. Na, K-ATPase pump in activated human lymphocytes: On the mechanisms of rapid and long-term increase in K influxes during the initiation of phytohemagglutinin-induced proliferation. Biochimica et Biophysica Acta. 1998;**1368**:61-72. DOI: 10.1016/s0005-2736(97)00164-8

[51] Grinstein S, Smith JD, Rowatt C, Dixon SJ. Mechanism of activation of lymphocyte Na^+/H^+ exchange by concanavalin A. A calcium- and protein kinase C-independent pathway. Journal of Biological Chemistry. 1987;**262**:15277-15284

[52] Chien EJ, Chang CP, Lee WF, Su TH, Wu CH. Non-genomic immunosuppressive actions of progesterone inhibits PHA-induced alkalinization and activation in T cells. Journal of Cellular Biochemistry. 2006;**99**:292-304. DOI: 10.1002/jcb.20858

[53] Baumgarten E, Brand MD, Pozzan T. Mechanism of activation of pyruvate dehydrogenase by mitogens in pig lymphocytes. Biochemical Journal. 1983;**216**:359-367. DOI: 10.1042/bj2160359

[54] Douglas SD. Electron microscopic and functional aspects of human lymphocyte response to mitogens. Immunological Reviews. 1972;**11**:39-59. DOI: 10.1111/j.1600-065x.1972.tb00045.x

Section 3

Special Theories Protocols and Techniques

Chapter 15

A Protocol for Minimal Single Protein Labeling with CyDye Fluors for Live Cell Internalization Assays

Antonio J. Castro, Alfonso Clemente and Juan de Dios Alché

Abstract

Individual proteins chemically labeled with fluorescent dyes can be localized and tracked in real-time experiments in order to get insights about the site and molecular mechanism of action. Here, we have adapted a protocol that was originally developed for two-dimensional fluorescence difference gel electrophoresis (2D-DIGE) applications, to label proteins with CyDye fluors for single-molecule internalization assays in living cells. This "minimal labeling" method offers a number of advantages including specificity and known stoichiometry, simplicity, high reproducibility, and sensitivity and allows multiplexing while minimizing perturbations of the biological system. Moreover, since only a single lysine (Lys) residue per protein molecule is labeled, this method is also quantitative. To validate experimentally our protocol, we carried out the fluorescent labeling of IBB1, a major soybean protease isoinhibitor of the Bowman-Birk family that is currently being investigated as colorectal chemopreventive agent. Then, we analyzed the *in vivo* internalization dynamics of the labeled IBB1 protein in human colorectal adenocarcinoma HT29 cells.

Keywords: Bowman-Birk inhibitors, CyDye fluors, fluorescence microscopy, HT29 cells, IBB1, live cell internalization assay, minimal fluorescent labeling

1. Introduction

Fluorescence-based assays allow protein quantification and measuring protein interactions, enzymatic activity, and conformational changes. Individual proteins chemically labeled with fluorescent dyes can also be localized and tracked in real-time experiments throughout the cell in order to get insights about the site and molecular mechanisms of their action [1]. The choice of the fluorescent marker will largely depend on the nature of the downstream assay to be performed with the labeled protein. Overall, an ideal fluorescent label should be specific, bright, and stable and should not interfere with the physicochemical and conformational properties of the protein. Unfortunately, according to our knowledge, none of the existing fluorescent tags accomplish all these criteria.

To analyze the dynamics of labeled proteins throughout the cell, we should combine live cell imaging and single-molecule methods. For this purpose, different fluorescent tags and labeling methods are available, including genetically encoded fluorescent proteins (e.g., GFP) [2], direct labeling with organic fluorophores (e.g., Cy3) [3], tag-mediated peptide and protein labeling methods (e.g., SNAP-tag) [4], and quantum dots (Qdots) [5]. For live cell internalization assays, organic fluorophores are preferable since they have a short lifetime (<5 ns) and they are brighter and more photostable than other fluorescent labels [6]. They also have a small size, thus interfering less with the biological system. Fluorescein (FITC) and rhodamine (TRITC) dyes have been often used for this purpose [7, 8]. The main disadvantages of these organic fluorophores are the low specificity of the labeling reaction and the difficulty of defining its stoichiometry. Additionally, the unbound dye could interfere with subsequent applications, so it should be removed before performing downstream procedures.

Cyanines are synthetic fluorescent dyes belonging to polymethine group that yield brighter and more stable fluorescence than FITC and TRITC [9]. At protein level, cyanines have been used to track single proteins in live cell assays [10], as well as in molecular interaction studies by fluorescence resonance energy transfer (FRET) [11]. Moreover, cyanine dyes constitute the basis of the two-dimensional fluorescence difference gel electrophoresis (2D-DIGE) technology [12]. This method was developed to multiplex different protein samples and minimize gel-to-gel quantitative variations. CyDye DIGE fluors are available as minimal and saturation labeling dyes. The former are used when the amount of sample is not a limiting factor, while saturation labeling dyes are intended for 2D-DIGE applications with scarce biological samples (e.g., those collected by laser capture microdissection). Minimal dyes can also be used for specific labeling of cell-surface proteins prior to 2D-DIGE analysis [13]. In this paper, we have tailored the 2D-DIGE minimal labeling protocol in order to tag single proteins for being used in live cell internalization assays. This protocol is suitable for both proteins purified from their natural sources and recombinant proteins expressed in heterologous systems (e.g., *Escherichia coli, Pichia pastoris,* etc.). Moreover, this method overcomes the problems of specificity and stoichiometry of the labeling reaction. To validate experimentally our protocol, we carried out internalization assays of IBB1, a major soybean protease isoinhibitor of the Bowman-Birk family, which is currently being investigated as colorectal chemopreventive agent [14, 15].

2. Description and experimental design

2D-DIGE CyDye minimal labeling dyes have an NHS ester reactive group that specifically binds to the ε amino group of lysine (Lys) residues by an amide linkage. Lysine represents about 3.6 ± 0.6% and 7.0 ± 0.6% of the total amino acid content in plant and animal proteins, respectively [16]. Accordingly, a general labeling procedure would likely result in labeling most of Lys residues. However, the 2D-DIGE minimal labeling protocol ensures that only 1–2% of the available Lys is labeled. Thus, this protocol facilitates successful labeling of an individual protein at a specific location (ε amino group of Lys) and with a well-defined stoichiometry (one Lys per protein molecule, 1:1). Moreover, this modification only adds just about 500 Da to the protein total molecular weight, thus minimizing perturbations of the biological system.

Overall, the protocol comprises three different steps (**Figure 1**). Firstly, the protein of interest is resuspended in the reconstitution buffer and labeled with the CyDye fluorophore of choice. This phase ends when the labeling is stopped by adding an excess of L-lysine (Lys) to the reaction mixture. When carrying out multiplex experiments with other fluorescent labels, it is important to select the appropriate CyDye fluors in order to visualize independently each fluorescent signal when imaging. Secondly,

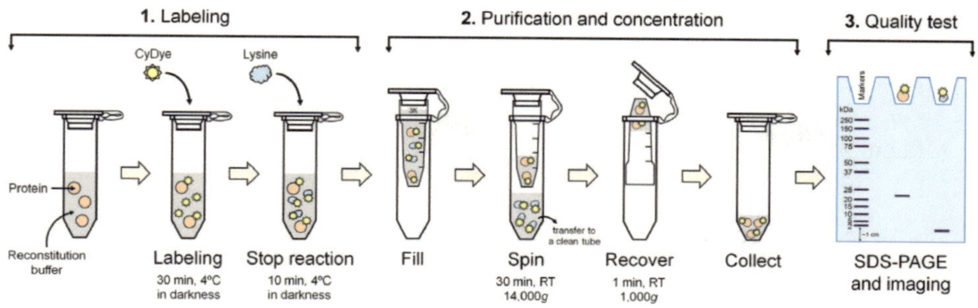

Figure 1.
Workflow for the minimal labeling of individual proteins with CyDye DIGE fluors.

the excess of free CyDye and exchange buffers is removed for sample cleanup. At this point, it is also possible to concentrate the Cy-labeled protein if necessary, by filtration through a size exclusion membrane of suitable cutoff. Finally, in the last step, the quality of the labeling reaction is verified by SDS-PAGE. The full standard protocol, which is accomplished in just a few hours, is described in detail below.

3. Labware chemicals and instruments

3.1 Reagents and solutions

Prepare all solutions using ultrapure water (18 MΩ cm at 25°C) and analytical grade reagents:

- 99.8% anhydrous N,N-dimethylformamide (DMF).
- Amersham™ CyDye™ DIGE Fluor minimal labeling kit (catalog no. 25-8010-65, GE Healthcare Bio-Sciences AB, Uppsala, Sweden). Alternatively, individual Cy™2, Cy™3, and Cy™5 minimal dyes are also available (see **Note 1**).
- 1× reconstitution buffer: 30 mM Tris-HCl (pH 8.5), 7 M urea, 2 M thiourea, and 4% (w/v) CHAPS (see **Note 2**).
- L-Lysine monohydrochloride.
- pH test strips, pH 4.5–10.0, resolution: 0.5 pH units.
- pH adjustment solution: 1× reconstitution buffer at pH 9.5 (see **Note 2**). Alternatively, 50 mM sodium hydroxide can be used.
- 12% polyacrylamide precast (or handmade) mini gels, 10-well, 30 μL.
- Prestained protein standards, broad range (e.g., Precision Plus Protein™ Dual Xtra, catalog no. 1610377, Bio-Rad).
- 2× gel loading buffer: 120 mM Tris-HCl (pH 6.8), 4% (w/v) SDS, 20% (v/v) glycerol, and 120 mM DTT (see **Note 3**).
- 1× electrophoresis running buffer: 25 mM Tris-HCl, 192 mM glycine, and 0.2% (w/v) SDS.

3.2 Materials and instruments

- Protein low-binding microcentrifuge 1.5 mL tubes.
- Set of variable volume (0.5–10, 10–100, and 100–1000 μL) single-channel pipettes.
- Amicon® Ultra-0.5 mL, Ultracel® 3 K filter device (catalog no. UFC500308. Millipore Corp., Burlington, MA, USA).
- Refrigerated centrifuge equipped with a fixed-angle rotor that can accommodate microcentrifuge 1.5 mL tubes, rated for 14,000 *g*.
- Vortex mixer.
- Mini-PROTEAN®-type vertical electrophoresis system.
- Image scanner equipped with a standard set of lasers (488, 532, and 635 nm) and emission filters.

4. Protein minimal fluorescent labeling procedure

4.1 Reconstitution of CyDye fluors

CyDye fluors are delivered lyophilized in opaque, conical microcentrifuge 1.5 mL tubes, so they have to be reconstituted before its use as follows:

a. Remove the CyDye from the freezer and leave for 5 min at room temperature.

b. Spin the microcentrifuge tube at 12,000 *g* for 10 s to collect the dye at the bottom.

c. Reconstitute the CyDye by adding the appropriate volume of DMF to obtain a stock solution at a final concentration of 1 mM. For example, in the case of a 5 nmol size pack, add 5 μl of DMF (see **Note 4**).

d. Replace the cap and vortex vigorously for 30 s.

e. Spin the microcentrifuge tube for 30 s at 12,000 *g*.

f. The reconstituted CyDye is now ready for protein labeling (see **Note 5**).

4.2 Minimal fluorescent labeling

Proceed as quickly as possible in order to minimize the time of handling, and also avoid exposure to light, which can quench the fluorescent properties of CyDye fluors. In addition, keep protein samples on ice to minimize degradation by proteases, and always use protein low-binding plastic microcentrifuge tubes as many proteins are able to bind to glassware:

a. Prepare a CyDye working solution to a final concentration of 400 μM by adding one volume of CyDye stock solution to 1.5 volumes of DMF.

b. Resuspend the lyophilized protein in the reconstitution buffer to a final concentration of 1 μg/μL.

c. Check that the pH remains at values near to 8.5 by spotting 1 μL on a pH indicator strip (see **Note 6**).

d. Take an aliquot of protein sample, and mix with the suitable volume of CyDye working solution. For optimal labeling, use 400 pmol of CyDye per 50 μg of protein (see **Note 7**).

e. Mix well by vortexing and centrifuge at 14,000 *g* for 10 s.

f. Incubate the mixture on ice for 30 min in the dark.

g. Add 1 μl of 10 mM L-lysine, mix by pipetting, and incubate for 10 min to stop the labeling reaction.

4.3 Purification and concentration of CyDye-labeled proteins

After labeling, the remaining dye (i.e., the CyDye bound to free L-Lys) should be removed from the protein solution before proceeding with downstream internalization experiments. Optionally, the labeled protein can be also concentrated at this step if necessary:

a. Place the Amicon Ultra-0.5 mL 3 K filter device into a microcentrifuge tube (see **Note 8**).

b. Add up to 500 μL of CyDye-labeled protein sample to the filter device.

c. Centrifuge at 14,000 *g* for 30 min. The final volume and concentration factor of the concentrate will depend on the spin time (see **Table 1** from [17]).

d. Transfer the first filtrate to a clean 1.5 mL polypropylene tube, and store at −20°C in the dark until use (see **Note 9**).

e. Dilute the concentrate with the desired cell culture medium to a final volume of 500 μL (see **Note 10**).

f. Spin at 14,000 *g* for 10–30 min as above and discard the filtrate.

g. Repeat Steps 5–6 thrice.

Dye	Color	Excitation max. (nm)	Emission max. (nm)	Laser (nm)	Emission filter
Cy2	Green	489	506	[1,2]Blue (488)	[1]520BP40; [2]530nmBP
Cy3	Red	550	570	[1,2]Green (532)	[1]580BP30; [2]605nmBP
Cy5	Blue	649	670	[1,2]Red (635)	[1]670BP30; [2]695nmBP

[1]*For Amersham™ Typhoon™ 9400 imager (GE Healthcare Bio-Sciences).*
[2]*For PharosFX™ Molecular© Imager (Bio-Rad).*

Table 1.
Spectral properties of CyDye™ fluors and lasers and emission filters used to scan fluorescent gels.

h. To recover the concentrated protein, place the filter device upside down in a new clean 1.5 mL tube supplied by the manufacturer.

i. Spin for 2 min at 1000 *g* to transfer the concentrated protein sample from the filter device to the tube (see **Note 11**).

4.4 Quality control of the labeling reaction

Before its use in internalization experiments, it is important to check that the protein was correctly labeled and purified. For this purpose, we will run a small sample of the labeled protein on 1-D SDS-PAGE along with a purity control sample as follows:

a. Make a 12% SDS-PAGE gel following standard protocols [18]. Alternatively, an equivalent precast gel can be used.

b. Mix a volume of the CyDye-labeled protein solution (~2 μg) with an equal volume of 2× gel loading buffer into a clean 1.5 mL tube by vigorously pipetting. Do not exceed the maximum capacity (30 μL) of the gel well.

c. Heat the protein sample at 95°C for 5 min to ensure full reduction of the labeled protein.

d. Make serial dilutions of the purity control sample (see **Section 4.3, d**) by adding different volumes (e.g., 5, 10, and 15 μL) to equal volumes of 2× gel loading buffer. It is not necessary to heat these samples.

e. Load the labeled protein sample along with the purity control samples in successive lanes on the gel.

f. Add the prestained protein standards according to the manufacturer's instructions at both ends of the gel, in order to track the electrophoretic migration of samples.

g. Run the samples at 180 V until the 2 kDa marker has nearly reached the bottom of the gel (leave about 1 cm).

h. Scan the gel at 100 μm resolution using the appropriate laser and emission filter (**Table 1**).

5. Experimental validation of the protocol

To validate experimentally our protocol, we carried out live cell internalization assays of IBB1, a major soybean protease isoinhibitor of the Bowman-Birk family with an ability to inhibit both trypsin- and chymotrypsin-like enzymes. This protein was purified from a commercially available BBI mixture of soybean isoinhibitors and tagged with CyDye fluors as described above. Then, its internalization in human colorectal adenocarcinoma HT29 cells was monitored using fluorescence microscopy.

5.1 Purification of soybean IBB1 protease inhibitor

A major Bowman-Birk isoinhibitor from soybean, called IBB1, was isolated from a commercial sample containing a mixture of IBB1 and IBBD2, as well as other inhibitors from the Kunitz family (catalog no. T9777, Sigma-Aldrich), as previously described [19]. Briefly, the commercial extract was resuspended to a final concentration of 0.7 mg/mL in 10 mL of elution buffer (25 mM sodium acetate buffer, pH 4.4). Aliquots of 2 mL were filtered through a Minisart® 0.2 μm pore size syringe filter (catalog no. 17821 K, Sartorius, Gottingen, Germany) and loaded on a Mono S 5/50 GL cation exchange column (catalog no. 17516801, GE Healthcare Bio-Sciences) connected to an ÄKTA™ FPLC apparatus (GE Healthcare Bio-Sciences). Fractionation was carried out using a linear gradient of 0–0.22 M NaCl in 25 mM sodium acetate buffer (pH 4.4) at a flow rate of 1 mL/min. The elution was monitored at 280 nm, and 0.5 mL fractions were collected.

In order to identify that fraction containing the pure IBB1 protein, trypsin (TIA) inhibitory activity of eluted samples was spectrophotometrically assayed in 96-well flat-bottom microtiter plates by using N_{α}-benzoyl-DL-arginine *p*-nitroanilide hydrochloride (BAPNA, catalog no. B4875, Sigma-Aldrich) as specific substrate [19]. Chymotrypsin inhibitory activity (CIA) evaluation was measured using a modified small-scale quantitative assay with N-benzoyl-L-tyrosine ethyl ester (BTEE, catalog no. B6125, Sigma-Aldrich) as previously described [20, 21]. The purity and identity of the IBB1 fraction containing both TIA and CIA were further confirmed by SDS-PAGE denaturing electrophoresis and peptide mass fingerprinting (PMF) analysis, following standard protocols. Finally, the IBB1 sample was dialyzed extensively against distilled water during 48 h at 4°C using a Spectra/Por™ 3, MWCO 3.5 kD, dialysis membrane (catalog no. 734-0687, Spectrum™ Labs, New Brighton, MN, USA), quantified using the standard BCA assay [22], and freeze-dried in a LyoQuest-55 lyophilizer (catalog no. 61644, Telstar, Terrassa, Spain) until use.

5.2 Minimal labeling of IBB1 protein

Two IBB1 samples were labeled with CyDye fluors Cy2 and Cy5, respectively, using the minimal labeling protocol described above. After purification and concentration steps by filtration, each labeled IBB1 protein was suspended in ~50 μL of Gibco® DMEM culture medium (catalog no. A1443001, Thermo Fisher Scientific, Waltham, MA, USA). To check the quality of labeling, approximately 2 μg of Cy2- and Cy5-labeled IBB1 proteins was analyzed by SDS-PAGE on a 12% polyacrylamide gel along with the control samples (i.e., $Lys_{\text{-Cy2}}$ and $Lys_{\text{-Cy5}}$) according to standard procedures and visualized in a PharosFX™ Molecular© Imager (Bio-Rad) using the appropriate laser and emission filter for each fluorochrome (see **Table 1**).

A single protein band with an apparent molecular weight of about 12 kDa was visible after scanning the gel at 488 nm ($IBB1_{\text{-Cy2}}$) and 635 nm ($IBB1_{\text{-Cy5}}$), proving the success of the minimal labeling reaction (**Figure 2**). When the gel was scanned at 488 nm, the fluorescent signal from IBB1-Cy5 was not detected. The signal from IBB1-Cy2 was also abolished when the gel was scanned at 635 nm, thus confirming the specificity of the labeling reaction (data not shown). Moreover, no fluorescent signal corresponding to the $Lys_{\text{-Cy2}}$ and $Lys_{\text{-Cy5}}$ bands were detected in IBB1 samples as compared with control samples (**Figure 2**), indicating that labeled IBB1 samples were pure and, thus, suitable for being used in live cell internalization experiments.

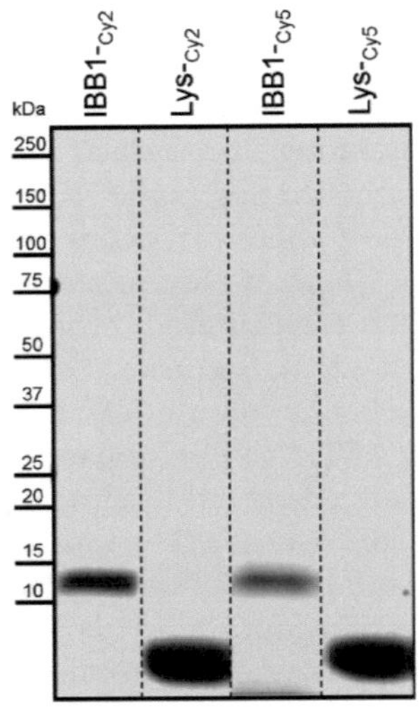

Figure 2.
Quality control by SDS-PAGE of IBB1 labeling with Cy2 and Cy5 dyes. Protein standards (kDa) are displayed on the left.

5.3 Live cell internalization experiments

5.3.1 IBB1 internalization assays

Human colorectal adenocarcinoma HT29 cells were cultured in DMEM medium, supplemented with 5% (v/v) fetal bovine serum, 2 mM glutamine, and 1% (v/v) antibiotic-antimycotic solution (catalog no. A5955, Sigma-Aldrich), as previously described [23]. After trypsinization, chambered µ-slides with a glass coverslip bottom (catalog no. 80444, Ibidi, Gräfelfing, Germany) were inoculated at a density of 25,000 HT29 cells per well in 700 µL of DMEM culture medium and incubated at 37°C under 5% CO2 in humidified air for 24 h to allow the cells to adhere to the chamber bottom. The culture medium was then supplemented with 50 µg of Cy5-labeled IBB1. Cell internalization dynamics was monitored using a C1 confocal laser microscope (Nikon Corp., Tokyo, Japan). Z-series images of HT29 cells were recorded at different time (min) intervals by exciting the sample with a red diode (633 nm) and processed with the software EZ-C1 Gold v2.10 build 240 (Nikon). Three independent experiments were carried out.

At $t_{0\ min}$, no fluorescent signal was observed inside HT29 cells (data not shown). The Cy5-labeled IBB1 protein crossed the plasma membrane of HT29 cells after a few minutes and was gradually accumulated, forming fluorescent patches randomly distributed across the cytoplasm (**Figure 3**). Negative controls of HT29 cells cultured with the unlabelled IBB1 protein did not show any fluorescent signal (data not shown).

5.3.2 Multiplex experiments

The minimal labeling protocol is suitable for multiplex experiments, either combining up to three different proteins labeled with Cy2, Cy3, and Cy5, respectively, or using CyDye fluors together with other fluorescent labels. Here, two different combinations of labels were tested. Firstly, nuclei of colon cancer cells were stained by adding to the culture medium 1 µL of Hoechst 33342 dye solution (catalog no. H1399, Thermo Fisher Scientific). HT29 cells were incubated for

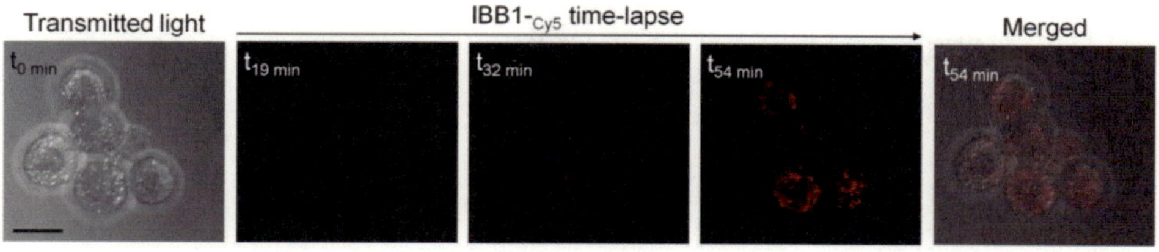

Figure 3.
Internalization dynamics of IBB1 protein into human colorectal adenocarcinoma HT29 cells. Fifty μg of IBB1-$_{Cy5}$ was added to the culture medium at the onset of culture, and HT29 cells were imaged at different time intervals. Cy5 fluorescence images are reproduced with permission from [24]. Bar = 50 μm.

30 min, and the culture medium was then supplemented with IBB1-$_{Cy2}$. Secondly, in addition to Hoechst, the fluorescent probe FM 4-64 (catalog no. T3166, Thermo Fisher Scientific), an endocytic tracer [25], was added to the growth medium (final concentration = 5 μg/mL) at the same time with IBB1-$_{Cy2}$. After 90 min of culture, the medium was supplemented with 0.35 μL of 50 mM brefeldin A (BFA), a potent inhibitor of exocytosis [26], and incubated for 30 min before imaging.

Cell internalization dynamics of IBB1-$_{Cy2}$ was monitored using an Eclipse Ti-U microscope (Nikon Corp.) equipped with a *p*E-300^{white} illumination system (CoolLED Ltd., Andover, UK) and a DFK 72AUCO2 camera (The Imaging Source, Bremen, Germany). Hoechst, Cy2, and FM 4-64 fluorescence images were obtained using a UV, blue, and green LED light, respectively. Images were merged with the Image-Pro Plus v6.0 software (Media Cybernetics, Rockville, MD, USA). For each combination of fluorescent labels, three independent experiments were carried out.

After the first 3 hours of culture, HT29 cells showed a very intense green fluorescent labeling from IBB1, forming patches randomly distributed across the polarized cytoplasm (green fluorescence image in **Figure 4**). However, the fluorescent signal of IBB1 did not overlap with Hoechst-specific labeling (blue fluorescence image corresponding to DNA in **Figure 4**), indicating that the protein did not internalize into the cell nucleus after this period of time.

On the other hand, we observed that the fluorescent signal from the endosome marker FM 4-64 (red fluorescence image in **Figure 5**) was localized in the plasma membrane and the cytoplasm of HT29 cells after 120 min of culture as expected. This signal clearly overlapped with labeling from IBB1 in the cytoplasm (green fluorescence image in **Figure 5**), giving a yellowish fluorescence in the merged image. There was no overlapping between nuclei (blue fluorescence image in **Figure 5**) and either IBB1 or FM4-64 labeling. These results suggest that IBB1 is internalized into the cytoplasm of HT29 cells through one of the existing endocytosis pathways.

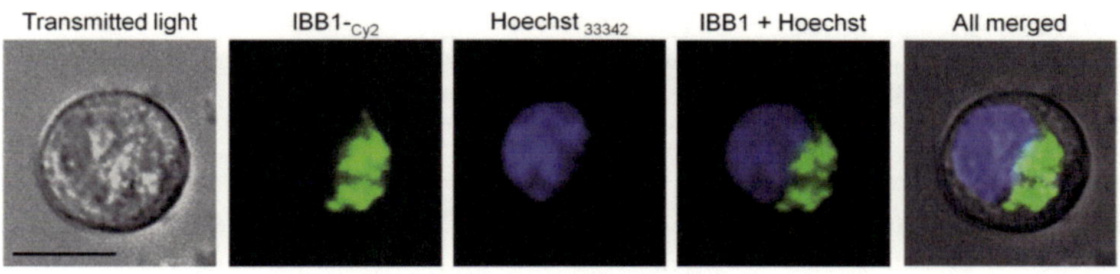

Figure 4.
Internalization of IBB1 (green) into the cytoplasm of human colorectal cancer HT29 cells. Nuclei were stained with Hoechst 33342 (blue). HT29 cells were imaged after 180 min of culture. Bar = 10 μm.

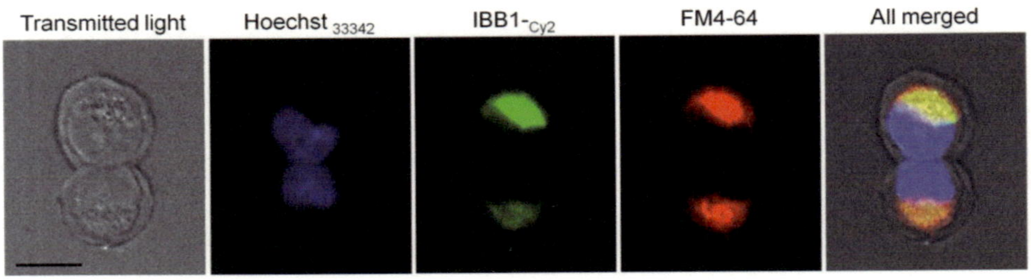

Figure 5.
Mechanism of IBB1 (green) internalization into human colorectal cancer HT29 cells. Nuclei were stained with Hoechst 33342 (blue). The endocytic tracer FM 4-64 (red) was added to the culture medium at the onset of the experiment. HT29 cells were imaged after 120 min of culture. Bar = 10 μm.

6. Troubleshooting guide

Users of the minimal fluorescent labeling protocol might encounter several problems, whose potential causes and their corresponding remedies are summarized in **Table 2**.

Probable causes	Remedies
Problem #1: low sample recovery in the concentrate	
1a. Adsorption of the protein to the filter device surfaces	1a. Remove the concentrate immediately after centrifugation
1b. Protein precipitation due to sample overload on the filter device	1b. Dilute the sample in the reconstitution buffer to a final concentration of 1 μL/mL before centrifugal spin
Problem #2: the Cy-labeled protein is not detected on the 1-D gel	
2a. DMF of poor quality	2a. Only use 99.8% anhydrous DMF
2b. DMF has been opened for longer than 3 months	2b. DMF must be less than 3 months old from day of opening
2c. The lifetime of the reconstituted CyDye has been exceeded	2c. Once reconstituted, the CyDye is stable and usable until the expiry date or for 2 months, whichever is sooner
2d. The CyDye has been exposed to light or left out of the freezer for a long time	2d. Make ready-to-use aliquots, and always store them at −20°C in the dark until use
2e. The pH in the labeling mixture has dropped below 8.0	2e. Adjust the pH to 8.5 by adding small volumes of the pH adjustment buffer (**Note 6**)
2 f. The protein solution concentration is less than 1 mg/mL	2 f. Concentrate the protein solution using a suitable filter device, and check again the pH and concentration of the sample
2 g. Incorrect CyDye to protein ratio	2 g. For optimal labeling, use 400 pmol of CyDye per 50 μg of protein
2 h. Insufficient protein sample loading on the gel	2 h. It is recommended to load between 1 and 5 μg of protein
Problem #3: detection of more than one fluorescent bands in the labeled protein sample on the 1-D gel	
3a. More than one Lys residue per molecule has been labeled with the dye (ladder effect)	3a. Decrease CyDye to protein ratio until only a single band is visible
3b. Presence of multimeric forms of the protein due to insufficient reduction	3b. Increase the concentration of DTT in the gel loading buffer
3c. Presence of protein fragments due to proteolysis	3c. Minimize protease activity by keeping protein samples at 4°C throughout the protocol
Problem #4: presence of free dye in the protein sample	
4a. Insufficient purification of the labeled protein	4a. Increase the number of filtration cycles in the protocol

Table 2.
Minimal fluorescent labeling with CyDye fluors of single proteins: troubleshooting guide.

Acknowledgements

This work was supported by MINECO-AEI (ERDF co-financed grants AGL2017-84298-P to AJC and AGL2017-83772-R to AC) and Junta de Andalucía (ERDF co-financed grant CA8313, AT17_5247 to JDA). We also thank Adrian Sabio, Anna Rodríguez, and Elena Lima-Cabello for their technical support.

Conflict of interest

The authors have no competing interests to declare.

Notes

Note 1. Caution! CyDye fluors are harmful in contact with the skin and if inhaled or swallowed. It is strongly recommended to wear lab gloves, mask, and coat when handling them to avoid staining clothing and the skin.

Note 2. Make aliquots and store them at −20°C until use.

Note 3. Important! Do not add bromophenol blue to the gel loading buffer. This compound is visible when scanning the gel at 635 nm, so it can interfere when using Cy5.

Note 4. Important! DMF must be less than 3 months old from day of opening. DMF is degraded over time, forming amine compounds that will compete with the protein for the CyDye.

Note 5. Important! This stock solution must be stored at −20°C and in the dark, being stable and usable until the expiry date or for 2 months, whichever is sooner.

Note 6. Critical step! There is little or no fluorescent labeling when pH values drop below 8.0. If pH of the protein solution has fallen to values below 8.0 after adding the lysis buffer, adjust the pH to 8.5 by adding a few drops of pH adjustment solution.

Note 7. Labeling reactions can be scaled up using more protein and CyDye flour.

Note 8. The Ultracel membrane contains trace amounts of glycerin. If this compound interferes with upstream analysis, pre-rinse the filter device with ultrapure water. If interference continues, rinse with 0.1 N NaOH followed by a second spin of ultrapure water. **Caution!** Do not allow the membrane to dry out once wet.

Note 9. This first filtrate contains the remaining CyDye, bound to free Lys, to be used for quality control of the labeling reaction (Section 4.4).

Note 10. The lysis buffer contains chemical reagents that are toxic for cells (e.g., urea, CHAPS, etc.). For its removal, add the same culture medium used for growing cells in your internalization assays.

Note 11. Alternatively, labeled protein samples can be stored for at least 3 months at −80°C in the dark.

Author details

Antonio J. Castro[1*], Alfonso Clemente[2] and Juan de Dios Alché[1]

1 Plant Reproductive Biology and Advanced Imaging Laboratory, Department of Biochemistry, Cell and Molecular Biology of Plants, Estación Experimental del Zaidín (CSIC), Granada, Spain

2 Gastrointestinal Health Laboratory, Department of Physiology and Biochemistry of Animal Nutrition, Estación Experimental del Zaidín (CSIC), Granada, Spain

*Address all correspondence to: antoniojesus.castro@eez.csic.es

References

[1] Toseland CP. Fluorescent labeling and modification of proteins. Journal of Chemical Biology. 2013;**6**:85-95. DOI: 10.1007/s12154-013-0094-5

[2] Chalfie M, Tu Y, Euskirchen G, Ward WW, Prasher DC. Green fluorescent protein as a marker for gene expression. Science. 1994;**263**(5148): 802-805. DOI: 10.1126/science.8303295

[3] Puljung MC, Zagotta WN. Labeling of specific cysteines in proteins using reversible metal protection. Biophysical Journal. 2011;**100**(10):2513-2521. DOI: 10.1016/j.bpj.2011.03.063

[4] Hinner MJ, Johnsson K. How to obtain labeled proteins and what to do with them. Current Opinion in Biotechnology. 2010;**21**:766-776. DOI: 10.1016/j.copbio.2010.09.011

[5] Michalet X, Pinaud FF, Bentolila LA, Tsay JM, Doose S, Li JJ, et al. Quantum dots for live cells, in vivo imaging, and diagnostics. Science. 2005;**307**(5709): 538-544. DOI: 10.1126/science.1104274

[6] Ha T, Tinnefeld P. Photophysics of fluorescent probes for single-molecule biophysics and super-resolution imaging. Annual Review of Physical Chemistry. 2012;**63**:595-617. DOI: 10.1146/annurev-physchem-032210-103340

[7] Maxfield FR, Schlessinger J, Shecter Y, Pastan I, Willingham MC. Collection of insulin, EGF, and α2-macroglobulin in the same patches on the surface of cultured fibroblasts and common internalization. Cell. 1978;**14**:805-810. DOI: 10.1016/0092-8674(78)90336-7

[8] Yavelow J, Scott CB, Mayer TC. Fluorescent visualization of binding and internalization of the anti-carcinogenic Bowman-Birk type protease inhibitors in transformed fibroblasts. Cancer Research. 1987;**47**:1602-1607

[9] Ernst LA, Gupta RK, Mujumdar RB, Waggoner AS. Cyanine dye labeling reagents for sulfhydryl groups. Cytometry. 1989;**10**(1):3-10. DOI: 10.1002/cyto.990100103

[10] Okamoto Y, Ninomiya H, Miwa S, Mas T. Cholesterol oxidation switches the internalization pathway of endothelin receptor type A from caveolae to clathrin-coated pits in Chinese hamster ovary cells. Journal of Biological Chemistry. 2000;**275**:6439-6446. DOI: 10.1074/jbc.275.9.6439

[11] Meyer BH, Martinez KL, Segura JM, Pascoal P, Hovius R, George N, et al. Covalent labeling of cell-surface proteins for in-vivo FRET studies. FEBS Letters. 2006;**580**:1654-1658. DOI: 10.1016/j.febslet.2006.02.007

[12] Unlü M, Morgan ME, Minden JS. Difference gel electrophoresis: A single gel method for detecting changes in protein extracts. Electrophoresis. 1997;**18**(11):2071-2077. DOI: 10.1002/elps.115018113

[13] Hagner-McWhirter A, Winkvist M, Bourin S, Marouga R. Selective labelling of cell-surface proteins using CyDye DIGE Fluor minimal dyes. Journal of Visualized Experiments. 2008;**21**. DOI: 10.3791/945

[14] Clemente A, Marín-Manzano MC, Jiménez E, Arques MC, Domoney C. The anti-proliferative effects of TI1B, a major Bowman-Birk isoinhibitor from pea (*Pisum sativum* L.), on HT29 colon cancer cells are mediated through protease inhibition. British Journal of Nutrition. 2012;**108**:S135-S144. DOI: 10.1017/S000711451200075X

[15] Utrilla MP, Peinado MJ, Ruiz R, Rodríguez-Nogales A, Algieri F, Rodríguez-Cabezas ME, et al. Pea (*Pisum sativum* L.) seed albumin extracts show anti-inflammatory effect

in the DSS model of mouse colitis. Molecular Nutrition & Food Research. 2015;**59**:807-819. DOI: doi.org/10.1002/mnfr.201400630

[16] Gorissen SHM, Crombag JJR, Senden JMG, Waterval WAH, Bierau J, Verdijk LB, et al. Protein content and amino acid composition of commercially available plant-based protein isolates. Amino Acids. 2018;**50**:1685-1695. DOI: 10.1007/s00726-018-2640-5

[17] Amicon® Ultra-0.5 Centrifugal Filter Devices: User Guide. Burlington, MA, USA: Millipore Corporation; 2011

[18] Gallagher SR. SDS-polyacrylamide gel electrophoresis (SDS-PAGE). Current Protocols Essential Laboratory Techniques. 2012;**6**:7.3.1-7.3.28. DOI: 10.1002/9780470089941.et0703s06

[19] Clemente A, Moreno FJ, Marín-Manzano MC, Jiménez E, Domoney C. The cytotoxic effect of Bowman-Birk isoinhibitors, IBB1 and IBBD2, from soybean (*Glycine max*) on HT29 human colorectal cancer cells is related to their intrinsic ability to inhibit serine proteases. Molecular Nutrition & Food Research. 2010;**54**:396-405. DOI: 10.1002/mnfr.200900122

[20] Clemente A, Jiménez E, Marín-Manzano MC, Rubio LA. Active Bowman-Birk inhibitors survive gastrointestinal digestion at the terminal ileum of pigs fed chickpea-based diets. Journal of the Science of Food and Agriculture. 2008;**88**:523-531. DOI: 10.1002/jsfa.3115

[21] Clemente A, MacKenzie DA, Jeenes DJ, Domoney C. The effect of variation within inhibitory domains on the activity of pea protease inhibitors from the Bowman-Birk class. Protein Expression and Purification. 2004;**36**:106-114. DOI: 10.1016/j.pep.2004.03.015

[22] Walker JM. The bicinchoninic acid (BCA) assay for protein quantitation. In: Walker JM, editor. The Protein Protocols Handbook. 2nd ed. Totowa, NJ: Humana Press; 2009. pp. 11-15. DOI: 10.1007/978-1-59745-198-7_3

[23] Clemente A, Gee JM, Johnson IT, Mackenzie DA, Domoney C. Pea (*Pisum sativum* L.) protease inhibitors from the Bowman-Birk class influence the growth of human colorectal adenocarcinoma HT29 cells in vitro. Journal of Agricultural and Food Chemistry. 2005;**53**:8979-8986. DOI: 10.1021/jf051528w

[24] Soria M, Porras P, Moya FJ, Sánchez N, Rodríguez B, Ruiz ML, et al. Internalization of the anti-carcinogenic IBB1, a major Bowman-Birk isoinhibitor from soybean (*Glycine max*), in HT29 colon cancer cells. igh School Students for Agricultural Science Research. 2014;**3**:59-68

[25] Bolte S, Talbot C, Boutte Y, Catrice C, Read ND, Satiat-Jeunemaitre B. FM-dyes as experimental probes for dissecting vesicle trafficking in living plant cells. Journal of Microscopy. 2004;**214**:159-173. DOI: 10.1111/j.0022-2720.2004.01348.x

[26] Helms JB, Rothman JE. Inhibition by brefeldin A of a Golgi membrane enzyme that catalyses exchange of guanine nucleotide bound to ARF. Nature. 1992;**360**(6402):352-354. DOI: 10.1038/360352a0

Chapter 16

Fluorogenic Polyfunctional Coumarin-Based Chemosensors for Multianalyte Detection

Alexander Dubonosov and Vladimir Bren

Abstract

Fluorogenic sensors capable of selective interaction with analyte, which leads to a change in the position or intensity of the fluorescence band, allow to detect ions or molecules *in situ* and *in vivo* and possess high sensitivity and efficiency. Currently, they are widely used in organic, biological, and medical chemistry and environmental sciences for express monitoring of the ionic composition of the medium. They represent a serious alternative to the bulky, expensive, non-transportable technical devices traditionally used for this purpose, such as atomic absorption, atomic emission, and XRF spectrometers. Polyfunctional sensors capable of independent detection of two or more kinds of "guests" from a multiple mixture of cations, anions, or molecules due to specific spectral responses via the same or different channels constitute a rapidly developing area of chemosensory science. This specific feature is associated with the presence of two or more coordination centers in their molecules, or the capability of one center to selectively respond to various analytes with individual spectral changes. Coumarin (2*H*-chromene-2-one) core is one of the most versatile frameworks for the design of fluorogenic polyfunctional chemosensors for multianalyte detection. In this chapter, we report on the review of sensing properties of this group of chemosensors based on functionalized coumarin derivatives, including their applications in bioimaging.

Keywords: coumarin, polyfunctional chemosensor, fluorescence, cations, anions, amino acids, bioimaging

1. Introduction

Chemosensor is a molecule of abiotic origin capable of selective interaction with analyte causing corresponding changes in the physical properties of the initial system (absorption spectra, fluorescence spectra, etc.) [1]. If a change occurs in spectral characteristics, the chemosensor relates to an optical type. There are two main types of optical chemosensors according to their mechanisms of action: chromogenic and fluorogenic [2–4]. In the case of chromogenic chemosensors after binding of analyte, there occurs a change in the electronic absorption spectra of the initial compounds. If this change can be seen with the human eye, we are dealing with a "naked-eye" chemosensor. Fluorogenic chemosensors can change their fluorescence spectrum after the interaction of analyte with receptor. It is highly desirable that this process is also accompanied by a "naked-eye" effect—contrast change in the color of emission. Chromogenic and fluorogenic chemosensor systems are widely used in organic, biological, and medical chemistry and environmental sciences for monitoring cations and anions. They represent a real alternative to the bulky, expensive, non-transportable technical devices, such as atomic absorption, atomic emission, and XRF spectrometers, that are traditionally used for this purpose. Of special efficiency are fluorogenic sensors, which use fluorescence for detection of various analytes, allow measurements *in situ* and *in vivo,* and are distinguished by highest sensitivity and selectivity. Recently a new scientific area in chemosensorics arose, which is associated with the design of poly- and bifunctional sensors capable of independent detection of two or more kinds of ions-"guests" due to the specific spectral responses via the same or different channels [5–8]. Using a single molecule possessing different reactions against multiple analytes is cost-effective and useful for practical applications.

Herein we report on the review of spectral, fluorescent, and sensing properties of new representatives of this group of chemosensors based on functionalized coumarin derivatives, including their applications in bioimaging.

Coumarins (derivatives of 2*H*-chromene-2-one) constitute a comprehensive group of extensively studied heterocyclic compounds in organic, physical, and medical chemistry [9, 10]. Some coumarins have been selected as privileged scaffolds for drug design [11–13], and a number of antitumor, antiproliferative, antioxidant, antifungal, anti-inflammatory, and antiviral agents have been obtained on their basis [14–19]. As a rule, substituted coumarins possess fluorescence in the visible part of the spectrum, as well as other useful photophysical properties. They are widely used in laser dyes, light-emitting devices, and solar cells. In addition, 2*H*-chromene-2-one is considered one of the most versatile frameworks for design of fluorescent, chemo- and biosensor systems [20–23].

There are several excellent reviews devoted entirely or partially to coumarin chemosensors [24–28], but polyfunctional coumarin-based sensors for multianalyte detection until now have not been considered.

2. Sensing of multiple metal cations

Fluorescent polyfunctional sensors for detection of metal cations must contain a metal chelating or binding fragment attached to a coumarin core capable of absorbing and emitting light. The formation of complexes with ions should cause a change in the electronic structure or molecular conformation, which should result in an increase or decrease in the emission intensity.

A fluorescent sensor **1** (**Figure 1**) demonstrates a high selectivity toward Al^{3+} and Zn^{2+} in the presence of many various metal cations. Aluminum is the third most

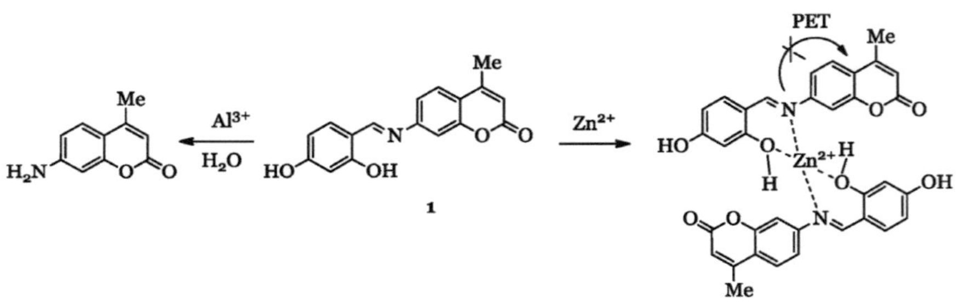

Figure 1.
Detection of Al^{3+} and Zn^{2+} by sensor **1**.

common element in the earth's crust. It is used in food additives and cookware, although its cations are highly toxic and may be associated with Parkinson's and Alzheimer's diseases, microcytic anemia, dialysis dementia, and osteomalacia. Zinc is the second most abundant *d*-metal cation, which plays a crucial role in gene transcription, regulation of metalloenzymes, and transmission of nerve signals. However, it has some toxicity and its excess in living cells can cause neurodegenerative disorders, epilepsy, and seizures.

A new emission band at 427 nm (an increase in intensity ~500 times) in the presence of Al^{3+} in ethanol-water mixture appears due to hydrolysis of imine **1**. The detection limit (LOD) was calculated to be 3.7×10^{-6} M. Detection of Zn^{2+} leads to a substantial initial fluorescence intensity enhancement at 489 nm due to the inhibition of PET process (photoinduced electron transfer) [29].

Coumarin **2** exhibits a significant enhancement of fluorescence intensity upon detection of Zn^{2+} (535 nm, 270-fold) or Al^{3+} ions (518 nm, 230-fold) in MeOH/H_2O mixture with LODs 3.75×10^{-8} and 1.14×10^{-8} M, respectively [30]. It has been shown that the sensing mechanism is based on inhibition of ICT process (intramolecular charge transfer) (**Figure 2**).

Compound **2** possesses a low toxicity and was used for fluorescent bioimaging of Zn^{2+} and Al^{3+} cations in PC12 cells.

Coumarin-crown compound **3** exhibits a high selectivity for the detection of Al^{3+}, Cu^{2+}, and Mg^{2+} in ethanol [31] (**Figure 3**). Copper(II) cations play an important role in biological systems. Their lack can lead to anemia and low white cell amount while an excess is accountable for neurodegenerative, Alzheimer's, and Wilson's diseases. Magnesium(II) cations are among the most abundant divalent cations in living cells. They are responsible for the formation of bone tissue, enzymatic biochemical reactions, cell proliferation, and DNA conformation stabilization, whereas their excessive concentration in the cytosol can lead to diabetes, hypertension, epilepsy, and Alzheimer's disease.

While copper(II) is identified by color change of solution from a slight yellow to orange, Al^{3+} and Mg^{2+} ions cause a significant fluorescence enhancement at 592 nm and 547 nm with low detection limits of 0.31 μM and 0.23 μM, respectively.

Chemosensor **4** was developed for dual detection of Fe^{3+} and Zn^{2+} ions [32] (**Figure 4**). Iron(III) is the most common *d*-metal cation in the human organism, it plays a significant role in many enzymatic reactions and in specialized transport and storage of proteins. Its lack can cause anemia, diabetes, hemochromatosis, and Parkinson's disease. Further development of this work led to the obtaining of a polyfunctional sensor **5** for Fe^{3+}, Zn^{2+}, and Cu^{2+} cations [33] (**Figure 4**). Formation of complex with Fe(III) ion and **4** or **5** is accompanied by a contrast color change from colorless to deep yellow.

Upon interaction with Zn^{2+} in CH_3OH/H_2O mixture, the emission intensity at 484 nm increases by five times compared to other metal ions. The LOD was found to be ~10^{-6} M. Since Cu^{2+} is a paramagnetic ion, its presence in the solution causes

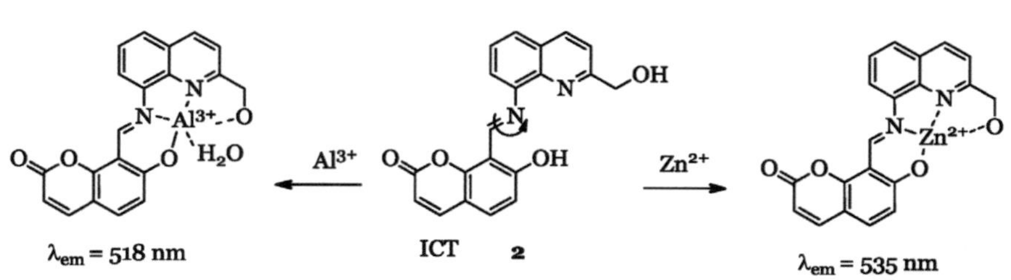

Figure 2.
Proposed scheme of detection Zn^{2+} and Al^{3+} by sensor 2.

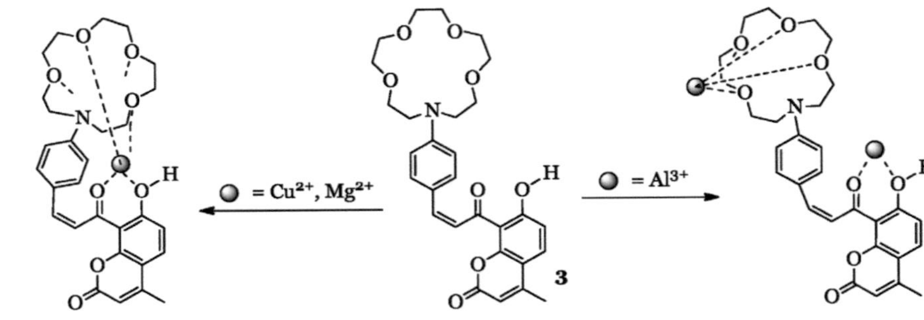

Figure 3.
Possible binding mechanisms of Cu^{2+}, Mg^{2+}, and Al^{3+} by coumarin 3.

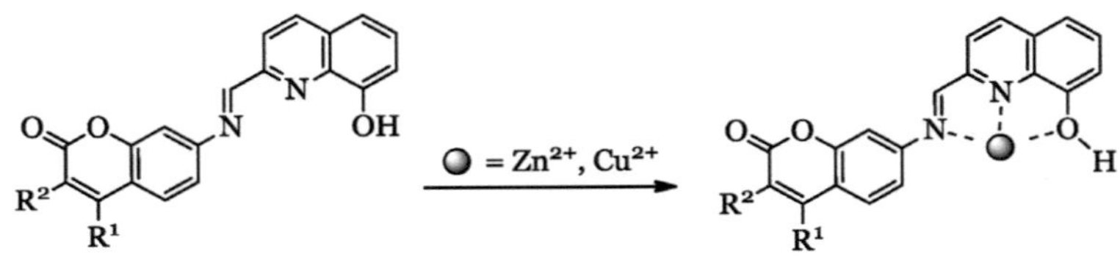

Figure 4.
Complexation of coumarins 4 and 5 with Zn^{2+} and Cu^{2+}.

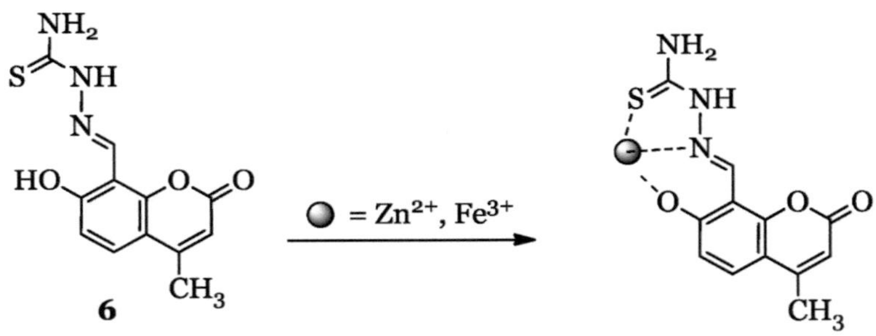

Figure 5.
Sensing of Fe^{3+} and Zn^{2+} by coumarin **6**.

a substantial quenching of initial fluorescence of **4** and **5**. The detection limit was calculated and found to be ~10^{-5} M.

A similar approach was used in design of a dual chemosensor **6** [34] (**Figure 5**). While Fe^{3+} ions in EtOH/H_2O solution cause only a visible color change from colorless to brown, the addition of Zn^{2+} ion resulted in 45-fold enhancement in the fluorescence intensity at 473 nm. The LOD was found to be 0.6×10^{-8} M.

Application of compound **6** as bioimaging fluorescent sensor for detection of Zn^{2+} in human cancer cells was also observed by fluorescent cell imager (**Figure 6**).

Chemosensor 7, as well as **4** and **5**, in water (1% EtOH) solution demonstrates the quenching of the fluorescent properties when adding Cu(II) ions and significantly increases the emission intensity in the presence of Zn(II) ions [35] (**Figure 7**).

When Cu^{2+} and Zn^{2+} were monitored by sensor 7, the LODs were 141 nM and 72 nM, respectively. About 85% of cells survive upon addition of 80 μM of 7 indicating its hypotoxicity and possibility of using it for cell imaging (**Figure 8**).

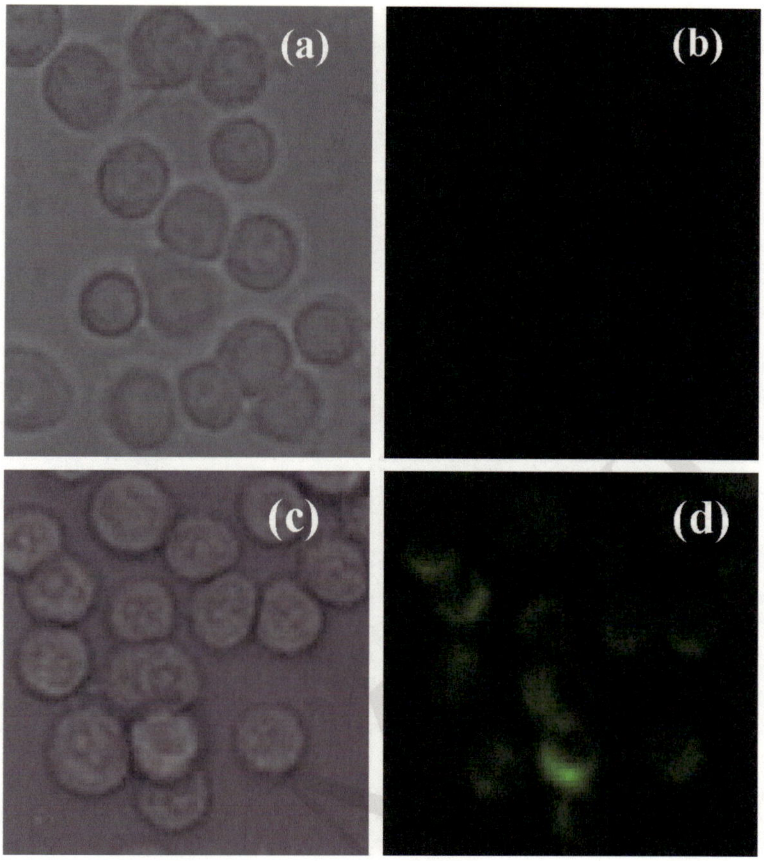

Figure 6.
*Fluorescence microscopic images of cancer cells treated with coumarin **6** (5 μM) in bright field (a) and merged field (b), pretreated with **6** (5 μM) followed by addition of 10 μM of Zn^{2+} in bright field (c) and merged field (d) [34].*

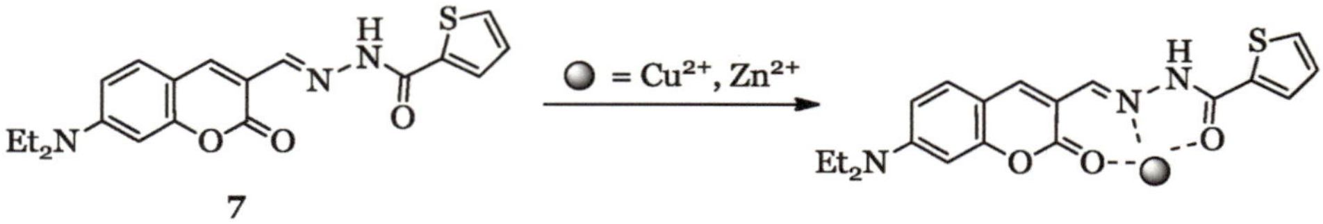

Figure 7.
Binding modes of Cu^{2+} and Zn^{2+} by coumarin 7.

Coumarin-naphthalene chemosensor **8** in the CH_3OH/H_2O mixture acts as a chemodosimeter for Ag^+ ions and a fluorescent sensor in relation to Cu^{2+} ions [36]. Silver is a potentially toxic and potentially carcinogenic element. It is known that Ag^+ can react with proteins in the body, block thiol groups of enzyme systems, and inhibit tissue respiration. Excessive concentration of Ag^+ ions in the body can lead to brain damage. Compound **8** possesses fluorescence at 480 nm, and its intensity is enhanced upon the addition of Cu^{2+} ions. However, the fluorescence intensity is quenched upon addition of Ag^+ ions due to irreversible desulfurization (**Figure 9**).

The detection limits of **8** are 8.1×10^{-9} M and 44.0×10^{-9} M for Cu^{2+} and Ag^+ ions, respectively. Compound **8** represents a safe and nontoxic to live cells

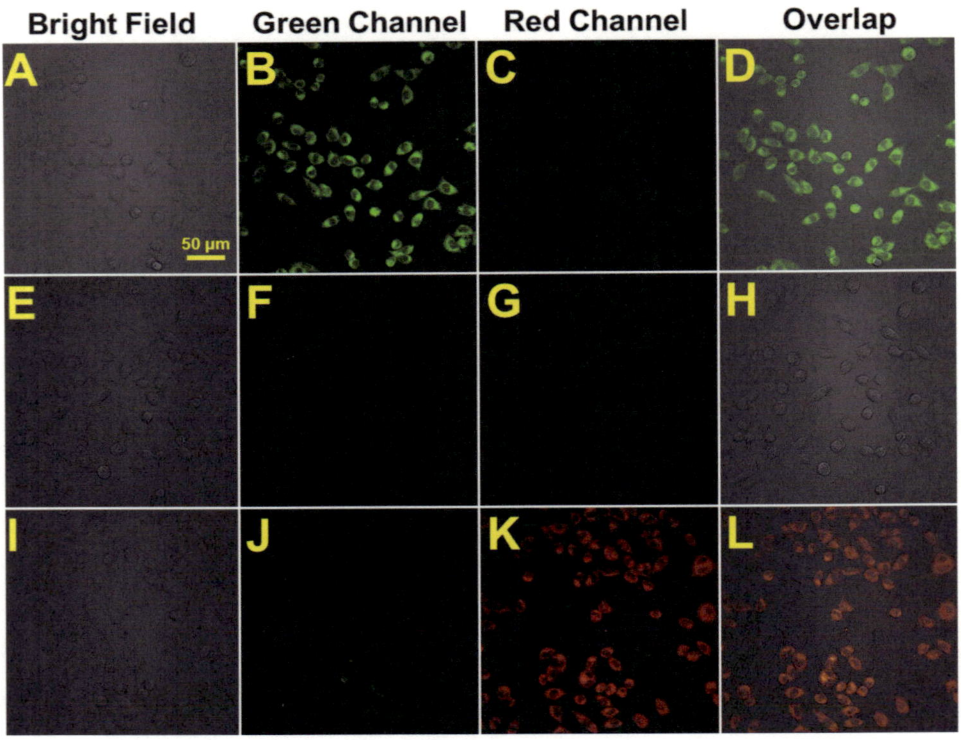

Figure 8.
Relative confocal fluorescence images of MRC-5 cells under different conditions with **7**. *MRC-5 cells treated with 5 μM of 7 (A–D), then further incubated with 5 μM of Cu(II) (E–H) and Zn(II) (I–L) ions for 10 min. Fluorescence images from left to right: Bright field, green channel, red channel, and overlap [35].*

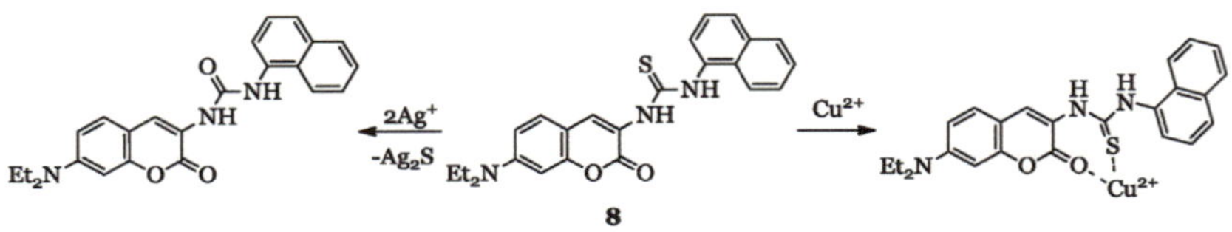

Figure 9.
Chemodosimeter and chemosensor properties of **8**.

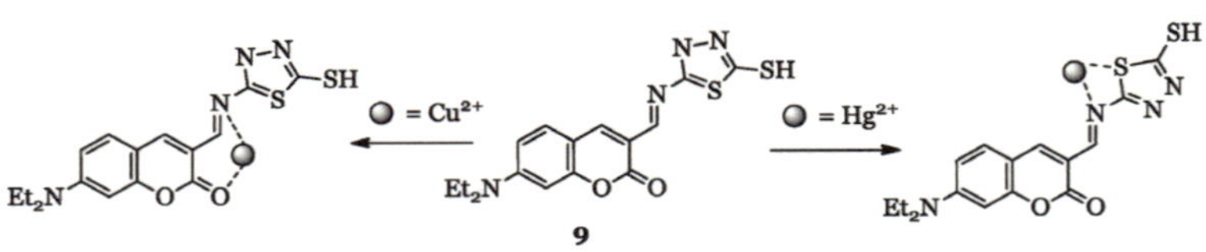

Figure 10.
The binding modes of **9** *with Cu^{2+} and Hg^{2+} ions.*

biosensor. Only weak emission was observed in human osteosarcoma cells U-2 OS when exposed to **8**, while strong blue and green fluorescence was observed upon addition of Cu^{2+} ions.

A dual-function coumarin chemosensor **9** could monitor Cu^{2+} and Hg^{2+} in CH_3CN/H_2O mixture [37]. The addition of Cu(II) cations changes the solution color from yellowish-brown to yellowish-green, while Hg^{2+} causes the fluorescence intensity enhancement at 572 nm (**Figure 10**).

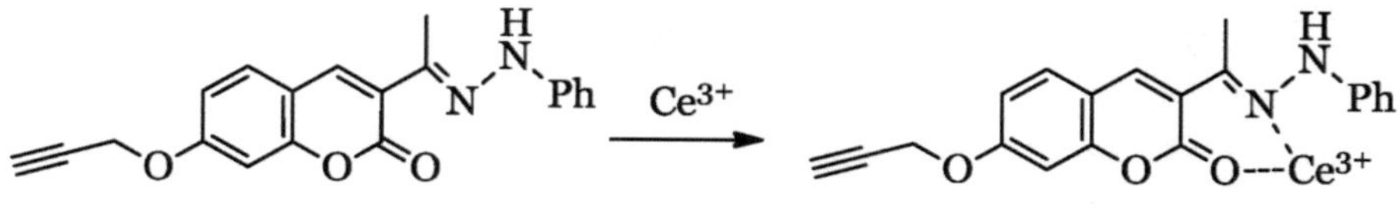

Figure 11.
Complexation of **10** *with* Ce^{3+}.

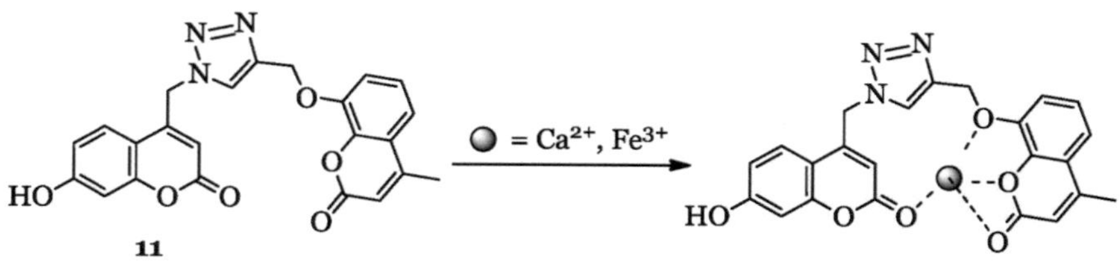

Figure 12.
Sensing of Ca^{2+} *and* Fe^{3+} *by* **11**.

The detection limit for Hg^{2+} was calculated to be 2.96 × 10^{-7} M.

Coumarin chemosensor **10** for detection of Ag^{+} and Ce^{3+} ions in an aqueous solution was developed [38]. Cerium is the most abundant of the lanthanides and refers to the conventionally toxic rare earth ultramicroelements. In the presence of Ag^{+}, the color of the solution of **10** in EtOH/H2O changes from pale yellow to brown. The addition of Ce^{3+} ion results in the substantial emission intensity enhancement at 350 nm (**Figure 11**).

The detection limit of Ce^{3+} ion by the sensor **10** was estimated to be 2.07 × 10^{-7} M.

Selective fluorescent coumarin-triazole chemosensor **11** toward Ca^{2+} and Fe^{3+} ions was synthesized [39]. Calcium(II) plays an important role in bone formation and as a second messenger in neurotransmitter release from neurons, in contraction of all muscle cell types and in fertilization. Inhibition of the PET process in **11** with Ca^{2+} ions leads to the fluorescence intensity increase at 473 nm, whereas complexation with Fe^{3+} causes its almost complete quenching (**Figure 12**).

The limit of detection was found to be 0.14 μM for Ca^{2+} and 0.25 μM for Fe^{3+}.

3. Sensing of multiple anions

A very small number of fluorogenic polyfunctional coumarin-based chemosensors for multianalyte detection has been created so far. This is due to the fact that the recognition of anions is in principle a very difficult problem, since charges of anions are more diffused than those of cations, which leads to rather weak electrostatic interactions between anions and receptor part of the sensor. As a result, the receptors connected with the coumarin core must have the ability to either form of hydrogen bonds with anions up to complete deprotonation, or to nucleophilic addition reactions. Anions play an important role in medicine, biology, and industry. A deficiency of fluoride ions can cause gum disease and osteoporosis, and an excess leads to fluorosis due to its nephrotoxic action. Both excess and deficiency of bromide and iodide anions affect the functioning of the thyroid gland and can cause serious diseases. Acetate anion is involved in various metabolic processes. Cyanide ion is highly toxic to humans even in small concentrations due to its strong interaction with cytochrome-oxidase.

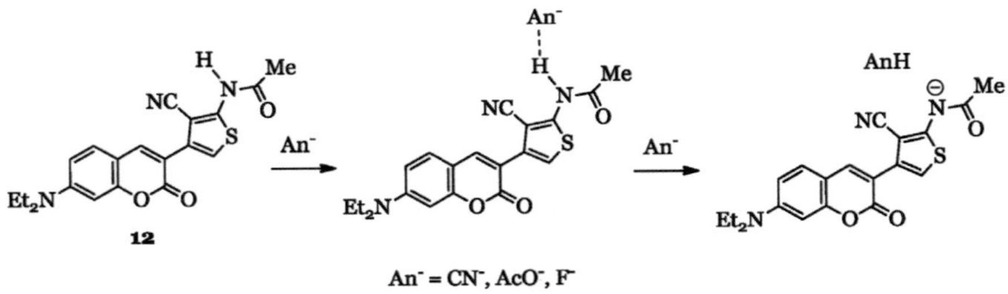

Figure 13.
Binding mode of coumarin **12** *with* CN^-*,* F^-*, and* AcO^- *in DMSO.*

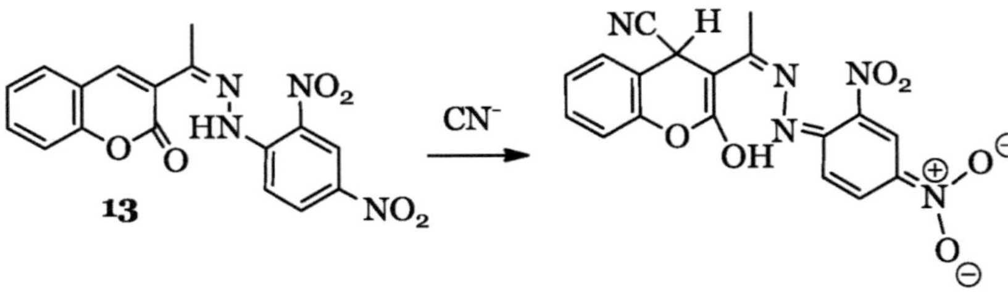

Figure 14.
The possible mode of binding for **13** *and* CN^-*.*

Coumarin chemosensor **12** is capable of detecting CN^-, F^-, and AcO^- in the presence of other ions-competitors [40]. Only cyanide anions caused a significant increase in the fluorescence intensity at 506 nm. However, only emission quenching was observed upon addition of CN^-, F^-, and AcO^- to the solution of **12** in DMSO. According to ^{1}H NMR titration data, this is due to the removal of the NH proton and the formation of the anionic form of **12** (**Figure 13**).

The fluorogenic and chromogenic chemosensor **13** in acetonitrile showed a change in the solution color upon addition of F^- and AcO^- ions from a yellow-green to red and orange, respectively [41]. In an aqueous medium, **13** selectively reacted with cyanide anion via a nucleophilic addition reaction, and the nonfluorescent solution turned a fluorescent blue-green at 495 nm (**Figure 14**).

The addition of cyanide anions to an aqueous solution of **13** containing blood plasma caused a significant fuorescence enhancement at 450 nm by ~7.7 times and a change in the nonfluorescent color of solution to blue. The LOD to detect cyanide anion in blood plasma was found to be 0.37 mM.

Coumarin-thiazole chemosensor **14** for CN^-, F^-, and AcO^- ions was synthesized [42]. Compound **14** demonstrated a color change in DMSO upon addition of these anions from yellow to deep red in the visible region and yellow fluorescence with CN^-.

^{1}H NMR and DFT calculation data correspond to the deprotonation mechanism, while for CN^- it is simultaneously supplemented by the addition reaction (**Figure 15**).

Coumarin **15** with the structure analogues to **14** was designed and synthesized for the selective detection of fluoride and cyanide anions [43]. Chemosensor **15** in acetonitrile selectively reacts with F^- via deprotonation mode, accompanied by a change in the color of the solution from colorless to deep red and a significant enhancement in the intensity of a yellow fluorescence. However, in an aqueous medium a substantial increase in the emission intensity at 506 nm was registered only in the presence of CN^- ions (**Figure 16**).

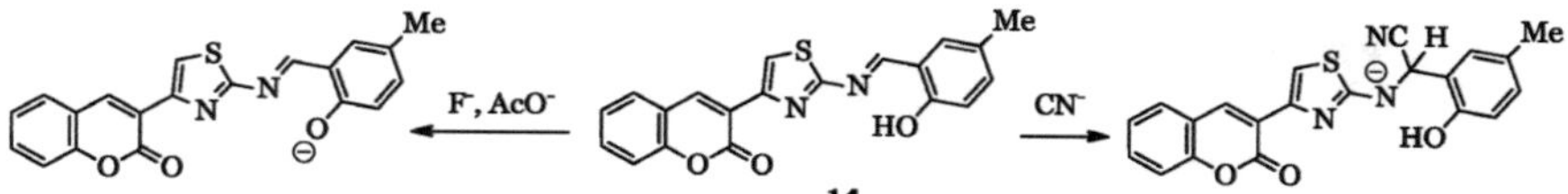

14

Figure 15.
Detection mechanisms of **14** *with CN⁻, AcO⁻, and F⁻.*

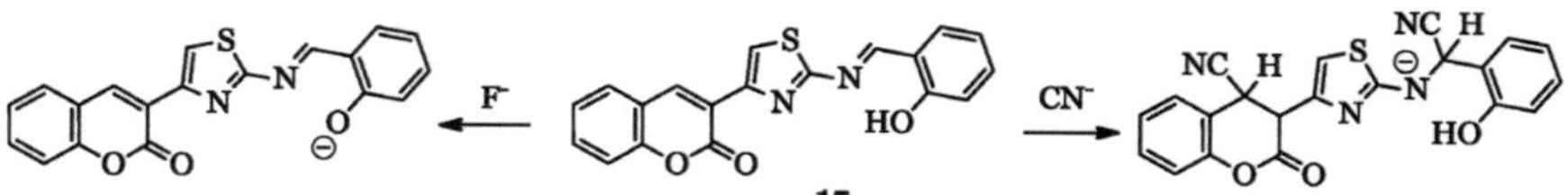

15

Figure 16.
Binding modes of **15** *with F⁻ and CN⁻.*

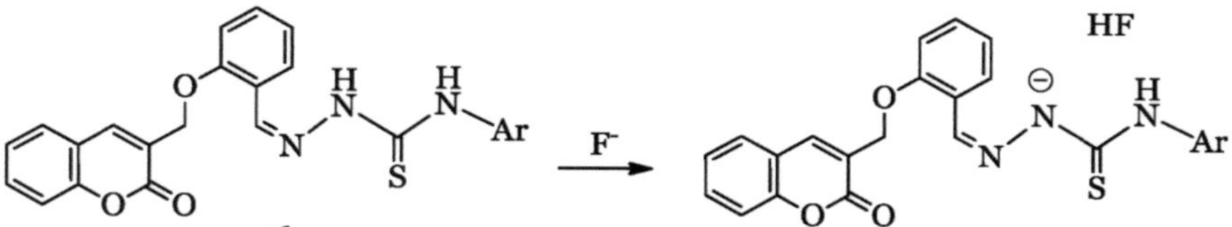

16

Figure 17.
Sensing of F⁻ by **16***.*

The LOD of fluoride ions in organic medium is 0.72 μM, while for cyanide ions in aqueous environment the LOD is 2.7 μM.

Coumarin thiosemicarbazones **16** act via naked-eye and fluorescence mechanisms [44] (**Figure 17**). They were found to be selective chemosensors for F^- with 1:1 receptor-anion ratio due to the appearance of a new emission band at 452 nm upon the addition of TBAF.

4. Sensing of metal cations and anions

As a rule, polyfunctional coumarin sensors for detection of metal cations and anions should include sites of various nature for the detection of these types of ions. Another displacement approach is based on the initial *in situ* formation of a complex with a cation, which then interacts with the anion, releasing the original sensor.

Diethylamine coumarin derivatives **17** and **18** were designed and synthesized to detect cyanide and copper(II) ions [45, 46]. The sensing mechanism is associated with the formation of a covalent bond between cyanide anions and 4-C carbon atom of coumarin (**Figure 18**). Red fluorescence of **17** at 669 nm is completely quenched, which is clearly visible to the naked eye. Complexation of **17** with Cu^{2+} exhibits color change from red to maroon and decreases the fluorescence intensity. LODs are 0.018 μM and 0.004 μM for CN^- and Cu^{2+}, respectively. Compound **18** also recognizes cyanide anions based on nucleophilic addition and copper(II) cations based on coordination reaction (**Figure 18**). However, in this case CN^- causes an obvious enhancement of fluorescence at 473 nm.

The detection limits of the compound **18** are 0.0071 μM (CN^-) and 0.014 μM (Cu^{2+}).

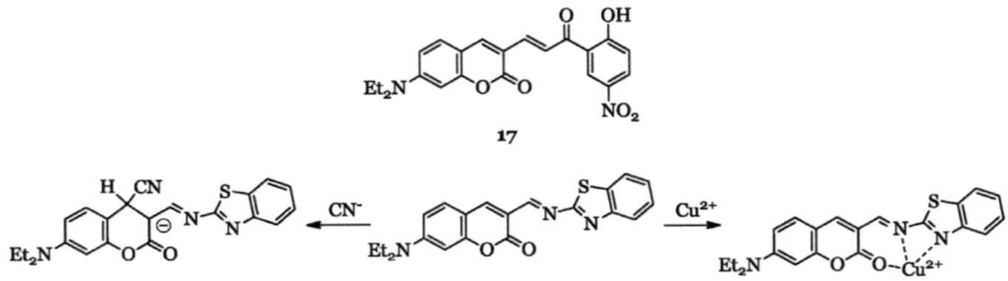

Figure 18.
Structures of **17** *and* **18** *and the proposed sensing mechanism of* **18**.

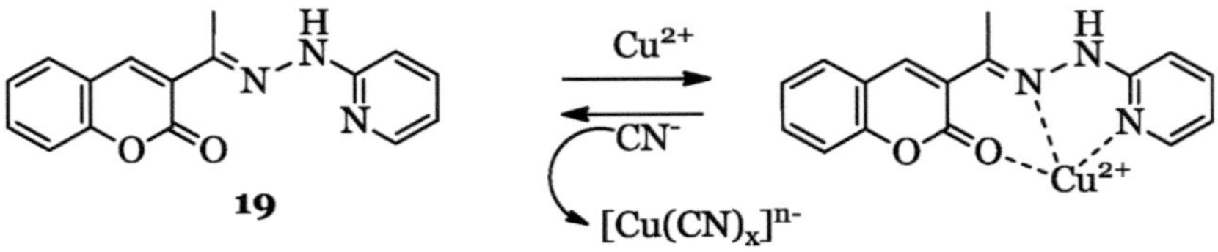

Figure 19.
Sensing of Cu^{2+} *and* CN^{-} *by* **19**.

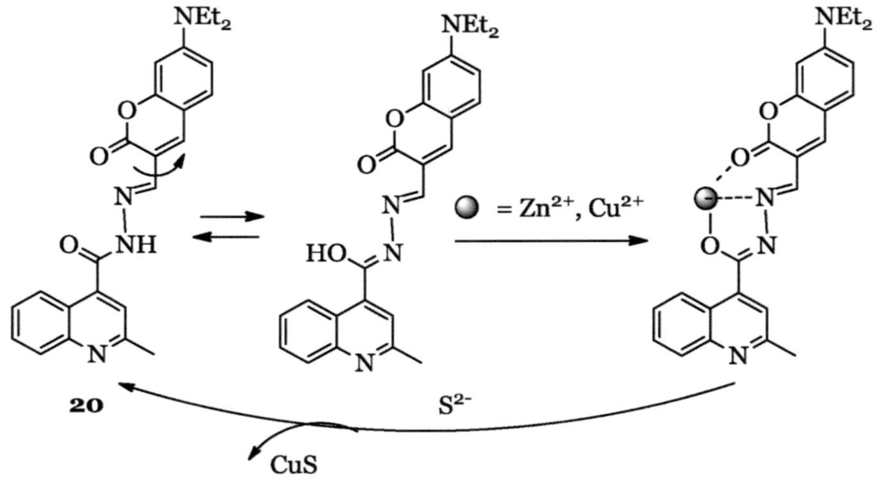

Figure 20.
Sensing of Zn^{2+}, Cu^{2+}, *and* S^{2-} *by* **20**.

Coumarin-based chemosensor **19** was obtained for selective fluorescent recognition of Cu^{2+} in MeOH/H_2O mixture and subsequent detection of CN^- via displacement approach [47]. Compound **19** demonstrates strong emission at 448 nm, which is selectively quenched upon addition of Cu^{2+} due to the formation of the complex (**Figure 19**) with the LOD of 3.76×10^{-7} M.

This *in situ* prepared chelate easily reacts with CN^- to form a very stable complex $[Cu(CN)_x]^{n-}$ and the initial fluorescence is restored with the LOD of 1.35×10^{-6} M, which is much lower than the WHO limit of CN^- (1.9 μM) for drinking water.

Coumarin **20** exhibits a significant increase in fluorescence intensity at 514 nm in the presence of Zn^{2+} ions, which is associated with the cessation of C=N isomerization process [48]. The detection limit reached at 5.76×10^{-7} M. As expected, the

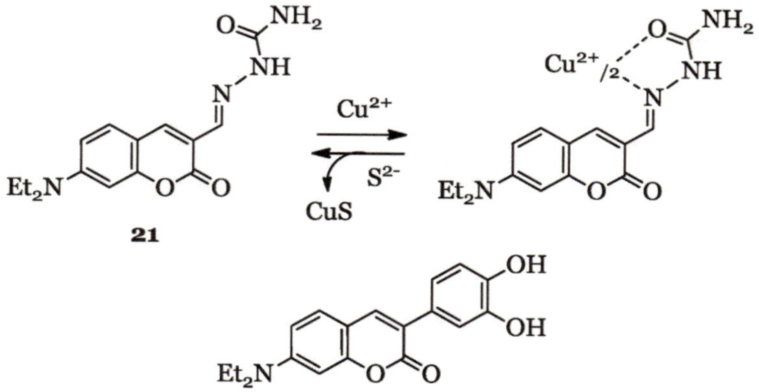

Figure 21.
*Structures of **21** and **22** and the proposed sensing mechanism of **21**.*

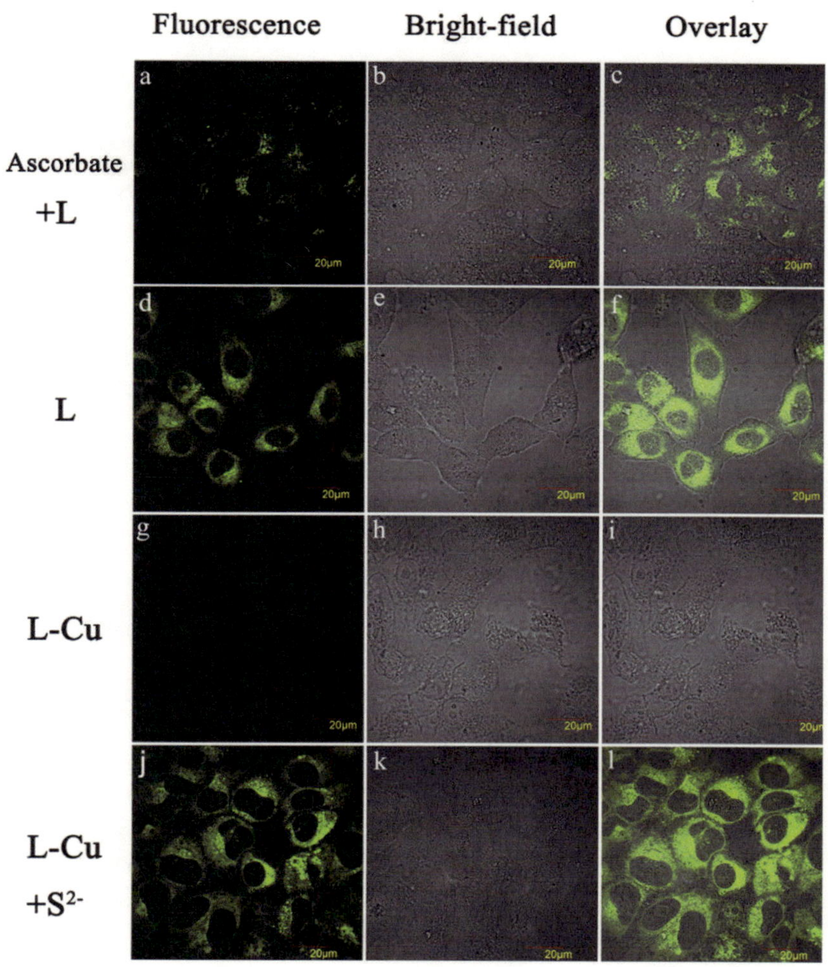

Figure 22.
*Confocal fluorescence imaging of A375 cells. Cells incubated with ascorbate (1 mM) for 3 h and stained with **21** (L, 10 μM) for 30 min (a–c); cells treated with **21** (10 μM) for 30 min (d–f); cells treated with **21** (10 μM) and cu^{2+} (20 μM) for 30 min (g–i); cells treated with **21**-Cu (10 μM) and S^{2-} (20 μM) for 30 min (j–l) [49].*

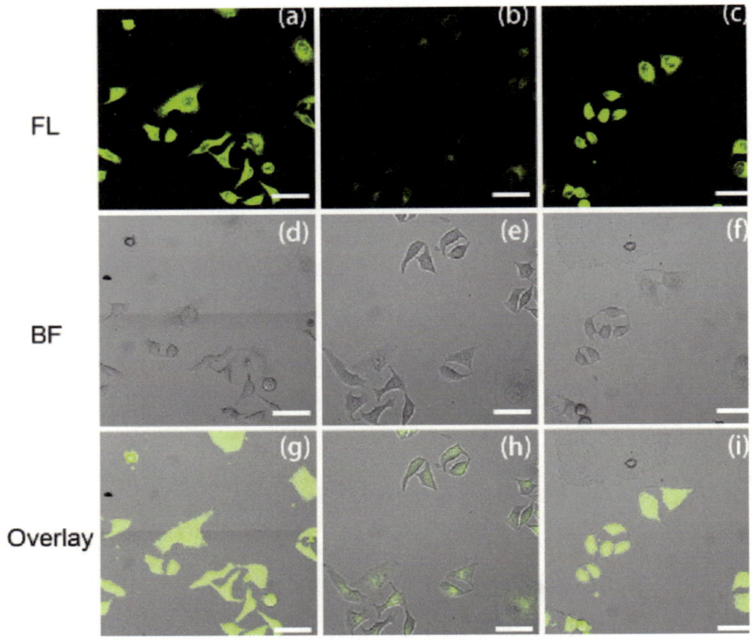

Figure 23.
Confocal fluorescence images of HeLa cells. (a) Cells incubated with **22** *(10 mM) for 30 min. (b) Cells incubated with* **22** *(10 mM) for 30 min, further incubated with* Cu^{2+} *(20 mM) for 30 min. (c) Cells incubated with* **22** *(10 mM) for 30 min, further incubated with* Cu^{2+} *(20 mM) for 30 min, and then incubated with* S^{2-} *(40 mM). (d-f) Bright-field pictures. (g-i) Overlapped pictures. Scale bar, 40 mm [50].*

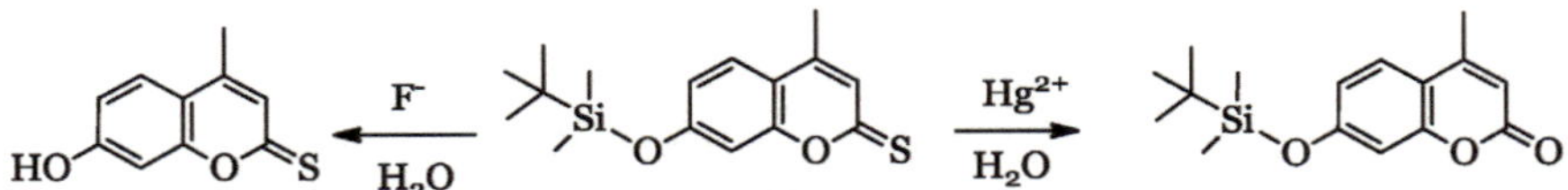

23

Figure 24.
Irreversible chemodosimeter sensing of Hg^{2+} *and* F^- *by* **23**.

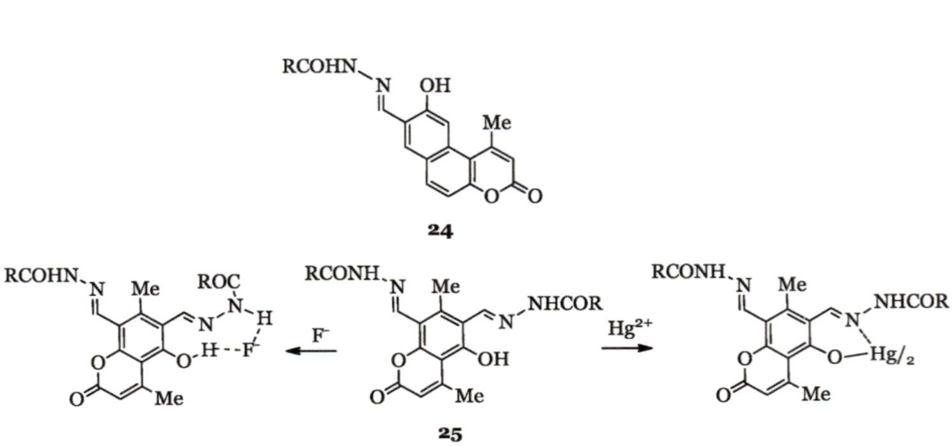

25

R = Ar, Het

Figure 25.
Structures of **24** *and* **25** *and a tentative scheme of sensing* Hg^{2+} *and* F^- *ions by the bifunctional chemosensors* **25**.

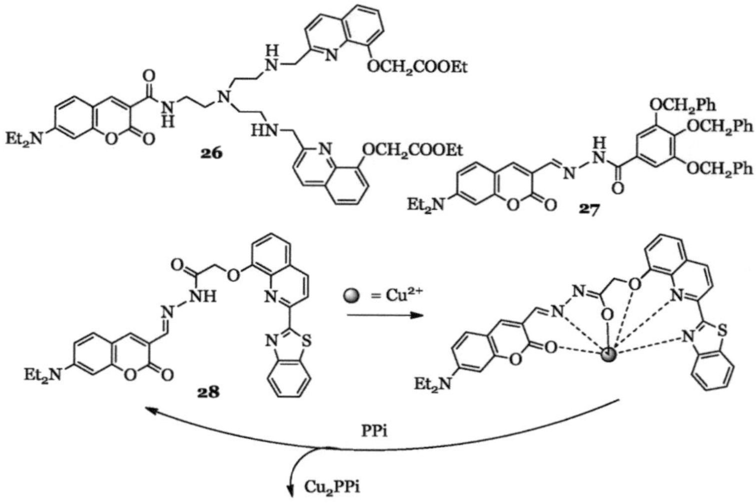

Figure 26.
Structures and sensing mechanism of **26–28.**

Cu(II) cations almost completely quenched this fluorescence due to their inherent paramagnetic properties (**Figure 20**). The LOD reached at 3.1×10^{-7} M.

Thus, prepared *in situ* complex with copper regenerates the initial fluorescence of **20** upon addition of S^{2-} even in the presence of F^-, Cl^-, Br^-, and I^- due to the formation of a very stable CuS. The LOD for S^{2-} was measured to be 1.9×10^{-5} M.

Further development of this approach has been applied in the design of chemosensory systems **21** and **22** suitable for intracellular biology applications. Coumarin **21** demonstrates an intensive emission at 516 nm [49, 50]. The fluorescence intensity decreases (~14-fold) upon addition of Cu^{2+} and reappears in the presence of S^{2-} anions (**Figure 21**). The LOD toward Cu(II) was found to be 2×10^{-8} M, which is lower than the most of the values reported in literature, and toward S^{2-}—6×10^{-8} M.

For the purpose of Cu^{2+} and S^{2-} biovisualization, confocal fluorescent imaging was performed using A375 cells. It is clearly visible in the dark field images that green fluorescence is significantly quenched by Cu^{2+} and restored after subsequent treatment by S^{2-} (**Figure 22**). The A375 cells were viable and maintained good shape in the entire process of this experiment, which means that **21** can successfully cross the cell membrane.

Similar results were obtained for **22**. HeLa cells were incubated with **22** (10 mM) at 37°C for 30 min and displayed bright green fluorescence (**Figure 23**). After incubation with Cu^{2+} for another 30 min, the emission of cells decreased. Upon addition of S^{2-} anions, the fluorescence intensity was restored. This indicates that **22** represents a potent candidate for sensing intracellular Cu^{2+} cations and S^{2-} anions in living cells.

The chemodosimeter approach was exploited for detection of Hg^{2+} and F^- ions by a simple coumarin derivative **23** [51] (**Figure 24**).

Upon addition of Hg^{2+} and F^- ions, **23** underwent desulfurization and desilylation to induce an increase in the fluorescence intensity at 491 nm and 526 nm, respectively.

Aroylhydrazones **24** and bis-aroylhydrazones of coumarin **25** display the properties of bifunctional fluorescent and colorimetric naked-eye chemosensors for mercury(II) cations and fluoride anions detection [52–54] (**Figure 25**).

The addition of Hg^{2+} ions in acetonitrile solution of **25** allows to observe a distinct naked-eye effect with the change of color from pale-yellow to bright-orange.

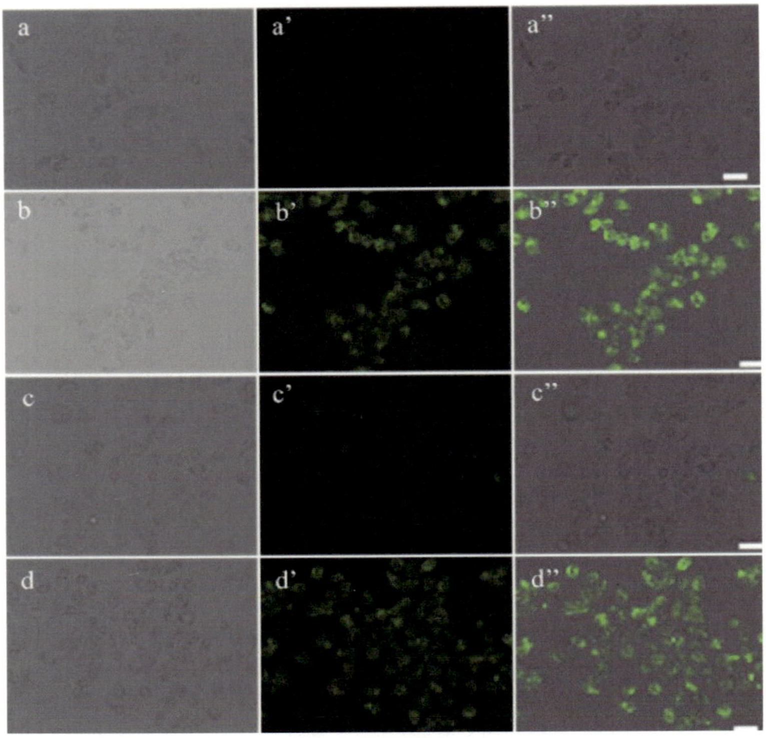

Figure 27.
Bright-field (a, b, c, d), fluorescence (a', b', c', d'), and confocal fluorescence microscope (a", b", c", d") images of HeLa cells: Blank cell (a, a', a"); cells treated with 5 μM **28** *(b, b', b"), then treated with 50 μM Cu^{2+} (c, c', c") and further treated with 50 μM PPi (d, d', d"). the scale bar was 50 μm [57].*

Simultaneously, the initial fluorescence is almost completely quenched. The LOD of mercury(II) cation sensing is 2.7 μM. In the presence of fluoride, cyanide, and acetate anions, a new absorption maximum in the visible spectral region appears. Furthermore, the formation of complex of **25** with fluoride anions is accompanied by the decrease in the relative intensity of the initial fluorescence I_0/I in ≈22 times.

Coumarin-based chemosensors **26–28** with complex chemical architecture were designed and synthesized for selective sequential recognition of Cu^{2+} and pyrophosphate anion (PPi) [55–57]. PPi is the main product of adenosine triphosphate hydrolysis in living cells, which is involved in important metabolic processes. The structures of **26–28** and the sensing mechanism are shown in **Figure 26.** These compounds demonstrate on–off fluorescence quenching toward the Cu^{2+} cation due to the formation of complexes that show off–on fluorescence enhancement upon addition of PPi over many competitive anions.

Chemosensor **28** showed sequential on–off–on fluorescent bioimaging of Cu^{2+} and PPi in HeLa cells. After the addition of **28**, the intense green fluorescence appeared (**Figure 27**). Cells incubated with Cu(II) cations efficiently quenched this emission, which was restored when treated with PPi. These data indicate that the sensor **28** possesses good cell permeability and can be used for bioimaging in live cells. The LOD for Cu^{2+} is 0.06 μM and for PPi it is 0.01 μM.

5. Sensing of metal cations and amino acids

Polyfunctional coumarin sensing of amino acids usually includes the initial detection of the appropriate metal cations, and in the second stage, the obtained *in situ* complex interacts with amino acid via displacement approach. A more complex problem is the creation of chemosensors capable of forming covalent bonds with the analyzed amino acid.

Amino acids are part of macromolecular proteins and represent essential substances for the growth and development of the human body. Cysteine (Cys) is of great importance in age defying, skin whitening, detoxifying, and improving immunity. Its deficiency causes premature senility, skin lesions, and uremia, while its excess can lead to senile dementia, neural tube defects, and osteoporosis. Histidine (His) is extremely important for the absorption of Fe^{2+} cations, vasodilation, and lowering blood pressure. The lack of His increases the risk of developing epilepsy, rheumatoid arthritis, and red cell aplasia, although its excessive content is associated with chronic kidney disease and Alzheimer's disease. Arginine (Arg) plays a vital role in cell replication, wound healing, and protein synthesis.

A simple coumarin sensor **29** selectively detects Hg^{2+} and Cys in an aqueous solution [58] (**Figure 28**). The addition of Hg(II) leads to a hypsochromic shift of the fluorescence emission band, while Cys almost completely quenches the emission of **29**. The latter process is seen by the naked eye under UV irradiation. The detection limit of **29** toward Cys is 8 μmol/L.

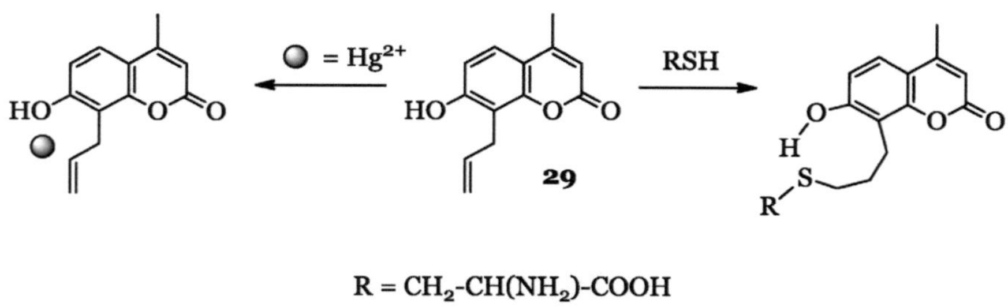

Figure 28.
Sensing mechanism of **29**.

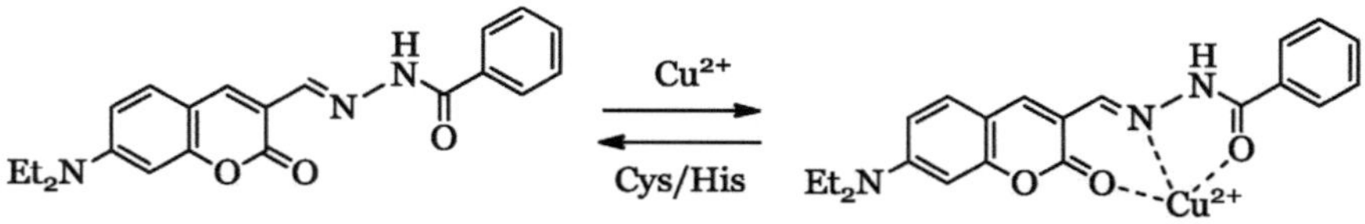

30

Figure 29.
Sensing mechanism of **30**.

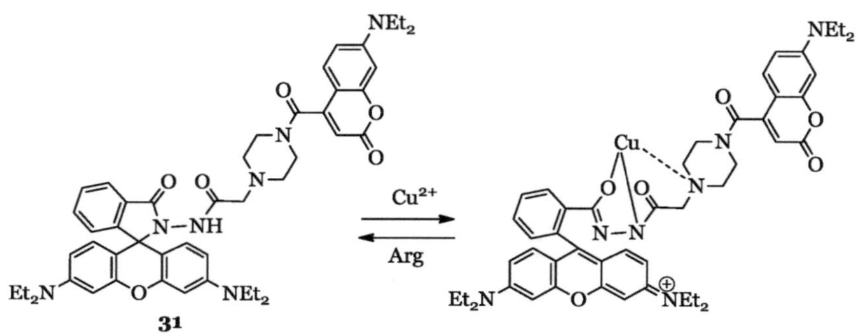

31

Figure 30.
Sequential detection of Cu^{2+} and Arg by **31**.

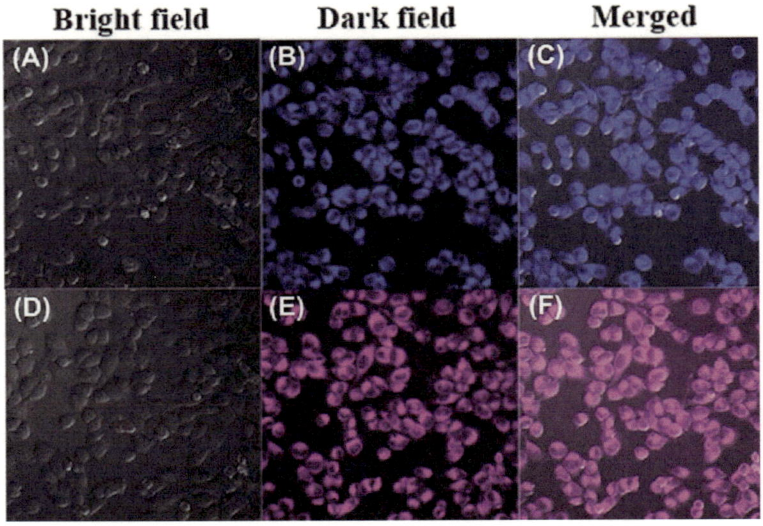

Figure 31.
Confocal fluorescence images of HeLa cells incubated with **31** *(20 mM) for 30 min (A–C) and then treated with* Cu^{2+} *(2 mM) for another 30 min (D–F). Images were obtained using an excitation of 405 nm and emission channels of (B) at 430–530 nm and (E) at 550–650 nm; (C and F) merge images of (A, B and D, E); (A and D) bright-field images of the cell culture [60].*

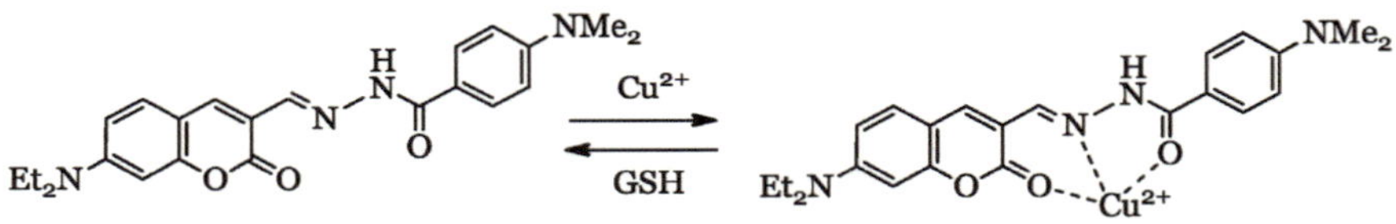

Figure 32.
Sequential detection of Cu^{2+} *and GSH by* **32.**

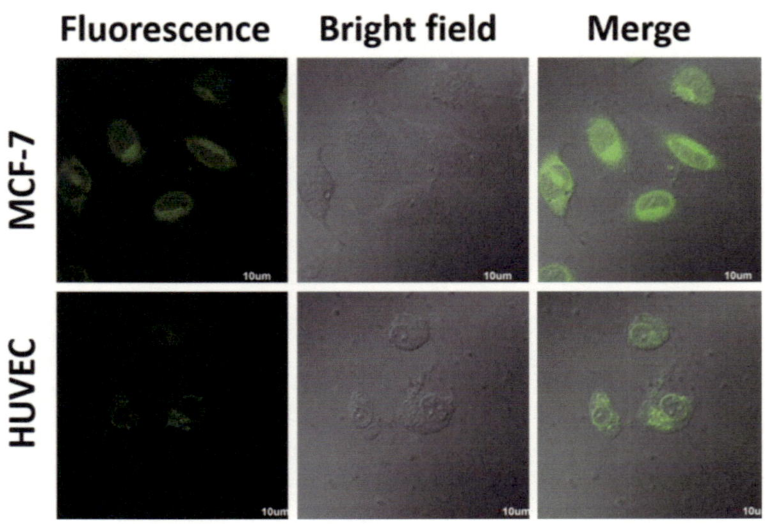

Figure 33.
Comparison of endogenous GSH level in MCF-7 and HUVEC cells after incubation with **32**-Cu^{2+} *complex. Left: Fluorescence images; middle: Bright-field images; right: Merged images [61].*

Coumarin **30** possesses a strong green emission at 527 nm [59]. Upon the addition of Cu^{2+}, **30** displays a significant fluorescence quenching. After the addition of Cys or His, the initial fluorescence is recovered due to the liberation of **30** (**Figure 29**).

Living A549 cells incubated with **30** exhibit notable emission. This fluorescence is quenched almost completely upon addition of Cu^{2+}. Further incubation of cells with Cys and His leads to the restoration of the initial fluorescence.

With the addition of Cu^{2+}, the solution of coumarin-rhodamine hybrid **31** in CH_3CN—H_2O mixture exhibits a naked-eye change from pale yellow to pink [60] (**Figure 30**). The fluorescence color converses from blue to pink (new maxima appear at 490 and 615 nm, which correspond to the emission of coumarin and rhodamine B moieties, respectively). The LOD for Cu^{2+} is 0.47 mM. The **31**-Cu^{2+} complex sequentially detects Arg with the restoration of blue fluorescence. The LOD for Arg is 0.33 mM.

The HeLa cells were incubated with **31** (20 mM) for 30 min, and a strong blue emission of the coumarin group was observed. Upon addition of Cu^{2+}, the HeLa cells exhibit strong pink fluorescence. These data show that **31** is cell permeable and can be applied to fluorescence imaging of intracellular Cu^{2+} (**Figure 31**).

Coumarin **32** was prepared for the detection of Cu^{2+} and glutathione (γ-glutamylcysteinylglycine, GSH) [61]. Overexpression of tumor biomarker GSH was found in many types of cancer. In the presence of Cu^{2}, **32** exhibits selective fluorescence quenching and color change from yellow to orange-red. When GSH was added to the solution, the initial fluorescence was recovered (**Figure 32**).

The LODs were calculated as 2.40×10^{-8} M and 1.29×10^{-7} M for Cu^{2+} and GSH, respectively.

MCF-7 and HUVEC cells were both incubated with **32**-Cu^{2+} complex for 30 min and then imaged under the same conditions (**Figure 33**). The fluorescence intensity in MCF-7 cells was above twofold higher than that in HUVEC cells, suggesting a higher GSH concentration in tumor cells. This is probably due to generation of additional GSH in tumor cells for resisting intrinsic oxidative stress.

Chemosensor **33** was designed for simultaneous detection of Cys, Hcy (homocysteine), and GSH [62]. Due to different binding mechanisms, compound **33** demonstrates enhancing of fluorescence intensity with 108-, 128-, and 30-fold at 457, 559, and 529 nm for Cys, Hcy, and GSH, respectively, through different excitation wavelengths (**Figure 34**).

For exogenous biothiols, the BEL-7402 cells were firstly pretreated with NEM and cellular biothiols and SH-containing proteins were deactivated. After incubation with **33**, no fluorescence could be observed (**Figure 35**).

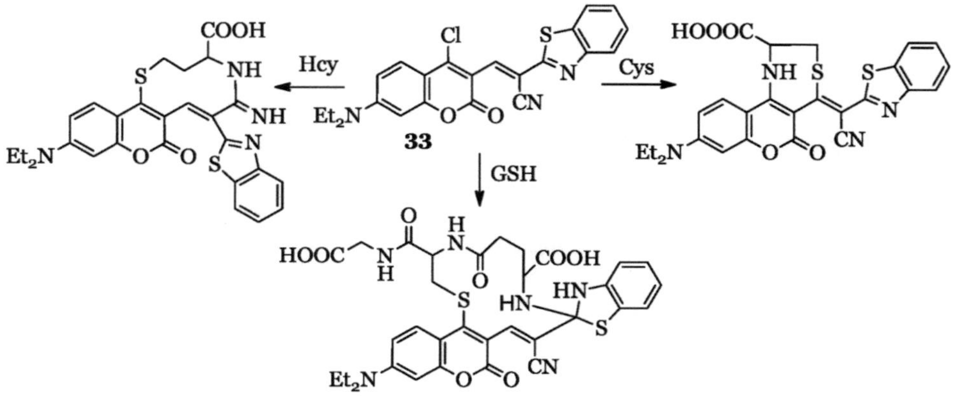

Figure 34.
Proposed mechanisms of bonding 33 with Cys, Hcy, and GSH.

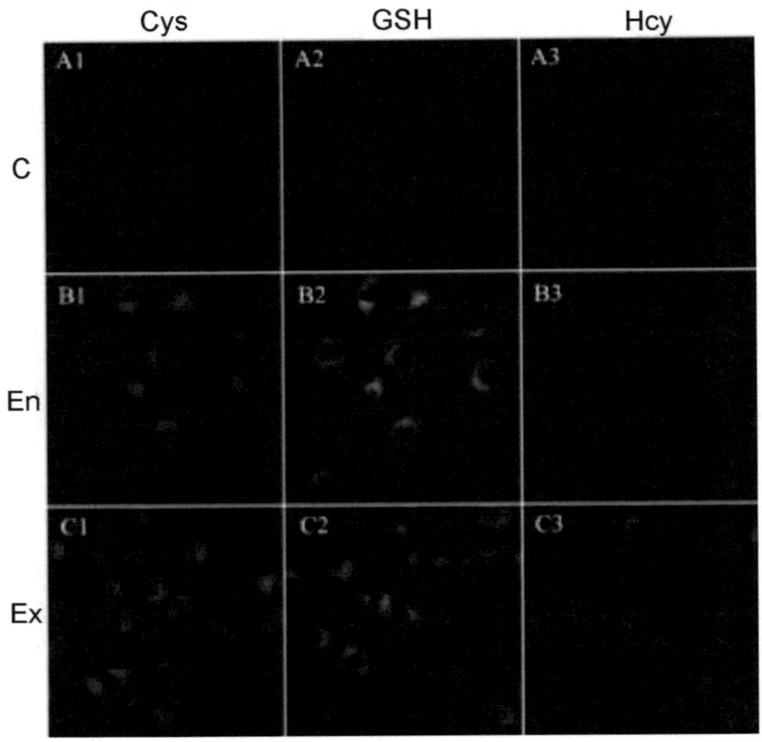

Figure 35.
Confocal fluorescence images of Cys, GSH, and Hcy in BEL-7402 cells. C: Control, en: Endogenous, ex: Exogenous. (A1–A3) cells were incubated for 30 min, then imaged. (B1–B3) cells were incubated with **33** *(2.5 μM) for 30 min, then imaged. (C1–C3) cells were pretreated with NEM (0.5 mM, 30 min), subsequently incubated with Cys/GSH/Hcy (500 μM, 30 min) and* **33** *(2.5 μM, 30 min), then imaged (λ_{ex} = 405 nm, λ_{em} = 421–475 nm for the blue channel; λ_{ex} = 458 nm, λ_{em} = 500–550 nm for the green channel; and λ_{ex} = 543 nm, λ_{em} = 552–617 nm for the red channel). Scale bar: 20 μm [62].*

After subsequent treatment with Cys, Hcy, and GSH, respectively, blue, red, and green fluorescence was observed from three different emission channels in living cells with high selectivity.

6. Conclusion

The design, synthesis, and investigation of fluorogenic polyfunctional coumarin chemosensors for multianalyte detection is an intriguing and extensively developing area of organic, medical, and biological chemistry. These sensors demonstrate high efficiency and selectivity combined with low cost and simplicity of analysis. Due to the limited size of the chapter, only sensors for the detection of metal cations, anions, and amino acids were considered, while sensors for proteins, DNA, RNA, etc. were ignored. Nevertheless, these data suggest that this group of polyfunctional chemosensors is extremely suitable for express analysis and bioimaging of various objects.

Acknowledgements

This research was financially supported by the Ministry of Science and Higher Education of the Russian Federation, project 0852-2020-2100-19. A. Dubonosov worked in the framework of the State assignment of the Southern Scientific Centre of the RAS No. 01201354239.

Conflict of interest

The authors declare "no conflict of interest."

Author details

Alexander Dubonosov[1*] and Vladimir Bren[2]

1 Federal Research Centre the Southern Scientific Centre of the Russian Academy of Sciences, Rostov on Don, Russian Federation

2 Institute of Physical and Organic Chemistry, Southern Federal University, Rostov on Don, Russian Federation

*Address all correspondence to: aled@ipoc.sfedu.ru

References

[1] Anslyn EV, Wang B. Chemosensors: Principles, Strategies, and Applications. Hoboken: Wiley; 2011. p. 540 DOI: 10.1002/97811180195802

[2] Singh R, Das G. Towards fluorogenic and chromogenic sensing of heavy metal ions in aqueous medium: A mini-review. In: Singh DK, Das S, Materny A, editors. Advances in Spectroscopy: Molecules to Materials. Singapore: Springer Nature; 2019. pp. 57-65

[3] Suganya S, Naha S, Velmathi S. A critical review on colorimetric and fluorescent probes for the sensing of analytes via relay recognition. ChemistrySelect. 2018;**3**(25):7231-7268. DOI: 10.1002/slct.201801222

[4] Liu Z, He W, Guo Z. Metal coordination in photoluminescent sensing. Chemical Society Reviews. 2013;**42**(4):1568-1600. DOI: 10.1039/c2cs35363f

[5] Yu L, Wang S, Huang K, Liu Z, Gao F, Zeng W. Fluorescent probes for dual and multi analyte detection. Tetrahedron. 2015;**71**(29):4679-4706. DOI: 10.1016/j.tet.2015.04.115

[6] Wu J, Kwon B, Liu W, Anslyn EV, Wang P, Kim JS. Chromogenic/fluorogenic ensemble chemosensing systems. Chemical Reviews. 2015;**115**(15):7893-7943. DOI: 10.1021/cr500553d

[7] Chhatwal M, Kumar A, Singh V, Gupta RD, Awasthi SK. Addressing of multiple-metal ions on a single platform. Coordination Chemistry Reviews. 2015;**292**:30-55. DOI: 10.1016/j.ccr.2015.02.009

[8] Singh H, Bhargava G, Kumar S, Singh P. Quadruple-signaling (PET, ICT, ESIPT, C=N rotation) mechanism-based dual chemosensor for detection of Cu^{2+} and Zn^{2+} ions: TRANSFER, INH and complimentary OR/NOR logic circuits. Journal of Photochemistry and Photobiology A. 2018;**357**:175-184. DOI: 10.1016/j.jphotochem.2018.02.030

[9] Penta S, editor. Advances in Structure and Activity Relationship of Coumarin Derivatives. Amsterdam: Elsevier-Academic Press; 2015. p. 190

[10] Calcio Gaudino E, Tagliapietra S, Martina K, Palmisano G, Cravotto G. Recent advances and perspectives in the synthesis of bioactive coumarins. RSC Advances. 2016;**6**:46394-46405. DOI: 10.1039/C6RA07071J

[11] Bräse S. Privileged Scaffolds in Medicinal Chemistry. Design, Synthesis, Evaluation. Cambridge: RSC; 2015. p. 476 DOI: 10.1039/9781782622246

[12] Anamika UD, Ekta JN, Sharma S. Advances in synthesis and potentially bioactive of coumarin derivatives. Current Organic Chemistry. 2019;**22**(26):2509-2536. DOI: 10.2174/1385272822666181029102140

[13] Grover J, Jachak SM. Coumarins as privileged scaffold for anti-inflammatory drug development. RSC Advances. 2015;**5**(49):38892-38905. DOI: 10.1039/C5RA05643H

[14] Emami S, Dadashpour S. Current developments of coumarin-based anti-cancer agents in medicinal chemistry. European Journal of Medicinal Chemistry. 2015;**102**:611-630. DOI: 10.1016/j.ejmech.2015.08.033

[15] Medina FG, Marrero JG, Macias-Alonso M, González MC, Córdova-Guerrero I, Teissier García AG, et al. Coumarin heterocyclic derivatives: Chemical synthesis and biological activity. Natural Product Reports. 2015;**32**(10):1472-1507. DOI: 10.1039/c4np00162a

[16] An R, Hou Z, Li JT, Yu HN, Mou YH, Guo C. Design, synthesis and biological evaluation of novel 4-substituted coumarin derivatives as antitumor agents. Molecules. 2018;**23**(9):E2281. DOI: 10.3390/molecules23092281

[17] Matos MJ, Vazquez-Rodriguez S, Fonseca A, Uriarte E, Santana L, Borges F. Heterocyclic antioxidants in nature: Coumarins. Current Organic Chemistry. 2017;**21**(4):311-324. DOI: 10.2174/1385272820666161017170652

[18] Kathiravan MK, Salake AB, Chothe AS, Dudhe PB, Watode RP, Mukta MS, et al. The biology and chemistry of antifungal agents: A review. Bioorganic and Medicinal Chemistry. 2012;**20**(19):5678-5695. DOI: 10.1016/j.bmc.2012.04.045

[19] Hassan MZ, Osman H, Ali MA, Ahsan MJ. Therapeutic potential of coumarins as antiviral agents. European Journal of Medicinal Chemistry. 2016;**123**:236-255. DOI: 10.1016/j.ejmech.2016.07.056

[20] Yamaji M, Hakoda Y, Okamoto H, Tani F. Photochemical synthesis and photophysical properties of coumarins bearing extended polyaromatic rings studied by emission and transient absorption measurements. Photochemical and Photobiological Sciences. 2017;**12**(4):555-563. DOI: 10.1039/c6pp00399k

[21] Al-Masoudi NA, Al-Salihi NJ, Marich YA, Markus T. Synthesis, and fluorescence properties of coumarin and benzocoumarin derivatives conjugated pyrimidine scaffolds for biological imaging applications. Journal of Fluorescence. 2015;**25**(6):1847-1854. DOI: 10.1007/s10895-015-1677-z

[22] Nazir R, Stasyuk AJ, Gryko DT. Vertically π-expanded coumarins: The synthesis and optical properties. Journal of Organic Chemistry. 2016;**81**(22):11104-11114. DOI: 10.1021/acs.joc.6b02094

[23] Wang ZS, Cui Y, Hara K, Dan-oh Y, Kasada C, Shinpo A. A high-light-harvesting-efficiency coumarin dye for stable dye-sensitized solar cells. Advanced Materials. 2007;**19**(8):1138-1141. DOI: 10.1002/adma.200601020

[24] Cao D, Liu Z, Verwilst P, Koo S, Jangjili P, Kim JS, et al. Coumarin-based small-molecule fluorescent chemosensors. Chemical Reviews. 2019;**119**(18):10403-10519. DOI: 10.1021/acs.chemrev.9b00145

[25] Katerinopoulos HE. The coumarin moiety as chromophore of fluorescent ion indicators in biological systems. Current Pharmaceutical Design. 2004;**10**(30):3835-3852. DOI: 10.2174/1381612043382666

[26] Formica M, Fusi V, Giorgi L, Micheloni M. New fluorescent chemosensors for metal ions in solution. Coordination Chemistry Reviews. 2012;**256**:170-192. DOI: 10.1016/j.ccr.2011.09.010

[27] Song Y, Chen Z, Li H. Advances in coumarin-derived fluorescent chemosensors for metal ions. Current Organic Chemistry. 2012;**16**(22):2690-2707. DOI: 10.2174/138527212804004544

[28] Sareen D, Kaur P. Strategies in detection of metal ions using dyes. Coordination Chemistry Reviews. 2014;**265**:125-154. DOI: 10.1016/j.ccr.2014.01.015

[29] Qin J, Fan L, Wang B, Yang Z, Li T. The design of a simple fluorescent chemosensor for Al^{3+}/Zn^{2+} via two different approaches. Analytical Methods. 2015;**7**(2):716-722. DOI: 10.1039/C4AY02351J

[30] Fu J, Chang Y, Li B, Wang X, Xie X, Xu K. A dual fluorescence probe for

Zn^{2+} and Al^{3+} through differentially response and bioimaging in living cells. Spectrochimica Acta A. 2020;**225**. Article 117493. DOI: 10.1016/j.saa.2019.117493

[31] Zhang Q, Ma R, Li Z, Liu Z. A multi-responsive crown ether-based colorimetric/fluorescent chemosensor for highly selective detection of Al^{3+}, Cu^{2+} and Mg^{2+}. Spectrochimica Acta A. 2020;**228**. Article 117857. DOI: 10.1016/j.saa.2019.117857

[32] Roy N, Dutta A, Mondal P, Paul PC, Singh TS. A new coumarin based dual functional chemosensor for colorimetric detection of Fe^{3+} and fluorescence turn-on response of Zn^{2+}. Sensors and Actuators B. 2016;**236**:719-731. DOI: 10.1016/j.snb.2016.06.061

[33] Roy N, Nath S, Dutta A, Mondal P, Paul PC, Singh TS. A highly efficient and selective coumarin based fluorescent probe for colorimetric detection of Fe^{3+} and fluorescence dual sensing of Zn^{2+} and Cu^{2+}. RSC Advances. 2016;**6**(68):63837-63847. DOI: 10.1039/C6RA12217E

[34] Wang L, Li W, Zhi W, Huang Y, Han J, Wang Y, et al. A new coumarin schiff based fluorescent-colorimetric chemosensor for dual monitoring of Zn^{2+} and Fe^{3+} in different solutions: An application to bio-imaging. Sensors and Actuators B. 2018;**260**:243-254. DOI: 10.1016/j.snb.2017.12.200

[35] He X, Xie Q, Fan J, Xu C, Xu W, Li Y, et al. Dual-functional chemosensor with colorimetric/ratiometric response to Cu(II)/Zn(II) ions and its applications in bioimaging and molecular logic gates. Dyes and Pigments. 2020;**177**. Article 108255. DOI: 10.1016/j.dyepig.2020.108255

[36] Kumar A, Mondal S, Kayshap KS, Hira SK, Manna PP, Dehaend W, et al. Water switched aggregation/disaggregation strategies of a coumarin-naphthalene conjugated sensor and its selectivity towards Cu^{2+} and Ag^{+} ions along with cell imaging studies on human osteosarcoma cells (U-2 OS). New Journal of Chemistry. 2018;**42**(13):10983-10988. DOI: 10.1039/c8nj01631c

[37] Xu P, Liu X, Zhao X, Zhu W, Fang M, Wu Z, et al. A dual-function chemosensor based on coumarin for fluorescent turn-on recognition of Hg^{2+} and colorimetric detection of Cu^{2+} in aqueous media. Journal of the Chinese Chemical Society. 2020;**67**(2):298-305. DOI: 10.1002/jccs.201900188

[38] Liu M, Xu Z, Song Y, Li H, Xian C. A novel coumarin-based chemosensor for colorimetric detection of Ag(I) ion and fluorogenic sensing of Ce(III) ion. Journal of Luminescence. 2018;**198**:337-341. DOI: 10.1016/j.jlumin.2018.02.047

[39] Puthiyedath T, Bahulayan D. A click derived triazole-coumarin derivative as fluorescence on-off PET based sensor for Ca^{2+} and Fe^{3+} ions. Sensors and Actuators B. 2018;**272**:110-117. DOI: 10.1016/j.snb.2018.05.126

[40] Yanar U, Babür B, Pekyilmaz D, Yahaya I, Aydiner B, Dede Y, et al. A fluorescent coumarin-thiophene hybrid as a ratiometric chemosensor for anions: Synthesis, photophysics, anion sensing and orbital interactions. Journal of Molecular Structure. 2016;**1108**:269-277. DOI: 10.1016/j.molstruc.2015.11.081

[41] Razi SS, Ali R, Srivastava P, Shahid M, Misra A. An efficient multichannel probe to detect anions in different media and its real application in human blood plasma. RSC Advances. 2014;**4**(43):22308-22317. DOI: 10.1039/c4ra02388a

[42] Şahin Ö, Özdemir ÜÖ, Seferoğlu N, Genc ZK, Kaya K, Aydiner B, et al. New platinum(II) and palladium(II) complexes of coumarin-thiazole Schiff base with a fluorescent chemosensor

properties: Synthesis, spectroscopic characterization, X-ray structure determination, in vitro anticancer activity on various human carcinoma cell lines and computational studies. Journal of Photochemistry and Photobiology B. 2018;**178**:428-439. DOI: 10.1016/j.jphotobiol.2017.11.030

[43] Padhan SK, Podh MB, Sahu PK, Sahu SN. Optical discrimination of fluoride and cyanide ions by coumarinsalicylidene based chromofluorescent probes in organic and aqueous medium. Sensors and Actuators B. 2018;**255**(2):1376-1390. DOI: 10.1016/j.snb.2017.08.133

[44] Islam M, Hameed A, Ayub K, Naseer MM, Hussain J, Alharthy RD, et al. Receptor-spacer-fluorophore based coumarin-thiosemicarbazones as anion chemosensors with "turn on" response: Spectroscopic and computational (DFT) studies. ChemistrySelect. 2018;**3**(26):7633-7642. DOI: 10.1002/slct.201801035

[45] Shan Y, Sun Y, Suna N, Guan R, Cao D, Wang K, et al. One diethylamine coumarin derivative with nitro substituted chalcone structure as chemosensor for cyanide and copper ions. Inorganic Chemistry Communications. 2015;**59**:68-70. DOI: 10.1016/j.inoche.2015.06.031

[46] Wang K, Feng W, Wang Y, Cao D, Guan R, Yu X, et al. A coumarin derivative with benzothiazole Schiff's base structure as chemosensor for cyanide and copper ions. Inorganic Chemistry Communications. 2016;**71**:102-104. DOI: 10.1016/j.inoche.2016.07.013

[47] Mukherjee S, Talukder S. A coumarin-based luminescent chemosensor for recognition of Cu^{2+} and its in-situ complex for CN-sensing via Cu^{2+} displacement approach. Journal of Fluorescence. 2017;**27**:1567-1572. DOI: 10.1007/s10895-016-1974-1

[48] Qin J, Yang Z. Design of a novel coumarin-based multifunctional fluorescent probe for Zn^{2+}/Cu^{2+}/S^{2-} in aqueous solution. Materials Science and Engineering C. 2015;**57**:265-271. DOI: 10.1016/j.msec.2015.07.064

[49] Feng Y, Yang Y, Wang Y, Qiu F, Song X, Tang X, et al. Dual-functional colorimetric fluorescent probe for sequential Cu^{2+} and S^{2-} detection in bio-imaging. Sensors and Actuators B. 2019;**288**:27-37. DOI: 10.1016/j.snb.2019.02.062

[50] Liu Z, Liu L, Li J, Qin Y, Zhao C, Mi C, et al. A new coumarin-based fluorescent probe for selective recognition of Cu^{2+} and S^{2-} in aqueous solution and living cells. Tetrahedron. 2019;**75**:3951-3957. DOI: 10.1016/j.tet.2019.05.057

[51] Gu L, Zheng T, Xu Z, Song Y, Li H, Xia S, et al. A novel bifunctional fluorescent and colorimetric probe for detection of mercury and fluoride ions. Spectrochimica Acta A. 2019;**207**:88-95. DOI: 10.1016/j.saa.2018.08.060

[52] Dubonosov AD, Bren VA, Minkin VI. Enolimine-ketoenamine tautomerism for chemosensing. In: Antonov L, editor. Tautomerism: Concepts and Applications in Science and Technology. Weinheim: Wiley-VCH; 2016. pp. 229-252. DOI: 10.1002/9783527695713.ch10

[53] Nikolaeva OG, Shepelenko EN, Tikhomirova KS, Revinskii YV, Dubonosov AD, Bren VA, et al. Bifunctional fluorescent and colorimetric "naked eye" aroylhydrazone chemosensors for Hg^{2+} and F^{-} ions detection. Mendeleev Communications. 2016;**26**(5):402-404. DOI: 10.1016/j.mencom.2016.09.012

[54] Nikolaeva OG, Karlutova OY, Dubonosov AD, Bren VA, Minkin VI. Synthesis and luminescence and ionochromic properties of

9-hydroxy-1-methyl-3-oxo-3H-benzo[f] chromene-8-carbaldehyde imines and hydrazones. Russian Journal of General Chemistry. 2020;**90**(2):196-201. DOI: 10.1134/S107036322002005X

[55] Meng X, Cao D, Hu Z, Han X, Li Z, Liang D, et al. A coumarin based highly selective fluorescent chemosensor for sequential recognition of Cu^{2+} and PPi. Tetrahedron Letters. 2018;**59**(49):4299-4304. DOI: 10.1016/j.tetlet.2018.10.048

[56] Zhao C, Chen J, Cao D, Wang J, Ma W. Novel coumarin-based containing denrons selective fluorescent chemosesor for sequential recognition of Cu^{2+} and PPi. Tetrahedron. 2019;**75**(13):1997-2003. DOI: 10.1016/j.tet.2019.02.024

[57] Li S, Cao D, Meng X, Hu Z, Li Z, Yuan C, et al. A novel fluorescent chemosensor based on coumarin and quinolinylbenzothiazole for sequential recognition of Cu^{2+} and PPi and its applicability in live cell imaging. Spectrochimica Acta A. 2020;**230**. Article 118022. DOI: 10.1016/j.saa.2019.118022

[58] Tao P, Chen D, Xu Z, Song Y, Li H, Xian C. A fluorescent probe for the dual detection of mercury ions and thiols based on a simple coumarin derivative. Coloration Technology. 2020;**136**(1):75-86. DOI: 10.1111/cote.12447

[59] Xie Y, Yan L, Li J. An on-off-on fluorescence probe based on Coumarin for Cu^{2+}, cysteine, and histidine detections. Applied Spectroscopy. 2019;**73**(7):794-800. DOI: 10.1177/0003702818821329

[60] Wang S, Ding H, Wang Y, Fan C, Liu G, Pu S. A colorimetric and ratiometric fluorescent sensor for sequentially detecting Cu^{2+} and arginine based on a coumarin-rhodamine B derivative and its application for bioimaging. RSC Advances. 2019;**9**(12):6643-6649. DOI: 10.1039/c8ra09943j

[61] Wang Z, Ding X, Huang Y, Yan X, Ding B, Li Q, et al. The development of coumarin Schiff base system applied as highly selective fluorescent/colorimetric probes for Cu^{2+} and tumor biomarker glutathione detection. Dyes and Pigments. 2020;**175**. Article 108156. DOI: 10.1016/j.dyepig.2019.108156

[62] Yin G, Niu T, Gan Y, Yu T, Yin P, Chen H, et al. A multi-signal fluorescent probe with multiple binding sites for simultaneous sensing of Cys, Hcy and GSH. Angewandte Chemie. 2018;**57**(18):4991-4994. DOI: 10.1002/anie.201800485

Chapter 17

Fluorescent Nanomaterials for Cellular Imaging

Kalyan Sundar Ghosh and Anchal Sharma

Abstract

This chapter will provide an overview of different nanomaterials, which are being used for nondestructive imaging of living entities such as cells and tissues. The chapter begins with the basics of fluorescence imaging followed by a discussion on the advantages of fluorescent nanomaterials as compared to commonly used molecular fluorophores and imaging probes. Specific features and applications of nanomaterials frequently exploited in bioimaging are summarized. These include fluorescent silicon-based nanomaterials, hydrogels, polymer dots, magnetic nanoparticles, fluorescent quantum dots, carbon dots and other carbon-based nanomaterials, noble metal nanoparticles, micelles, dendrimers, lipid nanoparticles, and so on. Specific examples on their applications in bioimaging including multimodal imaging and targeted imaging are illustrated.

Keywords: nanomaterials, fluorescence imaging, multimodal imaging

1. Introduction

In today's modern era, nanotechnology has emerged as one of the major discoveries, which become an inexorable part of our daily life. It has a wide range of applications in areas such as nanoelectronics, consumer products, biomaterials, and nanomedicines. This implies its interdisciplinary nature by combining different subject domains such as chemistry, biology, physics, material science, chemical, electrical, and mechanical engineering. Nanomaterials find remarkable biomedical applications due to: (i) their small size; (ii) fascinating chemical, physical, and optical properties; (iii) a highly loadable surface; and (iv) good biocompatibility [1]. On account of the resemblance in size of nanomaterials with biomolecules such as DNA, antibodies, oligonucleotides, glucose, proteins, and virus (a size range of 1–100 nm), various *in vitro* and *in vivo* studies involving nano and biomaterials are being reported [2]. Among different biomedical applications, the potential use of nanomaterials in bioimaging of cells and tissues has grabbed a significant interest of several researchers across the world. This technique has become an essential tool to investigate various biological processes occurring at the cellular and subcellular levels and gaining importance in both fundamental research and clinical diagnostics. Though myriad organic and inorganic imaging probes are already available, applications of nanomaterials as imaging probes are proliferating. The applications of nanomaterials in bioimaging cover from in-depth imaging of tissues to point of care (POC) testing [3]. There are numerous imaging modalities available in the medical field to access images of various cells and tissues. These include radio imaging, magnetic resonance imaging (MRI), CT imaging, electrochemical imaging, and even more futuristic imaging techniques such as laser ablation-inductively coupled plasma-mass spectroscopy (LA-ICP-MS) along with others. Some of the above methods are destructive in nature, require substantial preparation of samples, and possess limited resolution. These factors limit applications of these methods for imaging of cellular and subcellular structures in the body. This has paved the way for the development of imaging probes, which will have high resolution, can provide images in nondestructive way, and offer multimodal imaging to gain in-depth information. With the development of new functional materials, bioimaging using nanomaterials attracts substantial attention in biomedical research due to its versatility, selectivity, sensitivity, rich in contrast, and high-resolution properties obtainable in this technique. This also opens up a window to incorporate conventional imaging techniques with newly developed probes to record cellular events with high contrast and clarity. Earlier, optical imaging could be resoluted only to a limited range (~200 nm). But afterward with the advancement of fluorescence imaging methods such as photo-activated localization microscopy (PALM) and stimulated emission depletion (STED) microscopy, there was a substantial ramping up in the resolution [4, 5]. But still there are sort of limitations and those appear during the studies of animal models, which require high resolution and in-depth imaging. This clampdown can be curbed by using recently developed fluorescent nanomaterials imaging probes. Bioimaging using nanomaterials mostly adopts the following three techniques—(i) fluorescent nanomaterials are injected into the cells and the tissues to make them fluorescent; (ii) targeted bioimaging; and (iii) using nanomaterials as sensors to sense the cellular biochemical species, which are not fluorescent intrinsically. In targeted bioimaging, specific target sites are imaged using fluorescent nanomaterials. This is performed through the fabrication of suitably functionalized nanomaterials. The functional groups in those cases may be oligomers, receptors, ligands, and so on to detect a particular target site.

In this chapter, different types of fluorescent nanomaterials, which are being used as nanoprobes for fluorescence imaging to study cellular events, are discussed.

2. Fluorescence imaging

Cells, tissues, and organelles in the living beings are of such a tiny size that it is quite difficult to get clear and discernible images of them. This facilitates the need in microscopy to have certain kind of contrast in color of the entire cells and tissues and the localized areas of these tiny structures to get more information about the biological processes occurring in them. Fluorescence enables to solve both the issues of contrasting the overall structure and the localized area of that structure. Fluorescence is a type of luminescence that involves the absorption of radiations followed by emission of light of certain wavelength by the matter. The molecules, which have the ability to emit light on absorption, are called fluorophores. Fluorescence imaging is an invasive technique that enables to envision various biological processes occurring in living systems. It covers a wide range of observations, which includes protein expression, gene expression, and interactions occurring in cells and tissues. This technique basically involves the picturization of fluorescent proteins and dyes to understand various molecular mechanisms occurring in living cells, thus serving the purpose of being a potential tool for biochemical applications. In fluorescence microscopy, the samples are labeled with a fluorescent probe, and then, the sample is illuminated with a bright light. The labeled fluorophore absorbs and then emits at a certain wavelength. Therefore, on a black background, an image of high contrast can be obtained. Fluorescence imaging is the most widely used tool to study biological processes. There are two types of fluorescence imaging namely intrinsic and extrinsic fluorescence imaging. The first kind of imaging usually refers to the fluorescence emission by intrinsic fluorophoric species such as tryptophan residues in proteins, NADH in tissues, and so on. The second type of imaging is based on extrinsic fluorescence, which is exhibited by synthetic dyes and probes, nanomaterials, sensors, and so on. They basically help in the detection of species, which are not visible through direct fluorescence imaging. Various fluorescence imaging techniques are presently available to study the structures of cells and tissues in the body. These include fluorescent wide field microscopy, point scanning confocal microscopy, parallel confocal microscopy, two-photon microscopy, light sheet microscopy, total internal reflection fluorescence (TIRF) microscopy, and super resolution microscopy.

2.1 Fluorescence imaging over bright field

A bright field microscopy involves illumination of the sample with transmission of white light, which resulted in an image of high contrast. The image thus obtained is a dark image on white background. But this technique cannot differentiate transparent, translucent, and stained cells as well as the structure of the cells. On the contrary, fluorescence microscopy enables to study a specific area of cellular structure as the fluorophores have the ability to illuminate only the targeted area of activity. As a result, unwanted areas will have no or negligible fluorescence, which will be helpful in getting proper images and information of the targeted area.

2.1.1 Fluorescence microscopy imaging range

The imaging range for fluorescence microscopy is quite higher than the usual range that a human eye is able to see. This extended range can be obtained with the help of charge-coupled device (CCD) cameras, which accumulate light emitted by the sample [6].

2.1.2 Advantages of fluorescence imaging

The advantages of fluorescence imaging are as follows:

i. labeling only the desired area of interest in cellular physiology and investigating the biological processes;

ii. having the high resolving power as compared to conventional microscope, thus known as super resolution microscopy;

iii. enabling the identification of microelements of interest with the help of Raman spectroscopy;

iv. giving an insight into the structural dynamics using FRET, fluorescence anisotropy, and other fluorescence techniques; and

v. producing the high contrast images as compared to conventional microscope.

3. Fluorescent nanomaterials versus molecular fluorophores as imaging probes

Different nanomaterials are developed for imaging of cellular structures because of the advantages such as high contrast, imaging in targeted area, reduced cytotoxicity, and better selectivity [7]. For example, polyethylene glycol (PEG) coating on nanoparticles prevents undesirable interactions with other biomolecules [8]. The optical properties of intrinsically fluorescent nanoparticles are also not altered much on interactions with proteins [7]. In contrast to molecular imaging probes, nanomaterials have high retention time, which enables them to provide images for a prolong period [9]. Photostability of nanomaterials is usually superior to the conventional imaging probes. In fact, nanodiamonds do not show any photobleaching and can resist bleaching up to 100 MW cm^{-2} [10]. Another important advantage is the ease in internalization of nanomaterials into the cells. The internalization depends on the surface charges, and positive charge usually facilitates the internalization of nanomaterials. Thus, it can be easily targeted to the desired area of investigation, which is not feasible with common molecular imaging probes [11]. Also, to label a large number of cells at the same time to study any biological processes, nanomaterials are considered to be the simplest tool in high-throughput screening [12]. Most of the organic imaging fluorophores have size less than 1 nm as compared to the size of nanomaterials (1–100 nm). This provides a highly loadable surface for nanomaterials. Another important feature of nanomaterials is their use as multimodal imaging probes. Attachment of multiple components with nanomaterials enables them for application in more than one imaging techniques [13]. The nanomaterials also have the ability to show finer optical properties on account of surface plasmon resonance (SPR) that arises from collective electronic oscillations after interacting with the light of certain wavelength [14]. Nanoparticles can scatter absorbed light very efficiently and thus can be used as a major fluorescence imaging tool. Due to low quantum yield and low labeling ratio of dyes to targeted molecules in case of organic molecular fluorophores, low detection limit is difficult to be achieved. This limitation can be overcome by using fluorescent nanomaterials [15]. Another advantage of nanomaterials over traditional imaging probes is in theronaustics, which basically involves the combination of both therapeutic drugs and diagnostics [16]. Along with the binding of ligands and fluorophores on the surface of nanoparticles, drugs can also be incorporated in nanomaterials for simultaneous discernment and therapy [17].

3.1 Different types of nanomaterials frequently used in bioimaging and their features

Considering numerous exciting properties, different nanomaterials are developed for fluorescence imaging of cellular and subcellular structures. These include quantum dots, metal nanoparticles, magnetic nanoparticles, silica nanoparticles, nanodiamonds, hydrogels, and so on.

3.1.1 Fluorescent metallic nanoparticles

Use of metallic nanoparticles in molecular imaging in recent years has been keeping pace as it helps in attenuating the photophysical constraints of organic molecular fluorophores. These nanoparticles possess advantageous properties such as better photostability, improvised low detection limits, localized detection, high quantum yield, and clinical applicability [18, 19]. Suitably functionalized silver NPs are considered as efficient nano probe in bioimaging [20]. Metallic nanoparticles can also be used as imaging agents with high X-ray contrast because of their potency to absorb X-ray radiations. As a result, diseased tissue appeared with high contrast than the normal tissue in the image [21, 22]. Gold nanoparticles (AuNPs) have acquired notable attention owing to their low toxicity, good biocompatibility, and high absorption coefficient [23]. Gold nanorods and silver NPs are used in two-photon luminescence imaging of cancer cells [24, 25]. Molecularly targeted gold nanorods were established as a bright contrast agent for two-photon luminescence (TPL) imaging of cancer cells up to at a depth of 75 μm [24]. The TPL intensity of nanorod labeled cells was found to be much higher than the two-photon autofluorescence (TPAF) emission intensity from unlabeled cancer cells under deep imaging (**Figure 1**). The nanoclusters made from silver, gold, and copper show intrinsic fluorescence quite effectively, and their surface can be protected by coating them with alkanethiolate monolayers [26]. If their structures are modified effectively, then they can be used as agents in targeted imaging. Metallic nanoparticles enhance the fluorescence properties of fluorophores in their proximity on account of the interaction between the surface plasmon of the metal nanoparticles and the dipole moment of fluorophores [19]. Silver islands or silver particles deposited on a surface act as a scaffold in bioassays using metal-enhanced fluorescence (MEF) technique. Glass surface coated with silver island film can enhance the signal up to forty fold [27, 28]. Silver island films increase the sensitivity of the fluorescence signals of cell membrane as the fluorescence of the fluorophores enhances due to silver island films [29]. Coating of gold nanoclusters with ruthenium(II) polypyridyl provides the ability for DNA binding and thus leads to cellular uptake and imaging of cellular physiology [30].

3.1.2 Magnetic nanoparticles

Fluorescent magnetic nanoparticles play a major role in the biomedical applications and are widely used in bioimaging. Magnetic nanoparticles show various properties such as biocompatibility, superparamagnetic character, possibility of desired surface functionalization, and so on. They have been functionalized with various enzymes, proteins, and antibodies for improved targeting [31, 32]. Aminopropyl silica coated γ-Fe_2O_3 nanoparticles labeled with fluorescein can be used in both magnetic resonance and fluorescence imaging of cells and tissues [33].

3.1.3 Quantum dots

Quantum dots (QDs) are semiconducting nanocrystals with distinctive optical properties, which are quite unique from large sized particles because of quantum mechanical phenomenon. QDs are currently used extensively in fluorescence microscopy based on their high photostability, brightness, and tunable emission of light as compared to traditional organic and inorganic fluorophores [34–37]. Functionalization of quantum dots with protective ligands can alleviate toxicity and also reduces nonspecific binding with biomolecules. This also increases aqueous solubility and localized

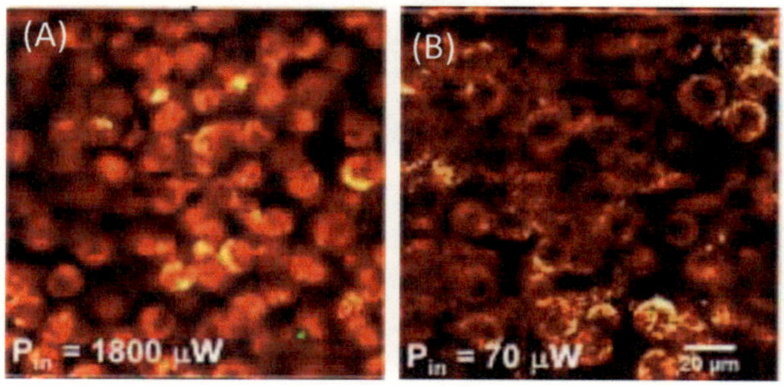

Figure 1.
Two-photon luminescence imaging of cancer cells embedded in a collagen matrix at 75 μm depth. (A) TPAF imaging of unlabeled cells and (B) TPL imaging of nanorods labeled cells. This figure is reprinted (adapted) with permission from Durr et al. [24]. Copyright © 2007, American Chemical Society.

targeting ability [36, 38–40]. On account of their small size, QDs have the tendency to get accumulated in liver or spleen, but this can be avoided by coating them with polyethylene glycol (PEG) [41]. A 3-mercaptopropanoic acid capped CdS QDs were used for the imaging of *Salmonella typhimurium* cells (**Figure 2A**) [42], and bioconjugated CdSe/ZnS QDs were used in multicolor imaging of HeLa cells [44]. ZnO QDs doped with Gd intensify the fluorescence, thus providing the QDs with the ability to act as nanoprobes for the rapid detection of cells [45]. Other QDs coated with polyethylene glycol and conjugated with peptides can target the nucleus and cytosol for imaging purpose [45–48]. Hybrid fluorescent nanocomposites such as silica-coated Fe_3O_4/CdTe quantum dots have been proven a promising agent for immune-labeling and fluorescence imaging of HeLa cells [49]. Bioimaging (**Figure 2B**) was also performed with the help of a TAT peptide conjugated fluorescent, radio-opaque, and paramagnetic CdS:Mn/ZnS QDs [43], and these conjugated QDs had efficiently crossed the blood-brain barrier.

Polymer-coated QDs bound with cancer-specific antibody offer the ability to image tumor targeted areas [50]. Fabrication of CdSe/ZnS QDs encapsulated in phospholipid micelles enables *in vivo* imaging of embryo [51]. The injected QDs had also exhibited greater stability without any toxicity. QDs conjugated with career peptide can be transfected into the cells for bioimaging in living organisms. An insect allatostatin has been identified that transfected into NIH 3T3 and A431 human squamous cells. QDs can be conjugated with allatostatin to make cellular entry easy for the purpose of imaging [52].

3.1.4 Hydrogels or hydrophilic polymers

Hydrogels are three-dimensional (3D) network of hydrophilic polymeric chains, which undergo swelling in water. The polymer chains are hitched in cross-link formation, and the structure contains a lot of water molecules inside [53]. The hydrophilicity is due to the presence of hydrophilic groups on its surface such as –COOH, $–SO_3H$, and $–NH_2$ [54]. Hydrogels are considered to be a potential candidate to act as an imaging agent for bioimaging. On account of their biodegradability, 3D crosslinking network, biocompatibility, sensitivity, nontoxicity, and quick gel to sol transformation in response to biological stimuli, they are being considered as promising fluorescent bioimaging probes [55, 56]. Hydrogels are fabricated using different procedures that involve metal-ligand complex formation, hydrogen bonding, host-guest complexation, and so on [57]. Fluorescent supramolecular metallohydrogel based on ruthenium(II) tris(bipyridine) complex had been used for cell imaging (**Figure 3**) [58].

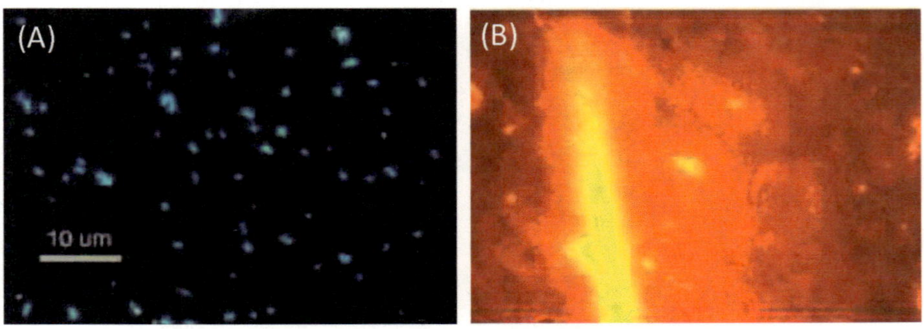

Figure 2.
(A) Fluorescence imaging of Salmonella typhimurium *cells by the aqueous CdS QDs. This figure is reprinted (adapted) with permission from Li et al. [42]. Copyright © 2007, American Chemical Society. (B) Fluorescence image of the branches of right middle cerebral artery of a cross section of rat brain after intra-arterial administration of TAT-conjugated QDs. This figure is reprinted (adapted) with permission from Santra et al. [43]. Copyright © 2005, American Chemical Society.*

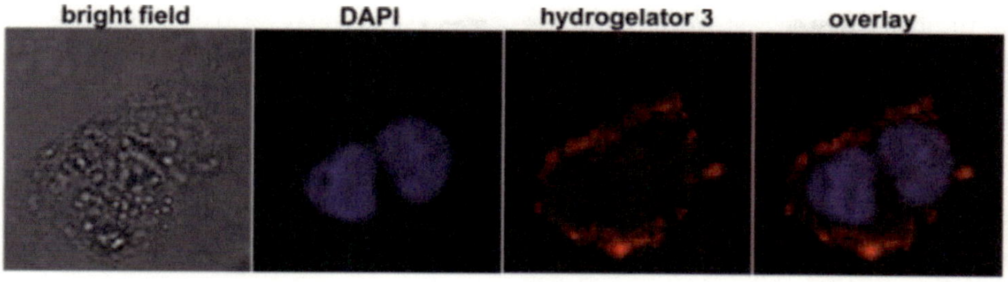

Figure 3.
Fluorescent image of HeLa cells incubated with metallohydrogel. From left to right: phase-contrast image, live cell stain DAPI (blue), luminescence emission of metallohydrogel (red), and the overlay image. This figure is reprinted (adapted) with permission from Zhang et al. [58]. Copyright © 2013, American Chemical Society.

Interactions between the hydrogels (based on gelatin cross linked with lysine diisocyanate ester) and the tissue were studied using multimodal imaging techniques [59]. Poly(ethylene glycol)-co-poly(ethylene oxide), which is also called pluronic hydrogel, was synthesized and proved to be a promising fluorescent nanoprobe for bioimaging [60].

3.1.5 Polymer dots

Polymer dot nanoparticles are fluorescent π-conjugated polymers having size of only few nanometers. They can be functionalized and encapsulated easily. They are highly fluorescent in both single-photon scanning microscopy and two-photon scanning microscopy [61, 62]. Conjugated polymers (CPs) acquire the maximum fluorescence brightness as compared to any other nanoparticles till date on account of high fluorescent decay rates, high quantum yield, high cross-sectional area (~10^{-14} cm^2) for absorption, and immoderate chromophore density [63]. Owing to small particle size, high cellular uptake through endocytosis, nontoxicity, and photostability, extraordinary emission rates endow the particle to act as efficient imaging probes. They have excellent fluorescence brightness that often widens to NIR region [64]. These semiconducting CPs have the tendency to get attached with peptides like chlorotoxin to image brain tumors [65]. The pH value inside the HeLa cells was demonstrated by imaging these cells using CPs with fluorescein, which is a pH indicator [66]. The fluorescence of anionic CPs is quenched by ions such as Cu(II) [67]. NIR-emissive CPs poly[2-methoxy-5-(2-ethylhexyloxy)-1,4-phenylenevinylene] doped with silicon 2,3-naphthalocyanine bis(trihexylsilyloxide) offers efficient FRET. A combination of bioluminescence resonance energy transfer (BRET) and FRET techniques was adopted with these CPs instead of their optical excitation, which does not provide enough in-depth information for imaging of cancer and lymph nodes [68]. In this case, the use of probes with longer life times helped in the separation of fluorescence signal from the background fluorescence and improved the signal-to-noise ratio. Considering this advantage, CPs were fabricated by encapsulating Ln complexes into the host polymer. Polyvinylcarbazole (PVK) (donor) was used as semiconducting host polymer incorporated with Eu complex (acceptor). These PVK/Eu complex CPs were found to show a luminescence lifetime of 509 μs. Cells labeled with these synthesized polymer dots were imaged and that showed very less background fluorescence as compared to the unlabeled ones [69]. When it comes to label receptors on the cells, monovalent or single chain CPs are considered to be an excellent labeling agent, for instance, alkyne terminated linear poly(p-phenylenevinylene) [70]. Furthermore, CPs have been establishing themselves as an efficient platform for the development of probes for multimodal imaging. Various polymer dots have been fabricated to combine two imaging techniques for molecular imaging. Conjugated polymers modified with BSA and functionalized with Au clusters act as a promising imaging probe that integrates two imaging techniques, that is, photoacoustic imaging and fluorescence imaging [71]. After injecting this fabricated polymer dots into an animal model, signals for both the imaging techniques were obtained. Thus, polymer dots incorporated with characteristics as per clinical point of view seem to be a promising nanoprobe agent as compared to the conventional imaging probes.

3.1.6 Silicon-based nanomaterials

Silicon nanoparticles (SiNPs) are being used in the manufacture of finest sensors and nanoprobes for various bioimaging applications. These nanoparticles are highly porous with high surface area to volume ratio and possess unique chemical, biodegradable, electrical, and optical properties [72, 73]. The product of biodegradation of silicon nanomaterials, for instance, orthosilicic acid is congenial with tissues in the body. Besides, silicon is present in human body as a trace element [74]. Owing to good biocompatibility, these nanoparticles are not cytotoxic, which are prerequisite for a good bioimaging agent. Fluorescent SiNPs are considered to be a consummate nanoprobe agent over traditional imaging probes due to high photostability, brightness, biocompatibility, and biodegradability [75]. The major limitation is the poor dispersibility of SiNPs in aqueous media, which constrained their use as a biological probe [76]. By taking into account the above limitation, luminescent SiNPs with hydrophilic ligands were synthesized that enhance the aqueous dispersibility and hence their utility in molecular imaging. Modified SiNPs were fabricated by encapsulating SiNPs in phospholipid micelle and used for *in vivo* imaging [77, 78]. Various other modified SiNPs were also synthesized with hydrophilic surface and were conjugated with proteins to use them for immunofluorescent cell imaging [79]. But due to large hydrodynamic radius (>50 nm), these modified SiNPs become less useful for *in vivo* imaging. Considering this, silicon nanowires (SiNWs) incorporated with glutaric acid were prepared with a size of <5 nm. They have excellent photostability and good aqueous dispersibility also [80, 81]. SiNPs capped with allylamine were also fabricated and used for labeling the HeLa cells. The image (**Figure 4**) showed brilliant bright SiNPs in the cytoplasm [82].

Commonly, two-step process is used to synthesize functional SiNPs. In the first step, SiNPs are synthesized, and then their surfaces are modified. To achieve this in one step, bottom-up approach was also used. In this method, hydrophilic molecules

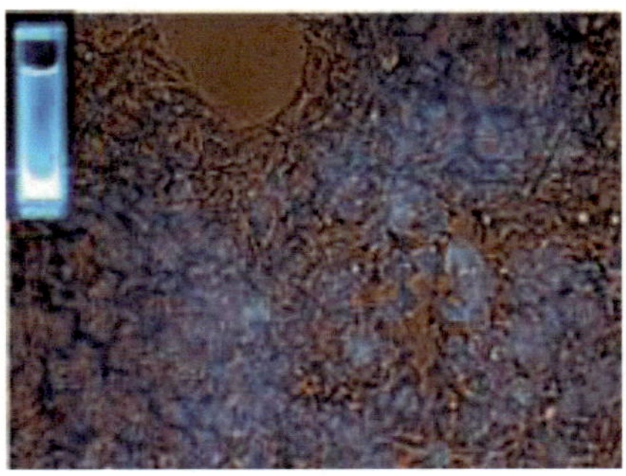

Figure 4.
Cell imaging using silicon quantum dots capped with allylamine. This figure is reprinted (adapted) with permission from Shiohara et al. [82]. Copyright © 2010, American Chemical Society.

containing silicon (e.g., 3-aminopropyl trimethoxysilane) were used as a source of silicon. The fluorescent SiNPs fabricated in this way were found to possess excellent aqueous dispersibility and high photostability over common organic fluorophores as well as II-VI QDs [83]. SiNWs coated with AuNPs are multicolored and are highly fluorescent nanomolecular beacons. They are highly sensitive for DNA detection and have the ability to detect up to ~50 pM [84]. Silica-based nanoparticles with fluorescence and magnetic properties can act as a nanoimage probe agent. Silica-coated iron oxide nanoparticles attached to the fluorescein dye have the ability to act as a multimodal probe agent to label bone marrow cells [85]. Organic fluorophores functionalized silica having gadolinium oxide (Gd_2O_3) NPs embedded in its shell were fabricated and then administered into mice and investigated further for its ability to act as a multimodal probe agent for both fluorescence microscopy and MRI (**Figure 5**). Gd_2O_3 core enhances the contrast of MRI, and the shell imparted excellent fluorescence thus suitable for multimodal imaging [86].

Another example for multimodal imaging is the nanoparticles having silica doped with dye-encapsulated in gadolinium silylated shell, which were used to label monocyte cells *in vitro* responsible for immune response in the body [87].

3.1.7 Carbon dots

Carbon dots or carbon-based quantum dots are small-sized semiconductor quasispherical luminescent nanoparticles with size below 10 nm and can also be as less than as 1 nm [88]. They were initially discovered from carbon nanotubes whose surface modification had resulted highly fluorescent emissions at different excitation wavelengths ranging between UV and near-IR [89]. C-dots are attributed with various characteristics that have grabbed the attention of several research groups.

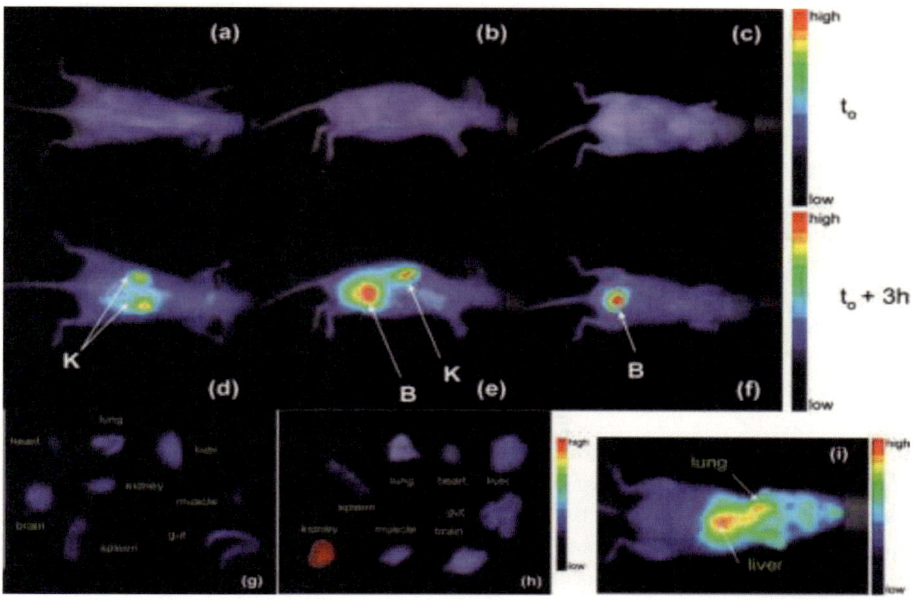

Figure 5.
Fluorescence reflectance imaging of a nude mouse (a–c) before and (d–f) 3 hours after the injection of Gd_2O_3-embedded silica NPs (K, kidneys; B, bladder). Fluorescence reflectance imaging of some organs after dissection (g) of a control mouse (no particles injected) and (h) of the nude mouse visualized on pictures (a–f). (i) Fluorescence reflectance imaging of a nude mouse after the injection of GadoSi$_2$C (particles without PEG). This figure is reprinted (adapted) with permission from Bridot et al. [86]. Copyright © 2007, American Chemical Society.

They are noncytotoxic and have sufficient photostability and chemical inertness with ease in surface modification [90, 91]. Their surface is full of defects, and after suitable modifications, the surface can be made highly fluorescent [92]. C-dots can be easily taken up by cells through a process called endocytosis and can image cells using both single and two-photon excitations [93]. The intracellular uptake can be enhanced by coupling them with peptides, which can translocate the active membrane. C-dots also undergo rapid renal excretion [80]. Aqueous dispersible C-dots processed from nanodiamonds using one-pot hydrothermal treatment were administered to NIH-3T3 cells, which exhibited different colored fluorescence (yellow and green) at different excitation wavelengths [94]. Highly luminescent C-dots fabricated from citric acid were used to image MC3T3 cells [95]. C-dots were also prepared through green synthesis from natural products, proteins, carbohydrates, biowastes, and food products [91]. The C-dots obtained from barbeque meat and surface modified with PEG resulted in high quantum yield and excellent fluorescence emission [96]. C-dot surface passivated with polyethyleneimine was used for labeling the HeLa cells, which were further viewed with green fluorescence. These quantum dots have also been proved to show astounding applications in theronaustics to attain both drug delivery and fluorescence imaging [97]. C-dot SiO_2 nanoparticles further functionalized with PEG molecules were fabricated that resulted in the augmentation of the brightness, biocompatibility, and stability. These synthesized nanoparticles were able to deliver anticancer drug doxorubicin inside the HeLa cells efficiently [98]. Doxorubicin was also loaded on C-dots synthesized from BSA and then administered into A549 cells. C-dots and the drug both were found localized in the cells and thus served the purpose for simultaneous bioimaging and drug delivery [99]. Carbogenic nanodots doped with iron oxide were fabricated for multimodal imaging (MRI and fluorescence microscopy). These nanoparticles were injected into rats and were uptaken by RAW264.7 cells and visualized in the cytoplasm. Fluorescence and MRI enhanced signals were observed because of T1 and T2 relaxation [100].

3.1.8 Various other carbon-based nanomaterials

Apart from C-dots, various other carbon containing nanomaterials such as graphene oxide, graphene, nanodiamonds, graphite oxide, and carbon nanotubes are available. The graphite oxide is sensitive to pH and is photostable like C-dots [101]. They can be synthesized in one-pot hydrothermal process. They give different colored fluorescence emission at different excitation wavelengths, and the photoluminescence is size-dependent. Nanographite oxide was taken up by A549 and was imaged in cell cytoplasm [102]. Graphene quantum dots (GQDs) are fabricated using hydrothermal treatment of graphene oxide. Both bottom-up and top-down methods were used for the synthesis of GQDs [103]. To enhance the fluorescence of GQDs, they are either reduced or surface passivated. They were further employed for the bioimaging of cell physiology, for instance, GQDs labeled MG-63 cells exhibited fluorescence at different excitation wavelengths (405 and 488 nm) [104]. Graphene is not often used for the purpose of imaging owing to its negligible fluorescence. Similarly, fullerenes too do not show any fluorescence and cannot be used for fluorescence imaging except fullerene C_{70}, which exhibits fluorescence and thus can be used as an efficient imaging probe [105]. Carbon nanotubes (CNTs) are known to be fluorescent in the near infrared region (NIR) with low quantum yield [106]. CNTs embedded in single-stranded DNA remain in active form up to 3 months [107]. On account of its fluorescence emission in NIR range, CNTs can be used for NIR fluorescence imaging of tumors in a targeted way [108]. Nanodiamonds contain nitrogen vacant spots and are being utilized extensively for bioimaging purposes based on their photoluminescence properties, photostability, and biocompatibility. They show red and green fluorescence at different excitation wavelengths [109].

3.1.9 Micelles, dendrimers, and lipid nanoparticles

NIR region is being extensively used for the imaging of cells and tissues due to its tendency to penetrate deeply and has an access to in-depth information. NIR dyes like indocyanine green (ICG) are being used for these days as image contrast agents. To enhance the specific targeting of ICG, it has been incorporated in several nanoparticles, but due to certain limitations, applications become restricted. Recently, ICG was formulated in micelles that include ICG and polycaprolactone (PCL). The ICG-PCL micelle has highly loadable surface and uniform size. They were found to have better fluorescence brightness, retention time, biocompatibility, no toxicity, and enhanced accumulation of ICG when compared with only ICG fluorescence imaging [110]. These micelles have also been proved to be useful for the surgery of tumors using intraoperative images. Another example is the fabrication of supramolecular micelles that involve perylenediimide and poly (D, L-lactide)-*b*-poly(ethyl ethylene phosphate), which is a block copolymer as precursors. Camptothecin drug was then loaded on the surface of synthesized supramolecular micelle and injected into mice having tumor cells. The fluorescence images suggest alleviation in the growth of tumor as compared to only drug [111]. Dendrimers are branched molecules having three-dimensional network and spherical morphology. Owing to their small size than normal nanoparticles, they can easily undergo endocytosis. They are highly fluorescent with very high molar absorbance. Thus, they can be considered as alternative of quantum dots for imaging [112]. Dendritic nanoprobes conjugated with cyanine dyes resulted in higher photostability, brightness, localization, and resolution for bioimaging [113]. Polyamide amine (PAMAM) dendrimer has been used widely for the fluorescence imaging of cancer cells. The receptor (sialoglycoprotein) present on cancer cells helped in the identification of this dendrimer by exhibiting blue emission. Dendrimers embedded with pH and NIR sensitive probes can be used to localize and target cancer cells [114]. Lipid NPs are being extensively used as drug delivery system and in bioimaging. They have been identified as excellent nanocarriers for the treatment of cancer. They can overcome various barriers in the physiological environment, thus enhancing the delivery of drug [115]. On account of their bioavailability, low-toxicity, and production in large scale, they are being widely used in bioimaging [116, 117]. The nonpolar lipid NPs are labeled with various probes and dyes (emitting in NIR region) for the purpose of bioimaging [118].

4. Conclusion and future directions

Advantages of fluorescence imaging of cellular events in the body have captured the attention of several research groups. This is on account of the advancement made in the bioimaging techniques. Various nanoparticles have been developed as remarkable imaging probes compared to the conventional imaging fluorophores, and many others are in the process. Cell is a biological entity through which the entire biological mechanism of an organism can be observed. Thus, such an image probe, which has the ability to go deep into the cells and helps in extracting the information out of it, was the necessity. Thus, various nanomaterials were identified that have been discussed in this chapter to be superior probes than the traditionally used fluorophores. Nanoparticles are often preferred to the conventional imaging probes for the intracellular biological activities owing to their ability to have an access to in-depth information. These nanoprobes have also been proved finer than other fluorophores owing to their high contrast, which can help to get brighter image. It makes easier to discern information with these highly fluorescent materials. Their high photostability aids in visualization of the image for a longer period of time. Better localizing ability of these materials assists to target only the affected cells and tissues and reduces nonspecific binding with other biomolecules. In addition to that, in-depth penetration and better endocytosis help in the easy uptake of these probes by the cells. Good biocompatibility and highly loadable surface add beneficial features in these probes. As a result, they can be used for both drug delivery and diagnosis by imaging. All these features make the nanomaterials more efficient than any other probes in the field of bioimaging. Still there are some limitations associated with them like toxicity that is a major issue that need to be addressed before its use in future. Its ability to combine various imaging modalities to act as a multimodal imaging probe is a remarkable feat that can bring revolution in cellular imaging.

Abbreviations

C-dots	carbon dots
CNTs	carbon nanotubes
CPs	conjugated polymers
GQDs	grapheme quantum dots
ICG	indocyanine green
NPs	nanoparticles
QDs	quantum dots
SiNPs	silicon nanoparticles
SiNWs	silicon nanowires

Author details

Kalyan Sundar Ghosh* and Anchal Sharma
Department of Chemistry, National Institute of Technology Hamirpur, Himachal Pradesh, India

*Address all correspondence to: kalyan@nith.ac.in

References

[1] Wilczewska AZ, Niemirowicz K, Markiewicz KH, Car H. Nanoparticles as drug delivery systems. Pharmacological Reports. 2012;**64**:1020-1037

[2] Narayan R, editor. Nanobiomaterials: Nanostructured Materials for Biomedical Applications. Amsterdam, Netherlands: Elsevier; 2017

[3] Syedmoradi L, Daneshpour M, Alvandipour M, Gomez FA, Hajghaseem H, Omidfar K. Point of care testing: The impact of nanotechnology. Biosensors & Bioelectronics. 2017;**87**:373-387

[4] Betzig E, Patterson GH, Sougrat R, Lindwasser OW, Olenych S, Bonifacino JS, et al. Imaging intracellular fluorescent proteins at nanometer resolution. Science. 2006;**313**:1642-1645

[5] Willig KI, Rizzoli SO, Westphal V, Jahn R, Hell SW. STED microscopy reveals that synaptotagmin remains clustered after synaptic vesicle exocytosis. Nature. 2006;**440**:935-939

[6] Salmon WC, Waters JC. CCD cameras for fluorescence imaging of living cells. In: Goldman et al., editors. Live Cell Imaging. 2nd ed. Cold Spring Harbor, NY, USA: CSHL Press; 2010

[7] Wolfbeis OS. An overview of nanoparticles commonly used in fluorescent bioimaging. Chemical Society Reviews. 2015;**44**:4743

[8] Suka S, Xua Q, Kima N, Hanesa J, Ensigna LM. PEGylation as a strategy for improving nanoparticle-based drug and gene delivery. Advanced Drug Delivery Reviews. 2016;**99**:28-51

[9] Sun X, Rossin R, Turner JL, Becker ML, Joralemon MJ, Welch MJ, et al. An assessment of the effects of shell crosslinked nanoparticle size, core composition, and surface PEGylation on *in vivo* biodistribution. Biomacromolecules. 2005;**6**:2541-2554

[10] Camejo SA, Adam MP, Besbes M, Hugonin JP, Jacques V, Greffet JJ, et al. Stimulated emission depletion microscopy resolves individual nitrogen vacancy centers in diamond nanocrystals. ACS Nano. 2013;**7**:10912-10919

[11] Behzadia S, Serpooshanb V, Taoa W, Hamalyc MA, Alkawareekd MY, Dreadene EC, et al. Cellular uptake of nanoparticles: Journey inside the cell. Chemical Society Reviews. 2017;**46**:4218-4244

[12] Nasir I, Fatih W, Svensson A, Radu D, Linse S, Cabaleiro Lago C, et al. High throughput screening method to explore protein interactions with nanoparticles. PLoS One. 2015;**10**:e0136687

[13] Liu Z, Kiessling F, Gatjens J. Advanced nanomaterials in multimodal imaging: Design, functionalization, and biomedical applications. Journal of Nanomaterials. 2010;**894303**

[14] Canovi M, Lucchetti J, Stravalaci M, Re F, Moscatelli D, Bigini P, et al. Applications of surface plasmon resonance (SPR) for the characterization of nanoparticles developed for biomedical purposes. Sensors. 2012;**12**:16421-16432

[15] Zhong W. Nanomaterials in fluorescence-based biosensing. Analytical and Bioanalytical Chemistry. 2009;**394**:47-59

[16] Chen F, Ehlerding EB, Cai W. Theranostic nanoparticles. Journal of Nuclear Medicine. 2014;**55**:1919-1922

[17] Madamsetty VS, Mukherjee A, Mukherjee S. Recent trends of the bio-inspired nanoparticles in cancer theranostics. Frontiers in Pharmacology. 2019;**10**:1264

[18] Zhang J, Fu Y, Zhao RY, Lakowicz JR. Metal nanoparticle fluorophore: A powerful fluorescence probe in single cell imaging. Proceedings of SPIE. 2010;**7571**

[19] Aslan K, Gryczynski I, Malicka J, Matveeva E, Lakowicz JR, Geddes CD. Metal-enhanced fluorescence: An emerging tool in biotechnology. Current Opinion in Biotechnology. 2005;**16**:55-62

[20] Pala R, Zeng Y, Pattnaik S, Busi S, Alomari N, Nauli SM, et al. Functionalized silver nanoparticles for sensing, molecular imaging and therapeutic applications. Current Nanomedicine. 2018;**8**:234-250

[21] Yu SB, Watson AD. Metal-based X-ray contrast media. Chemical Reviews. 1999;**99**:2353-2377

[22] Hainfeld JF, Slatkin DN, Focella TM, Smilowitz HM. Gold nanoparticles: A new X-ray contrast agent. The British Journal of Radiology. 2006;**79**:248-253

[23] Khan MS, Vishakante GD, Siddaramaiah H. Gold nanoparticles: A paradigm shift in biomedical applications. Advances in Colloid and Interface Science. 2013;**199-200**:44-58

[24] Durr NJ, Larson T, Smith DK, Korgel BA, Sokolov K, Yakar AB. Two-photon luminescence imaging of cancer cells using molecularly targeted gold nanorods. Nano Letters. 2007;**7**:941-945

[25] Jiang J, Gu H, Shao H, Devlin E, Papaefthymiou GC, Ying JY. Bifunctional Fe_3O_4–Ag heterodimer nanoparticles for two-photon fluorescence imaging and magnetic manipulation. Advanced Materials. 2008;**20**:4403-4407

[26] Luab Y, Chen W. Sub-nanometre sized metal clusters: From synthetic challenges to the unique property discoveries. Chemical Society Reviews. 2012;**41**:3594-3623

[27] Matveeva EG, Gryczynski Z, Lakowicz JR. Myoglobin immunoassay based on metal particle-enhanced fluorescence. Journal of Immunological Methods. 2005;**302**:26-35

[28] Matveeva E, Gryczynski Z, Malicka J, Gryczynski I, Lakowicz JR. Metal-enhanced fluorescence immunoassays using total internal reflection and silver island-coated surfaces. Analytical Biochemistry. 2004;**334**:303-311

[29] Zhang J, Fu Y, Liang D, Zhao RY, Lakowicz JR. Enhanced fluorescence images for labeled cells on silver island films. Langmuir. 2008;**24**:12452-12457

[30] Elmes RBP, Orange KN, Cloonan SM, Williams DC, Gunnlaugsson T. Luminescent ruthenium(II) polypyridyl functionalized gold nanoparticles; their DNA binding abilities and application as cellular imaging agents. Journal of the American Chemical Society. 2011;**133**:15862-15865

[31] Peng XH, Qian XM, Mao H, Wang AY, Chen ZG, Nie S, et al. Targeted magnetic iron oxide nanoparticles for tumor imaging and therapy. International Journal of Nanomedicine. 2008;**3**:311-321

[32] Hwu YSL, Lin YS, Josephrajan T, Hsu MH, Cheng FY, Yeh CH, et al. Targeted paclitaxel by conjugation to iron oxide and gold nanoparticles. Journal of the American Chemical Society. 2009;**131**:66-68

[33] García RS, Shelley Stafford ID, Gun'ko YK. Recent progress in synthesis and functionalization of multimodal fluorescent-magnetic nanoparticles for biological applications. Applied Sciences. 2018;**8**:172

[34] Alivisatos P. The use of nanocrystals in biological detection. Nature Biotechnology. 2004;**22**:47-52

[35] Gao XH, Yang LL, Petros JA, Marshall FF, Simons JW, Nie SM. In vivo molecular and cellular imaging with quantum dots. Current Opinion in Biotechnology. 2005;**16**:63-72

[36] Michalet X, Pinaud FF, Bentolila LA, Tsay JM, Doose S, Li JJ, et al. Quantum dots for live cells, in vivo imaging, and diagnostics. Science. 2005;**307**:538-544

[37] Smith AM, Duan HW, Mohs AM, Nie SM. Bioconjugated quantum dots for in vivo molecular and cellular imaging. Advanced Drug Delivery Reviews. 2008;**60**:1226-1240

[38] Mulder WJM, Koole R, Brandwijk RJ, Storm G, Chin PTK, Strijkers GJ, et al. Quantum dots with a paramagnetic coating as a bimodal molecular imaging probe. Nano Letters. 2006;**6**:1-6

[39] Mulder WJM, Strijkers GJ, Tilborg GAFV, Cormode DP, Fayad ZA, Nicolay K, et al. Nanoparticulate assemblies of amphiphiles and diagnostically active materials for multimodality imaging. Accounts of Chemical Research. 2009;**42**:904-914

[40] Pellegrino T, Manna L, Kudera S, Liedl T, Koktysh D, Rogach AL, et al. Hydrophobic nanocrystals coated with an amphiphilic polymer shell: A general route to water soluble nanocrystals. Nano Letters. 2004;**4**:703-707

[41] Gref R, Minamitake Y, Peracchia MT, Trubetskoy V, Torchilin V, Langer R. Biodegradable long-circulating polymeric nanospheres. Science. 1994;**263**:1600-1603

[42] Li H, Shih WY, Shih WH. Synthesis and characterization of aqueous carboxyl-capped CdS quantum dots for bioapplications. Industrial and Engineering Chemistry. 2007;**46**:2013-2019

[43] Santra S, Yang H, Holloway PH, Stanley JT, Mericle RA. Synthesis of water-dispersible fluorescent, radio-opaque, and paramagnetic CdS:Mn/ZnS quantum dots: A multifunctional probe for bioimaging. Journal of the American Chemical Society. 2005;**127**:1656-1657

[44] Jaiswal JK, Mattoussi H, Mauro JM, Simon SM. Long-term multiple color imaging of live cells using quantum dot bioconjugates. Nature Biotechnology. 2002;**21**:47-51

[45] Chen FQ, Gerion D. Fluorescent CdSe/ZnS nanocrystal-peptide conjugates for long-term, nontoxic imaging and nuclear targeting in living cells. Nano Letters. 2004;**4**:1827-1832

[46] Liu A, Peng S, Soo JC, Kuang M, Chen P, Duan H. Quantum dots with phenylboronic acid tags for specific labeling of sialic acids on living cells. Analytical Chemistry. 2011;**83**:1124-1130

[47] Yildiz I, McCaughan B, Cruickshank SF, Callan JF, Raymo FM. Biocompatible CdSe-ZnS Core-shell quantum dots coated with hydrophilic polythiols. Langmuir. 2009;**25**:7090-7096

[48] Ruan G, Agrawal A, Marcus AI, Nie SM. Imaging and tracking of Tat peptide-conjugated quantum dots in living cells: New insights into nanoparticle uptake, intracellular transport, and vesicle shedding. Journal of the American Chemical Society. 2007;**129**:14759-14766

[49] Sun P, Zhang HY, Liu C, Fang J, Wang M, Chen J, et al. Preparation and characterization of Fe_3O_4/CdTe magnetic/fluorescent nanocomposites and their applications in immuno-labeling and fluorescent imaging of cancer cells. Langmuir. 2010;**26**:1278-1284

[50] Papagiannaros A, Upponi J, Hartner W, Mongayt D, Levchenko T,

Torchilin V. Quantum dot loaded immunomicelles for tumor imaging. Journal of Medical Imaging. 2010;**10**:22

[51] Dubertret B, Skourides P, Norris DJ, Noireaux V, Brivanlou AH, Libchaber A. In vivo imaging of quantum dots encapsulated in phospholipid micelles. Science. 2002;**298**:1759-1762

[52] Walther C, Meyer K, Rennert R, Neundorf I. Quantum dot-carrier peptide conjugates suitable for imaging and delivery applications. Bioconjugate Chemistry. 2008;**19**:2346-2356

[53] Wichterle O, Lím D. Hydrophilic gels for biological use. Nature. 1960;**185**:117-118

[54] Hoffman AS. Hydrogels for biomedical applications. Advanced Drug Delivery Reviews. 2012;**64**:18-23

[55] Ahmed EM. Hydrogel: Preparation, characterization, and applications - A review. Journal of Advanced Research. 2015;**6**:105-121

[56] Sontyana AG, Mathew AP, Cho KH, Uthaman S, Park IK. Biopolymeric in situ hydrogels for tissue engineerin and bioimaging applications. Tissue Engineering and Regenerative Medicine. 2018;**15**:575-590

[57] Lim JYC, Lin Q, Xue K, Loh XJ. Recent advances in supramolecular hydrogels for biomedical applications. Materials Today. 2019;**3**:100021

[58] Zhang Y, Zhang B, Kuang Y, Gao Y, Shi J, Zhang XX, et al. A redox responsive, fluorescent supramolecular metallohydrogel consists of nanofibers with single-molecule width. Journal of the American Chemical Society. 2013;**135**:5008-5011

[59] Tondera C, Hauser S, Genge AK, Jung F, Neffe AT, Lendlein A, et al. Gelatin-based hydrogel degradation and tissue interaction in vivo: Insights from multimodal preclinical imaging in immunocompetent nude mice. Theranostics. 2016;**6**:2114-2128

[60] Chen M, Yin M. Design and development of fluorescent nanostructures for bioimaging. Progress in Polymer Science. 2014;**39**:365-395

[61] Wu C, Szymanski C, McNeill J. Preparation and encapsulation of highly fluorescent conjugated polymer nanoparticles. Langmuir. 2006;**22**:2956-2960

[62] Wu C, Szymanski C, Cain Z, McNeill J. Conjugated polymer dots for multiphoton fluorescence imaging. Journal of the American Chemical Society. 2007;**129**:12904-12905

[63] Pei QB, Yang Y. Efficient photoluminescence and electroluminescence from a soluble polyfluorene. Journal of the American Chemical Society. 1996;**118**:7416-7417

[64] Tang R, Feng X. Highly luminescent conjugated polymer nanoparticles for imaging and therapy. Canadian Chemical Transactions. 2013;**1**:78-84

[65] Wu C, Hansen SJ, Hou Q, Yu J, Zeigler M, Jin Y, et al. Design of highly emissive polymer dot bioconjugates for in vivo tumor targeting. Angewandte Chemie, International Edition. 2011;**50**:3430-3434

[66] Chan YH, Wu C, Ye F, Jin Y, Smith PB, Chiu DT. Development of ultrabright semiconducting polymer dots for ratiometric pH sensing.Analytical Chemistry. 2011;**83**:1448-1455

[67] Sun L, Hao D, Shen W, Qian Z, Zhu C. Highly sensitive fluorescent sensor for copper(II) based on amplified fluorescence quenching of a water-soluble NIR emitting conjugated polymer. Microchimica Acta. 2012;**177**:357-364

[68] Xiong L, Shuhendler AJ, Rao J. Self-luminescing BRET-FRET near-infrared dots for in vivo lymph-node mapping and tumour imaging. Nature Communications. 2012;**3**:1193

[69] Sun W, Yu J, Deng R, Rong Y, Fujimoto B, Wu C, et al. Semiconducting polymer dots doped with europium complexes showing ultranarrow emission and long luminescence lifetime for time-gated cellular imaging. Angewandte Chemie, International Edition. 2013;**52**:11294-11297

[70] Ye F, Wu C, Sun W, Yu J, Zhang X, Rong Y, et al. Semiconducting polymer dots with monofunctional groups. Chemical Communications. 2014;**50**:5604-5607

[71] Gao DY, Zhang PF, Liu YB, Sheng ZH, Chen HJ, Yuan Z. Protein-modified conjugated polymer nanoparticles with strong near-infrared absorption: A novel nanoplatform to design multifunctional nanoprobes for dual-modal photoacoustic and fluorescence imaging. Nanoscale. 2018;**10**:19742-19748

[72] Wang Y, Wang T, Da P, Xu M, Wu H, Zheng G. Silicon nanowires for biosensing, energy storage, and conversion. Advanced Materials. 2013;**25**:5177-5195

[73] Nishimura H, Ritchie K, Kasai RS, Goto M, Morone N, Sugimura H, et al. Biocompatible fluorescent silicon nanocrystals for single-molecule tracking and fluorescence imaging. The Journal of Cell Biology. 2013;**202**:967-983

[74] Park JH, Gu L, VonMaltzahn G, Ruoslahti E, Bhatia SN, Sailor MJ. Biodegradable luminescent porous silicon nanoparticles for in vivo applications. Nature Materials. 2009;**8**:331-336

[75] He Y, Fan CH, Lee ST. Silicon nanostructures for bioapplications. Nano Today. 2010;**5**:282-295

[76] Peng F, Su Y, Zhong Y, Fan C, Lee ST, He Y. Silicon nanomaterials platform for bioimaging, biosensing, and cancer therapy. Accounts of Chemical Research. 2014;**47**:612-623

[77] Erogbogbo F, Yong KT, Roy G, Xu X, Prasad PN, Swihart MT. Biocompatible luminescent silicon quantum dots for imaging of cancer cells. ACS Nano. 2008;**2**:873-878

[78] Erogbogbo F, Yong KT, Roy I, Hu R, Law WC, Zhao WW, et al. In vivo targeted cancer imaging, sentinel lymph node mapping and multi-channel imaging with biocompatible silicon nanocrystals. ACS Nano. 2011;**5**:413-423

[79] He Y, Su YY, Yang XB, Kang ZH, Xu TT, Zhang RQ, et al. Photo and pH stable, highly-luminescent silicon nanospheres and their bioconjugates for immunofluorescent cell imaging. Journal of the American Chemical Society. 2009;**131**:4434-4438

[80] Choi HS, Liu WH, Misra P, Tanaka E, Zimmer JP, Ipe BI, et al. Renal clearance of quantum dots. Nature Biotechnology. 2007;**25**:1165-1170

[81] He Y, Zhong YL, Peng F, Wei XP, Su YY, Lu YM, et al. One-pot microwave synthesis of water-dispersible, ultraphoto-and pH-stable, and highly fluorescent silicon quantum dots. Journal of the American Chemical Society. 2011;**133**:14192-14195

[82] Shiohara A, Hanada S, Prabakar S, Fujioka K, Lim TH, Yamamoto K, et al. Chemical reactions on surface molecules attached to silicon quantum dots. Journal of the American Chemical Society. 2010;**132**:248-253

[83] Zhong YL, Peng F, Bao F, Wang SY, Ji XY, Yang L, et al. Large-scale aqueous

synthesis of fluorescent and biocompatible silicon nanoparticles and their use as highly photostable biological probes. Journal of the American Chemical Society. 2013;**135**:8350-8356

[84] Su S, Wei XP, Zhong YL, Guo YY, Su YY, Huang Q, et al. Silicon nanowire-based molecular beacons for high-sensitivity and sequence-specific DNA multiplexed analysis. ACS Nano. 2012;**6**:2582-2590

[85] Lu C, Hung Y, Hsiao J, Yao M, Chung T, Lin Y, et al. Bifunctional magnetic silica nanoparticles for highly efficient human stem cell labeling. Nano Letters. 2007;**7**:149-154

[86] Bridot JL, Faure AC, Laurent S, Rivière C, Billotey C, Hiba B, et al. Hybrid gadolinium oxide nanoparticles: Multimodal contrast agents for in vivo imaging. Journal of the American Chemical Society. 2007;**129**:5076-5084

[87] Rieter WJ, Kim JS, Taylor KML, An H, Lin W, Tarrant T, et al. Hybrid silica nanoparticles for multimodal imaging. Angewandte Chemie, International Edition. 2007;**46**: 3680-3682

[88] Baker SN, Baker GA. Luminescent carbon nanodots: Emergent nanolights. Angewandte Chemie, International Edition. 2010;**49**:6726-6744

[89] Xu X, Ray R, Gu Y, Ploehn HJ, Gearheart L, Raker K, et al. Electrophoretic analysis and purification of fluorescent single-walled carbon nanotube fragments. Journal of the American Chemical Society. 2004;**126**:12736-12737

[90] Li H, Kang Z, Liu Y, Lee ST. Carbon nanodots: Synthesis, properties and applications. Journal of Materials Chemistry. 2012;**22**:24230

[91] Lim SY, Shen W, Gao Z. Carbon quantum dots and their applications. Chemical Society Reviews. 2015; **44**:362-381

[92] Fernando KAS, Sahu S, Liu Y, Lewis WK, Guliants EA, Jafariyan A, et al. Carbon quantum dots and applications in photocatalytic energy conversion. ACS Applied Materials & Interfaces. 2015;**7**:8363-8376

[93] Kong B, Zhu A, Ding C, Zhao X, Li B, Tian Y. Carbon dot-based inorganic-organic nanosystem for two-photon imaging and biosensing of pH variation in living cells and tissues. Advanced Materials. 2012;**24**:5844-5848

[94] Zhang X, Wang S, Zhu C, Liu M, Ji Y, Feng L, et al. Carbon-dots derived from nanodiamond: Photoluminescence tunable nanoparticles for cell imaging. Journal of Colloid and Interface Science. 2013;**397**:39-44

[95] Zhu S, Meng Q, Wang L, Zhang J, Song Y, Jin H, et al. Highly photoluminescent carbon dots for multicolor patterning, sensors, and bioimaging. Angewandte Chemie, International Edition. 2013;**125**:4045-4049

[96] Wang J, Sahu S, Sonkar SK, Tackett KN, Sun KW, Liu Y, et al. Versatility with carbon dots - From overcooked BBQ to brightly fluorescent agents and photocatalysts. RSC Advances. 2013;**3**:15604-15607

[97] Han B, Wang W, Wu H, Fang F, Wang N, Zhang X, et al. Polyethyleneimine modified fluorescent carbon dots and their application in cell labeling. Colloids and Surfaces, B: Biointerfaces. 2012;**100**:209-214

[98] Lai CW, Hsiao YH, Peng YK, Chou PT. Facile synthesis of highly emissive carbon dots from pyrolysis of glycerol; gram scale production of carbon dots/mSiO_2 for cell imaging and drug release. Journal of Materials Chemistry. 2012;**22**:14403

[99] Wang Q, Huang X, Long Y, Wang X, Zhang H, Zhu R, et al. Hollow luminescent carbon dots for drug delivery. Carbon. 2013;**59**:192-199

[100] Srivastava S, Awasthi R, Tripathi D, Rai MK, Agarwal V, Agrawal V, et al. Magnetic-nanoparticle-doped carbogenic nanocomposite: An effective magnetic resonance/ fluorescence multimodal imaging probe. Small. 2012;**8**:1099-1109

[101] Kochmann S, Hirsch T, Wolfbeis OS. The pH dependence of the total fluorescence of graphite oxide. Journal of Fluorescence. 2012;**22**:849-855

[102] Zhang X, Wang S, Liu M, Yang B, Feng L, Ji Y, et al. Size tunable fluorescent nano-graphite oxides: Preparation and cell imaging applications. Physical Chemistry Chemical Physics. 2013;**15**:19013-19018

[103] Luo P, Guan X, Yu Y, Li X, Yan F. Hydrothermal synthesis of graphene quantum dots supported on three-dimensional graphene for supercapacitors. Nanomaterials. 2019;**9**:201

[104] Fan Z, Li S, Yuan F, Fan L. Fluorescent graphene quantum dots for biosensing and bioimaging. RSC Advances. 2015;**5**:19773-19789

[105] Jeong J, Jung J, Choi M, Kim JW, Chung SJ, Lim S, et al. Color-tunable photoluminescent fullerene nanoparticles. Advanced Materials. 2012;**24**:1999-2003

[106] Liu Z, Yang K, Lee ST. Single-walled carbon nanotubes in biomedical imaging. Journal of Materials Chemistry. 2011;**21**:586-598

[107] Heller DA, Baik S, Eurell TE, Strano MS. Single-walled carbon nanotube spectroscopy in live cells: Towards long-term labels and optical sensors. Advanced Materials. 2005;**17**:2793-2799

[108] Yi H, Ghosh D, Ham MH, Qi J, Barone PW, Strano MS, et al. M13 phage-functionalized single-walled carbon nanotubes as nanoprobes for second near-infrared window fluorescence imaging of targeted tumors. Nano Letters. 2012;**12**:1176-1183

[109] Fu CC, Lee HY, Chen K, Lim TS, Wu HY, Lin PK, et al. Characterization and application of single fluorescent nanodiamonds as cellular biomarkers. Proceedings of the National Academy of Sciences of the United States of America. 2007;**104**:727-732

[110] Luo L, Yan L, Amirshaghaghi A, Wei Y, You T, Singhal S, et al. Indocyanine green-coated polycaprolactone micelles for fluorescence imaging of tumors. ACS Applied Bio Materials. 2020;**3**:2344-2349

[111] Sun M, Yin W, Dong X, Yang W, Zhao Y, Yin M. Fluorescent supramolecular micelles for imaging-guided cancer therapy. Nanoscale. 2016;**8**:5302-5312

[112] Mongin O, Krishna TR, Werts MHV, Caminade A-M, Majoral JP, Blanchard-Desce M. A modular approach to two-photon absorbing organic nanodots: Brilliant dendrimers as an alternative to semiconductor quantum dots. Chemical Communications. 2006;**8**:915-917

[113] Kim Y, Kim SH, Tanyeri M, Katzenellenbogen JA, Schroeder CM. Dendrimer probes for enhanced photostability and localization in fluorescence imaging. Biophysical Journal. 2013;**104**:1566-1575

[114] Devor A, Sakadzic S, Yaseen MA, Roussakis E, Tian P, Slovin H, et al. Functional imaging of cerebral oxygenation with intrinsic optical

contrast and phosphorescent probes. NeuroMethods. 2014;**85**:225-253

[115] Geszke-Moritz M, Moritz M. Solid lipid nanoparticles as attractive drug vehicles: Composition, properties and therapeutic strategies. Materials Science & Engineering, C: Materials for Biological Applications. 2016;**68**:982-994

[116] Dudhipala N, Puchchakayala G. Capecitabine lipid nanoparticles for anti-colon cancer activity in 1,2-dimethylhydrazine-induced colon cancer: Preparation, cytotoxic, pharmacokinetic and pathological evaluation. Drug Development and Industrial Pharmacy. 2018;**44**:1572-1582

[117] Marengo E, Cavalli R, Caputo O, Rodriguez L, Gasco MR. Scale-up of the preparation process of solid lipid nanospheres. Part I. International Journal of Pharmaceutics. 2000;**205**:3-13

[118] Gravier J, Navarro FP, Delmas T, Mittler F, Couffin AC, Vinet F, et al. Lipidots: Competitive organic alternative to quantum dots for in vivo fluorescence imaging. Journal of Biomedical Optics. 2011;**16**:096013

Chapter 18

Diffusion Theory for Cell Membrane Fluorescence Microscopy

Minchul Kang

Abstract

Since the discovery of fluorescent proteins and the development of DNA recombinant techniques, various fluorescence methods have significantly improved our understanding of cell biology at a molecular level. In particular, thanks, in large part, to technological advances in these fields, fluorescence techniques such as fluorescence recovery after photobleaching (FRAP), fluorescence correlation spectroscopy (FCS), and single-particle tracking (SPT) have become standard tools in studying cell membrane structure as well as the diffusion and interaction of biomolecules in the cell membrane. In this chapter, we will review some topics of the diffusion theory from both deterministic and probabilistic approaches, which are relevant to cell membrane fluorescence microscopy. Additionally, we will derive some basic equations for FARP and FCS based on the diffusion theory.

Keywords: diffusion theory, fluorescence recovery after photobleaching, fluorescence correlation spectroscopy, cell membranes

1. Introduction

Diffusion is an idealization of the random motion of one or more particles in space. Since diffusion is a dominant way for biological organisms to transport various molecules to desirable locations for cell signaling, the role of diffusion within biological systems is critical [1–3]. Therefore, to quantify the diffusion coefficient, a measure of diffusion rates, is essential to understand both the physiology and pathology of cells in terms of cell signaling time scales [1–3]. Moreover, the diffusion coefficients of proteins may also provide information on the landscape of the membrane environment where diffusion occurs [4–6]. However, quantifying the diffusion especially in live cell membranes is still challenging although a couple of tools are available including fluorescence recovery after photobleaching (FRAP) and fluorescence correlation spectroscopy (FCS) [7, 8]. Diffusion is quantified by a diffusion coefficient, D, which characterizes the proportionality in a linear relationship between mean squared displacement (MSD, $\langle x^2 \rangle$) of a Brownian particle and time [9, 10]. To determine the diffusion coefficients of biomolecules of interests, mathematical models for the diffusion process are compared with experimental data in FRAP and FCS analysis. In this chapter, we bridge the gap between experimental and theoretical aspects of FRAP and FCS by reviewing mathematical theories for FRAP and FCS.

 IntechOpen

2. Diffusion equation

2.1 Diffusion equation from the deterministic point of view

In 1855, Fick [11] published two cornerstone papers on diffusion, in which he proposed the fundamental laws describing the transport of mass due to the concentration gradient and an associated mathematical model. According to Fick's first law, the diffusive flux (J) is proportional to the concentration gradient of diffusants (du/dx) with a proportionality constant called a diffusion coefficient, D. In one-dimensional spatial dimension ($\mathbb{R}^1$), Fick's law can be represented as

$$J = -D\frac{du}{dx} \tag{1}$$

where $J(x,t)$ is the diffusion flux and $u(x,t)$ is the concentration of diffusants at the location x at time t. The diffusion coefficient can be calculated by the Stokes-Einstein equation [12, 13]:

$$D = \frac{k_B T}{6\pi\eta r} \tag{2}$$

where k_B is Boltzmann's constant, T is the absolute temperature, η is the dynamic viscosity, and r is the radius of the spherical particle. Assuming the conservation of mass in an infinitesimal interval $(x, x+\Delta x)$, we obtain

$$\begin{aligned} \frac{\partial}{\partial t}\{u(x,t)\Delta x\} &= J(x,t) - J(x+\Delta x, t); \\ \frac{\partial u}{\partial t} &= \frac{J(x,t) - J(x+\Delta x,t)}{\Delta x} \end{aligned} \tag{3}$$

where $u(x,t)\Delta x$ is the total number of molecules in the interval $(x, x+\Delta x)$ and $J(x,t) - J(x+\Delta x, t)$ is the difference of influx and efflux in and out of the interval (i.e., net change in the total number of molecules in the interval) as shown in **Figure 1**.

By combining Eqs. (1) and (3) and by taking the limit in $\Delta x \to 0$, we have Fick's second law that describes the diffusion process in a form of partial differential equation:

$$\frac{\partial u}{\partial t} = D\frac{\partial^2 u}{\partial x^2} \tag{4}$$

Eq. (4) is often referred to as the one-dimensional diffusion equation or heat equation. Similarly, two-dimensional ($\mathbb{R}^2$) and three-dimensional ($\mathbb{R}^3$) can be derived as

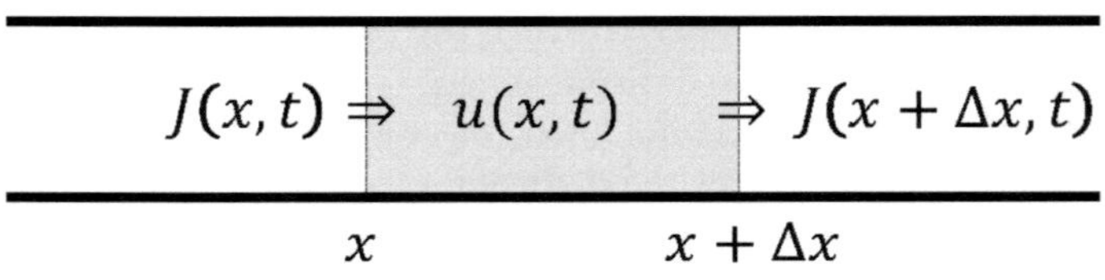

Figure 1.
The change in the number of molecules in an intestinal interval due to diffusion.

$$\frac{\partial u}{\partial t} = D\left(\frac{\partial^2 u}{\partial x^2} + \frac{\partial^2 u}{\partial y^2}\right)$$

$$\frac{\partial u}{\partial t} = D\left(\frac{\partial^2 u}{\partial x^2} + \frac{\partial^2 u}{\partial y^2} + \frac{\partial^2 u}{\partial z^2}\right) \tag{5}$$

In a more compact form, the diffusion equations are written using the Laplace operator, Δ:

$$u_t = D\Delta u \tag{6}$$

where $\Delta u = \frac{\partial^2 u}{\partial x^2}$ in $\mathbb{R}^1$, $\Delta u = \frac{\partial^2 u}{\partial x^2} + \frac{\partial^2 u}{\partial y^2}$ in $\mathbb{R}^2$, and $\Delta u = \frac{\partial^2 u}{\partial x^2} + \frac{\partial^2 u}{\partial y^2} + \frac{\partial^2 u}{\partial z^2}$ in $\mathbb{R}^3$. Importantly, the diffusion equation satisfies the following important properties:

1. **Property 1: Translation invariance.** If $u(x,t)$ is a solution of the heat equation, then for any fixed number x_0, the function $u(x - x_0, t)$ is also a solution.

2. **Property 2: Derivatives of solutions.** If $u(x,t)$ is a solution of the heat equation, then the partial derivatives of u also satisfy the heat equation.

3. **Property 3: Integrals and convolutions.** If $\Phi(x,t)$ is a solution of the heat equation, then $\Phi * g$ (the convolution of S with g) is also a solution where $\Phi * g(x,t) = \int_{-\infty}^{\infty} \Phi(x-y,t) g(y) dy$ provided that this improper integral converges. The improper integral $\Phi * g$ is called the convolution of Φ and g.

4. **Property 4: Dilation.** Suppose $a > 0$ is a constant. If $u(x,t)$ is a solution of the heat equation, then the dilated function $v(x,t) = u(\sqrt{a}x, at)$ is also a solution.

Based on these properties, we are now ready to solve the following initial value problem on $x \in \mathbb{R}^1$ for$0 \leq t < \infty$:

$$\begin{cases} u_t = Du_{xx} \\ u(x,0) = H(x) \end{cases} \text{where } H(x) = \begin{cases} 1, & x > 0 \\ 0, & x \leq 0 \end{cases} \tag{7}$$

where $H(x)$ is often referred to as the Heaviside function.

By **Property 4**, any solution $(u(x,t))$ is unaffected by the dilation $x \mapsto \sqrt{a}x$ and $t \mapsto at$ for any $a \in \mathbb{R}^1$. Since $\frac{x}{\sqrt{t}}$ is also unaffected by the dilations $\left(\frac{x}{\sqrt{t}} \mapsto \frac{\sqrt{a}x}{\sqrt{at}} = \frac{x}{\sqrt{t}}\right)$, we look for a solution in the form of $g\left(\alpha \frac{x}{\sqrt{t}}\right)$ for some constant α. Notice also that $g\left(\alpha \frac{x}{\sqrt{t}}\right)$ is also invariant under these dilations: $\left(\alpha \frac{\sqrt{a}x}{\sqrt{at}}\right) = g\left(\alpha \frac{x}{\sqrt{t}}\right)$. If we let $p = \alpha \frac{x}{\sqrt{t}}$ and choose $\alpha = \frac{1}{\sqrt{4D}}$, then by the chain rule, we have

$$0 = u_t - Du_{xx} = -\frac{p}{2t} g'(p) - \frac{\kappa}{4Dt} g''(p) = -\frac{1}{4t}\{g''(p) + 2pg'(p)\} \tag{8}$$

which reduces to an ordinary differential equation $g'' + 2pg' = 0$. This can be solved as.

$$g = C_2 + \int_0^p C_1 e^{-r^2} dr \text{ where } g(0) = C_2 \tag{9}$$

for arbitrary constants C_1 and C_2. Because as $t \to 0+, p \to \infty$, for $x > 0$

$$1 = \lim_{t \to 0+} u(x,t) = C_2 + \int_0^{\infty} C_1 e^{-r^2} dr = \frac{\sqrt{\pi}}{2} C_1 + C_2 \quad (10)$$

where we used a well-known identity (the error function integral):

$$\int_{-\infty}^{\infty} e^{-ax^2} dx = \sqrt{\frac{\pi}{a}}, \quad (11)$$

On the other hand, since as $t \to 0-, p \to -\infty$, for $x < 0$

$$0 = \lim_{t \to 0+} u(x,t) = C_2 + \int_0^{-\infty} C_1 e^{-r^2} dr = -\frac{\sqrt{\pi}}{2} C_1 + C_2 \quad (12)$$

which implies that $C_1 = \frac{1}{\sqrt{\pi}}$ and $C_2 = \frac{1}{2}$. Putting together, we have a solution

$$u(x,t) = \frac{1}{2} + \frac{1}{\sqrt{\pi}} \int_0^{x/\sqrt{4\kappa t}} e^{-r^2} dr \quad (13)$$

Define $\Phi(x,t) = u_x(x,t)$; then

$$\begin{aligned} \Phi(x,t) &= \frac{\partial}{\partial x} \left(\frac{1}{2} + \frac{1}{\sqrt{\pi}} \int_0^{x/\sqrt{4\kappa t}} e^{-r^2} dr \right) \\ &= \frac{1}{\sqrt{\pi}} e^{\frac{x^2}{4\kappa t}} \cdot \frac{1}{\sqrt{4\kappa t}} \\ &= \frac{1}{\sqrt{4\pi\kappa t}} e^{\frac{x^2}{4\kappa t}} \end{aligned} \quad (14)$$

By **Property 2, derivatives of solutions**, the function $\Phi(x,t) = \frac{1}{\sqrt{4\pi\kappa t}} e^{-\frac{x^2}{4\kappa t}}$ is also a solution to the diffusion equation. $\Phi(x,t)$ is called the (one-dimensional) heat kernel or the fundamental solution of the heat equation. The graphs of the heat kernel for different t are shown in **Figure 2**.

From **Figure 2**, we can see that the heat kernel $\Phi(x,t)$ has a "bell curve" graph of a normal distribution (Gaussian function) with $\sqrt{2Dt}$ as the standard deviation, which sometimes called the Gaussian root mean square width. Also, $\frac{1}{\sqrt{4\pi t}}$ modulates the amplitude of the Gaussian curves, and the amplitude blows up to ∞ as $t \to 0+$ and approaches 0 as $t \to \infty$, i.e.:

$$\lim_{t \to 0+} \Phi(x,t) = \begin{cases} 0 \text{ if } x \neq 0 \\ \infty \text{ if } x = 0 \end{cases}. \quad (15)$$

Also, from the error function integration (Eq. (14))

$$\begin{cases} \int_{-\infty}^{\infty} \Phi(x,t) dx = 1, \text{ for all } t \geq 0 \\ \int_{-\infty}^{\infty} \Phi(x-y,t) dx = 1, \text{ for all } t \geq 0 \end{cases} \quad (16)$$

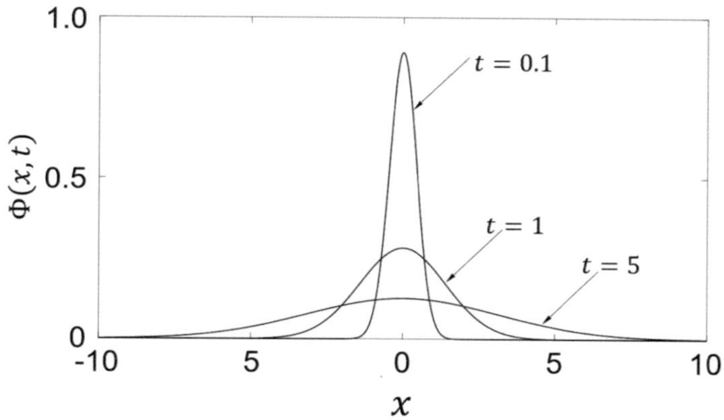

Figure 2.
The heat kernel graphs for different t.

Furthermore, it follows that (i) $\Phi(x-y,t)$ satisfies the heat equation (**Property 1: translation invariance**) and (ii) $(\Phi * \phi)(x,t) = \int_{-\infty}^{\infty} \Phi(x-y,t)\phi(y)dy$ satisfies the heat equation (**Property 3: integrals and convolutions**).

From the definition ($\Phi = u_x$), by differentiating Eq. (7) with respect to x, we see that $\Phi(x,t)$ satisfies

$$\begin{cases} \Phi_t = D\Phi_{xx} \\ \Phi(x,0) = u_x(x,0) = \dfrac{d}{dx}H(x) \end{cases} \tag{17}$$

Even though $H(x)$ is not differentiable due to discontinuity at $x = 0$, we can redefine differentiation in a broad sense (weak derivative) and under this weak derivative definition:

$$\frac{d}{dx}H(x) = \delta(x) = \begin{cases} 0 \text{ if } x \neq 0 \\ \infty \text{ if } x = 0 \end{cases} \tag{18}$$

where $\delta(x)$ is called the Dirac delta function. The Dirac delta function satisfies a few important properties:

1. $\lim_{t \to 0+} \Phi(x,t) = \delta(x)$

2. $\int_{-\infty}^{\infty} \delta(x)dx = 1$ and $\int_{-\infty}^{\infty} \delta(x-y)dx = 1$

3. $\int_{-\infty}^{\infty} \delta(x)f(x)dx = f(0)$ and $\int_{-\infty}^{\infty} \delta(x-y)f(x)dx = f(y)$

The third integration property is sometimes called the *sifting property* of the Dirac delta function. With these properties, we now can show (heuristically) $u(x,t) = (\Phi * \phi)(x,t)$ satisfies the following diffusion equation:

$$\begin{cases} u_t = Du_{xx} \\ u(x,0) = \phi(x) \end{cases} \tag{19}$$
$$\Rightarrow u(x,t) = (\Phi * \phi)(x,t)$$

To show $(\Phi * \phi)(x,t)$ satisfies the initial condition, we apply the sifting property of the Dirac delta function:

$$\begin{aligned} u(x,0) &= (\Phi * \phi)(x,0) \\ &= \int_{-\infty}^{\infty} \Phi(x-y,0)\phi(y)dy \\ &= \int_{-\infty}^{\infty} \delta(x-y)\phi(y)dy \\ &= \phi(x) \end{aligned} \tag{20}$$

In other words, this result (Eq. (19)) indicates that for any initial value problem, the solution can easily be found as a convolution of the heat kernel and initial data.

2.2 Diffusion equation from the stochastic point of view

In many biological systems, passive transports are often described by Brownian motion or diffusion that is observed in random drifting of pollen grains suspended in a fluid. Suppose a Brownian particle located at the position $x = 0$ when time $t = 0$ has moved randomly on a straight line during time Δt. Since the movement of a Brownian particle is random, the location of the Brownian particle at $t = \Delta t$ will be probabilistic. Especially, for smaller Δt elapsed, the Brownian particle will have a higher chance to be found near the starting location $x = 0$ similar to a normal (or Gaussian) probability distribution with zero mean and a small standard deviation. For this reason, the Brownian motion is often described mathematically by random variables in time, which is called a stochastic process (time-dependent random variable).

If we let X_t be a stochastic process in $\mathbb{R}^1$ describing the position of a fluorescence molecule at time t, i.e., "$X_t = x$" means that the location of a fluorescence molecule at time t is x, then the probability of the Brownian particle located within the interval $(0, \Delta x)$ at time t will be dependent on both Δx and the previous location:

$$\mathbb{P}\{X_t \in (0, \Delta x) | X_0 = 0\} \tag{21}$$

assuming the initial location is the origin ($X_0 = 0$). Bachelier [14] explicitly calculated this probability as

$$\mathbb{P}\{X_t \in (0, \Delta x) | X_0 = 0\} = \int_0^{\Delta x} \frac{1}{\sqrt{4\pi Dt}} \exp\left(-\frac{x^2}{4Dt}\right) dx \tag{22}$$

where D (μm/s^2) is a diffusion coefficient. The probability density function (the integrand) is the fundamental solution of heat equation (Eq. (14)) that is the normal distribution with standard deviation $\sigma = \sqrt{2Dt}$. Later, Einstein [12] showed that the probability density function of randomly moving particles (Brownian motion) satisfies the diffusion equation with a solution $\Phi(x,t)$ (Eq. (17)).

If $g(y)$ is the probability of a Brownian particle to be found at location y when $t = 0$, i.e., $\mathbb{P}\{X_0 = y\} = g(y)$, then the distribution of the Brownian particles can be determined by solving an initial value problem:

$$\begin{cases} \dfrac{\partial u}{\partial t} = D\dfrac{\partial^2 u}{\partial x^2}, \\ u(x,0) = g(x) \end{cases} \tag{23}$$

which has the solution

$$(\Phi * g)(x) = \int_{-\infty}^{\infty} \frac{1}{\sqrt{4\pi Dt}} \exp\left(-\frac{(x-y)^2}{4Dt}\right) g(y)dy. \tag{24}$$

as in Eq. (19).

2.3 Mean squared displacement

The spreading rate of diffusing particles is quantified by a diffusion coefficient, D, which characterizes a linear relationship between mean squared displacement ($\langle x^2 \rangle$) of a Brownian particle and time, where MSD is defined as

$$\begin{aligned}\langle x^2 \rangle &= \int_{-\infty}^{\infty} x^2\, \mathbb{P}\{X_t \in (0, \Delta x) | X_0 = 0\} dx \\ &= \int_{-\infty}^{\infty} x^2 \Phi(x,t) dx.\end{aligned} \tag{25}$$

For a diffusion process, MSD increases linearly in time with the rate of the diffusion coefficient:

$$\langle x^2 \rangle = 2nDt. \tag{26}$$

where n is the spatial dimension ($\mathbb{R}^n$) for a diffusion process. To derive this relation in 1D ($\mathbb{R}$), we consider $\frac{\partial}{\partial t}\langle x^2 \rangle$

$$\begin{aligned}\frac{\partial}{\partial t}\langle x^2 \rangle &= \frac{\partial}{dt}\int_{-\infty}^{\infty} x^2 \Phi(x,t) dx \\ &= \int_{-\infty}^{\infty} x^2 \frac{\partial}{\partial t}\Phi(x,t) dx \\ &= D\int_{-\infty}^{\infty} x^2 \frac{\partial^2}{\partial x^2}\Phi(x,t) dx\end{aligned} \tag{27}$$

where we used Eq. (17). Notice that by the product rule

$$D\int_{-\infty}^{\infty} \frac{\partial}{\partial x}\left(x^2 \frac{\partial}{\partial x}\Phi(x,t)\right) dx = D\int_{-\infty}^{\infty} 2x \frac{\partial}{\partial x}\Phi(x,t) dx + D\int_{-\infty}^{\infty} x^2 \frac{\partial^2}{\partial x^2}\Phi(x,t) dx \tag{28}$$

By solving for $D\int_{-\infty}^{\infty} x^2 \frac{\partial^2}{\partial x^2}\Phi(x,t) dx$

$$\begin{aligned}D\int_{-\infty}^{\infty} x^2 \frac{\partial^2}{\partial x^2}\Phi(x,t) dx &= D\int_{-\infty}^{\infty} \frac{\partial}{\partial x}\left(x^2 \frac{\partial}{\partial x}\Phi(x,t)\right) dx - \int_{-\infty}^{\infty} 2x \frac{\partial}{\partial x}\Phi(x,t) dx \\ &= D\left[x^2 \frac{\partial}{\partial x}\Phi(x,t)\right]_{-\infty}^{\infty} - D\int_{-\infty}^{\infty} 2x \frac{\partial}{\partial x}\Phi(x,t) dx \\ &= 0 - D\int_{-\infty}^{\infty} 2x \frac{\partial}{\partial x}\Phi(x,t) dx\end{aligned} \tag{29}$$

Next, by integration by parts

$$-D\int_{-\infty}^{\infty} 2x\frac{\partial}{\partial x}\Phi(x,t)dx = -D[2x\Phi(x,t)]_{-\infty}^{\infty} + D\int_{-\infty}^{\infty} 2\Phi(x,t)dx \\ = -0 + 2D \tag{30}$$

Finally, by putting all together

$$\frac{\partial}{\partial t}\langle x^2\rangle = 2D \\ \langle x^2\rangle = 2Dt, \tag{31}$$

for $\mathbb{R}^1$.

3. Fluorescence recovery after photobleaching

3.1 Principles of FRAP

Fluorescence recovery after photobleaching is a fluorescence-based biophysical tool developed in the 1970s to investigate the diffusion process in membranes of live cells. Discovery of the green fluorescent protein (GFP) and the invention of commercial confocal laser scanning microscopes (CLSMs) have broadened the accessibility of FRAP for many researchers in the field, and the applications of FRAP have become widely extended to the study of intracellular protein dynamics [15–18]. Over the four decades, there have been considerable advances in microscope technology. However, the basic principle of FRAP remains the same. In FRAP, fluorescently tagged molecules in a small region of interest (ROI) are irreversibly photobleached using a high-intensity laser source for a short period of time, and

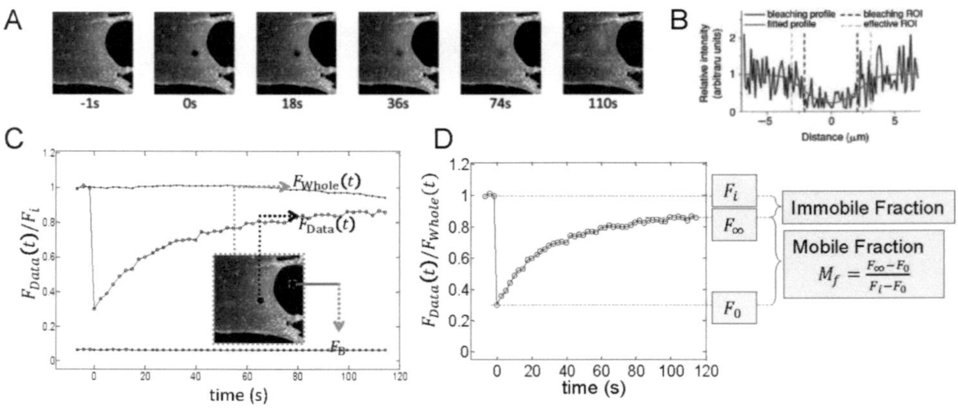

Figure 3.
Example of FRAP data. (A) Representative images from a FRAP experiment on Alexa488-CTxB. (B) A postbleach profile from the image for t = 0 shows a wider spreading radius (effective radius; r_e) than the bleaching spot radius (nominal radius; r_n) due to diffusion during photobleaching. (C) Mean fluorescence intensity (N = 13) from the bleaching ROI (∘, $F_{Data}(t)$), whole image (•, $F_{Whole}(t)$), and background (□) from a FRAP experiment of Alexa488-CTxB. The image in the inset shows the locations where $F_{Data}(t)$ (∘) and background (□) were measured. (D) In FRAP analysis, prebleach steady-state, postbleach initial, and postbleach steady-state fluorescence intensities are typically denoted as F_i, F_0, and F_i. These parameters can be used to calculate the mobile fraction (M_f) and the immobile fraction ($1 - M_f$) from the corrected FRAP data for photofading ($F_{Data}(t)/F_{Whole}(t)$) as indicated in the boxed equation.

then the exchange of fluorescence and photobleached molecules in and out of the bleached region is monitored using low-intensity laser excitation to track fluorescence recovery (**Figure 3A**). Due to the artifacts such as the diffusion during the photobleaching step (**Figure 3B**) and the photofading during the imaging step, FRAP data requires some corrections (**Figure 3C**). The diffusion during the photobleaching step can be corrected by using the experimentally measured postbleach profile as an initial condition for the FRAP model [19–21]. On the other hand, the photofading during the imaging step can be corrected by diving the raw FRAP data ($F_{Data}(t)$) by the fluorescence intensity from the whole image ($F_{Data}(t)$) (**Figure 3D**) [19–21]. Since different transport and reaction mechanisms may affect the curvature and the mobile fraction of a FRAP curve in various manners, kinetic parameters for underlying mechanisms can be obtained by comparing the FRAP curve to the corresponding theoretical FRAP models. For example, D can be measured by comparing a diffusion FRAP model with FRAP data for the best fitting D [19, 20].

3.2 Derivation of diffusion FRAP equation in $\mathbb{R}^1$

Quantitative FRAP analysis requires a mathematical description of fluorescence recovery for a given underlying transport/reaction kinetics through two different modes of CLSMs: photobleaching and photo-illumination. Although CLSMs use scanning laser for both photobleaching and photo-illumination, it has been reported for small bleaching spot size (we call this as the nominal radius of the laser); the scanning profile of CLSMs on a confocal plane is well approximated by a Gaussian function:

$$I_{r_n}(x) = \sqrt{\frac{2I_0^2}{\pi r_n^2}} \exp\left(-\frac{2x^2}{r_n^2}\right), \tag{32}$$

where r_n is the nominal radius, i.e., radius of a bleaching ROI (the half-width at e^{-2} laser intensity). I_{r_n} can be regarded as a photobleaching mode of CLSMs with a maximal laser intensity I_0. A bell-shaped profile of $I_{r_n}(x)$ defines total laser intensity I_0 with $\int_{-\infty}^{\infty} I_{r_n}(x)dx = I_0$ resulting from the error function integral (Eq. (11)). Since the high-intensity mode of laser ($I_{r_n}(x)$) causes photobleaching of fluorophores, for illumination, laser intensity has to be attenuated to a lower laser intensity level. Therefore, for an attenuation factor $\epsilon \ll 1$, a photo-illumination mode of CLSMs can be described as $\epsilon I_{r_n}(x)$. If we let $u(x,t)$ be the density of fluorophores (or fluorescent proteins) at a location x at time t, then fluorescence intensity at the position x at time t will be proportional to both the illumination laser intensity ($\epsilon I_{r_n}(x)$) and fluorophore density ($u(x,t)$). Assuming the linear proportionality, $f(x,t)$, the fluorescence intensity at a location (x,y) at time t can be described as

$$f(x,t) = q \cdot \epsilon I_{r_n}(x) u(x,t), \tag{33}$$

where the proportionality constant, q, is referred to as a quantum yield or quantum efficiency. When a CLSM system is used to photobleach fluorophores, its postbleach profile is not exactly the same as the laser profile in most cases due to diffusion occurring during the photobleaching step. Assuming the first-order photobleaching process with a photobleaching rate α, a governing equation for a photobleaching process of freely diffusing fluorescent proteins can be described as a reaction–diffusion equation:

$$\begin{cases} u_t = D\Delta u - \alpha I_{r_n}(x)u \\ u(x,0) = u_0 \end{cases} \tag{34}$$

where u_0 is the prebleach steady-state fluorescence intensity, which is regarded as a constant. Although the solution to Eq. (34) is hard to find, it is empirically proven [22] that a confocal postbleach profile can be described as a simple Gaussian function (constant minus Gaussian):

$$\varphi(x) = C_i\left(1 - K\exp\left(-\frac{2x^2}{r_e^2}\right)\right), \tag{35}$$

Note that different underlying kinetics for u yield a different FRAP equation. For free diffusion kinetics, the evolution of $u(x,t)$ can be described as the diffusion equation subject to the initial condition from a postbleach profile right after photobleaching.

$$\begin{cases} u_t = D\Delta u \\ u(x,0) = \varphi(x) \end{cases} \tag{36}$$

where $D(\mu m^2/s)$ is a diffusion coefficient and the Laplacian, $\Delta = \frac{\partial^2}{\partial x^2}$, in $\mathbb{R}^1$. The solution of the diffusion equation can be found as (Eq. (19))

$$u(x,t) = \Phi_D * \varphi$$

$$= \int \Phi_D(x - \overline{x}, t)\varphi(\overline{x})d\overline{x}$$

$$= \frac{C_i}{\sqrt{4\pi Dt}}\int \exp\left(-\frac{(x-\overline{x})^2}{4Dt}\right)\left[1 - K\exp\left(-\frac{2\overline{x}^2}{r_e^2}\right)\right]d\overline{x}$$

$$= \frac{C_i}{\sqrt{4\pi Dt}}\int \exp\left(-\frac{(x-\overline{x})^2}{4Dt}\right)d\overline{x} - \frac{C_iK}{\sqrt{4\pi Dt}}\int \exp\left(-\frac{(x-\overline{x})^2}{4Dt} - \frac{2\overline{x}^2}{r_e^2}\right)d\overline{x} \tag{37}$$

$$= C_i - \frac{C_iK}{\sqrt{4\pi Dt}}\int \exp\left(-\frac{(x-\overline{x})^2}{4Dt} - \frac{2\overline{x}^2}{r_e^2}\right)d\overline{x}$$

by Eq. (11) (error function integration).

The total fluorescence intensity from the region of interest can be found by integrating this local fluorescence intensity over the ROI:

$$F(t) = q\epsilon\int I_{r_n}(x)u(x,t)dx, \tag{38}$$

which is called a FRAP equation. To simplify Eq. (38) by using Eq. (37)

$$q\epsilon\int I_{r_n}(x)u(x,t)dx$$

$$= q\epsilon\int\left[\sqrt{\frac{2I_0^2}{\pi r_n^2}}\exp\left(-\frac{2x^2}{r_n^2}\right)\right]\left[C_i - \frac{C_iK}{\sqrt{4\pi Dt}}\int \exp\left(-\frac{(x-\overline{x})^2}{4Dt} - \frac{2\overline{x}^2}{r_e^2}\right)d\overline{x}\right]dx$$

$$= q\epsilon C_i\sqrt{\frac{2I_0^2}{\pi r_n^2}}\int \exp\left(-\frac{2x^2}{r_n^2}\right)dx - q\epsilon C_i K\sqrt{\frac{2I_0^2}{4\pi^2 r_n^2 Dt}}\iint \exp\left(-\frac{2x^2}{r_n^2}-\frac{(x-\overline{x})^2}{4Dt}-\frac{2\overline{x}^2}{r_e^2}\right)dx d\overline{x}$$

$$= q\epsilon C_i I_0 - q\epsilon C_i K\sqrt{\frac{2I_0^2}{4\pi^2 r_n^2 Dt}}\iint \exp\left(-\frac{2x^2}{r_n^2}-\frac{(x-\overline{x})^2}{4Dt}-\frac{2\overline{x}^2}{r_e^2}\right)dx d\overline{x}$$

$$= F_i - F_i K\sqrt{\frac{1}{2\pi^2 r_n^2 Dt}}\iint \exp\left(-\frac{2x^2}{r_n^2}-\frac{(x-\overline{x})^2}{4Dt}-\frac{2\overline{x}^2}{r_e^2}\right)dx d\overline{x} \tag{39}$$

where $F_i = q\epsilon C_i I_0$ is the prebleach fluorescence intensity due to fluorophore density C_i. If we let $\overline{x} = x + \theta\chi$ where $\theta = \sqrt{4Dt}$ $(d\overline{x} = \theta d\chi)$, then the integral term in Eq. (39) becomes

$$\iint \exp\left(-\frac{2x^2}{r_n^2}-\frac{(x-\overline{x})^2}{4Dt}-\frac{2\overline{x}^2}{r_e^2}\right)dx d\overline{x}$$

$$= \theta\iint \exp\left(-\frac{2x^2}{r_n^2}-\frac{\theta^2}{\theta^2}\chi^2-\frac{2(x+\theta\chi)^2}{r_e^2}\right)dx d\chi$$

$$= \theta\iint \exp\left(-\frac{2\left(r_e^2x^2+r_n^2(x+\theta\chi)^2\right)}{r_n^2r_e^2}-\chi^2\right)dx d\chi$$

$$= \theta\iint \exp\left(-\frac{2([r_e^2+r_n^2]x^2+2r_n^2\theta x\chi+\theta^2r_n^2\chi^2)}{r_n^2r_e^2}-\chi^2\right)dx d\chi \tag{40}$$

$$= \theta\iint \exp\left(-\frac{2[r_e^2+r_n^2]\left(x^2+2\frac{r_n^2\theta}{r_e^2+r_n^2}x\chi+\frac{\theta^2r_n^2}{r_e^2+r_n^2}\chi^2\right)}{r_n^2r_e^2}-\chi^2\right)dx d\chi$$

$$= \theta\iint \exp\left(-\frac{2[r_e^2+r_n^2]}{r_n^2r_e^2}\left\{\left(x+\frac{r_n^2\theta}{r_e^2+r_n^2}\chi\right)^2-\left(\left[\frac{r_n^2\theta}{r_e^2+r_n^2}\right]^2-\frac{\theta^2r_n^2}{r_e^2+r_n^2}\right)\chi^2\right\}-\chi^2\right)dx d\chi$$

$$= \theta\iint \exp\left(-\frac{2[r_e^2+r_n^2]}{r_n^2r_e^2}\left\{\left(x+\frac{r_n^2\theta}{r_e^2+r_n^2}\chi\right)^2-\left(\frac{r_n^4\theta^2-\theta^2r_n^4-\theta^2r_n^2r_e^2}{[r_e^2+r_n^2]^2}\right)\chi^2\right\}-\chi^2\right)dx d\chi$$

$$= \theta\iint \exp\left(-\frac{2[r_e^2+r_n^2]}{r_n^2r_e^2}\left\{\left(x+\frac{r_n^2\theta}{r_e^2+r_n^2}\chi\right)^2-\left(\frac{-\theta^2r_n^2r_e^2}{[r_e^2+r_n^2]^2}\right)\chi^2\right\}-\chi^2\right)dx d\chi$$

$$= \theta\iint \exp\left(-\left\{\frac{2[r_e^2+r_n^2]\left(x+\frac{r_n^2\theta}{r_e^2+r_n^2}\chi\right)^2}{r_n^2r_e^2}+\frac{2\theta^2}{r_e^2+r_n^2}\chi^2\right\}-\chi^2\right)dx d\chi$$

$$= \theta\int\left[\int \exp\left(\frac{2[r_e^2+r_n^2]\left(x+\frac{r_n^2\theta}{r_e^2+r_n^2}\chi\right)^2}{r_n^2r_e^2}\right)dx\right]\exp\left(-\left[\frac{2\theta^2}{r_e^2+r_n^2}+1\right]\chi^2\right)d\chi$$

$$= \theta\int\left[\int \exp\left(\frac{2[r_e^2+r_n^2]}{r_n^2r_e^2}x^2\right)dx\right]\exp\left(-\left[\frac{2\theta^2}{r_e^2+r_n^2}+1\right]\chi^2\right)d\chi$$

$$= \theta\int\left[\sqrt{\frac{\pi r_n^2r_e^2}{2(r_e^2+r_n^2)}}\right]\exp\left(-\left[\frac{2\theta^2}{r_e^2+r_n^2}+1\right]\chi^2\right)d\chi$$

$$= \theta\sqrt{\frac{\pi r_n^2 r_e^2}{2(r_e^2 + r_n^2)}} \int \exp\left(-\left[\frac{2\theta^2 + r_e^2 + r_n^2}{r_e^2 + r_n^2}\right]\chi^2\right) d\chi$$

$$= \theta\sqrt{\frac{\pi r_n^2 r_e^2}{2(r_e^2 + r_n^2)}}\sqrt{\frac{\pi(r_e^2 + r_n^2)}{2\theta^2 + r_e^2 + r_n^2}}$$

$$= \sqrt{\frac{\pi^2 r_n^2 r_e^2 (4Dt)}{2(8Dt + r_e^2 + r_n^2)}}$$

Therefore

$$F(t) = F_i - F_i K \sqrt{\frac{1}{2\pi^2 r_n^2 Dt}}\sqrt{\frac{\pi^2 r_n^2 r_e^2 (4Dt)}{2(8Dt + r_e^2 + r_n^2)}}$$
$$= F_i - F_i K \sqrt{\frac{r_e^2}{8Dt + r_e^2 + r_n^2}} \tag{41}$$
$$= F_i\left[1 - \frac{K}{\sqrt{1 + \gamma^2 + 2t/\tau_D}}\right]$$

where $\gamma = r_n/r_e$ and $\tau_D = r_e^2/(4D)$. If we consider the immobile fraction (**Figure 3D**), the FRAP equation for mobile fluorophores is found as

$$F(t) = F_i\left\{1 - \frac{K}{\sqrt{1 + \gamma^2 + 2t/\tau_D}}\right\}\mathbb{M} + (1 - \mathbb{M})F_0 \tag{42}$$

for the mobile fraction, $\mathbb{M}$ is defined as (**Figure 3D**)

$$\mathbb{M} = \frac{F_\infty - F_0}{F_i - F_0} \tag{43}$$

where F_i, F_0, and F_∞ are prebleach steady-state fluorescence intensity, postbleach initial fluorescence intensity ($F(0)$), and postbleach steady-state fluorescence intensity, respectively. The calculations for the 1D FRAP model can easily be extended to higher-dimensional cases. For example, a diffusion FRAP equation in 2D ($\mathbb{R}^2$) and 3D ($\mathbb{R}^3$) is found as

$$F(t) = F_i\left[1 - \frac{K}{1 + \gamma^2 + 2t/\tau_D}\right] \text{(2D)}$$
$$F(t) = F_i\left[1 - \frac{K}{(1 + \gamma^2 + 2t/\tau_D)\sqrt{1 + \gamma^2 + 2t/\tau_D}}\right] \text{(3D)} \tag{44}$$

4. Fluorescence correlation spectroscopy

4.1 Principles of fluorescence correlation spectroscopy

Fluorescence correlation spectroscopy is a standard bioengineering and biophysics technique for the study of molecular movements and interactions [23–25]. For FCS experiments, a laser beam is focused and stationed at a region of interest in the specimen (usually live cells). The illumination region formed by the focused laser is called a confocal volume, which is generally in the femtoliter range. As fluorescence molecules cross the confocal volume by diffusion or other transporting mechanisms, they emit fluorescence photons responding to the illumination laser (**Figure 4A**), and the fluctuations in the fluorescence signal, $F(t)$, is monitored as a function of time which is called raw FCS data. Since different FCS measurements from different cells can be quite different depending on the fluorescent protein expression level, the raw FCS data is first standardized by

$$\frac{\Delta F(t)}{\langle F\rangle_t} = \frac{(F(t) - \langle F(t)\rangle_t)}{\langle F\rangle_t} \tag{45}$$

where $F(t)$ is the fluorescence fluctuation in the confocal volume and $\langle F\rangle_t = \frac{1}{T}\int_0^T F(t)dt$ is the time average of the fluorescence fluctuation during observation time T. Notice that the mean of standardized data $(\Delta F(t)/\langle F\rangle_t)$ is zero. Next, the autocorrelation function of the standardized data is calculated by multiplying the standardized data, $\Delta F(t)/\langle F\rangle_t$, and the shifted standardized data by τ, $\Delta F(t+\tau)/\langle F\rangle_t$, and then taking the average over time:

$$G(\tau) = \left\langle \frac{\Delta F(t)}{\langle F\rangle_t} \cdot \frac{\Delta F(t+\tau)}{\langle F\rangle_t} \right\rangle_t \tag{46}$$

Notice that the autocorrelation has the maximum when $\tau = 0$ and converges to 0 as τ increases as $\Delta F(t)/\langle F\rangle_t$ and $\Delta F(t+\tau)/\langle F\rangle_t$ become independent for a large τ.

$$\begin{cases} G(0) = \left\langle \left(\frac{\Delta F(t)}{\langle F\rangle_t}\right)^2 \right\rangle_t > 0 \\ G(\tau) = \left\langle \frac{\Delta F(t)}{\langle F\rangle_t} \cdot \frac{\Delta F(t+\tau)}{\langle F\rangle_t} \right\rangle_t = \left\langle \frac{\Delta F(t)}{\langle F\rangle_t} \right\rangle_t \left\langle \frac{\Delta F(t+\tau)}{\langle F\rangle_t} \right\rangle_t = 0 \text{ for a large } \tau \end{cases} \tag{47}$$

An autocorrelation curve carries two crucial information. Since a large molecule will move slower than a light molecule, therefore the correlation decays at a longer time scale. On the other hand, the correlation amplitude is inversely proportional to

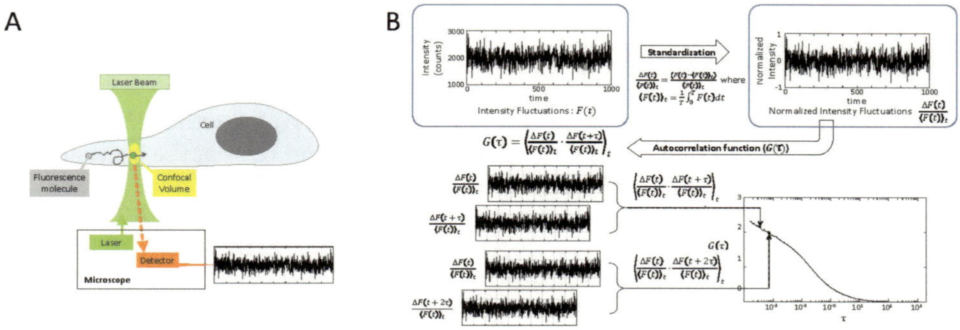

Figure 4.
Principles of fluorescence correlation spectroscopy analysis. (A) For FCS analysis for free diffusion, a static laser beam is focused on a specific region of interest. As the fluorescence molecules diffuse in and out of a certain domain, commonly called confocal volume (∼1 femtoliter), fluorescence intensities from the confocal volume fluctuate, yielding fluorescence time series. (B) The fluorescence time series data are processed into an autocorrelation curve by taking the average of the original time series data and the shifted time series data by τ to get an autocorrelation function (ACF) in τ. The ACF from the FCS data is next fitted to theoretical autocorrelation functions (ACFs) to determine underlying kinetic parameters, such as a diffusion coefficient.

the concentration of fluorophores due to the denominator for standardization. The information on the diffusion coefficient and concentration of fluorophores can be determined, once a mathematical model for $G(\tau)$ is developed.

Stationarity and ergodicity of the diffusion process play a pivotal role to derive an FCS equation in a closed, yet simple, form. A continuous-time dynamical system such as Brownian motion is called ergodic when all the accessible microstates such as the locations of a Brownian particle are equally probable over a long period, i.e., the statistical properties from the time average at a position are same as the ensemble (spatial) average at any moment. On the other hand, a stationary process is a stochastic process whose probability distribution and parameters are invariant by shifts in time. Stationary and ergodic properties of a diffusion process were proven mathematically [26].

If we let $n(x,t)$ be the fluorescence molecule density per unit area, the temporal average of $n(x,t)$ at a location x_0 and the spatial (ensemble) average of $n(x,t)$ can be defined as

$$\begin{cases} \langle n(x_0,t)\rangle_t = \lim\limits_{T\to\infty} \int_0^T n(x_0,t)dt \\ \langle n(x,t_0)\rangle = \mathbb{E}(n(x,t_0)) = \int_{-\infty}^{\infty} n(x,t_0)\mathbb{P}\{X_{t_0} = x\}dx. \end{cases} \tag{48}$$

Under stationarity and ergodicity of a diffusion process, we assume

$$\begin{aligned} \langle n(x_0,t)\rangle_t &= \lim_{T\to\infty} \frac{1}{T}\int_0^T n(x_0,t)dt \\ &= \int_{-\infty}^{\infty} n(x,t_0)\mathbb{P}\{X_{t_0} = x\}dx \text{ (Ergodikicity)} \\ &= \int_{-\infty}^{\infty} n(x,0)\mathbb{P}\{X_0 = x\}dx \text{ (Stationarity)} \\ &= \langle n(x,0)\rangle_x \end{aligned} \tag{49}$$

where $n(x,t_0)$ can be though as a snapshot of all the positions of Brownian particles at any fixed time t_0.

4.2 Derivation of diffusion FCS equation

For the fluorescence molecule density per unit area, $n(x,t)$, if we let $f(x,t)$ be the fluorescence intensities due to photons from fluorescent proteins at the location x at the time t, then $f(x,t)$ is proportional to $n(x,t)$. On the other hand, since more fluorescence photons can be generated under the higher laser intensity, $f(x,t)$ is also proportional to the laser intensity, $I(x)$. Therefore, $f(x,t)$ satisfies

$$f(x,t) = QI(x)n(x,t) \tag{50}$$

where Q is a proportionality constant for the product of the absorption cross section by the fluorescence quantum yield and the efficiency of fluorescence, and $I(x)$ is a function describing a Gaussian laser profile:

$$I(x) = \sqrt{\frac{2}{\pi\omega^2}} \exp\left(-2\frac{x^2}{\omega^2}\right) \tag{51}$$

where ω is the half-width of the beam at e^{-2}, which measures the size of a confocal volume (V).

A bell-shaped profile of $I(x)$ defines a unit confocal volume (V) with $|V| = \int_{-\infty}^{\infty} I(x)dx = 1$, resulting from the error function integral (Eq. (11)). Therefore, the fluorescent intensity (or the number of photons, $F(t)$) from the confocal volume is determined by

$$\begin{aligned} F(t) &= \int_{-\infty}^{\infty} f(x,t)dx \\ &= Q\int_{-\infty}^{\infty} I(x)n(x,t)dx \\ &= Q\int_{V} I(x)n(x,t)dx \end{aligned} \tag{52}$$

where we used the fact that the Gaussian laser profile defines the confocal volume in the last equality to switch the integration domain from V to $(-\infty, \infty)$.

Lastly, we will also assume the spatial and temporal independence of fluorescence intensities:

$$\begin{aligned} \langle f(x,t)f(y,t)\rangle_t &= \begin{cases} \left\langle (f(x,t))^2 \right\rangle_t \text{if } x = y \\ \langle f(x,t)\rangle_t \langle f(y,t)\rangle_t \text{if } x \neq y \end{cases} \\ &= \begin{cases} \left\langle (f(x,0))^2 \right\rangle \text{if } x = y \\ \langle f(x,0)\rangle \langle f(y,0)\rangle \text{if } x \neq y \end{cases}. \end{aligned} \tag{53}$$

This assumption hypothesizes that fluorescence intensities from different locations are not correlated but independent.

In FCS, to analyze the fluorescence fluctuations from the confocal volume (V), an autocorrelation function (ACF) of the variations in $F(t)$ is considered. The variations in the number of photons from the mean number of photons in a confocal volume (ΔF) are calculated by $\Delta F(t) = F(t) - \langle F \rangle_t$ where $F(t)$ and $\langle F \rangle_t$ are the fluorescence intensity in the confocal volume at time t and the mean fluorescence in the confocal volume, respectively. Therefore, by Eq. (52)

$$\begin{aligned} \Delta F(t) &= F(t) - \langle F \rangle_t \\ &= \int_{-\infty}^{\infty} f(x,t)dx - \lim_{T\to\infty} \frac{1}{T}\int_0^T f(x,t)dt \\ &= Q\int_{-\infty}^{\infty} I(x)n(x,t)dx - \lim_{T\to\infty} \frac{1}{T}\int_0^T \left[Q\int_{-\infty}^{\infty} I(x)n(x,t)dx \right] dt \\ &= Q\int_{-\infty}^{\infty} I(x)n(x,t)dx - Q\int_{-\infty}^{\infty} I(x)\langle n \rangle_t dx \\ &= Q\int_{-\infty}^{\infty} I(x)\Delta n(x,t)dx \end{aligned} \tag{54}$$

where $\Delta n(x,t) = n(x,t) - \langle n \rangle_t$, we used the identities $\int_{-\infty}^{\infty} I(x)dx = 1$.

Next, the autocorrelation function of the standardized fluorescence fluctuations, $\Delta F / \langle F \rangle_t$, is computed by

$$G(\tau) = \left\langle \frac{\Delta F(t)}{\langle F \rangle_t} \frac{\Delta F(t+\tau)}{\langle F \rangle_t} \right\rangle_t$$

$$\begin{aligned} \langle F \rangle_t^2 G(\tau) &= \lim_{T \to \infty} \int_0^T [\Delta F(t) \Delta F(t+\tau)] dt \\ &= \langle \Delta F(t) \Delta F(t+\tau) \rangle_t \\ &= \left\langle \left(Q \int_{-\infty}^{\infty} I(x) \Delta n(x,t) dx \right) \left(Q \int_{-\infty}^{\infty} I(x) \Delta n(x,t+\tau) dx \right) \right\rangle_t \\ &= Q^2 \int_{-\infty}^{\infty} \int_{-\infty}^{\infty} I(x) I(y) \langle \Delta n(x,t) \Delta n(x,t+\tau) \rangle_t \, dxdy \end{aligned} \tag{55}$$

where we used Eq. (50).

Notice that $n(x,t)$ satisfies the diffusion equation (Eq. (19)). Therefore, $\Delta n(x,t+\tau)$ also satisfies a diffusion equation in τ and x with initial time at t ($\tau = 0$):

$$\begin{cases} \dfrac{\partial}{\partial t} \Delta n(x,t+\tau) = D \dfrac{\partial^2}{\partial x^2} \Delta n(x,t+\tau) \\ \Delta n(x,t) = n(x,t) - \langle n \rangle_t \end{cases} \tag{56}$$

Consequently, the solution $\Delta n(x,t+\tau)$ is found as (Eq. (21))

$$\begin{aligned} \Delta n(x,t+\tau) &= \int_{-\infty}^{\infty} \Delta n(\overline{x},t) \frac{1}{\sqrt{4\pi D\tau}} \exp\left(-\frac{(x-\overline{x})^2}{4D\tau} \right) d\overline{x} \\ &= \int_{-\infty}^{\infty} \Delta n(\overline{x},t) \Phi(\tau, x-\overline{x}) d\overline{x}. \end{aligned} \tag{57}$$

Next, we use the ergodicity of a diffusion process to derive some essential properties of the double integral. Because diffusion is an ergodic process, the time average can be replaced by the ensemble average.

$$\begin{aligned} \langle \Delta n(x,t) \Delta n(y,t+\tau) \rangle_t &= \left\langle \Delta n(x,t) \int_{-\infty}^{\infty} \Delta n(\overline{x},t) \Phi(\tau, y-\overline{x}) d\overline{x} \right\rangle_t \\ &= \int_{-\infty}^{\infty} \langle \Delta n(x,t) \Delta n(\overline{x},t) \rangle_t \Phi(\tau, y-\overline{x}) d\overline{x} \\ &= \int_{-\infty}^{\infty} \langle \Delta n(x,0) \Delta n(\overline{x},0) \rangle \Phi(\tau, y-\overline{x}) d\overline{x} \\ &= \int_{-\infty}^{\infty} \left\langle (\Delta n(x,0))^2 \right\rangle \delta(x-\overline{x}) \Phi(\tau, y-\overline{x}) d\overline{x} \\ &= \sigma^2 \Phi(\tau, y-x) \end{aligned} \tag{58}$$

where $\sigma^2 = \left\langle (\Delta n(x,0))^2 \right\rangle$ is the variance of $n(x,0)$, or the mean square fluctuations of the fluorescence molecules, and $\delta(x-\overline{x})$ is the Dirac delta function defined as Eq. (18). In Eq. (58), the stationary and ergodic assumptions were used in the third line to convert the time average to the spatial average at $t = 0$.

$$\langle \Delta n(x,0)\Delta n(\overline{x},0)\rangle = \begin{cases} \left\langle (\Delta n(x,0))^2 \right\rangle \text{if } x = \overline{x} \\ \langle \Delta n(x,0)\rangle\langle \Delta n(\overline{x},0)\rangle \text{if } x \neq \overline{x} \end{cases}$$
$$= \begin{cases} \sigma^2 \text{if } x = \overline{x} \\ 0 \quad\quad \text{if } x \neq \overline{x} \end{cases}$$
$$= \sigma^2\delta(x-\overline{x}) \tag{59}$$

By plugging Eq. (58) back into Eq. (55)

$$\langle F\rangle_t^2 G(\tau) = Q^2\int_{-\infty}^{\infty}\int_{-\infty}^{\infty} I(x)I(y)\langle \Delta n(x,t)\Delta n(x,t+\tau)\rangle_t\, dxdy$$
$$= Q^2\int_{-\infty}^{\infty}\int_{-\infty}^{\infty} I(x)I(y)\sigma^2\Phi(\tau,y-x)dxdy \tag{60}$$
$$= Q^2\sigma^2\int_{-\infty}^{\infty}\int_{-\infty}^{\infty}\left[\frac{2}{\pi\omega^2}\exp\left(-\frac{2(x^2+y^2)}{\omega^2}\right)\right]\left[\frac{1}{\sqrt{4\pi D\tau}}\exp\left(-\frac{(x-y)^2}{4D\tau}\right)\right]dxdy$$
$$= Q^2\sigma^2\left(\frac{2}{\pi\omega^2}\cdot\frac{1}{\sqrt{4\pi D\tau}}\right)\int_{-\infty}^{\infty}\int_{-\infty}^{\infty}\exp\left(-\frac{2(x^2+y^2)}{\omega^2}-\frac{(x-y)^2}{4D\tau}\right)dxdy$$

If we substitute $y = x+\sqrt{4D\tau}\eta$ $(dy = \sqrt{4D\tau}d\eta)$, then

$$\langle F\rangle_t^2 G(\tau) = \frac{2Q^2\sigma^2}{\pi\omega^2\sqrt{4\pi Dt}}\int_{-\infty}^{\infty}\int_{-\infty}^{\infty}\exp\left(-\frac{2\left(x^2+\left(x+\sqrt{4D\tau}\eta\right)^2\right)}{\omega^2}-\eta^2\right)\sqrt{4D\tau}d\eta dy$$
$$= \frac{2Q^2\sigma^2}{\pi\sqrt{\pi\omega^2}}\int_{-\infty}^{\infty}\int_{-\infty}^{\infty}\exp\left(-\frac{2\left(x^2+\left(x+\sqrt{4D\tau}\eta\right)^2\right)+\omega^2\eta^2}{\omega^2}\right)d\eta dx \tag{61}$$
$$= \frac{2Q^2\sigma^2}{\pi\sqrt{\pi\omega^2}}\int_{-\infty}^{\infty}\int_{-\infty}^{\infty}\exp\left(-\frac{2\left(x^2+\left(x+\sqrt{4D\tau}\eta\right)^2\right)+\omega^2\eta^2}{\omega^2}\right)d\eta dx$$

where we used the fact

$$\int_{-\infty}^{\infty}\int_{-\infty}^{\infty}\exp\left(-\frac{2\left(x^2+\left(x+\sqrt{4D\tau}\eta\right)^2\right)}{\omega^2}-\eta^2\right)d\eta dy$$
$$= \int_{-\infty}^{\infty}\int_{-\infty}^{\infty}\exp\left(-\frac{4x^2+4\sqrt{4D\tau}x\eta+2(4D\tau)\eta^2+\omega^2\eta^2}{\omega^2}\right)d\eta dy \tag{62}$$
$$= \int_{-\infty}^{\infty}\int_{-\infty}^{\infty}\exp\left(-\frac{4\left(x^2+2\sqrt{D\tau}x\eta+\left[\sqrt{D\tau}\eta\right]^2\right)+\left[4D\tau+\omega^2\right]\eta^2}{\omega^2}\right)d\eta dy$$
$$= \int_{-\infty}^{\infty}\int_{-\infty}^{\infty}\exp\left(-\frac{4(x+\sqrt{D\tau}\eta)^2+\left[4D\tau+\omega^2\right]\eta^2}{\omega^2}\right)d\eta dy$$

Now, we can evaluate the inner integral in Eq. (46) using a substitution $z = x + \sqrt{D\tau}\eta$ for x

$$\int_{-\infty}^{\infty} \exp\left(-\frac{4(x+\sqrt{D\tau}\eta)^2}{\omega^2}\right)dx = \int_{-\infty}^{\infty} \exp\left(-\frac{4z^2}{\omega^2}\right)dz \tag{63}$$
$$= \frac{\omega\sqrt{\pi}}{2}$$

where we used Eq. (11). Back to Eq. (61)

$$\langle F\rangle_t^2 G(\tau) = \frac{2Q^2\sigma^2}{\pi\sqrt{\pi}\omega^2}\int_{-\infty}^{\infty}\left\{\int_{-\infty}^{\infty} \exp\left(-\frac{4(x+\sqrt{D\tau}\eta)^2}{\omega^2}\right)dx\right\} \exp\left(-\frac{(4D\tau+\omega^2)\eta^2}{\omega^2}\right)d\eta$$
$$= \frac{Q^2\sigma^2}{\omega\pi}\int_{-\infty}^{\infty} \exp\left(-\frac{(4D\tau+\omega^2)\eta^2}{\omega^2}\right)d\eta \tag{64}$$
$$= \frac{Q^2\sigma^2}{\omega\pi}\sqrt{\frac{\omega^2\pi}{4D\tau+\omega^2}}$$
$$= \frac{Q^2\sigma^2}{\omega\sqrt{\pi}}\frac{1}{\sqrt{1+\tau/\tau\_D}}$$

by the error function integration (Eq. (11)), where $\tau_D = \omega^2/(4D)$, which is a diffusion time.

If fluorescence molecules undergo Brownian motion, then the number of photons in a confocal volume changes in time due to random movements of fluorescence molecules in and out of the confocal volume. In FCS analysis, the number of photons (or fluorescence molecules) from a confocal volume at any moment t is assumed to follow a Poisson distribution, in which the probability for k fluorescence molecules (or photons) to be found in the confocal volume is

$$\mathbb{P}(F(t) = k) = \frac{e^{-\lambda}\lambda^k}{k!} \tag{65}$$

where $\lambda = \langle F\rangle_t$ is the average number of fluorescence molecules (or photons) in the confocal volume. This assumption is reasonable for a diffusion process since the arrival process of infinitely many identical independent diffusion processes was shown to be a Poisson process [27]. Importantly, the mean (or expectation) and variance of a Poisson random variable are known to be equal

$$\begin{cases} \mathbb{E}(F(t)) = \sum_{k=0}^{\infty} k\mathbb{P}\{F(t) = k\} = \langle F\rangle_t \\ \sigma^2 = \mathbb{E}\left(\left|F(t) - \langle F\rangle_t\right|^2\right) = \langle F\rangle_t \end{cases}. \tag{66}$$

Since we assumed that $F(t)$ follows the Poisson statistics that has equal variance and mean

$$\langle F\rangle_t^2 G(0) = \langle \Delta F(t)\Delta F(t+0)\rangle_t \tag{67}$$
$$= \left\langle (\Delta F(t))^2 \right\rangle_t$$

$$= \sigma^2$$
$$= \langle F \rangle_t$$

by Eq. (66). On the other hand, by Eq. (64)

$$G(0) = \frac{Q^2 \sigma^2}{\omega \sqrt{\pi}} \tag{68}$$

which indicates that

$$\frac{1}{\langle F \rangle_t} = \frac{Q^2 \sigma^2}{\omega \sqrt{\pi}} \tag{69}$$

By replacing the bulk parameters in Eq. (47) with $1/\langle F \rangle_t$

$$G(\tau) = \frac{1}{\langle F \rangle_t} \frac{1}{\sqrt{1 + \tau/\tau_D}} \tag{70}$$

As we saw, with a Poisson distribution assumption on $F(t)$, we can readily determine the average density of fluorescence molecules as well as the average number of fluorescence molecules in the confocal volume. Similar to FRAP equations, FCS equations in higher spatial dimensions can be found by similar calculations

$$G(\tau) = \frac{1}{\langle F \rangle_t} \frac{1}{\left(1 + \tau/\tau_{D_{xy}}\right)} (2D) \tag{71}$$
$$G(\tau) = \frac{1}{\langle F \rangle_t} \frac{1}{\left(1 + \tau/\tau_{Dxy}\right) \sqrt{1 + \tau/\tau_{D_z}}} (3D)$$

where $\tau_{D_{xy}} = \omega_{xy}^2/(4D)$ and $\tau_{D_z} = \omega_z^2/(4D)$ with ω_{xy} = the half-width of the beam at e^{-2} in $x/y-$ direction and ω_z = the half-width of the beam at e^{-2} in $z-$ direction.

5. Conclusion

Diffusion plays a crucial role within biological systems in many different temporal and spatial scales from various perspectives. It is a dominant way for biological organisms to transport multiple molecules to desirable locations for cell signaling. However, to quantify the molecular diffusion, especially in live cells, is still challenging although a couple of tools are available, including fluorescence recovery after photobleaching and fluorescence correlation spectroscopy. Although FRAP and FCS were originally developed to study biological diffusion processes, they are now being applied not only to a diffusion process but also to a broad range of biochemical processes, including binding kinetics and anomalous diffusion. Since the derivation of FRAP and FCS equations for many biochemical processes shares many common steps with the diffusion FRAP and FCS equations, it is essential to understand the mathematical theory behind the diffusion FRAP /FCS equation [18, 22, 25, 28–32]. In this study, we provide a simple and straightforward derivation of FRAP/FCS equation for free diffusion based on calculus-level mathematics, so that FRAP/FCS equations and its applications are accessible to a broad audience. Although the applications of these FRAP and FCS equations to cell membrane biophysics from experimental perspectives can be a very important topic, it is beyond the scope of this chapter and therefore will not be covered here. These topics are well documented in various references, and interested readers are referred to [20, 31, 33], and references therein. We hope that this tutorial is understandable as well as gives readers a solid theoretical foundation for FRAP and FCS, bridging the gap between experimental and theoretical aspects of FRAP and FCS.

Author details

Minchul Kang
Department of Mathematics, Texas A&M University-Commerce, Commerce, TX, USA

*Address all correspondence to: minchul.kang@tamuc.edu

References

[1] Murray J. Mathematical Biology: I. An Introduction (Interdisciplinary Applied Mathematics). 3rd ed. New York: Springer; 2007

[2] Murray J. Mathematical Biology II: Spatial Models and Biomedical Applications (Interdisciplinary Applied Mathematics). 3rd ed. New York: Springer; 2011

[3] Okubo A, Levin S. Diffusion and Ecological Problems, Modern Perspectives. 2nd ed. New York: Springer; 2002

[4] Ritchie K, Spector J. Single molecule studies of molecular diffusion in cellular membranes: Determining membrane structure. Biopolymers. 2007;**87**(2–3): 95-101

[5] Skaug MJ, Faller R, Longo ML. Correlating anomalous diffusion with lipid bilayer membrane structure using single molecule tracking and atomic force microscopy. The Journal of Chemical Physics. 2011;**134**(21):215101

[6] Das BB, Park SH, Opella SJ. Membrane protein structure from rotational diffusion. Biochimica et Biophysica Acta. 2015;**1848**(1 Pt B): 229-245. DOI: 10.1016/j.bbamem. 2014.04.002

[7] Axelrod D, Koppel DE, Schlessinger J, Elson E, Webb WW. Mobility measurement by analysis of fluorescence photobleaching recovery kinetics. Biophysical Journal. 1976; **16**(9):1055-1069

[8] Elson EL, Magde D. Fluorescence correlation spectroscopy. I. Conceptual basis and theory. Biopolymers. 1974;**13**: 1-27

[9] Crank J. The Mathematics of Diffusion (Oxford Science Publications). 2nd ed. USA: Oxford University Press; 1980

[10] Carslaw HS, Jaeger JC. Conduction of Heat in Solids (Oxford Science Publications). 2nd ed. USA: Oxford University Press; 1986

[11] Fick A. On liquid diffusion. Annalen Der Physik und Chemie Ergänzung. 1885;**94**:59

[12] Einstein A. Über die von der molekularkinetischen Theorie der Wärme geforderte Bewegung von in ruhenden Flüssigkeiten suspendierten Teilchen. Annalen der Physik. 1905; **322**(8):549-560

[13] von Smoluchowski M. Zur kinetischen theorie der brownschen molekularbewegung und der suspensionen. Annalen der Physik. 1906;**326**(14):756-780

[14] Bachelier L. Théorie de la spéculation. Annales Scientifiques de l'École Normale Supérieure. 1900;**17**(3): 21-86

[15] Carrero G, Crawford E, Hendzel MJ, de Vries G. Characterizing fluorescence recovery curves for nuclear proteins undergoing binding events. Bulletin of Mathematical Biology. 2004;**66**: 1515-1545

[16] Sprague BL, Pego RL, Stavreva DA, McNally JG. Analysis of binding reactions by fluorescence recovery after photobleaching. Biophysical Journal. 2004;**86**(6):3473-3495

[17] Houtsmuller AB. Fluorescence recovery after photobleaching: Application to nuclear proteins. Advances in Biochemical Engineering/ Biotechnology. 2005;**95**:177-199

[18] Kang M, Day CA, DiBenedetto E, Kenworthy AK. A quantitative approach to analyze binding diffusion kinetics by confocal FRAP. Biophysical Journal. 2010;**99**(9):2737-2747

[19] Kang M, Kenworthy A. Complex applications of simple FRAP on membranes. In: Faller R et al., editors. Biomembrane Frontiers. New York, USA: Humana Press; 2009

[20] Day CA, Kraft LJ, Kang M, Kenworthy AK. Analysis of protein and lipid dynamics using confocal fluorescence recovery after photobleaching (FRAP). Current Protocols in Cytometry. 2012. Chapter 2; Unit 2.19

[21] Kang M, Andreani M, Kenworthy AK. Normalizations, scaling, and photo-fading corrections for FRAP data analysis and their implications. PLoS One. 2015;**10**(5):e0127966

[22] Kang M, Day CA, Drake K, Kenworthy AK, DiBenedetto E. A generalization of theory for two-dimensional fluorescence recovery after photobleaching applicable to confocal laser scanning microscopes. Biophysical Journal. 2009;**97**(5):1501-1511

[23] Elson EL. Fluorescence correlation spectroscopy: Past, present, future. Biophysical Journal. 2011;**101**(12): 2855-2870

[24] Medina MA, Schwille P. Fluorescence correlation spectroscopy for the detection and study of single molecules in biology. BioEssays. 2002; **24**(8):758-764

[25] Lee K, Astudillo N, Kang M. A simple derivation of diffusion fluorescence correlation spectroscopy equations. Journal of Fluorescence. 2020. DOI: 10.1007/s10895-019-02476-z [Epub ahead of print]

[26] Derman C. Ergodic property of the Brownian motion process. Proceedings of the National Academy of Sciences of the United States of America. 1954; **40**(12):1155-1158

[27] Nadler B, Schuss Z. The stationary arrival process of independent diffusers from a continuum to an absorbing boundary is Poissonian. SIAM Journal on Applied Mathematics. 2001;**62**(2): 433-447

[28] Kang M, Kenworthy A. A closed-form analytic expression for FRAP formula for the binding diffusion model. Biophysical Journal. 2008;**95**(2):L13-L15

[29] Kang M, DiBenedetto E, Kenworthy A. Proposed correction to Feder's anomalous diffusion FRAP equations. Biophysical Journal. 2011; **100**(3):791-792

[30] Kang M, Day CA, Drake K, Kenworthy A, DiBenedetto E. Simplified equation to extract diffusion coefficients from confocal FRAP data. Traffic. 2012;**13**(12):1589-1600

[31] Kang M, Andreani M, Kenworthy A. Validation of normalizations, scaling, and photofading corrections for FRAP data analysis. PLoS One. 2015;**10**(5): e0127966

[32] Kang M, Day CA, Drake K, Kenworthy A. A novel computational framework for $D(t)$ from fluorescence recovery after photobleaching data reveals various anomalous diffusion types in live cell membranes. Traffic. 2019;**20**(11):867-880

[33] Chiantia S, Ries J, Schwille P. Fluorescence correlation spectroscopy in membrane structure elucidation. Biochimica et Biophysica Acta. 2009; **1788**(1):225-233

Chapter 19

Special Cuvettes for Spectrofluorimeters

Nikolai Vekshin

Abstract

Special cuvettes that dramatically increase the fluorescence intensity of diluted solutions and suspensions of isolated biomembranes and living cells are described. (1) Mirror cuvettes have a reflecting aluminum or silver layer, putting on their three external sides, having a slot-hole window at frontal side center. (2) Mirror microcuvettes have two reflecting sides. (3) Total internal reflection cuvettes have lateral reflecting side and two triangular prisms, attached outside; the cuvette works effectively with a parallel excitation light beam. (4) In cuvette with diagonal quartz plate, very small quantity of a sample is attached in the form of a thin layer. (5) Mirror cuvette with diagonal plate. All described cuvettes can be applied for measurements of spectra excitation and emission, excited-state lifetime, and polarization degree. Also, they can be used for photo-bleaching of biological chromophores.

Keywords: cuvettes, fluorescence, spectroscopy, excitation, emission, lifetime, polarization degree, photo-bleaching

1. Introduction

Fluorescence spectroscopy is one of the most high-sensitive methods, which allow to detect very low concentrations of substances (μM, nM, and even pM) and to distinguish one substance from another [1–5]. Fluorescence spectroscopy is used for studying the functioning of living cells [6–8]. It is intended also for detecting various dyes and substances in cells and membranes [6–9].

Fluorescence spectroscopy gives valuable information on spatial organization and behavior of biological structures and about appearance of intermolecular complexes, for example, DNA with proteins, enzymes with substrates, etc. It is applied in studies of conformation properties, mobility, and other properties of macromolecules, biological membranes, and living cells [6–9].

It is an extremely informative method, since it allows us to receive the data on the distances, orientations, and intermolecular interactions. An advantage of fluorescent spectroscopy is the absence of damage of a sample in the course of research. Also, it is possible to work with native biological preparations.

The measured intensity of fluorescence of a biological sample is determined by its optical density, lifetime of the excited state, fluorescence quantum yield, and effective collection of emission beam.

Fluorescence spectra can be detected by special techniques—spectrofluorimeters. In such devices, a scanning of excitation wavelength by the first monochromator (placed before a cuvette with a testified sample) leads to changes in the intensity of fluorescence according to the form of the absorption spectrum; that is, form and position of the excitation spectrum should coincide with the absorption spectrum. At fixed excitation wavelength, a scanning of the second monochromator (placed after a cuvette with a sample) gives the emission spectrum (named also as spectrum of fluorescence or emission).

Various fluorescence techniques are widely used to increase the sensitivity of the method [1–6]. One of the approaches to enhance the sensitivity of fluorescence analysis is the increase of the length of optical path of an exciting light in a sample. Usually, it is reached by consecutive reflections between concave mirrors, located in a common spectrofluorimeter near the cuvette [5] filled with a studied solution or suspension. However, only two passes of stimulating light in such systems take place, and also losses of light at media borders exist. Therefore, the gain (amplification factor) G is <3. Besides, parasitic reflections at borders and mirrors lead to penetration of exciting light into registration channel that leads to impossibility of correct measurements of weak fluorescence.

That is why special cuvettes which dramatically increase the fluorescence intensity of diluted solutions and suspensions of isolated biomembranes or living cells were invented [7, 8]. They will be described below.

2. Methods and results

2.1 Multipass cuvettes

To overcome the mentioned specified difficulties, the multipass cuvettes (**Figure 1**), allowing to raise G up to 4–10 times and to remove parasitic reflections, can be used [7, 8]. Such cuvettes allow to lower the concentration of substances and to receive an intensive fluorescence signal.

Mirror cuvettes are intended for measuring the fluorescence of weakly absorbing solutions and can be used to register the excitation and emission spectra in UV and visible region and lifetimes. The cuvettes provide for many-fold increase of fluorescence intensity due to multiple passage of exciting light through the solution being tested and due to additional fluorescence collected.

The stimulating light in multipass cuvettes is almost wholly absorbed by a solution or suspension, and, thus, there are almost no parasitic reflections at cuvette sides (as against usual cuvettes with external concave mirrors, all reflections here are useful). Penetration of stimulating light into registration channel, therefore, sharply decreases. It is especially essential at low intensities of emission

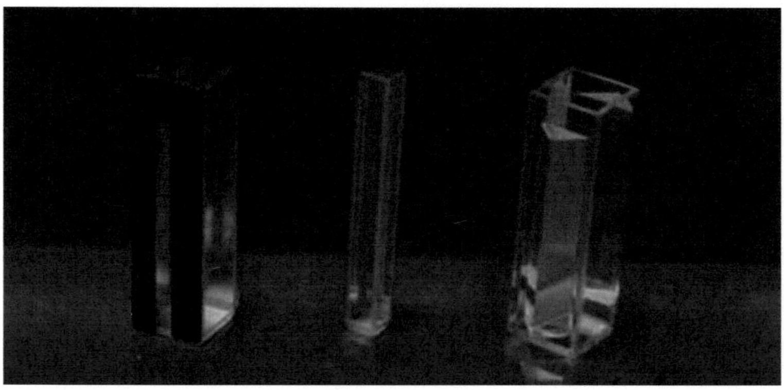

Figure 1.
Mirror cuvette, mirror microcuvette, and TIR cuvette.

of light-scattered suspensions of cells or membranes. Amplification of fluorescence intensity from a sample in multipass cuvettes allows also to raise the accuracy of measurements of fluorescence polarization degree and lifetime of excited state.

Mirror cuvettes represent themselves as a quartz 1-cm cell with a reflecting aluminum or silver layer, putting on their three external sides (outside protection is enamel), having a slot-hole window at frontal side center—for entrance of stimulating light (**Figure 2**). Mirror cuvettes may be applied in serial spectrofluorimeters without changes in their design. The cuvette can be used for recording emission or excitation spectra and for determining the excited-state lifetime of weakly absorbing solutions and suspensions.

Mirror microcuvettes are a quartz 0.4-cm cuvette with a reflecting layer, putting on their two external sides. This cuvette allows work with small volumes (0.15–0.2 ml) of solutions or suspensions. Mirror microcuvettes may be applied in spectrofluorimeters, using a special holder, which has a standard external size of 12.5 × 12.5 × 40 mm.

Total internal reflection (TIR) cuvettes contain a lateral reflecting side and two triangular prisms elements, attached outside to two transparent sides (**Figure 2**). A parallel excited light beam enters sideways in frontal side and, passing through solution (or diluted suspension) and back prism, undergoes total internal reflections at quartz/air borders and comes back, making 3–5 times passes through solution or suspension, having very low optical density (as rule, not more 0.01). The TIR cuvette works effectively only with a parallel excitation light beam, especially at laser excitation. It requires application a special holder, displacing the cuvette relating to exciting beam to the best position.

The gain (G) of fluorescence intensity in multipass cuvettes (in comparison with standard single-pass cuvettes) is determined by the expression:

$$G = \left(1 + \rho T + \rho^2 T^2 + \ldots \rho^n T^n\right)(1 + \rho). \quad (1)$$

Here ρ is the reflection factor of a mirror or prism, T is the light transmission of a solution or suspension at the excitation band at one beam pass, and n is the number of passes. The first member of the equation describes amplification of fluorescence intensity due to repeated passage of stimulating light, and the second one takes into account additional collection of emission due to the lateral reflecting wall. The more T (i.e., the less optical density), the bigger G, aspiring, however, to

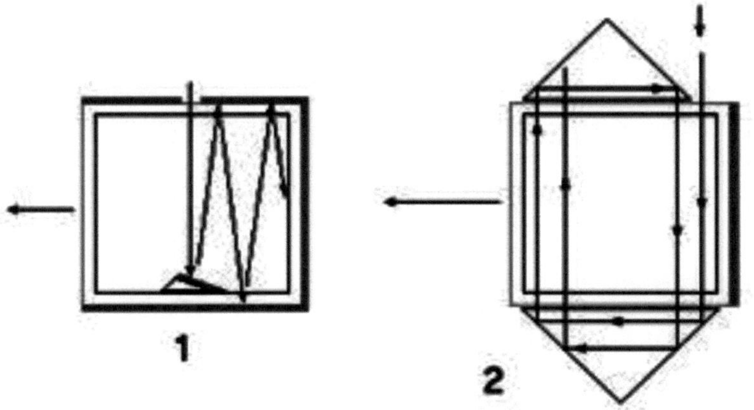

Figure 2.
Scheme of Mirror cuvette (1) and TIR cuvette (2); top view. Arrows show the direction of light beams. Mirror layers are shaded.

some limit. Really, the G value for the 1-cm mirror cuvette with aluminum covering reaches 4–7 (at UV region) and with silver covering reaches 6–9 (at visible region). For TIR cuvette, G reaches 10, if excitation is done by laser beam.

2.2 Spectral measurements

Emission spectra of 10-μM solution of ANS dye in ethanol in standard and mirror cuvettes are given in **Figure 3**. The intensity of fluorescence in mirror cuvettes is sometimes higher than that in the standard one. The form of the emission spectrum does not change.

Table 1 summarizes the calculated and experimental gain G, obtained for various samples (diluted solutions and suspensions) in different cuvettes.

2.3 Excited-state lifetime measurements

If an object is excited by short-time light flare, which has a duration less than the time of emission transition of molecules from S_1 into S_0, it is possible, using a stroboscopic multichannel detector or phase detector, to register the decay emission kinetics in nanosecond timescale.

In the presence of only one kind of radiating molecules, decay is described as [1–6]:

$$dN\ (t) = -A\ N\ (t)\ dt. \tag{2}$$

Here dN is the loss of the number of excited molecules during time t, A is Einstein's factor for spontaneous transitions from S_1 to S_0, and N is the number of molecules at S_1 level. The integration gives:

$$N\ (t) = -N\ exp\ (-A\ t). \tag{3}$$

Decay has exponential character. The decrease of decay intensity in *exp* times (~ in 2.7 times) is named as *lifetime* (τ) of the excited state. The lifetime is a quantitative characteristic of the average duration of existence of molecules in the excited state. For instance, perylene, consisting of five condensed aromatic rings, has τ ~ 5 ns, and pyrene, consisting of the same four aromatic rings, has τ more than 300 ns. Molecules of one kind in different conditions have various lifetimes. For example, NADH in water has τ = 0.5 ns, but in the mitochondrial NADH dehydrogenase 2 ns [9]. Generally speaking, τ does not depend on concentration of fluorescent molecules, if they do not form own aggregates.

Since the fluorescence intensity of a sample in multipass cuvette strongly increases, the excited-state lifetime measurements become more accurate.

Figure 4 demonstrates the lifetime distributions of tryptophan fluorescence of cod parvalbumin solution measured in a standard cuvette and in a multipass one. Cod parvalbumin was dissolved in a 25 mM of Tris-HCl and 1 mM of $CaCl_2$ with pH of 8.2; the absorbance at 295 nm was less than 0.1; excitation wavelength was 295 nm, and emission was detected at 320 nm; and monochromator slits were 5 nm. The UV excitation was made with the Orsay synchrotron (France).

Lifetime distributions were measured through correlation fluorescence spectroscopy. The 0.03 ns band, obtained with standard cuvettes (**Figure 4**, left), is an artifact, caused by a big noise of small fluorescence signal and also by some penetration of the exciting light into registering channel. It is prevented if multipass cuvettes are used (**Figure 4**, right).

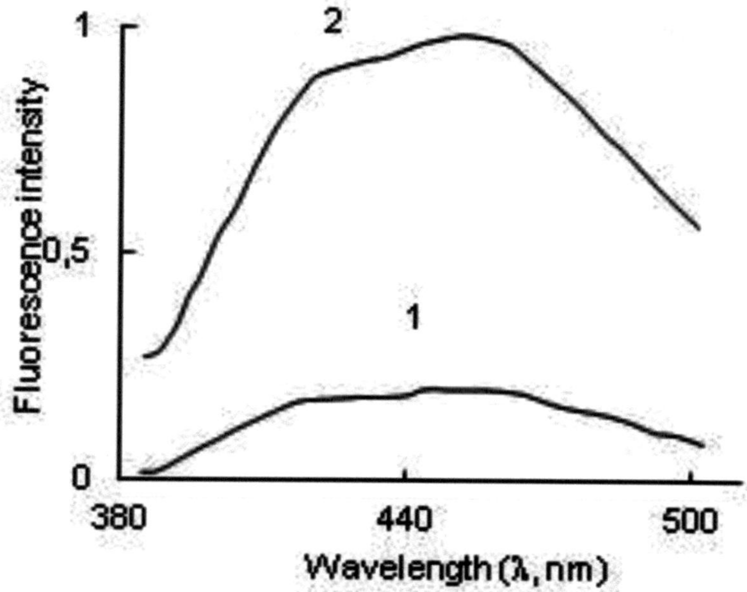

Figure 3.
Fluorescence spectrum of ANS (in ethanol) in standard cuvette (1) and in mirror cuvette (2). Excitation was 360 nm; slits were 5 nm. Spectrofluorimeter was "Perkin Elmer MPF-44B."

Cuvette	$G_{theor.}$	$G_{exper.}$
Standard 1-cm cuvette	1	1
Cuvette and two outer concave mirrors	3.3	3
Mirror cuvette with aluminum coating	9	6
Mirror cuvette with silver coating (excepting UV region)	17	8
Total internal reflection cuvette	20	10

Table 1.
Gain of fluorescence intensity, obtained with various cuvettes.

In the case of subnanosecond lifetimes, the measurements should be made using multipass cuvettes in all channels—for studied sample and for reference blank sample.

Table 2 presents the lifetime components of fluorescence of ribonuclease T_1, parvalbumin, and phospholipase A_2. Excitation was 295 nm (slits were 5 nm). RNAase T1 was dissolved in the 100 mM acetic buffer (pH 5.5). The calcium form of parvalbumin was used in the 25 mM Tris-HCl buffer with 1 mM $CaCl_2$ (pH 8.2). Pork phospholipase A_2 was in the 100 mM acetic buffer (pH 5.8) or in 90% glycerol. Detection of emission at 310 and 320 nm was done in mirror cuvettes to increase the intensity. Emission at other wavelengths was measured in standard 1-cm quartz cuvettes.

Multipass cuvettes are especially useful in phase modulation measurements of lifetimes, when, because of narrow slits of entrance monochromator and modulator, the intensity of exciting beam is too small. These cuvettes allow to raise the accuracy of phase modulation measurements. For example, a solution of ANS in

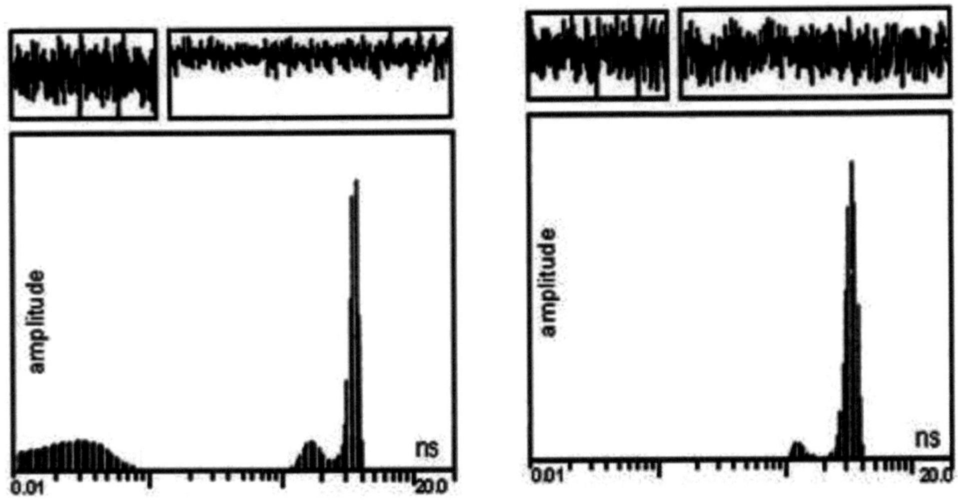

Figure 4.
Lifetime distributions of tryptophan fluorescence of cod parvalbumin in a standard 1-cm quartz cuvette (left figure) and in a multipass one (right figure).

Protein	λ_{em} (nm)	τ_1 (ns)	τ_2 (ns)	τ_3 (ns)	a_1	a_2	a_3	τ (ns)
Ribonuclease	310	—	—	4	—	—	1	4
	320	—	—	4	—	—	1	4
	340	—	—	4	—	—	1	4
	375	—	—	4	—	—	1	4
Parvalbumin	310	—	1.1	3.4	—	0.15	0.85	2.7
	320	—	1.3	3.3	—	0.07	0.93	3.2
	340	—	1.7	3.3	—	0.05	0.95	3.3
	375	—	—	3.4	—	—	1	3.4
Phospholipase	320	0.6	2.1	5.3	0.71	0.24	0.05	1.2
	350	0.7	2.7	7.2	0.63	0.32	0.05	1.7
	385	0.7	2.7	6.5	0.61	0.33	0.06	1.7
Phospholipase in glycerol	320	0.4	2.1	5.6	0.4	0.43	0.17	2
	350	0.8	2.7	6.3	0.19	0.54	0.27	3.3
	385	—	—	4	—	—	1	4

Table 2.
Lifetimes and amplitudes of tryptophan emission of ribonuclease T_1, parvalbumin, and phospholipase A_2.

ethanol in standard cuvettes gives fluctuations in lifetime > ± 0.5 ns (lifetime was 7.7 ns; it was measured by "SLM-4800" at frequency of 30 MHz; excitation was 370 nm; excitation slits were 1 nm), but in multipass cuvettes (at the same conditions), fluctuations were < ±0.1 ns.

Multipassing does not bring appreciable contribution in measurements of nanosecond lifetimes, since several passes of the beam lead to artifact delay only about of 0.1 ns. In the case of subnanosecond lifetimes, it is necessary to do measurements with differential mode, using multipass cuvettes in all channels and entering an amendment for artifact delay.

3. Polarization degree of fluorescence

Molecules of many organic substances consist of a flat chromophore and various groups attached to it. Their structures are optically anisotropic. Therefore, emission of motionless molecules is polarized (even at the nonpolarized excitation). The greatest contribution to the total emission is brought by those molecules, in which chromophores are located in a perpendicularly position to the excitation light flow. It means that even at a chaotic arrangement of molecules, for example, in suspensions of membranes or cells, the emission is partially polarized. The intensity of fluorescence in some direction can be presented as a sum of two light flows, F_{II} and $F_{\perp}$, polarized at right angle.

If to illuminate an object by linearly polarized light, the fluorescence becomes even more polarized. For example, it is possible to direct the light beam from a lamp on a sample through the polarizer (polarizing prism) and to put the analyzer (the same second element) sideways of the sample, passing through which the emission will get on the photomultiplier. The intensity of emission, measured at identical orientations of polarizer and analyzer, is named as "parallel" component. The intensity, registered at orientations, crossed under 90 degrees between the polarizer and the analyzer, is the "perpendicular" component. The degree of polarization at excitation by linearly polarized light is expressed as [6]:

$$P = (F\| - F\perp)/(F\| + F\perp). \quad (4)$$

Fluorescence polarization degree (P) of various samples in standard cuvettes and mirror cuvettes is presented in **Table 3**. A very small decrease of P in mirror cuvettes is due to depolarization of exciting light at reflections on mirror sides. This small artifact can be easily eliminated as an addition of constant value 0.005.

3.1 Photo-bleaching

Multipass cuvettes may be useful not only for fluorescence analysis but also for many-fold intensification of photo-bleaching or laser flash photolysis of chromophores or dyes in solutions or suspensions.

The energy of UV quantum, absorbed by a cell or biomembrane, is spent mainly for fluctuations, so - on instant strong local heating [7, 8].

If such local heating is too high, it can result to denaturation of proteins and their aggregation. For instance, the influence of UV irradiation on intensity of tryptophan fluorescence of α-crystalline (it is a protein from bovine eye lens) is shown in **Figure 5**. A solution of the protein, hermetically closed and thermostated at 10°C (here, thermo-aggregation is possible to be neglected) in mirror microcuvette, was continuously irradiated for 120 min with UV light of a 450-W xenon lamp through holographic monochromator, positioned at 280 nm with slits of 16 nm. The intensity of tryptophan fluorescence (excitation was 295 nm; emission was 340 nm; emission slits were 4 nm) was quickly measured during a course of irradiation. The observed decrease in fluorescence of α-crystalline, caused by photoinduced denaturation, is due mainly to the increase of light-scattering during aggregation. Light-scattering can be measured from optical density at 310 nm of the sample in a standard compartment of spectrophotometer or, by the other way, from intensity of scattering light, detected from the sample by the photomultiplier of a spectrofluorimeter under a right angle. In the second case, both monochromators have to be established at an identical wavelength of 310 nm (at slits = 1 nm).

Sample (solution or suspension)	P in standard cuvette	P in mirror cuvette
Tryptophan residues of sarcoplasmic reticulum (ex. 295 nm, em. 320 nm)	0.36	0.355
Tryptophan residues of mitochondrial suspension	0.29	0.285
ANS in bovine albumin	0.28	0.275
7-Aminoactinomycin in DNA	0.31	0.305

Table 3.
Fluorescence polarization degree of various samples.

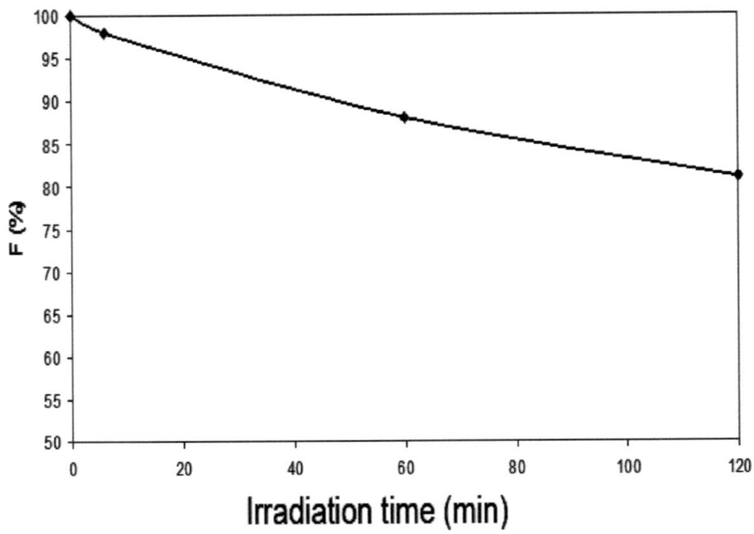

Figure 5.
Influence of UV radiation on the intensity of tryptophan fluorescence of α-crystalline (0.7 mg/ml) during photoaggregation of the protein at 10°C. irradiation was at ~280 nm in mirror microcuvette.

Fluorescent kinetics of photo-bleaching of flavin mononucleotide (FMN) in aqueous solution and in diluted suspension of proto-mitochondria [9] after illumination by blue light of the mercury lamp SVD-120A in a mirror cuvette is given in **Figure 6**.

Photo-bleaching is induced by photo-destruction of FMN, namely, by oxidation of its triplet state by molecular oxygen. The absorption band of FMN, detected at 450 nm, was "burns out" [7, 9].

3.2 Cuvette with diagonal plate

Cuvette with transparent diagonal quartz plate can be applied for many-fold increase of fluorescence signal and for many-time using of a single sample, adsorbed on the plate. A schematic diagram of a cuvette with the diagonal transparent quartz plate is shown in **Figure 7**. The sample in the form of a smear or film is located on the side of the plate facing the exciting light.

A sample of cells or isolated organelles is attached on the surface of a transparent quartz plate, which is placed inside a standard cuvette at a right angle to the exciting light. A sample has to be as a smear or film. The cuvette should be filled by water (if cells cannot be desorbed from the plate) or hydrophobic solvent like hexane (if cells are not tightly fixed on the plate).

A transparent quartz plate is used for many-fold increase of the sensitivity of the fluorescence analysis. The sample (matter of inquiry) is placed exactly at the center of the surface of the plate. The size of the attached smear (film of organelles or cells or other samples) should correspond to the light spot, formed by the exciting lens (**Figure 7**).

To get the best results, the object under study (cells, cellular organelles, or nuclear DNA, etc.) is applied to the surface of the plate in the form of a thin smear (with drying from water within 4–5 min). Fluorescent dyes, dissolved in a solution, are deposited as microdroplets (1–4 μl) to the sample. After that, the sample is dried for 1–2 min. The plate with the dried sample is placed into the cuvette filled by liquid, which weakly interacted with sample, for example, hexane, perfluorodecalin or, in opposite case, for water-insoluble samples, by isotonic aqueous medium or water.

In all experiments, the plate was placed along a diagonal of the cuvette, as shown in **Figure 7**, to exclude the penetration of the reflected artifact light into the recording emission channel. Virtually, any films of samples (stained cells or

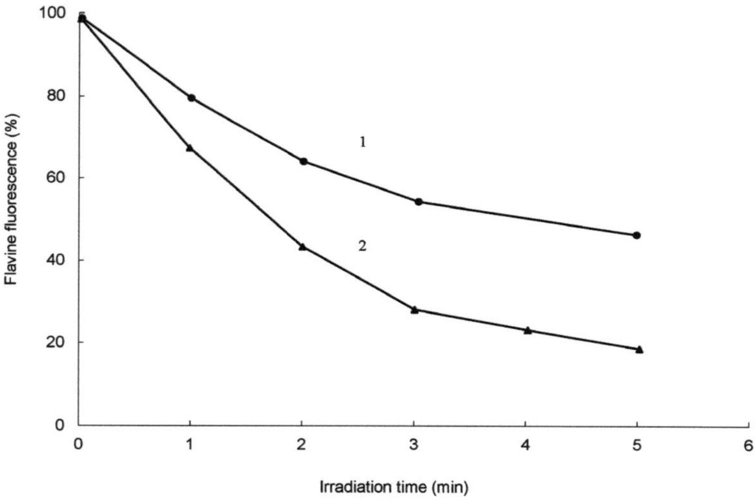

Figure 6.
Dependence of photodestruction of proto-mitochondrial flavins (1) and free FMN (2) at the time of blue irradiation. It was measured by fall in the intensity of flavin fluorescence, appeared in the burning band at 450 nm, and detected as decrease in fluorescence intensity at 525 nm.

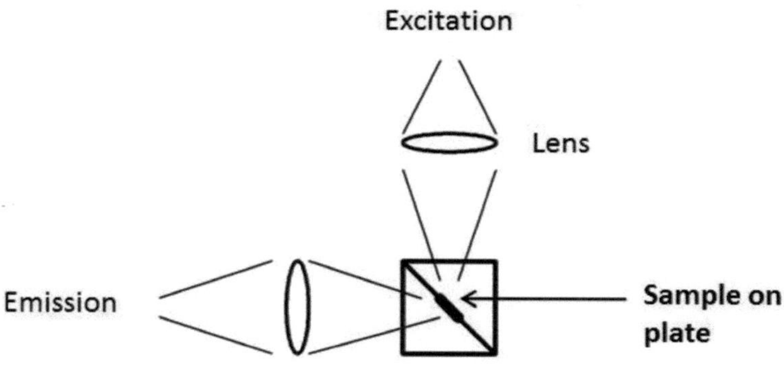

Figure 7.
Cuvette with the diagonal transparent quartz plate (top view). The sample in the form of a smear or film is located on the side of the plate facing the exciting light.

organelles, etc.) applied on the surface of the plate produce fluorescence signals that are several times higher than signals yielded by the same quantity of the studied substance in the solution filling the cuvette volume, although the sample on the plate is usually hypochromized (the extinction factor of chromophores is substantially decreased due to the sieve and screening effects).

High intensity of fluorescence of the layer on the plate arises due to several reasons. First, the whole studied substance is concentrated in the small square of layer, but, while being in a solution, it is distributed over the whole volume of the cuvette. Second, the sample is attached on the plate at the same place, where the focus of both lenses is located (the area of the sample is equal to that of the light spot). Third, the excited molecules of the sample in the condensed phase are less subjected to deactivation (quenching) than in solutions.

The cuvette with transparent diagonal quartz plate can be applied for chemical treatments of a sample. For instance, **Table 4** presents the data on formation of pyrene excimers (excitation was 335 nm, and emission was 393 nm for monomers and 470 nm for excimers) before and after extraction of mitochondrial lipids by acetone from mitochondrial smear. Also, quenching of tryptophan fluorescence (excitation was 286 nm, and emission was 335 nm) of mitochondrial smear by pyrene (2.4 μM) is shown.

During the oxidation of succinate in respiratory chain of mitochondria, a consumption of oxygen, accompanied by increased τ_m and F_m of pyrene, takes place (**Table 5**) [8]. At exhaustion of oxygen, luminescence of pyrene leaves on a plateau. Mitochondria were 0.8 mg of protein per ml in the incubation solution: 10 mM Tris-HCl, 10 mM phosphate, 50 mM KCl, 150 mM sucrose, and pH = 7.5. Pyrene was 1 μM. Excitation was 336 nm, and emission was 393 nm (monomers) and 480 nm (excimers); slits were 2 nm.

In anaerobic conditions, the lifetime of pyrene monomers in mitochondria equals 157 ns (**Table 5**). In without-oxygen organic solvents, τ of pyrene is much more; for example, τ in cyclohexane is 370 ns.

The highly sensitive sensor was designed to measure the molecular oxygen content in solutions and diluted suspensions, placed in a quartz cuvette, which contains the diagonal plate with a pyrene, protected by a Teflon film (**Figure 8**). The sensor can be used in any standard spectrofluorimeter. Unlike analogs, this oxygen sensor is not inertial.

3.3 Mirror cuvette with diagonal plate

An additional multiple increase in the fluorescent signal is attained by the use of special mirror cuvette with diagonal plate. Such cuvette (**Figure 9**) is very useful when the emission intensity from a sample is too low.

In all experiments, the cuvette with the plate was filled with a solvent, the refractive index of which is close to that of quartz. The filling with a solvent is intended to (i) eliminate spurious light reflections at boundaries, (ii) reliably prevent the fluorescence channel from the incoming exciting light, (iii) prevent

Mitochondria	Quenching (%)	*Fe/Fm*
Native	27	0.47
Without lipids	23	0.1

Table 4.
Quenching of tryptophan fluorescence of mitochondrial smear by pyrene and pyrene monomer/excimer ratio before and after extraction of lipids from mitochondrial smear.

Oxygen (μM)	*Fm*	*Fe*	*Fe/Fm* τ_m (ns)
130	95	11.5	0.12105
~1	150	23	0.15157

Table 5.
Fluorescence of pyrene in membranes of mitochondria in aerobic and anaerobic conditions.

Figure 8.
Cuvette with the diagonal transparent quartz plate with plastic microcontainer. The sample—suspension of cells or solution of pyrene (oxygen sensor).

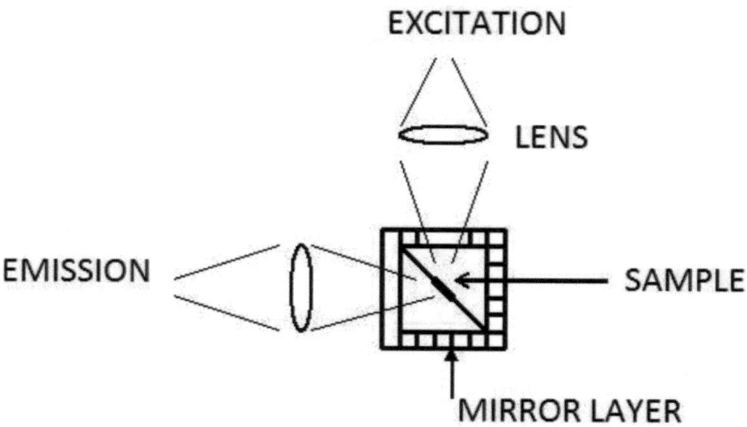

Figure 9.
Schematic diagram of mirror cuvette with the diagonal transparent quartz plate (top view). The sample in the form of a smear or film is located on the side of the plate facing the exciting light.

the sample from strong overheating by the exciting light, and (iv) allow measurement of the kinetic processes on the sample surface. The matter is that molecules of the studied substance are sufficiently mobile on the sample surface contacting the solution and make it possible to observe the kinetics of chemical and biochemical reactions using fluorescence spectroscopy methods.

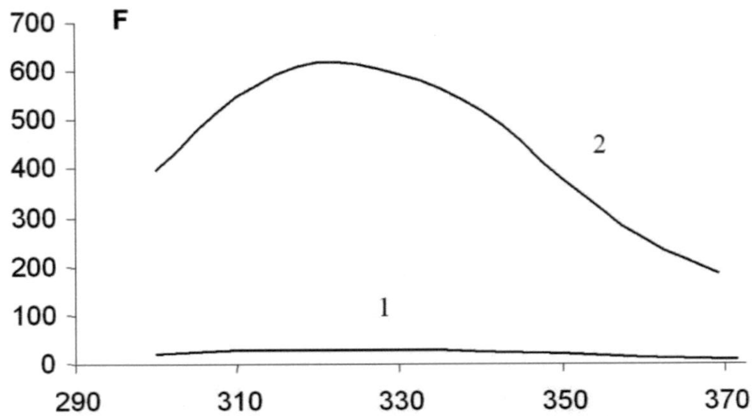

Figure 10.
Spectrum of the tryptophan fluorescence of mitochondrial proteins: (1) aqueous solution in the standard cuvette and (2) sample film on the surface of the quartz plate in the mirror cuvette.

Sample	Exciting wavelength (nm)	Maximum of fluorescence spectrum (nm)	Fluorescence intensity of		
			Dissolved sample in mirror cuvette	Sample layer on the plate in standard cuvette	Sample layer on the plate in mirror cuvette
Rhodamine B	540	570	3.5	4.1	13.8
Pyrene	335	390	3.3	3.8	11.9
Tryptophan	280	350	3.1	3.5	10.1
7-AAMD	550	660	3.6	3.8	12.9
7-AAMD in DNA	570	630	3.7	4.2	14.8
Mitochondria	286	340	3.4	4.7	15.2
Leucocytes	450	525	3.1	3.8	11.1

Note: concentrations <1 μM; fluorescence intensity of the dissolved sample in the standard cuvette was assumed to be 1; 7-AAMD is 7-aminoactinomycin D.

Table 6.
Fluorescence parameters and intensities (gain, G) of various samples on the plate, diagonally cutting in quartz 1-cm cuvette.

The use of the plate ensures another possibility; namely, the film sample on it can be used repeatedly, placing the plate (after rinsing) in the required medium. For example, a thin layer of mitochondria (from rat liver) safely lies on the plate even at a long (1–2 h) period of being in the isotonic water phase (the adhesion and stability of a thick layer are rather worse). In this case, the initial activity of enzymes is not lost. In the inert perfluorodecalin, this layer remains unchanged for many hours.

Figure 10 shows, for example, spectra of the tryptophan fluorescence of proteins of mitochondrial diluted suspension in the standard cuvette (curve *1*) and in the form of a film on the quartz plate, diagonally cutting into the mirror cuvette (curve *2*). This mirror cuvette ensures a higher fluorescence signal from the sample on the surface of the plate than that from the solution in the cuvette volume. The 1-cm mirror cuvette usually increases the output signal from the dissolved sample by a factor of 3.1–3.7 as compared with the standard cuvette. It additionally increases the signal by a factor of 3.5–4.7 for a sample layer on the plate, as compared with the standard cuvette with a diagonal plate. As a result, the total gain of the recorded fluorescence signal in the mirror cuvette with a diagonal plate proves to be at least an order of magnitude higher than that of the standard cuvette filled with a solution of the same substance (**Table 6**).

Similar results also were observed in experiments with dyes and aromatic hydrocarbons. **Table 6** summarizes relative fluorescence values of some substances, when they are placed on the plate's surface (comparing to the solution of a sample in the cuvette volume).

4. Conclusions

A number of special cuvettes which dramatically increase the fluorescence intensity of diluted solutions and suspensions are tested. All described cuvettes can be applied for measurements of spectra excitation and emission, excited-state lifetime, and polarization degree. These cuvettes enable a multiple increase in the sensitivity of fluorimetric measurements with common spectrofluorimeters. It is also possible to study, in particular, the kinetics of phenomena on the sample surface, to detect small concentrations of substances, e.g., slight quantities of proteins and DNA in solutions, and this information can be used for biomedical analyses, in crime detection, etc. Also, in principle, the described cuvettes could be applied for photo-bleaching experiments.

Author details

Nikolai Vekshin
Institute of Cell Biophysics of FIC of PNCBI of RAS, Moscow, Russia

*Address all correspondence to: nvekshin@rambler.ru

References

[1] Holler FJ, Skoog DA, Crouch SR. Principles of Instrumental Analysis. Cengage Learning Publisher; 2006

[2] Valeur B. Molecular Fluorescence: Principles and Applications. Wiley-VCH Publisher; 2012

[3] Sharma A, Schulman S. Introduction to Fluorescence Spectroscopy. Wiley Publisher; 1999

[4] Gauglitz G, Vo-Dinh T. Handbook of Spectroscopy. Wiley-VCH Publisher; 2003

[5] Levshin LV, Saletskij AM. Luminescence and Its Measurements (in Russian). Moscow: MGU; 1989

[6] Lakowicz J. Principles of Fluorescent Spectroscopy. Berlin: Springer; 2006

[7] Vekshin NL. Fluorescence Spectroscopy of Biomacromolecules. Lambert Academic Publisher; 2012

[8] Vekshin NL. Photonics of Biopolymers. Berlin: Springer; 2002

[9] Vekshin NL. Biophysics of Mitochondria. Berlin: Springer; 2019

Chapter 20

Pore-Forming Proteins: Fluorescent Dyes to Study the Channel Functionality and Biophysical Properties

Guilherme Teixeira and Robson Faria

Abstract

Large conductance channels, when activated transiently, alter the plasma membrane permeability permitting the passage of molecules with the size of until 1000 Da. These pore-forming proteins are found in the plasma membrane of mammals and invertebrate cells such as the P2X7 receptor, transient receptor potential vanilloid 1 (TRPV1), transient potential melastin receptor 8 (TRPM8), pannexin-1, and hemichannel proteins. Other proteins may be secreted for organisms creating nonselective pathways to large molecules and proteins. Some examples are α-hemolysin and gramicidin. In both cases, an efficient manner to evaluate the functionality is using fluorescent dyes. Thus, we discuss the uses of the fluorescent dyes as tools to study the functionality and some biophysical properties of the proteins described above, among others.

Keywords: fluorescent dyes, ion channels, pore-forming proteins

1. Introduction

Techniques involving fluorescence have been used and improved in studies involving cells, the visualization, and research of their diverse biochemical structures and functions. Detection of proteins, recognition of organelles, nucleic acids, membrane permeabilization, and ions present in cells are widely explored through the feasibility of fluorescent dyes according to their physical-chemical characteristics such as excitation and emission lengths, electrical charges, chemical structures, and molecular weights [1, 2].

Ion channel-forming ionotropic receptors are extremely important proteins in maintaining the potential of cell membranes. In some cases, they are able to form pores (high conductance channels) in the plasma membrane, allowing the passage of certain ions of varying charges and different physical characteristics according to the concentration and electrochemical gradient. Thus, there are a large number of high conductance ion channels capable of opening pores, changing the permeability of the plasma membrane. As examples, we can mention the P2X7 purinergic receptor, transient receptor potential vanilloid (TRPV), voltage-dependent anion channel (VDAC), and others. When stimulated, they can allow ions to pass through the formed pores and molecules with up to 1000 Da to the intracellular medium [3–7]. Thus, certain fluorescent dyes can be used to observe these pores of high ionic conductance, thus being able to further explore about the biophysical and chemical characteristics of these membrane proteins.

2. Pore-forming ionic channels in mammals

The formation of pores in the plasma membrane is a phenomenon often caused by defense proteins in mammalian cells as an immune response to infections by pathogens, tumors, or danger signs. As a response, several proteins are activated, forming pores such as perforin, gasdermin, complement system, mixed lineage kinase domain-like pseudokinase (MLKL), and granulysin, thus mediating cytotoxic and antimicrobial immune activities aimed at preventing and protecting the body from harmful agents [8, 9].

In addition to cellular defense proteins, ion channels are another class of proteins involved in the formation of pores in the plasma membrane, being responsible for the transit of ions such as selective to Ca^{2+}, N^{+}, K^{+}, and Cl^{-} and acting in a physiological and pathophysiological way. There are a large number of ion channels addressed in the literature related to the pore opening. Among them, the P2X7 receptor has been highlighted. The P2X7 receptor, which is the most studied among P2X, is a cationic channel activated by adenosine-5′-triphosphate (ATP) [10, 11]. This channel is responsible for several intracellular processes, such as apoptosis, NLRP3 inflammasome (one of the IL-1β study pathways), inflammation, and pain. When activated, P2X7 leads to an increase in ion flow that can intensify with a decrease in the influx of cations from divalent charges such as Ca^{2+} [12]. Electrophysiology (EP) studies addressed in 2005 by Faria et al. [4] about the high pore conductance of the P2X7 receptor tested on 2BH4 and peritoneal macrophage cells observed unitary conductance values corresponding to 400 pS. In addition to these data, experiments using high-molecular-weight cationic and anionic fluorescent dyes also confirm the high pore conductance. Regarding the permeabilization through the pore, fluorophores tested were able to penetrate the cell regardless of its load, ethidium bromide (1^{+}), propidium iodide (2^{+}), and Lucifer Yellow (2^{-}).

However, other channels in the literature also correlate with the pore of the P2X7 receptor showing high conductance and permeability to fluorophores through the membrane. As highlights are the TRP receptors as TRPV 1–4, with notoriety for TRPV1, which is the most studied subtype of TRPs. This subtype is expressed in several cells of the organism such as the liver, lung, and nociceptive neurons. TRPV1 is a nonselective cationic channel with high permeability for Ca^{2+}, with a unitary conductance of 35–77 pS [13, 14]; however, the unitary conductance of the pore formed was not record until now. Although the ionic channel may be activated for distinct stimuli (capsaicin, low pH, and heat), this receptor only permits the fluorescent dye uptake after capsaicin activation and low pH conditions. Other TRP subtypes related to high conductance channel formation are the transient potential ankhirin type-1 (TRPA1) cationic channels with a conductance of 87–100 pS and transient potential melastin receptor 8 (TRPM8) with a conductance of 40–83 pS, according to [14, 15]. Pannexin-1 [16, 17], connexin hemichannel [18], calcium homeostasis modulator 1 (CALHM1) [19], voltage-dependent anion channel (VDAC1 highlighted) [6, 20], maxi anion [21], P2X (P2X2 and P2X4) [11], and maitotoxin-induced pore [22] are some other different types of channels with unitary conductance values greater than 200 pS and permeable to molecules of until 1000 Da [4, 10, 23]. More details about these pore-forming proteins and some techniques used to record their functionality are shown in **Table 1**.

Pore-forming proteins	EF	PLP	FC	FPR	CA	CPA
P2X7R	[4]	[24]	[25]	[26]	[26]	[4]
TRPA1	[27]	[28]	N/A	[27]	[27]	[27]
Maxi anion	[17]	N/A	N/A	[17]	[17]	[17]
TRPV1	[15]	[59]	N/A	[71]	[24]	[15]
Plasma VDAC	[29]	[30]	N/A	N/A	[29]	[29]
Connexin	[31]	[32]	[33]	[31]	[31]	[31]
Pannexin	[34]	[16]	[35]	[17]	[17]	[17]
Maitotoxin	[22]	[36]	N/A	[22]	[37]	[37]
CALHM1	[31]	N/A	[38]	[31]	[31]	[31]
TRPM8	[39]	[40]	[41]	N/A	[40]	[42]

Table taken from Ref. [49] and adapted.

Table 1.
Pore-forming proteins and some assessment techniques as electrophysiology (EP), planar lipid bilayer (PLP), flow cytometry (FC), fluorescent plate reader (FPR), colorimetric assay (CA), and cell permeabilization assay (CPA) for functional detection of the high conductance.

Although they have similarities to conductance, these different channels have particularities related to biophysical properties, having different functionalities and mechanisms. Some channels exhibit a distinct selectivity for certain ions, some having preference for anions such as Cl^-, PO_4^{3-}, and glutamate in the case of VDAC1 and maxi anion or cations like Ca^{2+} for CALHM1 [6, 19, 43].

These channels are the subject of studies for their involvement in several cellular processes, such as intracellular signaling response, release of cytokines, inflammation, cell death, pain, tumors, and metabolism regulation. Interestingly, the same functions are performed for these large conductance channels with distinct cell types or tissues. We can highlight the role of TRPs (TRPV1 and TRPM8), P2X7 receptor, connexin 43, and pannexin-1 [11, 44–48].

Based on the study of these channels, several molecular and cellular biology techniques are employed to obtain the best understanding of the functions that these receptors perform. Among these assays, the cell death assay, flow cytometry, intracellular calcium quantification, cytokine assay, and cell permeabilization stand out. However, for the best performance of these techniques, fluorescent dyes are used according to their physical-chemical characteristics and molecular mass, serving as a strategy to obtain better results and performance of experimental tests [49]. For additional information about some of the techniques mentioned above, more details involving their respective methodologies can be found in the reviews published below [50–52].

3. Fluorescent dyes and their use in the study of pores by ion channels

Fluorescence is a kind of luminescence in which the electron can capture energy by entering an excited state for a short period and returning to its normal state by emitting light. The first discoveries involving this form of light were in the middle of the sixteenth century by physician and botanist Nicolás Monardes through a wood (*Lignum nephriticum*) from Mexico used to treat kidney diseases [53, 54]. Over the years, the term "fluorescence" was coined by physicist George Gabriel Stokes, in addition to presenting a list of fluorescent substances, confirming new advances in the area. With the advancement of science and the years, the first fluorescent dye was synthesized, fluorescein, an anionic dye produced by German chemist Adolf von Baeyer, a molecule that would be used in several laboratory methods in the future [49, 55].

Fluorescent dyes are widely used in various specialties within science, among them; the staining of different structures in tissues and cells stands out. A vast number of laboratory techniques of molecular and cellular biology in broad fields of research use these molecules, based on their chemical and physical properties according to the method worked, thus being an excellent resource in the investigation and visualization of the topic addressed [56, 57]. Among many techniques, the opening function of the high conductance pores mediated through ion channels can be evaluated in experimental tests for the capture of intracellular dye [4, 49]. Many dyes are impermeable to the membrane due to their molecular weight and physical characteristics such as positive charge (cationic dyes) and negative charge (anionic dyes). However, due to the opening of the pores with a subsequent increase in permeability, it is possible to investigate these dyes through the capture by fluorescence reading. The interaction of fluorescent dyes in the intracellular medium occurs according to the electrical charge of the target structure to be marked. Several structures that make up the intracellular environment, such as ions, nucleic acids, and organelles, each are having a specific electrical character, thus enabling the connection with fluorophores after permeating through the high conductance pores [1, 58].

In 1987, one of the first works was carried out with the use of fluorescent dyes to determine pore permeability, in which there were significant results about the subject. Steinberg et al. [58] in a study using macrophages of the J774 strain found that the presence of extracellular ATP was able to induce increased cation flow through an unidentified membrane receptor. In order to better analyze the P2X7 receptor permeability, five fluorescent anionic dyes impermeable to the membrane with varying were tested. The authors tested Lucifer Yellow (457 Da, 2^-) being the Ca^{2+} pentavalent Fura-2 indicator (832 Da, 5^-), 6-carboxyfluorescein (376 Da, 2^-), trypan blue, and Evans blue with similar charge and molecular weight (960.81 Da, 4^-). The first three dyes showed permeability through the influence of ATP. However, trypan blue and Evans blue were not permeable to the cell and may suggest a specific conductance by the receptor through the different molecular weights of the fluorophores used by the authors. This data led to the exploration and better understanding of P2X pores, mainly from the P2X7 receptor [11, 12]. However, the data about the P2X7 receptor pore permeability to cationic fluorescent dyes with charge higher 2^+ (propidium iodide) was not investigated.

In another study, Faria et al. [59] evaluated the dye uptake involving proteins P2X7, TRPV1, TRPA1, and maxi anion in astrocytes. The selectivity of these pores for cationic dyes (ethidium bromide and propidium iodide) and anionics (fluorescein and calcein) was measured. The TRPV1 and TRPA1 receptors were selective for cationic ones allowing their permeability and maxi anion for anionic ones, suggesting a possible preference of these pores concerning the electric charge of the fluorophore. This aspect did not occur with the P2X7 receptor, already reported, and another study about the subject [4]. However, the data about the maximal size of cationic dyes able to permeate for TRPV1 and TRPA1 channels or anionic for maxi anion channels was not investigated. Additionally, the permeability to compounds with the higher charge was not discussed.

One of the points to be noted regarding the capture of dyes in the investigation of pore-forming proteins is the low diversity of fluorophores used in these tests. In most experiments, "standard" dyes are used, among which are the cationic ethidium bromide (394 Da, 1^+), propidium iodide (668.39 Da, 2^+), YO-PRO-1 (629 Da, 2^+), and DAPI (350.25 Da, 2^+) which mostly interact with nucleic acids. Anionics such as Lucifer Yellow (457 Da, 2^-) Fluo-4 (736.63 Da, 4^-), Fura-2 (832 Da, 5^-), and fluorescein (332.31 Da, 1^-) are usually ion markers or intracellular structures [27, 59, 60]. A study by McCoy et al. [42] on the permeabilization of large molecules mediated by the TRPM8 receptor, the cationic dye PO-PRO-3 (605.30 Da, 2^+), was used to measure the dilation of the pore. However, it is not common to use fluorophores other than those mentioned above, having a great scenario to be explored in ionic conductance tests. Another impacting factor regarding dye testing is related to the electrical charge in which the molecule stands out. As seen previously, mainly in cationic dyes, the charges vary between the charge of 1^+ and 2^+, a fact little investigated, and the search for tests in dyes with more tri or tetravalent charges is relevant in the evaluation of these pores. However, pores similar to dyes also have charges, suggesting varied possibilities for molecular interaction of pores with the physicochemical properties of fluorophores, a fact that is not thoroughly investigated in scientific researches [49].

We consider it relevant to emphasize that some of the pore-forming proteins mentioned have no reports of the use of fluorescent dyes such as pl-VDAC. Other proteins possess basically the same ones studied at the P2X7 receptor, such as pannexin-1, maitotoxin, and maxi anion [22, 34, 59].

The only exception would be connexins, such as connexin 43, which is due to its prominent regulatory function in several vital processes, such as cardiac synchronism. However, these proteins have too much variability of homotypic and heterotypic channels with unitary conductance and permeability by size and charge of the variable molecule. Thus, the biophysical characterization of these channels can be models to generate protocols for the study of the dyes above in the other high unit conductance channels [61].

There is a plurality of fluorophores described in the literature still used in the evaluation of pore formation functionality [62], given a large field to be explored and researched regarding the use of these dyes in order to understand more clearly some specific characteristics of these pores. The following tables (**Tables 2** and **3**) contain a total of 24 different examples of fluorescent dyes with varied molecular weights, structures, and different charges (mono-, di-, tri-, or tetravalent), cutoff, within their physical-chemical characteristics, in which can be tested in future experiments to deepen the understanding about the biophysical properties of these high ion conductance channels. We hope that the data contained in these tables can contribute to the purpose of studies of the pore-forming channels, which may serve as a tactic and add discoveries to the scientific field, following more knowledge of cell physiology.

A detail to note is that the table corresponding to cationic lacks more examples of tetravalent dyes. The proteins addressed in the previous topic are reported to allow the permeabilization of molecules of up to 1000 Da. However, there are other types of tetravalent fluorophores found in the literature that exceed the permeable molecular weight limit for these pores [2]. BoBo 1 (1202.66 Da), BoBo 3 (1254.73 Da), ToTo-1 (1302.78 Da), and ToTo 3 (1354.85 Da) are some other examples of charged dyes (4^+) [62]. Thus, the lack of other viable tetravalent dyes to the corresponding permeability values of the described proteins prevents the exploration and evaluation of the pores with these molecules. Therefore, we consider it necessary to investigate other fluorescent dyes with this type of charge, aiming at possible future experiments on the molecular pore interaction with this level of net charge [56, 63].

Anionic dyes	MW	Charge	Molecular structure
PromoFluor 840	788.96	1⁻	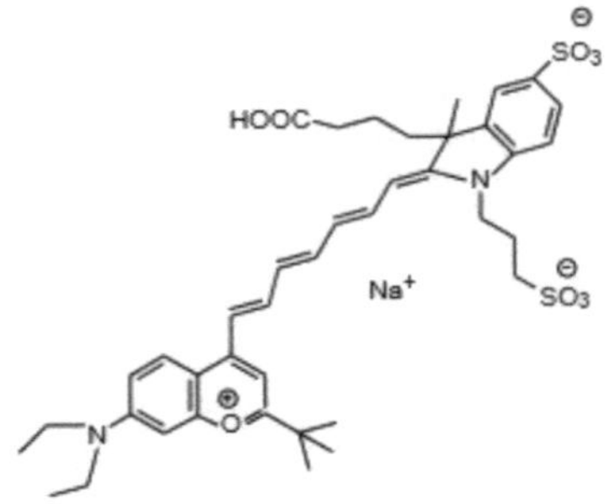
Yakima Yellow	654.30	1⁻	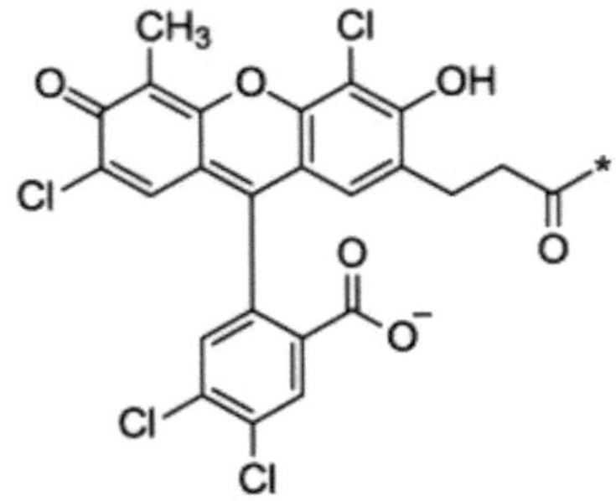
5-Carboxyfluorescein	376.32	1⁻	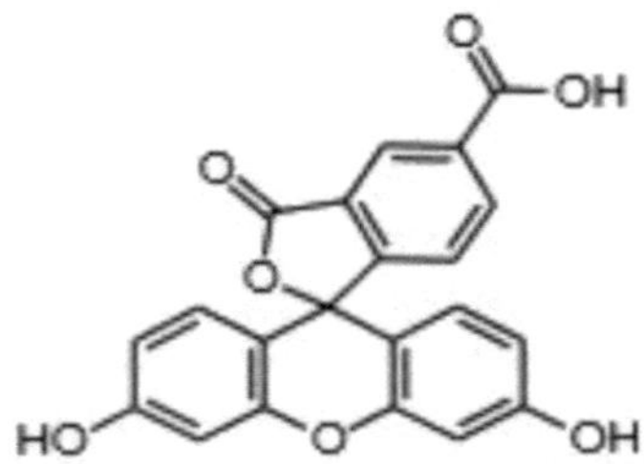
PromoFluor 680	852.96	2⁻	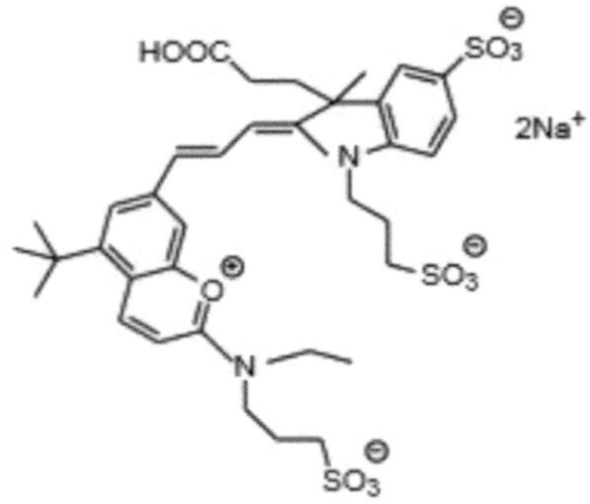
JOE	602.34	2⁻	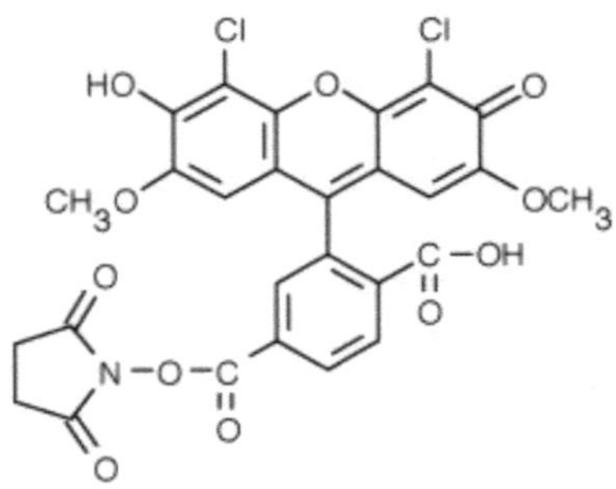

Anionic dyes	MW	Charge	Molecular structure
Dichlorofluorescein	**401.2**	2^-	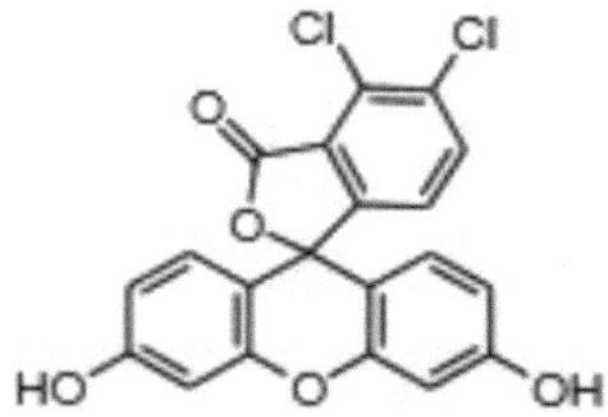
PromoFluor 405	**695.59**	3^-	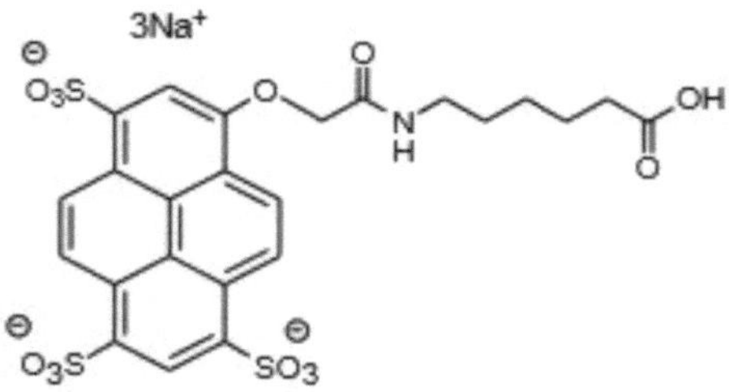
Methoxypyrenetrisulfonic acid trisulfonate (MPTS)	**538.4**	3^-	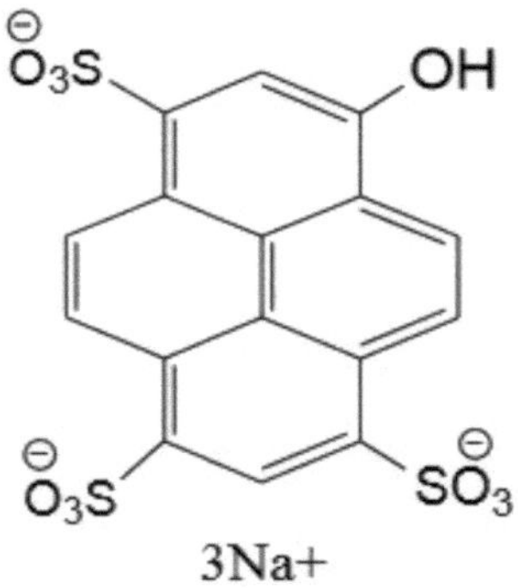
8-Aminopyrene-1,3,6-trisulfonic acid (APTS)	**523.38**	3^-	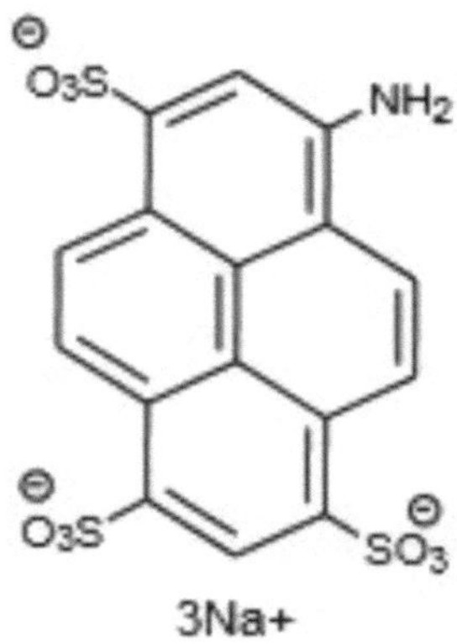
Tetrakis(4sulfophenyl) porphine (TSPP)	**935**	4^-	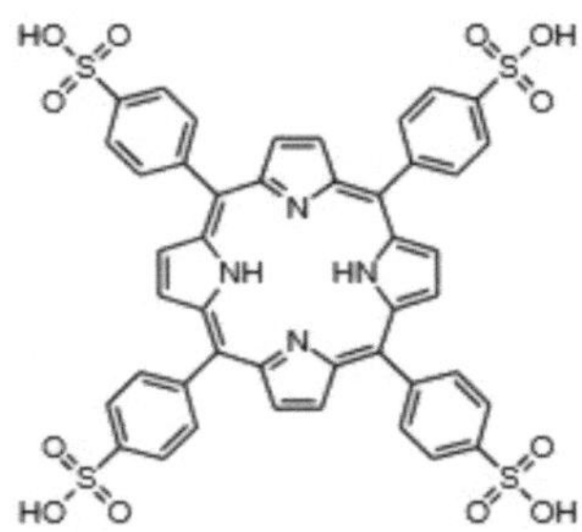

Anionic dyes	MW	Charge	Molecular structure
Tetrakis(4carboxyphenyl) porphine (TCPP)	790.8	4^-	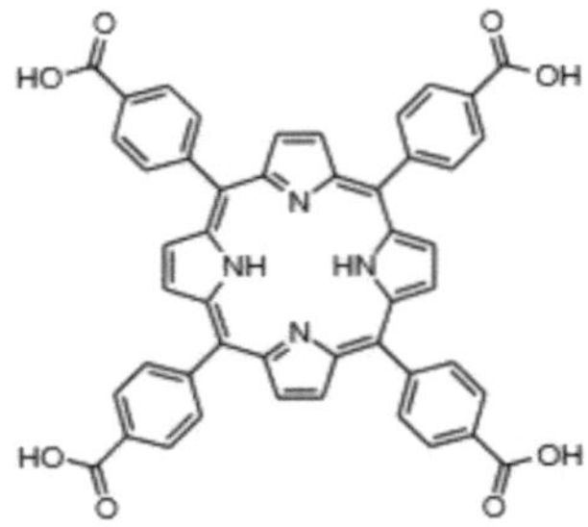
Quin-2	693.87	4^-	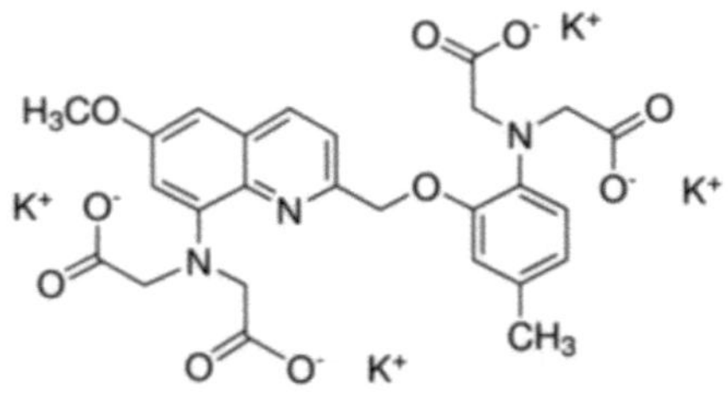

Table 2.
Fluorescent anionic dyes.

Cationic dyes	MW	Charge	Molecular structure
QSY9 carboxylic acid succinimidyl ester	951.43	1^+	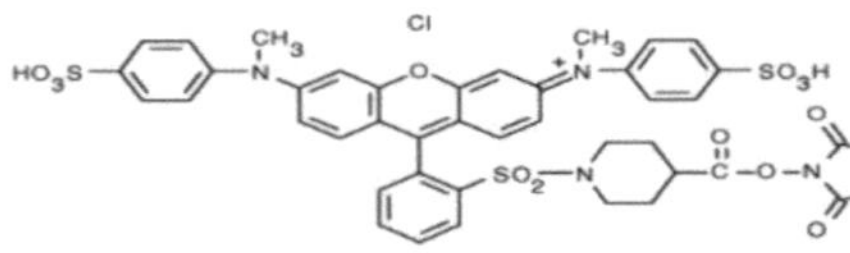
PyMPO	564.4	1^+	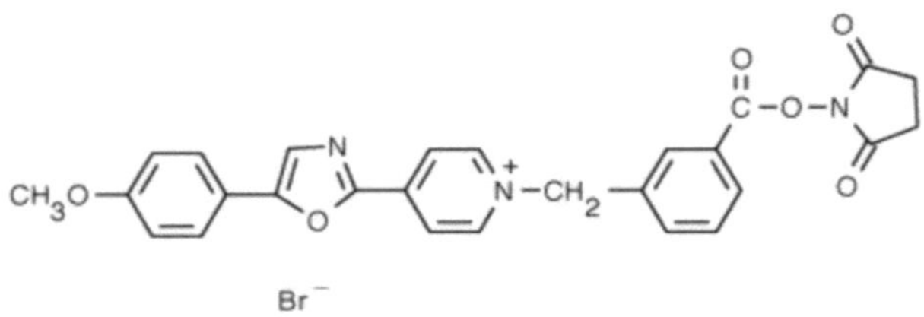
Hexidium iodide	497.42	1^+	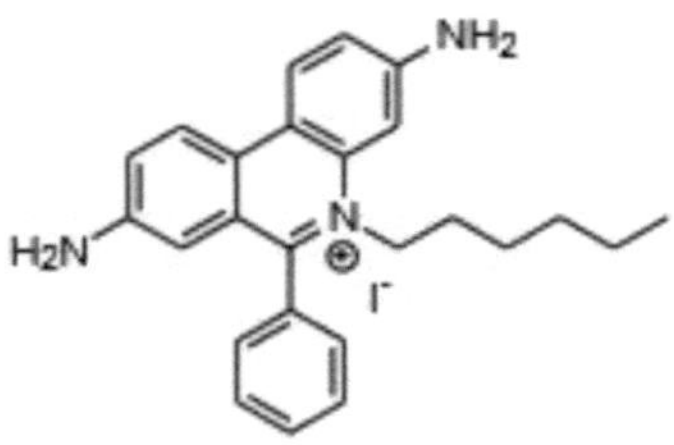
TO-PRO-1	645.38	2^+	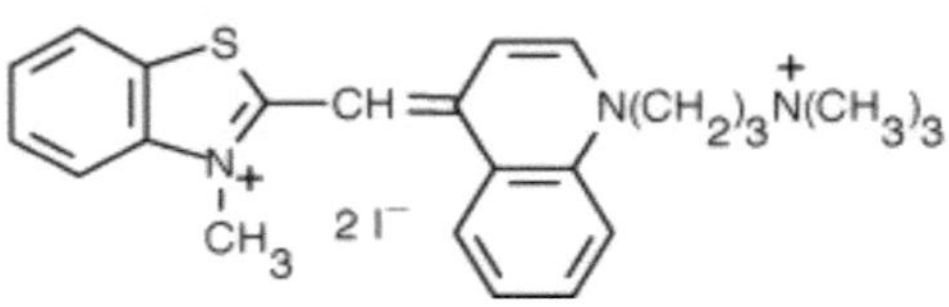

Cationic dyes	MW	Charge	Molecular structure
JO-PRO-1	**630.31**	2^+	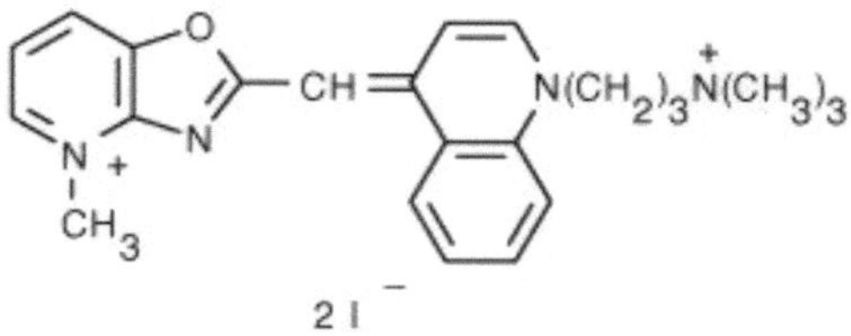
BO-PRO-1	**595.32**	2^+	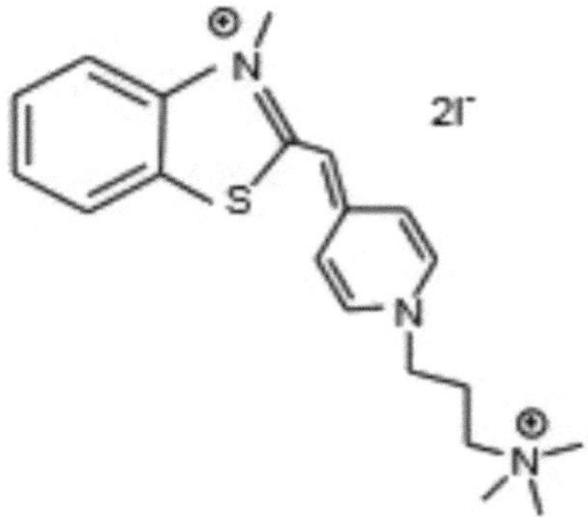
EvaGreen	**509.73**	2^+	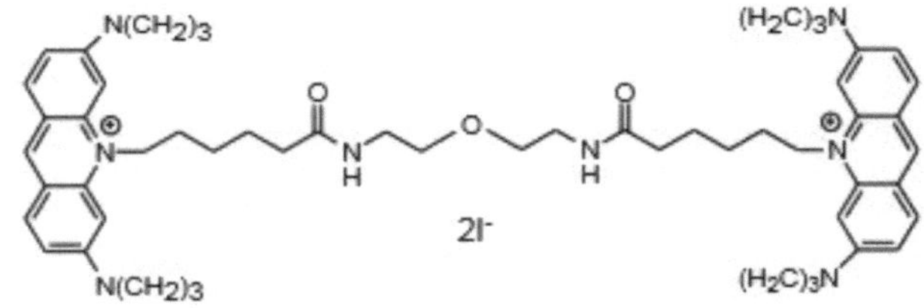
Hoechst 769121	**596.96**	3^+	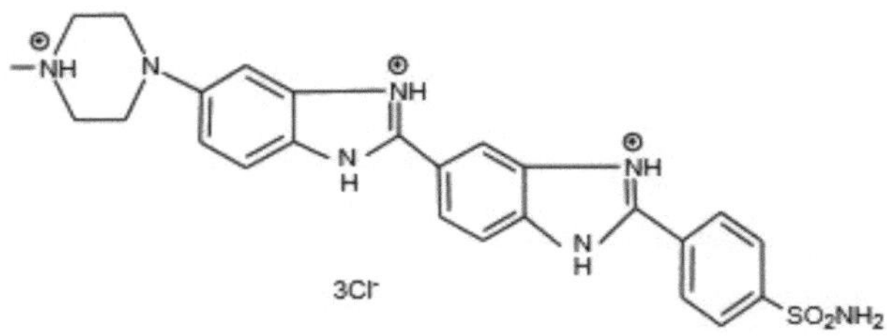
Hoechst 33342	**561.93**	3^+	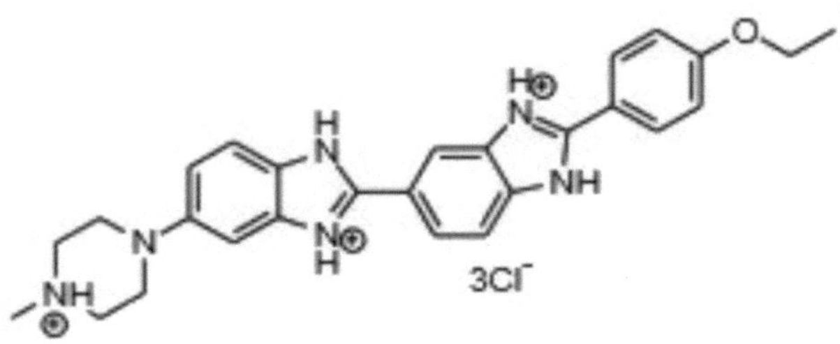
Hoechst 34580	**560.96**	3^+	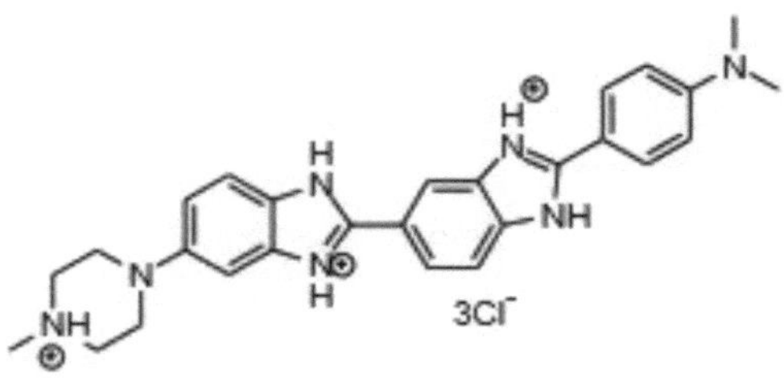

Cationic dyes	MW	Charge	Molecular structure
Hoechst 33258	533.88	3^+	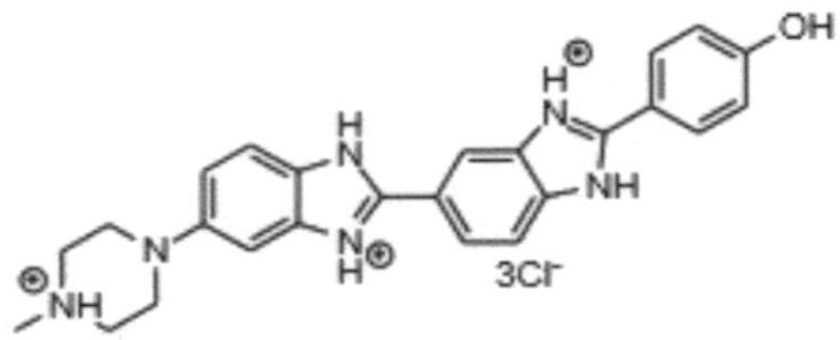
EthD-1	856.77	4^+	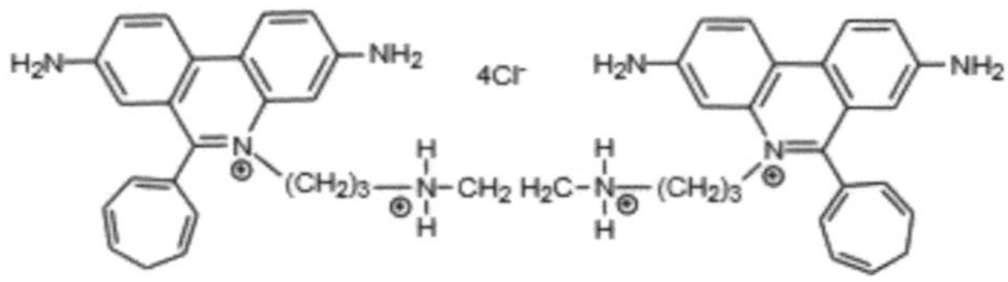

Table 3.
Fluorescent cationic dyes.

4. Conclusion

Scientific advances with methods involving fluorescence, such as fluorimetry and colorimetry, have led to the significant discoveries in different horizons of biological knowledge, with the application of fluorophores in these methodologies having an immense importance in the purpose of evaluating different topics involving the study of cells. Studies on the permeabilization of the plasma membrane through high conductance pore-forming proteins are a topic widely addressed in laboratory research, with the use of fluorescent dyes being a useful device in the evaluation and investigation of the functionality of ionotropic receptors. Through the tables presented above, several anionic and cationic dyes with different charge values (mono-, di-, tri-, and tetravalent) can be used by researchers in numerous experiments, aiming to check these pores through their physical-chemical properties and conductance. We hope that this review will motivate further research on the applicability, mechanisms, and biophysical properties of these high ion conductance channels, aiming at new knowledge and discoveries for the cellular biological perspective.

Acknowledgements

We are grateful to the Coordination of Superior Level Staff Improvement (CAPES), Finance Code 001, Oswaldo Cruz Institute (IOC), and Postgraduation of Sciences and Biotechnology (PPBI), and Federal Fluminense University (UFF).

Conflict of interest

The authors declare no conflict of interest.

Notes/thanks/other declarations

We are grateful to the Coordination of Superior Level Staff Improvement (CAPES), Finance Code 001, National Council for Scientific and Technological

Development, Foundation Carlos Chagas Filho for Research Support of the State of Rio de Janeiro, Oswaldo Cruz Institute (IOC), Postgraduation and Cellular and Molecular Biology (BCM) from IOC, and Postgraduation of Sciences and Biotechnology (PPBI) from Federal Fluminense University (UFF).

Author details

Guilherme Teixeira and Robson Faria*
Laboratory of Toxoplasmosis and Other Protozoans, Oswaldo Cruz Institute (IOC), Fiocruz, Rio de Janeiro, Brazil

*Address all correspondence to: robson.xavier@gmail.com

References

[1] Giepmans BNG, Adams SR, Ellisman MH, Tsien RY. The fluorescent toolbox for assessing protein location and function. Science. 2006;**312**:217-224

[2] Petit JM, Denis-Gay M, Ratinaud MH. Assessment of fluorochromes for cellular structure and function studies by flow cytometry. Biology of the Cell. 1993;**78**:1-13

[3] Adams DJ. Ionic channels in vascular endothelial cells. Trends in Cardiovascular Medicine. 1994;**4**:18-26

[4] Faria RX, DeFarias FP, Alves LA. Are second messengers crucial for opening the pore associated with P2X 7 receptor? American Journal of Physiology-Cell Physiology. 2005;**288**:C260-C271

[5] Hille B. Ionic selectivity of Na and K channels of nerve membranes. Membranes (Basel). 1975;**3**:255-323

[6] Shoshan-Barmatz V, Mizrachi D. VDAC1: From structure to cancer therapy. Frontiers in Oncology. 2012;**2**:1-34

[7] Tominaga M, Tominaga T. Structure and function of TRPV1. Pflügers Archiv - European Journal of Physiology. 2005;**451**:143-150

[8] Brogden KA, Ackermann M, McCray PB, Tack BF. Antimicrobial peptides in animals and their role in host defences. International Journal of Antimicrobial Agents. 2003;**22**: 465-478

[9] Liu X, Lieberman J. Knocking 'em Dead: Pore-Forming Proteins in Immune Defense. Annual Review of Immunology. 2020;**38**:455-485

[10] Alves LA, Ferreira LB, Pacheco PF, Mendivelso EAC, Teixeira PCN, Faria RX. Pore forming channels as a drug delivery system for photodynamic therapy in cancer associated with nanoscintillators. Oncotarget. 2018;**8**: 25342-25354

[11] Burnstock G. Purine and pyrimidine receptors. Cellular and Molecular Life Sciences. 2007;**64**:1471-1483

[12] North RA. Molecular physiology of P2X receptors. Physiological Reviews. 2002;**82**:1013-1067

[13] Cui M, Honore P, Zhong C, Gauvin D, Mikusa J, Hernandez G, et al. TRPV1 receptors in the CNS play a key role in broad-spectrum analgesia of TRPV1 antagonists. The Journal of Neuroscience. 2006;**26**:9385-9393

[14] Ferreira LGB, Faria RX. TRPing on the pore phenomenon: What do we know about transient receptor potential ion channel-related pore dilation up to now? Journal of Bioenergetics and Biomembranes. 2016;**48**:1-12

[15] Diver MM, Cheng Y, Julius D. Structural insights into TRPM8 inhibition and desensitization. Science. 2019;**365**:1434-1440

[16] Mou L, Ke M, Song M, Shan Y, Xiao Q, Liu Q, et al. Structural basis for gating mechanism of Pannexin 1 channel. Cell Research. 2020;**30**:452-454

[17] Pelegrin P, Surprenant A. Pannexin-1 mediates large pore formation and interleukin-1β release by the ATP-gated P2X7 receptor. The EMBO Journal. 2006;**25**:5071-5082

[18] Stout CE, Costantin JL, Naus CCG, Charles AC. Intercellular calcium signaling in astrocytes via ATP release through connexin hemichannels. The Journal of Biological Chemistry. 2002; **277**:10482-10488

[19] Ma Z, Siebert AP, Cheung KH, Lee RJ, Johnson B, Cohen AS, et al.

Calcium homeostasis modulator 1 (CALHM1) is the pore-forming subunit of an ion channel that mediates extracellular Ca^{2+} regulation of neuronal excitability. Proceedings of the National Academy of Sciences of the United States of America. 2012;**109**:E1963-E1971

[20] Sabirov RZ, Dutta AK, Okada Y. Volume-dependent ATP-conductive large-conductance anion channel as a pathway for swelling-induced ATP release. The Journal of General Physiology. 2001;**118**:251-266

[21] Islam MR, Uramoto H, Okada T, Sabirov RZ, Okada Y. Maxi-anion channel and pannexin 1 hemichannel constitute separate pathways for swelling-induced ATP release in murine L929 fibrosarcoma cells. American Journal of Physiology-Cell Physiology. 2012;**303**:C924-C935

[22] Schilling WP, Sinkins WG, Estacion M. Maitotoxin activates a nonselective cation channel and a P2Z/P2X7-like cytolytic pore in human skin fibroblasts. American Journal of Physiology-Cell Physiology. 1999;**277**:C755-C765

[23] Alves LA, De Melo Reis RA, De Souza CAM, De Freitas MS, Teixeira PCN, Neto Moreira Ferreira D, et al. The P2X7 receptor: Shifting from a low- to a high-conductance channel—An enigmatic phenomenon? Biochimica et Biophysica Acta - Biomembranes. 2014;**1838**:2578-2587

[24] De Souza CAM, Teixeira PCN, Faria RX, Krylova O, Pohl P, Alves LA. A consensus segment in the M2 domain of the hP2X 7 receptor shows ion channel activity in planar lipid bilayers and in biological membranes. Biochimica et Biophysica Acta, Biomembranes. 2012;**1818**:64-71

[25] Jursik C, Sluyter R, Georgiou JG, Fuller SJ, Wiley JS, Gu BJ. A quantitative method for routine measurement of cell surface P2X7 receptor function in leucocyte subsets by two-colour time-resolved flow cytometry. Journal of Immunological Methods. 2007;**325**:67-77

[26] Faria RX, de Jesus Hiller N, Salles JP, Resende JALC, Diogo RT, von Ranke NL, et al. Arylboronic acids inhibit P2X7 receptor function and the acute inflammatory response. Journal of Bioenergetics and Biomembranes. 2019;**51**:277-290

[27] Banke TG, Chaplan SR, Wickenden AD. Dynamic changes in the TRPA1 selectivity filter lead to progressive but reversible pore dilation. American Journal of Physiology-Cell Physiology. 2010;**298**:C1457-C1468

[28] Sinica V, Zimova L, Barvikova K, Macikova L, Barvik I, Vlachova V. Human and mouse TRPA1 are heat and cold sensors differentially tuned by voltage. Cell. 2019;**9**:57-80

[29] Elinder F, Akanda N, Tofighi R, Shimizu S, Tsujimoto Y, Orrenius S, et al. Opening of plasma membrane voltage-dependent anion channels (VDAC) precedes caspase activation in neuronal apoptosis induced by toxic stimuli. Cell Death and Differentiation. 2005;**12**:1134-1140

[30] Böhm R, Amodeo GF, Murlidaran S, Chavali S, Wagner G, Winterhalter M, et al. The structural basis for low conductance in the membrane protein VDAC upon β-NADH binding and voltage gating. Structure. 2020;**28**:206-214

[31] Gaete PS, Lillo MA, Lopez WI, Liu Y, Harris AL, Contreras JE. Novel biophysical properties of molecular permeation in CALHM1 and connexin channels. Biophysical Journal. 2020;**118**:270a

[32] Carnarius C, Kreir M, Krick M, Methfessel C, Moehrle V, Valerius O,

et al. Green fluorescent protein changes the conductance of connexin 43 (Cx43) hemichannels reconstituted in planar lipid bilayers. The Journal of Biological Chemistry. 2012;**287**:2877-2886

[33] Xu C, Fan Z, Shan W, Hao Y, Ma J, Huang Q, et al. Cyclic stretch influenced expression of membrane connexin 43 in human periodontal ligament cell. Archives of Oral Biology. 2012;**57**:1602-1608

[34] Iglesias R, Locovei S, Roque A, Alberto AP, Dahl G, Spray DC, et al. P2X7 receptor-pannexin1 complex: Pharmacology and signaling. American Journal of Physiology-Cell Physiology. 2008;**295**:C752-C760

[35] Chekeni FB, Elliott MR, Sandilos JK, Walk SF, Kinchen JM, Lazarowski ER, et al. Pannexin 1 channels mediate "find-me" signal release and membrane permeability during apoptosis. Nature. 2010;**467**:863-867

[36] Treviño CL, De La Vega-Beltrán JL, Nishigaki T, Felix R, Darszon A. Maitotoxin potently promotes Ca^{2+} influx in mouse spermatogenic cells and sperm, and induces the acrosome reaction. Journal of Cellular Physiology. 2006;**206**:449-456

[37] Caillaud A, Yasumoto T, Diogène J. Detection and quantification of maitotoxin-like compounds using a neuroblastoma (Neuro-2a) cell based assay. Application to the screening of maitotoxin-like compounds in *Gambierdiscus* spp. Toxicon. 2010;**56**:36-44

[38] Moreno-Ortega AJ, Buendia I, Mouhid L, Egea J, Lucea S, Ruiz-Nuño A, et al. CALHM1 and its polymorphism P86L differentially control Ca^{2+} homeostasis, mitogen-activated protein kinase signaling, and cell vulnerability upon exposure to amyloid β. Aging Cell. 2015;**14**: 1094-1102

[39] Kühn FJP, Knop G, Lückhoff A. The transmembrane segment S6 determines cation versus anion selectivity of TRPM2 and TRPM8. The Journal of Biological Chemistry. 2007;**282**:27598-27609

[40] Asuthkar S, Velpula KK, Elustondo PA, Demirkhanyan L, Zakharian E. TRPM8 channel as a novel molecular target in androgen-regulated prostate cancer cells. Oncotarget. 2015; **6**:17221-17236

[41] Zeng J, Wu Y, Zhuang S, Qin L, Hua S, Mungur R, et al. Identification of the role of TRPM8 in glioblastoma and its effect on proliferation, apoptosis and invasion of the U251 human glioblastoma cell line. Oncology Reports. 2019;**42**:1517-1526

[42] McCoy DD, Palkar R, Yang Y, Ongun S, McKemy DD. Cellular permeation of large molecules mediated by TRPM8 channels. Neuroscience Letters. 2017;**639**:59-67

[43] Sabirov RZ, Merzlyak PG. Plasmalemmal VDAC controversies and maxi-anion channel puzzle. Biochimica et Biophysica Acta, Biomembranes. 2012;**1818**:1570-1580

[44] Hofmann F, Navarrete M, Álvarez J, Guerrero I, Gleisner MA, Tittarelli A, et al. Cx43-gap junctions accumulate at the cytotoxic immunological synapse enabling cytotoxic T lymphocyte melanoma cell killing. International Journal of Molecular Sciences. 2019;**20**: 1-16

[45] North RA. P2X receptors. Philosophical Transactions of the Royal Society of London. Series B, Biological Sciences. 2016;**371**:1-7

[46] Samivel R, Kim DW, Son HR, Rhee YH, Kim EH, Kim JH, et al. The role of TRPV1 in the CD4+ T cell-mediated inflammatory response of allergic rhinitis. Oncotarget. 2016;**7**:148-160

[47] Valdebenito S, Barreto A, Eugenin EA. The role of connexin and pannexin containing channels in the innate and acquired immune response. Biochimica et Biophysica Acta, Biomembranes. 2018;**1860**:154-165

[48] Zhang L, Barritt GJ. TRPM8 in prostate cancer cells: A potential diagnostic and prognostic marker with a secretory function? Endocrine-Related Cancer. 2006;**13**:27-38

[49] Ferreira L, Pereira L, Faria R. Fluorescent dyes as a reliable tool in P2X7 receptor-associated pore studies. Journal of Bioenergetics and Biomembranes. 2015;**47**:283-307

[50] Gee KR, Brown KA, Chen WNU, Bishop-Stewart J, Gray D, Johnson I. Chemical and physiological characterization of fluo-4 Ca^{2+}-indicator dyes. Cell Calcium. 2000;**97**:97-106

[51] Kepp O, Galluzzi L, Lipinski M, Yuan J, Kroemer G. Cell death assays for drug discovery. Nature Reviews Drug Discovery. 2011;**10**:221-237

[52] Picot J, Guerin CL, Le Van Kim C, Boulanger CM. Flow cytometry: Retrospective, fundamentals and recent instrumentation. Cytotechnology. 2012; **64**:109-130

[53] Lakowicz JR. Principles of Fluorescence Spectroscopy. 3rd ed. Boston, MA: Springer; 2006

[54] Valeur B, Berberan-Santos MN. A brief history of fluorescence and phosphorescence before the emergence of quantum theory. Journal of Chemical Education. 2011;**88**:731-738

[55] Baeyer A. Ueber eine neue Klasse von Farbstoffen. Berichte der Dtsch Chem Gesellschaft. 1871;**4**:555-558

[56] Khadria AS, Senes A. Fluorophores, environments, and quantification techniques in the analysis of transmembrane helix interaction using FRET. Biopolymers. 2015;**104**:247-264

[57] Lemasters JJ, Ramshesh VK. Imaging of mitochondrial polarization and depolarization with cationic fluorophores. Methods in Cell Biology. 2007;**80**:283-295

[58] Steinberg TH, Newman AS, Swanson JA, Silverstein SC. ATP4-permeabilizes the plasma membrane of mouse macrophages to fluorescent dyes. The Journal of Biological Chemistry. 1987;**262**:8884-8888

[59] Faria RX, Reis RAM, Ferreira LGB, Cezar-de-Mello PFT, Moraes MO. P2X7R large pore is partially blocked by pore forming proteins antagonists in astrocytes. Journal of Bioenergetics and Biomembranes. 2016;**48**:309-324

[60] Rissiek B, Stabernack J, Cordes M, Duan Y, Behr S, Menzel S, et al. Astrocytes and microglia are resistant to NAD+-mediated cell death along the ARTC2/P2X7 axis. Frontiers in Molecular Neuroscience. 2020;**12**:1-8

[61] Johnson RG, Le HC, Evenson K, Loberg SW, Myslajek TM, Prabhu A, et al. Connexin hemichannels: Methods for dye uptake and leakage. The Journal of Membrane Biology. 2016;**249**:719-741

[62] Sabnis RW. Handbook of Fluorescent Dyes and Probes. Hoboken, New Jersey: John Wiley & Sons; 2015

[63] Stockert JC, Blazquez-Castro A. Fluorescence Microscopy in Life Sciences. Sharjah, UAE: Bentham Science Publishers; 2017